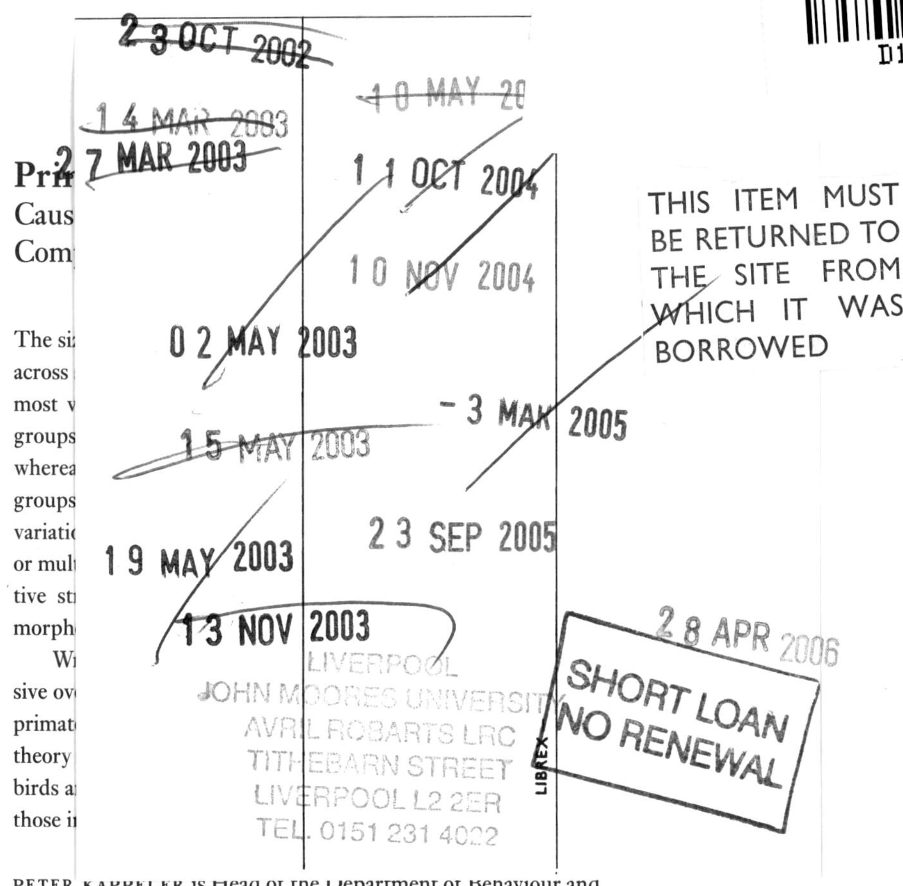

PETER KAPPELER is Head of the Department of Behaviour and Ecology at the Deutsches Primatenzentrum in Göttingen, Germany. His research focuses on the lemurs of Madagascar, and he is co-editor of *Lemur Social Systems and their Ecological Basis* (1993) with J.U. Ganzhorn.

Primate Males

Causes and Consequences of Variation in Group Composition

Edited by
Peter M. Kappeler

CAMBRIDGE
UNIVERSITY PRESS

PUBLISHED BY THE PRESS SYNDICATE OF THE UNIVERSITY OF CAMBRIDGE
The Pitt Building, Trumpington Street, Cambridge, United Kingdom

CAMBRIDGE UNIVERSITY PRESS
The Edinburgh Building, Cambridge CB2 2RU, UK http://www.cup.cam.ac.uk
40 West 20th Street, New York, NY 10011-4211, USA http://www.cup.org
10 Stamford Road, Oakleigh, Melbourne 3166, Australia
Ruiz de Alarcón 13, 28014 Madrid, Spain

First published 2000

Printed in the United Kingdom at the University Press, Cambridge

Typeface Ehrhardt 9/12pt. *System* QuarkXPress® [SE]

A catalogue record for this book is available from the British Library

Library of Congress Cataloguing in Publication data
Primate males : causes and consequences of variation in group
composition / edited by Peter M. Kappeler.
p. cm.
Includes bibliographical references.
ISBN 0 521 65119 0. – ISBN 0 521 65846 2 (pbk.)
1. Primates–Behavior. 2. Primates–Variation. 3. Social
behavior in animals. I. Kappeler, Peter M.
QL737.P9P67258 2000
599′.15–DC21 99-15472 CIP

ISBN 0 521 65119 0 hardback
ISBN 0 521 65846 2 paperback

To Jakob and Josef – my dearest males

Contents

List of contributors ix
Preface xi

I Introduction 1
1 Primate males: history and theory 3
PETER M. KAPPELER

II Comparative perspectives on male–female association 9
2 Multi-male breeding groups in birds: ecological causes and social conflicts 11
NICHOLAS B. DAVIES

3 Males in macropod society 21
PETER J. JARMAN

4 Social counterstrategies against infanticide by males in primates and other mammals 34
CAREL P. VAN SCHAIK

III Variation in male numbers: taxon-level analyses 53
5 Causes and consequences of unusual sex ratios among lemurs 55
PETER M. KAPPELER

6. The number of adult males in callitrichine groups and its implications for callitrichine social evolution 64
ECKHARD W. HEYMANN

7 From binding brotherhoods to short-term sovereignty: the dilemma of male Cebidae 72
KAREN B. STRIER

8 The number of males in guenon groups 84
MARINA CORDS

9 Socioecology of baboons: the interaction of male and female strategies 97
ROBERT A. BARTON

10 Variation in adult sex ratios of red colobus monkey social groups: implications for interspecific comparisons 108
THOMAS T. STRUHSAKER

11 **The number of males in langur groups: monopolizability of females or demographic processes?** 120
ELISABETH H.M. STERCK & JAN A.R.A.M. VAN HOOFF

12 **Costs and benefits of the one-male, age-graded, and all-male phases in wild Thomas's langur groups** 130
ROMY STEENBEEK, ELISABETH H.M. STERCK, HAN DE VRIES & JAN A.R.A.M. VAN HOOFF

13 **Male dispersal and mating season influxes in Hanuman langurs living in multi-male groups** 146
CAROLA BORRIES

14 **Rethinking monogamy: the gibbon case** 159
VOLKER SOMMER & ULRICH REICHARD

15 **Causes and consequences of variation in male mountain gorilla life histories and group membership** 169
DAVID P. WATTS

IV Behavioral aspects of male coexistence 181

16 **Relationships among non-human primate males: a deductive framework** 183
JAN A.R.A.M. VAN HOOFF

17 **Collective benefits, free-riders, and male extra-group conflict** 192
CHARLES L. NUNN

18 **Dominance, egalitarianism, and stalemate: an experimental approach to male–male competition in Barbary macaques** 205
SIGNE PREUSCHOFT & ANDREAS PAUL

V Evolutionary determinants and consequences 217

19 **The evolution of male philopatry in neotropical monkeys** 219
THERESA R. POPE

20 **Models of outcome and process: predicting the number of males in primate groups** 236
JEANNE ALTMANN

21 **Why are male chimpanzees more gregarious than mothers? A scramble competition hypothesis** 248
RICHARD W. WRANGHAM

22 **Male mating strategies: a modeling approach** 259
ROBIN I.M. DUNBAR

VI Conclusions 269

23 **Understanding male primates** 271
MICHAEL E. PEREIRA, TIMOTHY H. CLUTTON-BROCK & PETER M. KAPPELER

References 278

Index 314

Contributors

ALTMANN, JEANNE
Department of Ecology and Evolutionary Biology, Princeton University, Princeton, NJ 08544–1003, USA
altj@princeton.edu

BARTON, ROBERT A.
Department of Anthropology, University of Durham, Durham DH1 3HN, UK
R.A.Barton@durham.ac.uk

BORRIES, CAROLA
Department of Anthropology, Suny, Stony Brook, NY 11794–4364, USA
akoenig@motes.cc.sunysb.edu

CLUTTON-BROCK, TIMOTHY H.
Department of Zoology, University of Cambridge, Downing Street, Cambridge CB2 3EJ, UK
thcb@hermes.cam.ac.uk

CORDS, MARINA
Schermerhorn Hall, Department of Anthropology, Columbia University, 1200 Amsterdam Avenue, New York, NY 10027–5529, USA
mc51@columbia.edu

DAVIES, NICHOLAS B.
Department of Zoology, University of Cambridge, Downing Street, Cambridge CB2 3EJ, UK
amj22@hermes.cam.ac.uk

DE VRIES, HAN
Ethologie en Socio-oecologie, Universiteit Utrecht, PO Box 80086 3508 TB Utrecht, The Netherlands

DUNBAR, ROBIN I.M.
ESRC Research Centre in Economic Learning and Social Evolution, School of Biological Sciences, University of Liverpool, Liverpool L69 3BX, UK
rimd@liverpool.ac.uk

HEYMANN, ECKHARD W.
Abteilung Verhaltensforschung/Ökologie, Deutsches Primatenzentrum, Kellnerweg 4, 37077 Göttingen, Germany
eheyman@gwdg.de

JARMAN, PETER J.
University of New England, Department of Ecosystem Management, Armidale, NSW 2351, Australia
pjarman@metz.une.edu.au

KAPPELER, PETER M.
Abteilung Verhaltensforschung/Ökologie, Deutsches Primatenzentrum, Kellnerweg 4, 37077 Göttingen, Germany
pkappel@gwdg.de

NUNN, CHARLES L.
Department of Biology, University of Virginia, Charlottesville, VA 22903–2477, USA
charlie.nunn@virginia.edu

PAUL, ANDREAS
Institut für Anthropologie, Universität Göttingen, Bürgerstrasse 50, 37073 Göttingen, Germany

PEREIRA, MICHAEL E.
Department of Biology, Bucknell University, Lewisburg, PA 17837, USA
mpereira@bucknell.edu

POPE, THERESA R.
Department of Biological Anthropology & Anatomy, Duke University, Box 90383, Durham, NC 27708, USA
trpope@acpub.duke.edu

PREUSCHOFT, SIGNE
Living Links Center, Emory University, Yerkes RPRC, 954 North Gatewood Road, Atlanta, GA 30322, USA
signe@rmy.emory.edu

REICHARD, ULRICH
Max-Planck-Institut für evolutionäre Anthropologie, Inselstrasse 22, 04103 Leipzig, Germany
reichard@eva.mpg.de

SOMMER, VOLKER
Department of Anthropology, University College, Gower Street, London WC1E 6BT, UK
V.Sommer@ucl.ac.uk

STEENBEEK, ROMY
Ethologie en Socio-oecologie, Universiteit Utrecht, PO Box 80086, 3508 TB Utrecht, The Netherlands
R.steenbeek@biol.ruu.nl

STERCK, ELISABETH H.M.
Ethologie en Socio-oecologie, Universiteit Utrecht, PO Box 80086, 3508 TB Utrecht, The Netherlands
sterck@neuretp.biol.ruu.nl

STRIER, KAREN B.
Department of Anthropology, University of Wisconsin, Madison, WI 53706–1393, USA
kbstrier@facstaff.wisc.edu

STRUHSAKER, THOMAS T.
Department of Biological Anthropology & Anatomy, Duke University, Box 90383, Durham, NC 27708, USA
tomstruh@acpub.duke.edu

VAN HOOFF, JAN A.R.A.M.
Ethologie en Socio-oecologie, Universiteit Utrecht, PO Box 80086, 3508 TB Utrecht, The Netherlands
J.A.R.A.M.vanHooff@biol.ruu.nl

VAN SCHAIK, CAREL P.
Department of Biological Anthropology & Anatomy, Duke University, Box 90383, Durham, NC 27708, USA
vschaik@acpub.duke.edu

WATTS, DAVID P.
Department of Anthropology, Yale University, PO Box 208277, New Haven, CT 06520, USA
david.watts@yale.edu

WRANGHAM, RICHARD W.
Department of Anthropology, Harvard University, Peabody Museum, 11 Divinity Avenue, Cambridge, MA 02138, USA
wrangham@husc.harvard.edu

Preface

This is a book about primate males. More specifically, its focus is on the causes and consequences of variation in the number of males per group within and across taxa. This variation lies at the heart of basic problems in primate socioecology, such as grouping patterns, reproductive strategies, social relationships, and dispersal.

Questions about the causes and consequences of variation in the number of males per group have a long history in studies of primate socioecology. In fact, they played a central role in the first systematic attempts to make sense out of the varied primate social systems and to understand their adaptive basis. Seminal papers by Crook & Gartlan, and Eisenberg and colleagues used the number of males as a variable to organize the rapidly growing information about primate social systems. The resulting categories were used to examine correlations with ecological variables, an approach that had been successfully applied earlier to birds.

Subsequent studies by Clutton-Brock, Harvey and others used the distinction between single-male and multimale species to explain successfully variation in other traits, but there were no breakthroughs on the question of the adaptive origins of the social systems *per se*. Despite, or perhaps because of, growing dissatisfaction with these largely correlational studies, most researchers moved on to other questions, and, as Harvey & Harcourt put it retrospectively in 1984, 'the question of single-male versus multimale breeding systems remained a matter of speculation'.

Following important papers by Wrangham and van Schaik, the focus began to shift toward questions about female behavior in the early 1980s, perhaps also because females were practically ignored in the first classificatory approaches. Important papers by Ridley, Andelman, Wrangham, and Jeanne Altmann revisited the issue in the late 1980s, but it was not until recently that new questions, methods, and data generated renewed interest in primate males and their behavior. Most recently, carefully controlled comparative studies re-examined potential determinants of the number of males in primate groups, albeit with somewhat conflicting conclusions.

We were therefore confronted with a situation characterized by partly unresolved old, fundamental questions, a wealth of data on the diversity of primate social systems from many new field studies, as well as important theoretical and methodological advances. Time was therefore ripe for a basic inventory and some organizing questions. An international conference with focused contributions by the recognized authorities on various taxa and topics seemed to be an appropriate way to assess and organize the existing information. This conference took place in December 1997, hosted by the German Primate Center (DPZ). For four days, 19 invited speakers and more than 200 other participants discussed 47 oral and poster presentations. Following the conference, 21 contributions were submitted in written form, and each one was subjected to rigorous peer review. These contributions constitute a representative sample of the contributions to the conference, encompassing specific case studies, comprehensive reviews, theoretical modeling, as well as studies of non-primate societies that provide important comparative perspectives on general principles.

The conference, as well as the resulting volume, would not have been possible without the help of many people and organizations. The first Göttinger Freilandtage were made possible by generous grants from the Deutsche Forschungsgemeinschaft (DFG), the Wenner-Gren Foundation for Anthropological Research, the Niedersächsisches Ministerium für Wissenschaft und Kultur, and the Sparkasse Göttingen. Michael Lankeit generated the idea of the Göttinger Freilandtage and subsequently supported their first installment in many ways. All members of the Abteilung Verhaltensforschung/Ökologie of the DPZ, in particular Andreas Koenig, Carola Borries, Dietmar Zinner, Eckhard Heymann, Christoph Knogge, Stephanie Heiduck, Julia Ostner, Oliver Schülke, and Ulrike Walbaum helped beyond the call of duty with logistical and organizational problems. Other members of the DPZ, especially Rudi Ilse, Erhard Neuhaus, Michael Schwibbe, and Silke Singer provided much appreciated help with various practical problems. Tracey Sanderson's advice and encouragement were crucial in realizing this volume. Ulrike Walbaum had the nerves to compile the joint reference list, and Carel van Schaik, Volker Sommer and Michael Pereira provided helpful suggestions about the organization of this book. I thank all of them. This book is dedicated to my father and son, who, like the rest of my family, supported and inspired me in many ways during the evolution of this volume.

Göttingen, January 1999 Peter M. Kappeler

Part I
Introduction

1 • Primate males: history and theory

PETER M. KAPPELER

INTRODUCTION

The integration of behavioral and ecological studies that generate and test hypotheses about the adaptive bases of diversity in animal social systems is a major goal of socioecology. Studies of primates continue to provide important contributions to this discipline because our comparatively well studied closest relatives exhibit a stunning diversity of grouping patterns and complex social behavior that continue to inspire the interest of numerous zoologists, anthropologists and psychologists.

Across groups, populations and species, the number of adult males per group is the most variable aspect of the social system of those primates that live in permanent groups of more than two adults. This volume features a combination of case studies and synthetic reviews that examine questions revolving around causes of variation in male numbers and its consequences for male and female social behavior. In this introductory chapter, I will review salient earlier studies with the goal of sketching a theoretical framework for the other contributions to this volume.

CAUSES OF VARIATION IN THE NUMBER OF MALES PER GROUP

Single and multi-male primate groups have been recognized as fundamental types of reproductive units since the earliest comparative analyses of primate social systems (Crook & Gartlan, 1966; Eisenberg *et al.*, 1972). These influential studies used the number of males as a variable to organize the rapidly growing information about primate social systems. The resulting categories or grades were used to examine correlations with ecological variables, an approach that had been successfully applied earlier to birds (Crook, 1968), and later to other mammals (Jarman, 1974). These categories were, at least implicitly, treated as adaptive, species-specific traits.

The first of these attempts by Crook & Gartlan (1966) to order information about primate social systems was structured after Crook's thesis (1964), in which he successfully correlated the diversity in social organization and behavior of more than 90 species of weaver birds with broad characteristics of their habitat. As in the earlier study, group characteristics were considered as a target of selection. A new element was introduced to the analysis of primate social systems by postulating a progressive shift between grades that culminated in the presumed condition of early humans. Males played the prominent role in this discussion of different grades. One-male groups characterized the members of the highest grade, which all live in open grassland and savannah habitats. Under these harsh ecological conditions, 'the presence of several large males, only functional in mating and playing no part in rearing young, results in the consumption of much food not used in maintaining the species' (Crook & Gartlan, 1966). The occurrence of predominantly multi-male groups in the next lower grade was attributed to an apparently increased predation risk, even though many of these species also live in open country. Here, 'the increased size and aggressive nature of males are considered to have been pre-adaptive to their role in troop defence', however (Crook & Gartlan, 1966).

Sexual selection theory was also implicated in explanations of both types of groups. Increased competition among males was recognized as an inevitable consequence of a multi-male structure, due to excellent visibility, group cohesion and the seasonality of mating: 'The exclusive possession of a harem probably increases inter-male competition for females further and the occurrence of all-male population units indicates a considerable degree of exclusion of potential reproductives from breeding'.

Subsequently, Crook (1970) acknowledged the existence of intraspecific variation in social organization and discussed the then-available information on baboons, langurs and vervets. He continued to link social systems, in particular one-male and multi-male groups, to ecological factors, such as diet, food dispersion and predation. Gartlan (1968), in a later analysis of social dominance, continued to view the group as 'an adaptive unit, the actual form of which is determined by ecological pressures', but he conceded 'that there is no evidence of simple, linear social evolution from the arboreal Lemuriformes to the anthropoid apes'.

Thus, cause and effect of variation in the number of males were not yet clearly identified, and different questionable functional explanations for species in very similar habitats were provided by this first analysis. In retrospect, these analyses were tremendously important for two reasons. First, they clearly indicated that ecology and behavior are intricately linked. Second, they stimulated a new wave of primate field studies, which continue to provide a valuable basis for our current knowledge of primate behavior and ecology (Terborgh & Janson, 1986).

An important paper by Eisenberg *et al.* (1972) refined and broadened the general approach initiated by Crook and Gartlan. They refined the classification of multi-male species by introducing the category of age-graded male troops, which are defined by the absence of fully adult males of equivalent age. Functionally, age-graded male groups were considered 'a variation on the uni-male theme,' brought about by intermediate levels of tolerance by adult males toward several younger males, which were allowed to mature longer in their natal group. Multi-male groups, in contrast, were characterized by the presence of several adult males of equivalent age, which, presumably as a result of their increased tolerance, show affiliative and cooperative behavior among each other. Importantly, these authors stressed that such permanently bisexual groups with several males are unusual among mammals. Eisenberg *et al.* not only differentiated between different types of groups with variable numbers of functionally adult males, they also suggested that there may be a greater range of causes underlying this variability than the antipredator benefit of multi-male groups and the adaptation to extreme environments of uni-male groups.

Eisenberg *et al.* (1972) also rejected the notion of species-specific social structures. Instead, they emphasized intraspecific variation and recommended using modal categories for interspecific comparisons. They continued to define grades, based on group composition, feeding ecology and habitat use, but they emphasized parallel evolution, rather than linear increase of social complexity. Correlations between social system and ecological parameters continued to be stressed, however. Exceptional populations or species were explained with a new factor: phylogenetic inertia (see also DiFiore & Rendall, 1994), even though this undermined the socioecological approach focusing on adaptations.

A number of different male roles in various societies were discussed, such as leadership, defense against neighboring groups and protection against predators, but females and their reproductive strategies were still largely neglected. Females were discussed in detail only in connection with their role in infant care. In addition, they were assigned an important function in modifying social structure: 'In an evolutionary sense, the number of males in a given troop will depend on what advantage the males are to the reproducing females.' What the exact nature of these determinants of social systems beyond ecological ones and their underlying mechanisms were, was not clearly identified, however.

This early phase of primate socioecology characterized by collecting and classifying information was followed by a set of quantitative comparative studies (Terborgh & Janson, 1986). Following Jorde & Spuhler (1974), Clutton-Brock, Harvey and their collaborators (Clutton-Brock, 1974; Clutton-Brock & Harvey, 1977a, 1977b; Clutton-Brock *et al.*, 1977; Harvey *et al.*, 1978) used the distinction between single-male and multi-male species successfully to explain variation in life history and sexually selected traits, but there were no breakthroughs on the question of the adaptive origins of the social systems *per se*.

Using a much larger data base from the increasing number of field studies, Clutton-Brock & Harvey (1977a) recognized that social organization is not entirely independent of taxonomic affiliation. In contrast to previous analyses, they studied not merely group composition, but also group size and other population parameters, an approach that further emphasized the continuous nature of the observed variation. Single-male and multi-male groups could still be associated with habitat type, but no longer with particular diets. Clutton-Brock & Harvey (1977a) also clearly identified the individual as the appropriate level of analysis, as exemplified by their cost/benefit analysis of adding an extra male to an initially single-male group. In further contrast to previous studies, they concluded that reproductive costs and benefits are more important than energetic ones, and that different factors are probably involved in different species in determining the number of males.

Despite, or perhaps because of, growing dissatisfaction with these largely correlational studies, most researchers moved on to other questions. Following important papers by Wrangham (1979, 1980) and van Schaik (1983; van Schaik & van Hooff, 1983), the focus began to shift toward questions about female behavior in the early 1980s, partly because it is more closely linked to ecological conditions, and perhaps also because females were largely ignored by the first socioecological studies. Nevertheless, some studies from that period contributed important arguments to the discussion about the number of males. Predation risk as the ultimate cause of multi-male groups had been questioned before (Eisenberg *et al.*, 1972), but van Schaik & van Hooff (1983) showed that cooperative defense could only be a consequence of the presence of multiple males and not a cause of it. More importantly, monopolization of females was clearly identified as the key factor responsible for the distinction between single-male and multi-male groups (Clutton-Brock & Harvey, 1977a; Wrangham, 1979; van Schaik & van Hooff, 1983). Genetic relatedness among males and the costs of being solitary were also identified as important potential determinants of male behavior (van Schaik & van Hooff, 1983).

Important papers revisited the issue of variation in the number of males per group in the late 1980s. First, Ridley (1986) refined the key concept of male monopolization potential by examining the effects of the length of the breeding season on group composition. The important idea, which goes back to Trivers (1972), was that 'In a species with a short breeding season, in which several females may become sexually receptive at the same time, a single male may not be able to monopolize and mate with them all.' As a result, a male has little to gain from excluding other males, and a multi-male system results. Conversely, a single strong, powerful male might be able to monopolize each female of a group in species with a long breeding season. Using a new method that controls for evolutionary dependence among species values, Ridley (1986) showed in a comparative test that the number of males and the length of the breeding season are not independent of each other and that they are correlated in the predicted direction. Ridley argued that the relationship is a causal one and that the length of the breeding season may be partly under female control.

In the same year, Andelman (1986) presented an analysis which identified female numbers and dispersion as important determinants of male monopolization potential. Among Cercopithecines, groups with up to five females are single-male, whereas those with ten or more females are generally multi-male. In groups with intermediate numbers of females (six to ten), both social systems occur. In the latter groups, several factors, such as group cohesion, predation risk and male coalitions, which vary in their relative importance, are responsible for variation within and between species with very similar female group sizes (see also Rowell, 1988a). In contrast to Ridley's study, Andelman found no evidence for a relationship between birth synchrony and social system among the Cercopithecines. A second important result of this analysis was the observation that adult sex ratios among multi-male Cercopithecines were relatively invariant, which argued against the long-held notion that multi-male groups present an adaptation to high predation pressure. This result also reinforced the general argument that the number of males, beyond the single-male/multi-male dichotomy, strongly depends on female group size.

This argument was forcefully repeated in a critique of Ridley's (1986) analysis. Jeanne Altmann (1990) identified several methodological flaws in Ridley's study, most of them concerned with the reliability and classification of individual observations. More importantly, she extended Andelman's (1986) argument about the importance of female group size for the number of males to all primates. She also stressed the importance of considering long-term advantages and population sex ratios in studies of male reproductive strategies. In extending Ridley's suggestion about female control of reproductive synchrony, Altmann proposed that females may synchronize their reproduction in order to increase the likelihood of the presence of several males, which in turn would reduce their risk of infanticide and the intensity of reproductive competition with other females.

Since then, three studies examined the proposed key determinants of the number of males in sophisticated quantitative analyses. First, a carefully controlled comparative study demonstrated that predation risk may increase the number of adult males (van Schaik & Hörstermann, 1994). Controlling for the number of females per group, it could be shown that the presence of monkey-eating eagles tends to increase the number of males in howler and colobus monkeys on average from one to two, whereas ecologically similar langurs, which are not exposed to such predators, tend to live in single-male groups. This result is especially interesting because these taxa represent the majority of primates with intermediate numbers of females, where variation in the number of males is most pronounced (Andelman, 1986).

Finally, two recent comparative studies re-examined the relative importance of spatial and temporal distribution of females on male monopolization potential, using methods that control for phylogenetic dependencies of social systems among taxa (DiFiore & Rendall, 1994; Kappeler, 1999a). Mitani *et al.* (1996a) found that the number of males in primate groups, in particular the qualitative difference between single-male and multi-male groups, is positively associated with the number of females, and not with temporal distribution of their receptive periods. They noted several exceptions of species in which a small number of females is associated with several males and suggested that males may gain benefits in these species that offset the potential costs incurred through increased male–male competition. Using the same data set, Nunn (1999) confirmed the effect of female group size, but he also demonstrated that temporal overlap of female receptive periods predicts the number of males after controlling for the number of females. A complete assessment may therefore have to await the availability of additional data, in particular on group-living lemurs, which are characterized by small group size, the presence of several males and extremely short breeding seasons (Kappeler, 1997a).

CONSEQUENCES OF VARIATION IN THE NUMBER OF MALES PER GROUP

Variation in the number of males is also interesting because it has important consequences for social behavior at several levels. For example, the switch from single-male to multi-male groups has profound consequences for the resulting mating system, as well as the associated reproductive strategies of both sexes. It also affects social relationships, in particular the organization of male co-residence in multi-male groups and male–female relations.

Relationships among males are dominated by reproductive competition. Males are inherently less tolerant of each other than females because their reproductive success is limited by access to a nonshareable resource: fertile females. Whenever possible, they should therefore attempt to monopolize access to this resource (Trivers, 1972). Superior size, strength, weaponry and aggressiveness promote this endeavor, and stunning sexual dimorphism in these traits is one consequence of these selective pressures (Clutton-Brock *et al.*, 1977; Clutton-Brock, 1985; Plavcan & van Schaik, 1992). Whenever monopolization of females is not possible, behavioral and physiological mechanisms that promote reproductive skew have apparently evolved to facilitate male coexistence (Bercovitch, 1991; Dixson, 1997). Formalized dominance relationships are of paramount importance in this context (Bernstein, 1976; de Waal, 1986, 1989a). Delayed maturity, group transfers and alternative reproductive strategies are important mechanisms available to subordinate males to alleviate the consequences of their disadvantaged position (Cowlishaw & Dunbar, 1991; Dunbar & Cowlishaw, 1992; Alberts & Altmann, 1995a, 1995b).

Cooperation among males is primarily expected when they live with relatives, so that they can accrue inclusive fitness benefits (van Hooff & van Schaik, 1994). Communal defense of groups of females or their ranges and male bonding are indeed observed in species with male philopatry (van Hooff & van Schaik, 1994), sometimes even within groups (Watts, 1998a). Mutual or reciprocal benefits can also promote short-term cooperation between unrelated males (Noë & Sluijter, 1990). However, affiliative behavior between non-relatives is also commonly observed, but still poorly understood (Rowell, 1988a). Several of these topics have been studied in some detail (de Ruiter & van Hooff, 1993; van Hooff & van Schaik, 1994), but long-term studies of all aspects of male relationships are still scarce (see, for example, Altmann *et al.*, 1997), so that a number of questions remain unanswered. For example, do age-specific patterns of male mortality differ among single-male and multi-male species? What are the maturational and sexual strategies employed by adolescent males? Why do some males stay in their natal groups whereas others leave and join established groups or found new ones? Why are all-male bands so rare among primates, and what are the relationships among their residents? How common are solitary males and how do they try to optimize their survival and reproductive success?

Variation in the number of resident males also has far-reaching consequences for intersexual relationships. Questions about the female perspective are particularly relevant here (Smuts & Smuts, 1993), i.e., which social costs and benefits can females expect in single-male and multi-male groups? Female choice of mates is constrained by definition in stable single-male groups. Females in single-male groups also experience a high potential risk of infanticide in the case of group takeovers (Watts, 1989; Sommer, 1994). On the other hand, the risk or intensity of other forms of sexual coercion may be reduced, compared to multi-male groups (Smuts & Smuts, 1993). Females in multi-male groups have potentially more male assistance in rearing young (Wright, 1990) and have opportunities to establish friendships with individual males that support them in conflicts (e.g., Smuts, 1985). Thus, females are clearly not just passive bystanders that, depending on male competitiveness or tolerance, happen to end up in single-male or multi-male groups, but rather have fundamental interests of their own at stake.

Several recent studies indicate that female primates are very sensitive to the risk of infanticide and that they adjust their behavior and even physiology accordingly (Sterck, 1997, 1998; Steenbeek *et al.*, 1999), and females in other species may form alliances in response to this sort of conspecific threat (Treves & Chapman, 1996). In general, females can reduce this risk in multi-male groups, which they should prefer, if given a choice (Altmann, 1990). Females may also benefit through increased opportunities for mate choice in a multi-male group. Why females in some multi-male groups also engage in copulations with extra-group males remains to be studied in detail, however.

Males and females are therefore likely to have a conflict of interest over the number of males in a group. Which mechanisms do females have at their disposal to win this evolutionary battle? Resource characteristics permitting, they could form such large groups that defensibility by a single male becomes impossible. Critical female group size with respect to defensibility may also vary according to visibility (Rowell, 1988a) and group cohesion (van Schaik & van Hooff, 1983). Synchronization of fertile periods within groups is another theoretically possible way to reduce male monopolization potential (Ims, 1990). Variation in other aspects of female reproductive behavior and physiology suggests that females can modify them to make matings with several males possible. Sexual swellings and mating calls are common among multi-male species, which also tend to have longer follicular phases (van Schaik *et al.*, 1999). Finally, in species in which females are free to migrate, they can transfer into groups with the optimal number of males (Sterck, 1997). Thus, females could take an active role in influencing basic aspects of group composition, but these questions have only begun to be studied in detail (van Schaik, 1996; Sterck *et al.*, 1997).

CONCLUSIONS

Thus, the causes and consequences of variation in the number of males per group are far from completely understood. Much progress has been made in the past 30 years toward understanding the selective factors that determine the composition of primate groups. The focus has shifted from group to individual adaptations, from concentrating exclusively on males to considering reproductive strategies of both sexes, and from looking for only ecological correlates to acknowledging the importance of social factors and sex-specific interests in structuring group composition. Similarly, the list of known or suspected mechanisms underlying variation in group size and composition has grown in parallel.

It was not until recently, however, that new questions, e.g., about mechanisms of sperm competition and aspects of sexual coercion (Dixson, 1991; Smuts & Smuts, 1993; Harcourt & Gardiner, 1994; Harcourt, 1995), new methods, such as DNA-fingerprinting and comparative phylogenetic analyses (Cheverud *et al.*, 1986; Ely & Kurland, 1989; Martin *et al.*, 1992; Sillén-Tullberg & Møller, 1993; de Ruiter *et al.*, 1994; Garber, 1994; Purvis, 1995; Goldberg & Wrangham, 1997) and new data on variability in social organization within and among species (Gautier-Hion *et al.*, 1988; Kappeler & Ganzhorn, 1993; Davies & Oates, 1994; McGrew *et al.*, 1996; Norconk *et al.*, 1996), permitted each contributor to this volume to take a fresh look at a particular taxon or a specific question in this context based on the developments outlined in this chapter. The final chapter attempts to summarize and evaluate their new results, conclusions and questions about primate males (see also Pereira, 1998; Kappeler, 1999c).

Part II
Comparative perspectives on male–female association

Socioecology has been a comparative discipline from its beginnings. It is therefore only appropriate to open this volume on primate males with reviews of non-primate taxa to provide a broad comparative perspective.

Studies of birds have always set standards in socioecology. The work by Nick Davies and his colleagues on the social and mating systems of dunnocks and other little brown birds is no exception. In his lucid review, Davies outlines causes for variation in the number of males in these birds and explores its behavioral consequences with great rigor. This work continues to provide one of the most intuitive examples of the origin and nature of sexual conflict. As in other taxa, males attempt to monopolize mating access, whereas females try to solicit matings, and therefore help in raising their young, from several males. The small groups of these birds provide excellent opportunities to study the resulting behavioral responses, as well variation in skew of male reproductive success. Shared territory defense by males, the importance of ecological factors for variation in social systems, and certain aspects of female sexual behavior are identified by Davies as fascinating parallels with some primates, so that these studies should continue to inspire students of primate behavior to illuminate these factors in the same detail.

Insightful comparative information can also be obtained from Peter Jarman's study of kangaroos and their relatives. These marsupials, with life histories completely different from those of eutherian mammals, provide instructive null models for analyses of variation in social organization. Group-living macropods are described as large mammals that live in essentially open societies. Variation in group composition, or, more specifically, male numbers, appears completely irrelevant to these animals. The reproductive biology of marsupials may provide an important part of an explanation of these differences with primates and other eutherian mammals. Because lost dependent offspring can be immediately replaced by activating a dormant embryo, infanticide cannot evolve as an adaptive male reproductive strategy, a factor that appears to be a powerful determinant of male–female relations in primates. Nevertheless, other aspects of males' reproductive behavior, such as roving in search of unpredictably distributed estrous females and fierce fighting for temporary monopolization of these females, reflect observations familiar to us from eutherian species. Jarman's comparison between diverse but unrelated groups of mammals therefore generates some caution towards sweeping generalizations about mammalian socioecology. It also provides a convincing example of a strong, and previously neglected, link between life history traits and social behavior.

This is a theme taken up by Carel van Schaik in the final chapter of this section. He builds on the results of previous studies that demonstrated that the risk of infanticide is a powerful predictor for the occurrence of male–female association in primates. This idea is based on the assumption that males can act as protectors of vulnerable infants. Van Schaik develops the relative lengths of gestation and lactation as logical predictors of the risk of infanticide and shows that this variable can also predict intersexual association in other mammalian taxa reasonably well. Exceptional species indicate that other factors must be at work, however, so that this first test should provide mammalogists with specific questions for focused research. An important conclusion emerging from these analyses is that primate females should prefer to live in groups with several males to reduce the risk of infanticide. Van Schaik's discussion of several potential counterstrategies that males and females could employ to decide this sexual conflict in their favor should stimulate additional work on this fascinating topic.

2 • Multi-male breeding groups in birds: ecological causes and social conflicts

NICHOLAS B. DAVIES

INTRODUCTION

Two routes to multi-male groups

Thirty years ago, ornithologists would have had little to contribute to this volume. Lack (1968) had concluded that most birds are monogamous because 'each male and each female will leave most descendants if they share in raising a brood'. Some species were known to be polygamous and Lack followed John Crook's (1964) lead in linking species' differences in breeding systems to differences in ecological factors such as food and predation.

The early 1970s heralded a revolution in thinking about breeding systems. Parker (1970) and Trivers (1972) pointed out that it would pay a male to adopt a 'mixed strategy', both seeking extra-pair matings and guarding paternity. Selection on males to enhance success in sperm competition leads to selection on females to encourage or discourage multiple mating, so producing social conflicts which can be just as important as ecological factors in determining mating systems (Maynard Smith, 1977; Parker, 1979, 1984). Sexual reproduction was no longer viewed as a harmonious venture in which males and females cooperated to maximize their joint success, but rather as a battleground where it often paid individuals to desert offspring or otherwise exploit their mates. The key realization was that it did not make sense to ask which social system was best adapted to particular ecological conditions. Rather, social systems should be viewed as outcomes of individual behavior.

Inspired by these new ideas, field workers began more detailed studies which focused on adaptive decision making by individuals and the consequences for social systems. Aided by new molecular techniques for measuring parentage, it was discovered that extra-pair paternity was widespread, even in socially monogamous species (Westneat *et al.*, 1990; Birkhead & Møller, 1992; Gowaty, 1996). Furthermore, there was often variability in mating systems within a population, arising as different outcomes of sexual conflict (Alatalo *et al.*, 1981; Oring, 1982; Davies, 1989; Szekely *et al.*, 1996).

At about the same time as Parker and Trivers were changing our views about mating systems, cooperative breeding was beginning to be understood for the first time due to Hamilton's (1964) ideas on the evolution of altruism. In some species of birds, the young remain on their natal territory prior to dispersal or prior to inheriting the breeding role at home, and they help their parents to raise further broods. Such 'helping at the nest' can increase the helper's indirect fitness, through either improved production of sibs or improved survival of parents through reducing their workload. Helping can also increase the helper's direct fitness (personal reproduction) if the helper gains paternity or maternity in the home nest, or if it increases the helper's future reproduction, for example through experience or by acting as a payment to the breeders for permission to remain at home (Brown, 1987; Emlen & Wrege, 1989; Mulder & Langmore, 1993; Komdeur, 1996).

The last 30 years, therefore, have led to the realization that bird breeding systems can often be complex. Two routes to multi-male groups can be recognized.

(*a*) *Shared matings*. Several males may share matings with a single female (polyandry) or several females (polygynandry). The males may be close relatives (e.g. father and sons or sibs) as in acorn woodpeckers (*Melanerpes formicivorus*: Koenig & Mumme, 1987) or unrelated, as in dunnocks (*Prunella modularis*: Davies, 1992), Smith's longspurs (*Calcarius pictus*: Briskie, 1992), some pukeko (*Porphyrio porphyrio*) populations (Jamieson *et al.*, 1994), Galapagos hawks (*Buteo galapagoensis*: Faaborg *et al.*, 1995) and brown skuas (*Catharacta lonnbergi*: Millar *et al.*, 1994). The groups can arise through coalitions of male relatives staying on their natal territory to breed or through related or unrelated coalitions taking over other territories (both occur in acorn woodpeckers and pukekos). Alternatively, multi-male groups may arise when single males join other unrelated males. This occurs in dunnocks, where a male helps to feed chicks only if he gained a share of the matings (Davies *et al.*, 1992), and in pied kingfishers (*Ceryle rudis*), where a male may help even in the absence of a mating share because this increases the chance of him mating with the female in future attempts (Reyer, 1990; see Price, 1990, for a similar case in a primate).

(*b*) *Helpers on the natal territory*. In most birds, females disperse more than males, so helping at the nest more often involves young males staying on their natal territory and helping to raise sibs (Brown, 1987).

These two routes merge if males inherit the breeding vacancy on their natal territory when their mother dies. For example, in acorn woodpeckers (Koenig & Mumme, 1987) and striped-backed wrens (*Campylorhynchus nuchalis*: Rabenold *et al.*, 1990; Piper & Slater, 1993), young males avoid incest with their mothers but compete for matings if their mother is replaced by a new female. Nevertheless, it is still useful to distinguish the two routes because it helps focus on the various types of fitness gain in multi-male groups, including both present and future fitness through direct and indirect means (Brown, 1987). In some bird species, populations may include multi-male groups formed through both routes (Table 2.1).

The aim of this chapter is not to provide a comprehensive review but rather to focus on some well-studied species to raise general issues which may also apply to primates. Birds differ from primates in two obvious ways. First, males play more of an active role in helping to rear offspring, and second, females are more mobile and less often in defendable groups. Where polygyny occurs, it is usually through male monopolization of resources which females require (food, nest sites) rather than of females themselves (Emlen & Oring, 1977). Nevertheless, there are some striking parallels between birds and primates in the selective pressures giving rise to multi-male groups, and in their consequences for social relationships and individual reproductive success. Table 2.2 summarizes some of the main themes that will emerge. I begin by discussing the simplest case of a multi-male group (two males with one female) and then consider polygynandry (several males with several females) and finally polygynandry together with non-breeding helpers.

Table 2.1. *Paternity, chick feeding, and group structure in multi-male groups of white-browed scrubwrens*

Group structure	*n*	Percentage of cases in which the beta male: shared paternity	helped feed nestlings
Beta male with mother and father	44	0	50
Beta male with mother and new male	2	0	0
Beta male with father and new female	38	19	78
Beta male with unrelated male and female	16	46	65

Note: Multi-male groups (two males and one female) of white-browed scrubwrens (*Sericornis frontalis*) in the Canberra Botanic Garden, Australia, include: groups formed (i) through beta males joining unrelated pairs and gaining shared paternity, and (ii) through natal philopatry, where young males may gain indirect fitness through helping their parents or direct and indirect fitness through sharing paternity with their father breeding with a new female. From Magrath & Whittingham (1997) and Whittingham *et al.* (1997).

DUNNOCK POLYANDRY AND POLYGYNANDRY

'Dun' means brown, 'ock' signifies little, and the dunnock is the archetypal little brown bird which feeds on the ground, often under bushes and hedges, searching for small insects and seeds. Colleagues and I have studied a color-ringed population of about 80 breeding adults in the Cambridge University Botanic Garden (Davies, 1992). Both males and females are territorial and various mating systems emerge depending on how the male territories overlap the female territories. A single male may defend one female territory (monogamy) or two adjacent female territories (polygyny). Alternatively, two males may jointly defend one female territory (polyandry) or two or three adjacent female territories (polygynandry). The two males that share a territory are not close relatives and there is usually a clear dominance order, with the alpha (usually older) male able to displace the beta male (often a first-year) from feeding sites and the vicinity of the female. Occasionally a third male (gamma) is involved. During a ten-year study, 55% of breeding groups ($n = 254$) were multi-male groups (30% polyandry, 25% polygynandry).

Multi-male groups form in two ways (Fig. 2.1). First, a beta male may join a monogamous pair or a polygynous trio to form, respectively, a polyandrous or polygynandrous group. The resident alpha male always tries to chase the beta male away and sometimes beta males can be seen wandering from territory to territory. Eventually, however, they manage to force themselves in somewhere; there are no 'floater' males. Second, adjacent male territories may merge when a

Table 2.2. *Some of the main themes emerging from studies of multi-male breeding groups of birds*

Selective pressures	Ecological factors	Social conflicts
1. Territory acquisition and defense	Habitat saturation Territory quality Intruder pressure	
2. Exclusive mate guarding	Habitat (e.g., visibility) Territory size Intruder pressure Female synchrony	1. Stay versus leave 2. Paternity sharing 3. Parental effort (kinship among breeders, helpers and offspring affects these outcomes)
3. Cooperative brood care	Harsh environment Food availability	

single female wanders between them (to form a polyandrous trio) or when a male of a monogamous pair expands to dominate a monogamous pair next door (to form a polygynandrous quartet). Two males were more likely to monopolize more than one female when neighboring female territories were smaller. For example, an experimental supply of food to females led to a reduction in their territory size and an increase in polygynandry (Davies & Lundberg, 1984).

Although two males may eventually agree to share a territory, and cooperate in defense against neighbors, once the female has completed her nest and begins to solicit matings, the alpha male guards her closely, following her wherever she goes, and he tries to prevent the beta male from mating. However, beta males often gain a share of the matings, not only through their own persistence but also because a female actively avoids the alpha male's guarding and encourages the beta male to mate by soliciting to him. Beta males are more likely to gain shared matings on larger territories with denser vegetation, where alpha males often lose their female and so she can hide away with the beta male. In polygynandry, beta males are more likely to be successful if the two females have synchronous fertile periods, because whenever the alpha male associates with one female, the beta male has free access to the other (Davies & Hatchwell, 1992). Therefore, the outcome of sexual conflict over mate guarding depends partly on ecological factors, partly on female synchrony, and partly on the relative competitive abilities of the alpha and beta males and females.

DNA-fingerprinting showed that shared matings often led to shared paternity of a brood, with a beta male's paternity share increasing in proportion to his mating share (Burke *et al.*, 1989; Davies *et al.*, 1992). Beta males helped to feed the young only if they gained a share of the matings. There was no indication that males could distinguish their own-sired young. For example, a male would help to feed the brood if he obtained a mating share even in cases in which he got no paternity. Therefore, males apparently used mating share as an indirect cue to paternity chances. To understand the conflicts, it is helpful to separate the two types of multi-male group.

(*a*) *Polyandry*. If both males shared matings, then both helped the female to feed the young. The help from two males exceeded that of one male, so the chicks were better fed and as a result females raised more young. Temporary removal experiments during the female's fertile period, which varied the alpha:beta mating share, showed that each male increased his provisioning rate with an increased share of the mating time. The total effort of the two males was maximal at an equal share of the work (Hatchwell & Davies, 1990), which occurred when matings were shared equally between them (Davies *et al.*, 1992). Observations of female behavior suggested that the females did, indeed, promote an equal mating share. A female solicited at a greater rate to whichever male had gained the least access, which was usually the beta male. She was thus able to boost his copulation share, so that beta males who gained just 20% of the mating access time could nevertheless perform 50% of the matings (Davies *et al.*, 1996). Females laid a larger clutch when both males mated compared to cases where the alpha male monopolized matings, so they varied clutch size in relation to the amount of male help they expected (Davies & Hatchwell, 1992). An interesting parallel here is with callitrichid primates, where twinning evolved after

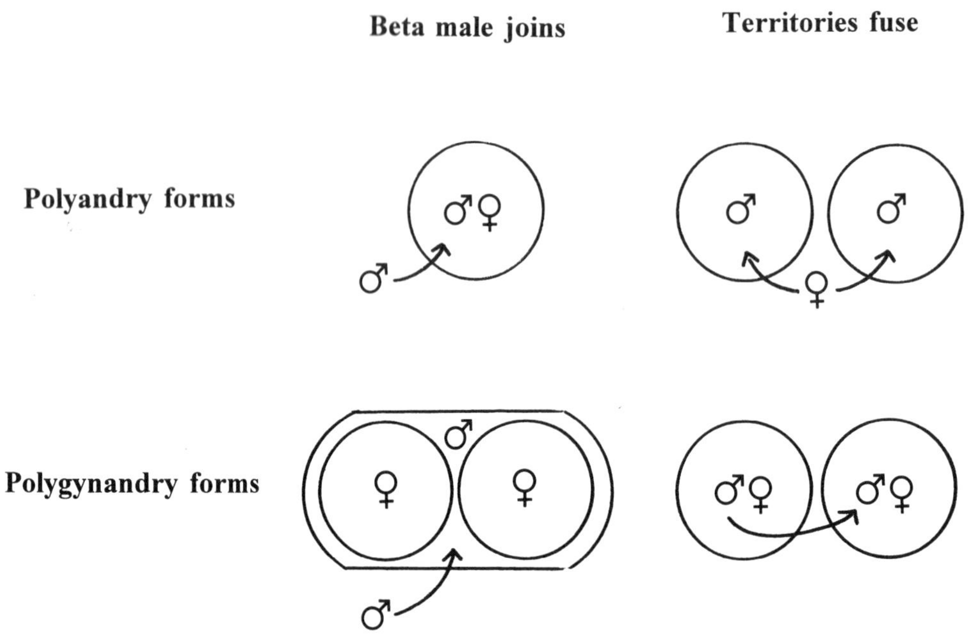

Beta male joins
Territories fuse
Polyandry forms
Polygynandry forms

Fig. 2.1. In dunnocks, multi-male groups usually involve two (unrelated) males. Polyandry may form by a beta male joining a monogamous pair or when a female wanders between two neighboring male territories. Polygynandry may form by a beta male joining a polygynous trio, or when a male of a monogamous pair invades a neighboring pair's territory.

pair-bonding, so females increased brood size only when they were assured of the potential of male help (Dunbar, 1995a).

(*b*) *Polygynandry*. If two female dunnocks had overlapping fertile periods, the alpha male usually spent time with both. As a result, both males mated with both females because the beta male was able to mate with whichever female was unguarded at the time. When the nestlings hatched, each male then usually helped full time at a different nest, choosing the brood belonging to the female with whom he had gained the greater mating share (Davies & Hatchwell, 1992). A game theory model predicts that this is often the stable way for the two males to divide their effort (Sozou & Houston, 1994). Now we have a single male helping at a nest where he has part paternity. In contrast to polyandry, there was no tendency for a male to reduce his parental effort in relation to a reduced mating share (Davies & Hatchwell, 1992). The likely explanation for the different male reactions is as follows. In polyandry, when a male loses paternity it is to another male helping at the same nest. Any reduction in his effort due to paternity loss will be fully compensated for by an increase in effort by the other male, who has gained paternity at his expense. In polygynandry, however, when each male helps at a different nest, any reduction in male effort is not compensated for fully by the female. In this case, a male with part paternity may be forced to provision at a full rate to save the lives of the chicks in the brood which are likely to be his own.

(*c*) *Variable mating systems*. The variable mating system of the dunnock can be interpreted as the different outcomes of sexual conflict. These stem from two sources. First, the increased production of young arising from a second male's help does not compensate the alpha male for shared paternity, so while the female promotes shared matings, the alpha male attempts to monopolize exclusive paternity. Second, a female suffers reduced reproductive success from shared male care but the combined output of two females exceeds that from one. So a female tries to drive other females away while a male attempts to gain a second female. Polygynandry can be viewed as a stalemate to the conflict: a male cannot drive a second male away and neither female can evict the other (Davies, 1992). Thus multi-male dunnock groups arise despite the alpha male's attempts to prevent them from forming. They are promoted by: (i) beta male persistence, (ii) the female encouragement of shared paternity, and (iii) the difficulty of exclusive mate guarding by alpha males in dense habitat or when there are two synchronously fertile females. These three factors may also apply to the formation of multi-male groups in primates (e.g. Ridley, 1986; Dunbar, 1988; Srivastava & Dunbar, 1996).

ALPINE ACCENTOR POLYGYNANDRY

A congener of the dunnock, the alpine accentor (*Prunella collaris*) inhabits high mountain tops. Our study in the French Pyrenées (Davies *et al.*, 1995, 1996; Hartley *et al.*, 1995) and those by Nakamura (1990, 1998a, 1998b) in the Japanese Alps and Heer (1996) in the Swiss Alps show that it breeds in large polygynandrous groups. An alpha male and from one to five unrelated subordinate males share a large range within which from two to five unrelated females breed. In the Pyrenées, the diet was mainly small invertebrates, especially taxa which dominate the 'fallout fauna' which is swept up to the mountain tops by winds from the valleys below. The birds' large ranges were necessary to exploit these prey, which were scarcer and more patchily distributed in both space and time than down by the treeline where dunnocks breed.

Each female in the group had her own nest but their fertile periods were often staggered, so that the first females began to breed two to three weeks before the last female. There was no tendency for females within a group either to avoid overlap or to synchronize their breeding. Rather, differences in timing were probably related to their ability to find sufficient food and suitable nest sites (Davies *et al.*, 1995). The behavioral conflicts were the same as in the dunnocks. The alpha male attempted to guard females while a female encouraged matings from subordinate males too. A male was more likely to help feed the brood if he had gained a share of the matings, and paternity of a brood was often split among two or three males (Hartley *et al.*, 1995; Heer, 1996). Females gained increased male help from sharing the matings, with up to four males helping at one nest (Nakamura, 1998a). However, when several females had synchronous broods, different males tended to help full time at different nests, and a male helped at the nest of the female with whom he had gained a greater mating share, just as in dunnock polygynandry (Hartley *et al.*, 1995). Males preferred to compete for matings than to help feed chicks, so when other females were fertile, a female with young often had to care for them alone, at least for part of the time. As with the dunnocks, females fledged more and heavier chicks with more male help, but this did not compensate the alpha male for shared paternity, which explains the behavioral conflicts (Nakamura, 1998a).

In summary, the large polygynandrous groups of alpine accentors, combined with female asynchrony, provide the potential for an alpha male to monopolize the matings with several females in turn. This produces potential polygyny costs for females, who thus gain by soliciting to subordinate males too to maximize male help with parental care. Although the behavioral and reproductive conflicts were similar to those in dunnocks, there were two interesting differences.

1. Whereas each female dunnock had an exclusive territory, in the Pyrenées, female alpine accentors had overlapping ranges and often competed directly with each other, for example when the whole group fed together in a snow clearing. Female alpine accentors showed two behavior patterns which reflected such direct competition. First, they sang during their fertile period, especially when the males were associating with other females, and their song attracted males for mating (Langmore *et al.*, 1996). Second, they developed bright red cloacas to signal their fertility (Nakamura, 1990), reminiscent of the sexual swellings of female primates which breed in multi-male groups (Dixson, 1983; Pagel, 1994b). Females solicited up to 50 times per hour and increased their solicitation rate when other females in the group were fertile (Davies *et al.*, 1996). Like song, this display seemed to represent an attempt to outsignal female rivals and so gain male attention. In primates, too, females are more likely to initiate the copulations when there is increased competition among females (Dunbar, 1978, 1988) or when females gain from mating with multiple males (Janson, 1984).
2. Whereas a female dunnock showed equal preference for alpha and beta males, a female alpine accentor preferred mixed matings but with a bias in favor of the alpha male. We found no evidence that alpha males were better sires: in broods of mixed paternity, chicks sired by alpha males were no heavier than those sired by subordinate males (Davies *et al.*, 1996). Instead, we suggested that the female preference maximized male help because, unlike dunnocks, alpha and subordinate male alpine accentors reacted differently to reduced mating share. When just one male helped at a nest, subordinate males did not reduce their provisioning effort in relation to a reduced

mating share, provided they gained more than a critical minimum share. Alpha males, however, did reduce their effort as paternity chances were reduced, perhaps because they had greater opportunity costs from helping, in the form of lost opportunities of mating with other females in the group. Under these conditions, it pays a female to give a minimum share to a subordinate male so he crosses the helping threshold, and most to the alpha male to maximize his help (Davies *et al.*, 1996). Alpha male alpine accentors may behave differently from alpha male dunnocks because they have up to five females, compared to dunnocks which usually have just one or two, so they may demand greater paternity payoffs before they are willing to invest in paternal care and so forgo mating opportunities elsewhere.

FEMALE PROMOTION OF MULTI-MALE GROUPS TO MAXIMIZE MALE HELP OR MINIMIZE MALE HARASSMENT

Recent game theory models by Harada & Iwasa (1996) and Houston *et al.* (1997) provide a useful framework for summarizing the dunnock and alpine accentor results. They also include the possibility of 'negative' male help in the form of infanticide, which makes the argument more relevant to primates. Although female primates may gain help from males with whom they have copulated, in the form of either infant carrying (Goldizen, 1987a, 1989) or better access to food (Janson, 1984), the main advantage to multiple mating may often be to give males sufficient paternity chances to avoid infanticide (Hrdy, 1977, 1979; Borries, 1997).

The models involve a two-step game in which the female chooses a distribution of paternity among the males and the males then choose their level of parental effort (negative or positive), which depends both on their paternity and on the effort of the other participants. The male reactions can then influence the evolution of the female's decision concerning paternity. The main conclusion is that the female's best option varies with the form of the function linking male care to paternity. Consider the case where a female has the opportunity to mate with two males and she has complete control over the distribution of matings. In Fig. 2.2a, where the male responses are concave down and symmetrical, the female does best to split paternity equally between them, as in dunnocks, and the result is cooperative polyandry. In Fig. 2.2b, where the male responses are concave down but asymmetric, it pays the female to bias the paternity, as in alpine accentors, and again the outcome is cooperative polyandry. In Fig. 2.2c, where male responses are convex up, females do best to give all the matings to one male so there is monogamy. Alternatively, if a second male poses a threat of infanticide, it pays the female to give him a small share of the paternity and the outcome is social monogamy with extra-pair mating (Fig. 2.2d).

Hartley & Davies (1994) removed female dunnocks from some territories to give neighboring females with two males the opportunity to mate with a third male. They found that females often refused to give a third male a mating share and suggested that this was because shared paternity among three or more males may lead to less total male effort than from two males. However, Houston *et al.* (1997) show that the optimal number of males is very sensitive to the fitness function linking male effort with paternity and male effort with current and future fitness. As a result, it is difficult to make precise predictions.

Two further points are worth making to end this section. First, in birds extra-pair matings may also be driven by female choice for better genetic sires for their offspring (Hasselquist *et al.*, 1996; Kempenaers *et al.*, 1997). In some cases, female behavior may reflect a trade-off between multiple mating for genetic benefits and the need for paternal care. For example, female superb fairy wrens (*Malurus cyaneus*) allocate all the paternity to their mate if they have no helpers, but with helpers they are apparently freed from the constraints of relying on their mate for help and often give most of their matings to high-quality males on other territories (Mulder *et al.*, 1994). Second, the sexual conflict over the formation of multi-male breeding groups, which dominates the lives of dunnocks and alpine accentors, will change to sexual cooperation if the extra production of young from multi-male help offsets the cost of paternity sharing. This seems to occur in saddle-backed tamarins (*Saguinus fuscicollis*), for which lone pairs are unlikely to raise offspring and polyandry appears to pay both males and females (Goldizen, 1987a, 1989).

MULTI-MALE GROUPS FOR COOPERATIVE TERRITORY DEFENSE

In primates, it may pay males to form strong affiliative associations with other males in order to increase their access to groups of females (van Hooff & van Schaik, 1992; Strier, 1994b). In birds, too, males may cooperate to acquire or defend high-quality territories to increase their access to mates and sometimes they share copulations amicably.

Acorn woodpeckers

A good example is the acorn woodpecker studied by Koenig & Mumme (1987) in coastal California. The birds live in

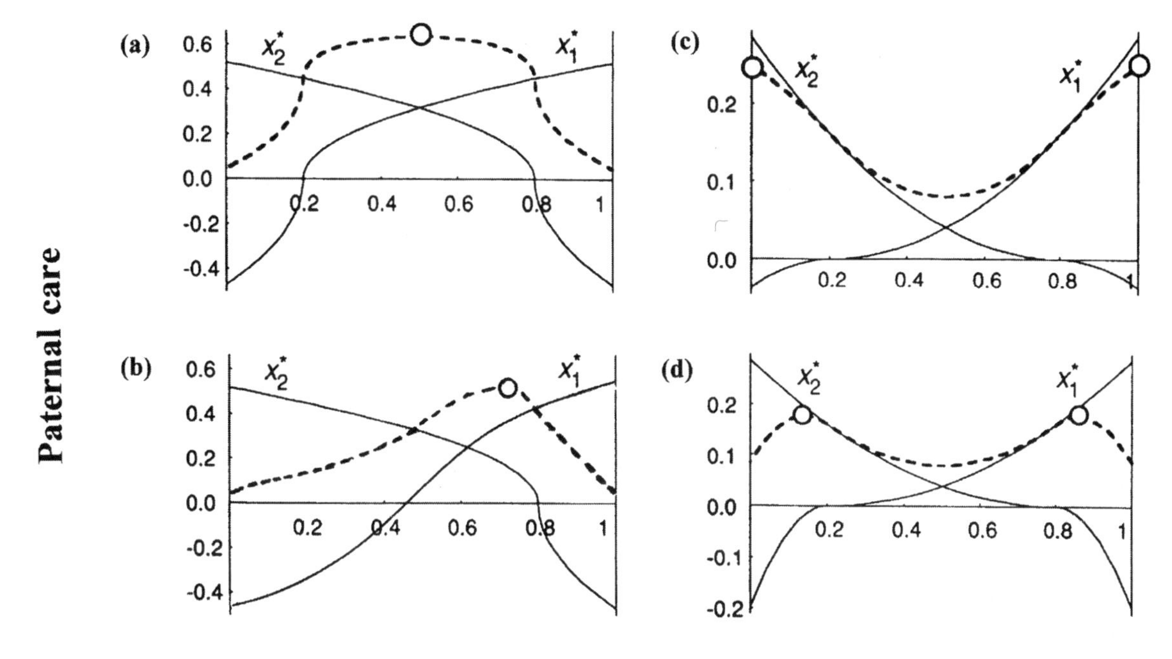

Fig. 2.2. How a female best shares matings between two males (X1 and X2) depends on how a male's parental care or harrassment varies in relation to his paternity. (a) The two males behave in the same way. Each demands a minimum paternity share, in this case 0.2. to refrain from infanticide (negative care). Thereafter, paternal care increases with paternity but at a decelerating rate. The female's best option to maximize the total help from both males (dotted line) is to share paternity equally between them. Dunnock cooperative polyandry fits this case well. (b) Male 1 demands a greater paternity share for a given parental effort so it pays the female to bias the paternity in his favor. This occurs in alpine accentor polygynandry. (c) If there are minimal risks of infanticide and male responses are accelerating, then it pays females to give all the matings to one male and only he provides help (hence genetic and social monogamy). (d) With more substantial infanticide risks and accelerating curves, it pays females to give a small share to one male (to avoid infanticide from him) and most to the other male (to maximize his help). The result is social monogamy (one male helps) with extra-pair mating. Figure modified from Harada & Iwasa (1996).

large polygynandrous groups and each group defends a permanent territory centered on large stores of acorns embedded in the bark of granary trees. The survival and reproductive success of the group members are dependent on these stores. The groups contain up to four co-breeding males, up to three co-breeding females, plus up to ten non-breeding helpers, who are previous offspring. The breeding males are relatives, often sibs or father and sons. The females, too, are relatives (from another family), either sibs or mother and daughters, and they all lay in one communal nest. The helpers remain on their natal territory for a while, either because there are no breeding vacancies or because the quality of the vacancy is poorer than the home territory. Delayed dispersal to await a better territory, or to inherit the natal territory, can often be better than immediate dispersal and breeding on a poorer territory (Stacey & Ligon, 1987; Koenig *et al.*, 1992).

When breeding vacancies occur, the result is a 'power struggle' in which up to 30 birds from as many as 15 neighboring groups arrive and compete to take over the territory. The contest may last several days as same-sex sibling coalitions battle for control of the granary. Hannon *et al.* (1985) removed breeding males from some territories and found that larger male coalitions tended to win the vacancy. Males in duos and trios survive better than single males, probably because they can take over and defend better quality territories. As a result, male lifetime reproductive success is greater in a multi-male group and this selects for male cooperation (Mumme *et al.*, 1988).

Takeovers by new males often spark off interesting conflicts in the group. Female offspring act as non-breeding helpers as long as their father remains in the group. When he dies, they are released from an 'incest taboo' and compete with their mother to breed with the new males (Koenig &

Table 2.3. *Comparison of multi-male groups of pukeko in two areas of New Zealand*

	South Island Otago[1]	North Island Shakespear[2]
Group composition	1–3 unrelated males + 1–2 females	3–7 related males + 1–2 females + previous offspring
Territories	Breeding season only Flocks in winter	Permanent year-round
Winters	Harsh	Mild
Adult mortality	High	Low
Breeding vacancies	Abundant, but high-quality territories scarce	Scarce
Young birds	Disperse	Remain on natal territory
Reproductive success per male	Higher in coalition than in pair	Higher in pair than in a group[3]
Reproductive skew	Low (equal share)	High (dominant males get most paternity)
Why do multi-male groups form?	Polyandry-threshold model	Habitat saturation (prisoner's dilemma)[3]

Notes:

[1] Jamieson *et al.* (1994); Jamieson (1997); Jamieson & Quinn (in preparation).

[2] Craig & Jamieson (1990).

[3] These conclusions refer to another study site on North Island where breeding density was lower than at Shakespear, and where the kinship among the breeders was not known: Craig (1984).

Pitelka, 1979). This competition often involves egg destruction as females attempt to increase their own share of the communal clutch (Mumme *et al.*, 1983). The male helpers are now faced with the prospect of helping their mother to raise half-sibs and may leave the group or be driven off by the new breeding males. In other species, young males may turn on their mother, drive her off, and attract new females to the territory, so attaining breeding status themselves (Emlen, 1997).

Pukekos

Competition to take over high-quality territories can also promote cooperation and amicable sharing of copulations among unrelated males, who may as a result enjoy greater lifetime success in a multi-male group than on their own (Galapagos hawks, Faaborg & Bednarz, 1990; Faaborg *et al.*, 1995; and perhaps also Tasmanian native hens (*Gallinula mortierii*, Goldizen *et al.*, 1998). Jamieson's study of pukeko near Otago, on the South Island of New Zealand, provides a particularly well-studied example (Jamieson *et al.*, 1994; Jamieson, 1997; Jamieson & Quinn, in preparation). Table 2.3 (left-hand side) summarizes the main points. Coalitions of two (sometimes three) unrelated males form in the winter flocks and cooperate to take over and defend the scarce, high-quality breeding territories which provide rich vegetation for food and safe nesting. Single males end up on poorer territories. In the male coalitions, males not only cooperate in territory defense but also share matings equally and do not interrupt each other's copulations (in marked contrast to dunnocks). In a multi-male group, as a result of gaining a better territory, each male has higher reproductive success and survival than single males. Thus, cooperation results from males crossing a 'polyandry threshold' (Gowaty, 1981; Davies, 1992, Fig. 14.1). Although it pays males to cooperate, there are no benefits to dominant females from sharing their nest with unrelated subordinate females (when there are two females in a group, both lay in the same nest). Dominant females try to evict subordinates but shared nesting often occurs because of the difficulties of evicting a persistent parasite and because of constraints on egg recognition. Males, however, benefit from two females so there is a conflict between the sexes here, as in dunnocks (Jamieson & Quinn, in preparation).

Another study by Craig of pukeko on North Island provides a fascinating contrast (Table 2.3, right-hand side). Here, high adult survival in the milder winters leads to habitat saturation and the formation of multi-male kin groups, rather like those of the acorn woodpecker. In a second study area on North Island, Craig (1984) found that males had higher reproductive success in monogamous pairs than in multi-male groups (the reverse of the Otago population). Nevertheless, most males accepted subordinate males on their territory. One explanation could be that subordinates force themselves on to pair territories (as in dunnocks). However, Craig suggested an alternative explanation under the assumption (not yet tested) that dominants can decide whether a subordinate settles. He argued that if all birds bred as pairs, it would pay a dominant male to accept a subordinate male. By so doing, it could expand its territory at the expense of neighboring pairs and each male may then enjoy increased success. The only way a neighbor can prevent this is by matching its defense forces and accepting a subordinate, too. The end result is rather like the prisoner's dilemma: each male would do best in a pair but under these conditions groups prosper, so the stable solution is for all males to breed in groups despite the fact that they do less well than in pairs.

The breeding systems of some neotropical primates are remarkably similar to the 'polygynandry with helpers' shown by acorn woodpeckers and pukekos. For example, in red howler monkeys (*Alouatta seniculus*) it pays males to form related or unrelated coalitions to take over the limited number of female groups. Furthermore, young males may remain in their natal troops, helping to defend the territory, until breeding vacancies arise either at home or nearby (Pope, 1990).

REPRODUCTIVE SKEW IN MULTI-MALE GROUPS

At first sight, the contrast between the two pukeko studies seems to support 'reproductive skew theory' (Emlen, 1982; Vehrencamp, 1983a, 1983b; Reeve & Ratnieks, 1993; Keller & Reeve, 1994). This has two main assumptions: (1) dominant males can control the mating share, and (2) they benefit from the presence of subordinates in the group, for example through shared brood care and territory defense. The theory predicts that dominants should then offer subordinates a 'staying incentive' in the form of a share of the matings. Three predictions are:

1. The share offered should be larger to unrelated males than to related males because the former will gain fitness only through shared paternity while related males also gain indirect fitness through their kinship with the dominant, so may stay with smaller direct gains. Therefore, unrelated coalitions should have lower reproductive skew than related coalitions.
2. The mating share offered to the subordinates will increase with an increased chance that the subordinate can leave and breed independently.

a)

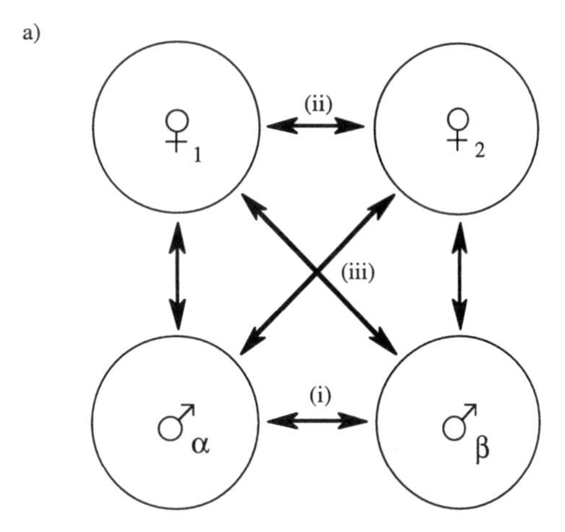

b)

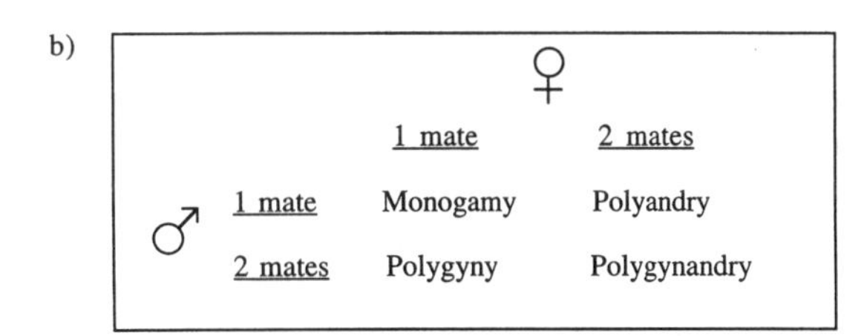

Fig. 2.3. (a) Interactions within the sexes (i and ii) may involve cooperation or conflict. These can create conflicts between the sexes (iii) if one sex gains from multiple mates but the other suffers a cost. (b) Different mating systems emerge depending on the ability of individuals to exclude rivals or form alliances with members of the same sex, and to monopolize members of the opposite sex. From Davies *et al.* (1996).

3. The offer should be larger to subordinates who are better competitors as a 'peace incentive' to prevent fights in the group.

As predicted, there is lower reproductive skew in the Otago population where unrelated males cooperate and where there are greater opportunities for independent breeding by subordinates (Jamieson *et al.*, 1994; Table 2.3). However, there are alternative explanations. Subordinates in related multi-male groups may get lower mating shares because they are poorer competitors or because they avoid inbreeding with female relatives. Furthermore, it seems unlikely that the assumptions of current reproductive skew theory will apply to many cases (Cant, 1998; Clutton-Brock, 1998). For example, dominant males are certainly unable to gain complete control of the mating share in dunnocks and alpine accentors, where subordinate males and females favor outcomes different from the dominant male's optimum. Dominant males in these two species also do not benefit from a subordinate's presence. Rather, multi-male groups occur despite the alpha male's attempts to prevent them. Models for reproductive skew need to consider such conflicting interests between three or more players (Davies, 1992; Jamieson & Quinn, in preparation).

Figure 2.3 summarizes the various interactions discussed in this chapter, and their outcomes in terms of mating systems. All the interactions may involve various degrees of conflict or cooperation. First, two males may agree to share matings if their cooperation is needed to defend the territory or females against other males, or if through their cooperation in brood care they can each raise more offspring than when working alone. Alternatively, the dominant male may attempt to monopolize the matings if the gain from the subordinate's assistance does not compensate him for shared paternity. Second, two females may cooperate if their alliance gains them access to better resources (Wrangham, 1980) or mates or reduces predation (Smuts & Smuts, 1993), but there will be conflicts if these benefits are outweighed by the costs of polygyny, such as shared male help or increased infanticide by competing females. Finally, these interactions within the sexes can create conflicts between them if the dominant male gains by excluding subordinate males but the female benefits from multiple paternity, or if a male benefits from polygyny but the females suffer. The various mating outcomes and degrees of reproductive skew may then reflect the different resolutions to these conflicts and will depend on which individuals can gain their best options despite the conflicting interests of others.

CONCLUSIONS

There are striking parallels between the formation of multi-male groups in birds and primates, and their consequences for social relationships and reproduction. In birds, multi-male groups often involve (1) shared matings with one female (polyandry) or several females (polygynandry), or (2) non-breeding males which act as helpers on their natal territory. Selective pressures promoting multi-male groups through male or female benefit, or both, include competition to acquire and defend territories, the difficulties of exclusive mate guarding, and the benefits of cooperative brood care. In dunnocks and alpine accentors, females encourage multi-male groups by soliciting matings from subordinate males while dominant males attempt to evict subordinates or limit their paternity. Females have greatest reproductive success if they share matings equally between two males (dunnocks) or bias the share toward the alpha male (alpine accentors), while a male does best if he gains exclusive paternity. The outcome of sexual conflicts may depend partly on ecological factors and also on how male parental effort (including negative effort in the form of infanticide) relates to paternity and how it affects current and future fitness. In birds, males are most likely to share copulations amicably in cases where it pays to cooperate to defend or acquire high-quality territories. Such cooperation occurs among male kin in acorn woodpeckers and among unrelated males in pukekos and Galapagos hawks. To understand how matings are shared in multi-male groups, current theory needs to be extended to consider the reproductive conflicts which occur between and within the sexes.

ACKNOWLEDGMENTS

The studies of dunnocks and alpine accentors were done together with Mike Bruford, Terry Burke, André Desrochers, Ian Hartley, Ben Hatchwell, Naomi Langmore, Arne Lundberg, Denis Nebel, Tim Robson and Jackie Skeer, and I thank them and the Natural Environment Research Council for funding the work. The summary of the pukeko studies was stimulated by discussions with Ian Jamieson and by a visit to Otago University, generously funded by a William Evans Fellowship. I also thank Peter Kappeler for helpful comments and for arranging the stimulating meeting which gave rise to this volume, and Ann Jeffrey for help with the manuscript.

3 • Males in macropod society

PETER J. JARMAN

INTRODUCTION

Most people are familiar with the large kangaroo species that are so often displayed in zoos, but know little about the diversity of other species of kangaroos, wallabies and rat-kangaroos that together compose the macropods. Indeed, most people do not know how diverse the marsupials are in general. Hence they fail to realize the opportunities that marsupials present to test, upon a second infra-class of Mammalia, the generalizability of correlations between ecology, life history and social organization derived from studies of eutherian ('placental') mammals.

The Eutheria and Metatheria separated some 104 million years ago (Kirsch *et al.*, 1997). While they share primary mammalian characteristics of morphology and physiology, they differ in metabolic rate (slightly) and in organization of the stages of embryonic and juvenile dependence on the mother. Nevertheless, the two infra-classes have followed remarkably similar evolutionary trajectories, evolving trophic specializations that converge in many details of morphology and behavior. Those convergences make one think that, all else being equal, a socioecological argument for the adaptiveness of a behavioral trait derived for one infra-class should hold true for comparable species in the other. Conversely, failure of radiations to converge in what appear to be similar situations should also be instructive (as argued by Kappeler & Heymann, 1996, for comparisons within the primates).

Comparisons require that the radiations be similarly rich; no generalizable comparison can be made between a single, possibly idiosyncratic species and a diverse array of species. The metatherian (marsupial) order Diprotodontia, containing 10 families and about 130 species, is as diverse as the eutherian order Primates, containing 11 families and about 180 species. The most recent revision (Kirsch *et al.*, 1997) divides the diprotodontian super-family Macropodoidea (colloquially called the macropodoids) into two families: Hypsiprymnodontidae, with one living species, the minute, rainforest-dwelling musky rat-kangaroo *Hypsiprymnodon moschatus*, so plesiomorphic that it bears two or three young at a birth and does not hop; and Macropodidae, containing all other living kangaroos, wallabies and rat-kangaroos. This family, the macropods, contains two subfamilies, Potoroinae (potoroines: ten species of potoroos, bettongs and rat-kangaroos) and Macropodinae (macropodines: 56 species of kangaroos and wallabies). The macropods form the most species-rich family in their order. They were even richer until a tribe of mainly large, short-faced, browsing kangaroos, the Sthenurini, became extinct (except for one small species) in the late Pleistocene period. Macropods show species richness similar to that of the great primate radiations, the Callitrichidae, Cebidae, Cercopithecidae, and the Madagascan lemurs (Table 3.1). Like the primates, and the artiodactyl Bovidae and Cervidae, they show one of the successful ways of being a large, herbivorous, modern mammal.

Table 3.1. *Richness of recently extant species and genera among some of the great radiations of herbivorous mammals*

Family	Genera	Species
Metatheria		
Macropodidae	15	66
Eutheria		
Artiodactyla		
Bovidae	46	120
Cervidae	16	36
Primates		
Callitrichidae	5	26
Cebidae	11	51
Cercopithecidae	16	90
Lorisidae	8	16
'Lemurs'	14	32

Note:
Numbers of genera and species are taken from Strahan (1995, with minor up-dating) for macropods, Macdonald (1984) for artiodactyls, and Kappeler & Heymann (1996) for primates. 'Lemurs' contain the Madagascan families Cheirogaleidae, Daubentonidae, Indridae, Lepilemuridae, and Lemuridae.

Previous comparisons have been made of community structure and dietary adaptations (Smith & Ganzhorn, 1996) and of social organization, behavior and life history (Winter, 1996), between lemuriform primates and arboreal marsupials; of mating systems, growth and sexual dimorphism amongst bovids, cervids and macropods (Jarman, 1983); and of spatial organization and behavior amongst many metatherian and eutherian radiations (Jarman & Kruuk, 1996).

This contribution aims to stimulate thought about the effects and adaptive value of male behavior in primate groups by looking at those characteristics in the macropods. The chapter first reviews the natural history of macropods, then uses as an example grouping, reproduction and male behavior in the eastern gray kangaroo (*Macropus giganteus*), a social and well-studied macropod species, before arguing the adaptiveness of macropod systems and comparing macropods and primates.

CHARACTERISTICS OF FAMILY MACROPODIDAE

Sizes and ecology

The 66 extant macropods range from 1 kg to a maximal size of 93 kg for very large males of the red kangaroo (*Macropus rufus*: Table 3.2; Jarman, 1989a). As described below, the smaller species are homomorphic, while the largest are strongly heteromorphic (*sensu* Jarman, 1983); the largest female macropods weigh less than half the heaviest male conspecifics.

Macropods are found in nearly all habitats from deserts to rainforest, from the coast to the alpine zone. Most species are terrestrial, but tree-kangaroo species in the genus *Dendrolagus* climb trees, and rock-wallaby species in the genus *Petrogale* species nimbly climb rocky cliffs and steep hillsides, and even into low trees. All the potoroine species hide from predators in a nest (and one in self-dug burrows) in the daytime, emerging at night to forage. Rock-wallabies take daytime refuge in caves, overhangs and interstices between boulders to escape from predators and extremes of radiant heat. Many small or medium-sized wallabies 'squat' (like hares, *Lepus* spp.) during the day, under cover of vegetation, emerging at night to feed. Tree-kangaroos roost in rainforest trees during the day and forage there at night; however, some also forage and move on the ground at night (Flannery *et al.*, 1996). Forest-dwelling, medium-sized macropods come out into clearings to forage only at night, some extending this foraging to dusk and dawn.

Only the largest macropod species do not regularly use a defined sheltering habitat or structure, but even they use shade of bushes or trees to escape the sun. The largest species are least restricted to nocturnal foraging, and feed well into the daytime; the eastern gray kangaroo forages for as much as 17 hours a day in winter (Clarke *et al.*, 1989). No species in the Macropodidae confines its activity to the daytime, although the plesiomorphic musky rat-kangaroo is diurnal, hiding in a nest at night (Dennis & Johnson, 1995).

Needing shelter, most medium-sized and some small macropods live in forested habitats. The largest species inhabit open communities: grasslands, shrublands, savannah and light woodland. Some small species also inhabit those open habitats, squatting or nesting in the shelter of shrubs or grass tussocks.

All extant macropods have narrow muzzles and tend to pluck food items (e.g. a leaf or blade of grass) one at a time. The potoroines are highly selective for plant storage organs, including underground sporocarps of fungi (truffles) (Dawson, 1989), and occasionally eat invertebrates (personal observation). Their food items are rich, scattered, and have to be individually sought and processed. Most macropodines take a mixed diet of grasses, forbs, dicot leaves or fallen fruits. Some of these items are patchily distributed in space and time. Only the largest kangaroos are pure grazers. In the right habitat, their food items can be well distributed, although their availability and quality vary in space and time.

Sociality and predation

The great majority of macropod species are solitary or at best weakly gregarious. Only some of the largest species regularly forage in groups (Croft, 1989; Jarman & Coulson, 1989; Jarman, 1991); they can be considered 'social' because animals interact and coordinate activities and direction of movement within the group. Even for social macropod species, mean foraging group size is small, usually less than ten animals (Jarman & Coulson, 1989). Groups are of open membership, and group flux (rate of change of members) can be high. Although a foraging group of kangaroos may change membership several times in an hour, most individuals in social species spend most of their lives accompanied by conspecifics.

Macropod society is normally 'open.' With few exceptions, any class of animal can be in a particular place and can join or leave a group, although some individuals temporarily attract or repel others. Reported instances of 'territoriality' consist of defense of access to a limited refuge resource. However, some male rock-wallabies and perhaps tree-

Table 3.2. *Characteristics of genera of macropods (Macropodidae) and the musky rat-kangaroo (Hypsiprymnodontidae)*

	Number of species	Weight range (kg)	Weight dimorphism	Breeding seasonality	Young per birth	Activity period	Sociality
Hypsiprymnodontidae							
Hypsiprymnodon	1	0.5	1.2	S	2(3)	D	A
Macropodidae							
Potoroinae							
Aepyprymnus	1	3.5	1.0	A	1	N	A
Bettongia	4	1–2	1.0	A	1	N	A
Caloprymnus	1	1	0.8	A	1	N	A
Potorous	4	1–2	1.0–1.1	A	1	N	A
Macropodinae							
Dendrolagus	6	5–15	1.0–2.0	A	1	N, N+C (D)	A
Dorcopsis	3	5–11	1.5–3.0?	A	1	N (D)	A
Dorcopsulus	2	1.5–3.5	0.7–1.0?	A	1	N?	A
Lagorchestes	4	0.9–4.5	1.0–1.2	A	1	N	A
Macropus	14	4–93	1.3–2.5	A(S)	1	N+C, N+D	A,S
Onychogalea	3	3.5–9	1.3–1.6	A	1	N	A
Petrogale	16	1–12	0.9–1.4	A	1	N(+C)	A, (S)
Setonix	1	2.5–4	1.2	S(A)	1	N(+C)	A, (S)
Thylogale	5	2.5–12	1.5–1.9	A	1	N	A
Wallabia	1	10–20	1.5	A	1	N+C (D)	A
Lagostrophus	1	1–2	1.0	S	1	N	A

Note:
Weight range (kg) is of adult animals of both sexes; however, weight dimorphism is based on the ratio of greatest recorded weight of male to that of female for each species. Breeding seasonality is either aseasonal (A) if young are born at any time of year, or seasonal (S) if it is known that there are some months each year in which young are not born. One species and one island subspecies in the genus *Macropus* and the island subspecies of *Setonix brachyurus* breed seasonally. Activity period is that part of the 24 hours in which animals move and forage; it is either D = diurnal, N = nocturnal, or C = crepuscular. Exceptions shown by one species in the genus, or some populations, are shown in brackets. Sociality is either A = asocial or S = social (most animals spend part of the day in groups).

kangaroos may defend, for weeks or months, access to the refuges regularly used by one or a small number of females. Horsup (1996) reports mutual grooming between a male and a female allied rock-wallaby (*Petrogale assimilis*) associating in long-term pairs; mutual grooming is highly exceptional between the sexes in macropods. In many species, a male may defend, for a few days leading up to estrus, the space within a few meters of a female. Those examples apart, pre-emptive defense of space is not a feature of macropod society.

Social macropods do not live in permanent groups whose members exclude non-members. Nor are groups a long-term expression of male or female mating strategy; a male neither gathers a harem of females that he defends against other males, nor excludes all other males from a fixed territory within which he has exclusive access to females. Short-term groups of males may form around a pro-estrous or estrous female, but such a reproductive group is ephemeral (lasting a few hours at most) and its members do not cooperate. Macropod groups are not reproductively based.

Lack of sociality does not mean that males do not interact. Indeed, because of the typical 'roving male' mating system, males of species that typically forage solitarily may meet and interact with all the males that their home range encompasses. Individual foraging ranges are never exclusive.

Macropods evolved with a suite of predators including marsupial 'lions,' Thylacoleonids, thylacine (*Thylacinus cynocephalus*) and its ecological replacement the dingo (*Canis lupus dingo*) (which invaded Australia only 3500 years ago) and wedge-tailed eagle (*Aquila audax*). All the small, solitary macropods are nocturnally active, spending the day in a nest or under cover, using crypsis to avoid eagles, to whom they are vulnerable if flushed from the nest. Some use physically secure shelter, such as hollow logs, as refuge from pursuing predators. .

In some of the larger, diurnal, social macropod species, the individual rate of being alert falls, and that of feeding rises, with increasing group size (Jarman, 1987). Dingoes are less able to approach (close and undetected) large groups than small (Jarman & Wright, 1993). This implies that grouping plays an antipredator role for the social macropod species, foraging in open country.

Macropods display a unique antipredator tactic. Females, when pursued by a cursorial predator, can relax the muscles of the pouch-opening, letting the pouch-young (if it is advanced enough to be detached from the nipple) fall to the ground. I have recorded this occurring in eastern gray kangaroo, western gray kangaroo (*Macropus fuliginosus*), red kangaroo, common wallaroo (*M. robustus*), red-necked wallaby (*M. rufogriseus*), swamp wallaby (*Wallabia bicolor*), and rufous bettong (*Aepyprymnus rufescens*). Ejection of pouch-young is a well known problem in handling captured macropod females of many species. Ejected pouch-young are easily caught by predators, aiding the mother's escape. Robertshaw & Harden (1985, 1986) recorded predation by dingoes upon swamp wallaby pouch-young and juveniles being so continuous that the locally typical summer peak in production of juveniles was replaced by year-round production. Such dingo predation can severely limit recruitment to the subadult population in swamp wallabies (Robertshaw & Harden, 1989) and eastern gray kangaroos (personal observation).

Despite this tactic of pouch-young ejection, on being alarmed by a predator, a female whose young is temporarily out of the pouch will, if she can, call her young and help it into the pouch before fleeing. She chooses carefully when and where to let her young out of the pouch, especially while it is small, and trains it to return upon her call (Stuart-Dick, 1987).

In the largest macropod species, large males can defend themselves effectively against attack by dingoes (Wright, 1993). However, there are no records of males attacking dingoes to defend females or juveniles.

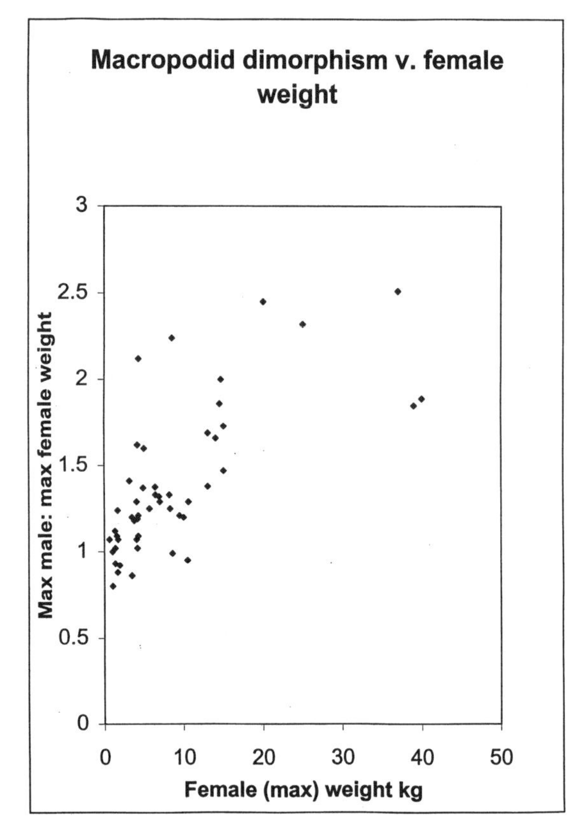

Fig. 3.1. Sexual size dimorphism and species size in the Macropodoidea. The ratio of maximum known male weight to maximum known female weight, plotted against maximum female weight (as an indicator of species size) for 52 macropod species. Data sources as in Jarman (1989a), with additional data from Flannery *et al.* (1996) and personal communications from several scientists.

Growth and dimorphism

The larger macropod species are all sexually dimorphic in weight (Fig. 3.1). Below a species' maximal female weight of about 5 kg (including all the extant potoroines), species cluster around homomorphy. Above that weight, with two exceptions, maximal male weight is usually much greater than maximal female weight. Nearly all species in which maximal female weight exceeds 15 kg are strongly heteromorphic, males being between 1.5 and 2.5 times heavier than maximal female weight. Some solitary species are nevertheless heteromorphic, e.g., bridled nailtail wallaby (*Onychogalea fraenata*), red-necked pademelon (*Thylogale thetis*), and swamp wallaby (Jarman, 1989a).

Dimorphism in growth and shape accompanies dimorphism in weight. The potoroines and the homomorphic macropodines appear to stop growing at an 'adult' body weight (but see below), but males in all the heteromorphic species grow either faster than females or for long after the age at which female growth stops (Jarman, 1989a). Males become bigger and heavier than females, and exaggerate arm and shoulder bone length and muscle development.

Larger macropods use their arms and forepaws to grapple the opponent during fighting; their other weapons are the hind feet, with which they kick the opponent. Heteromorphy has exaggerated the forelimb girdle, which plays only a small part in locomotion, but not the hind limbs, which, although weapons, are also the major locomotor organs.

Persistent male growth and progressive exaggeration of forelimb length and musculature differentiate males of successive ages or stages of development. They indicate likely rank in the mainly size-based hierarchies that prevail in macropod male society.

Reproduction

Details of the anatomy and physiology of reproduction in macropods can be found in Hume *et al.* (1989) and Seebeck & Rose (1989). This section gives a general picture of reproduction, relevant to the activities of males in the social organization and behavior of macropods.

Female reproduction

All macropod females can begin breeding before reaching maximal weight. Females of some smaller species (e.g., rufous hare-wallaby, *Lagorchestes hirsutus*, G. Lundie-Jenkins, personal communication; rufous bettong, personal observation) may conceive just before they are weaned! Females of the most heteromorphic species continue evident growth well past puberty; a red kangaroo female may first conceive when weighing 17 kg, yet continue to grow to 35 kg, double her weight at puberty.

All macropodid species bear a single young at a birth, after a gestation of 21–38 days (one or a few days less long than the estrous cycle). At birth, the young climbs unaided into the forward-facing pouch and attaches to a nipple. There it remains for 100–320 days. It detaches from the nipple and is able to leave and re-enter the pouch for the last days or months. Its mother finally prevents its re-entry a day or two before she gives birth to her next offspring, but it continues to put its muzzle into the pouch and to suck from the same elongated nipple that nourished it during pouch-life. Weaning occurs at widely varying ages. Males play no part in care for the young.

The mother comes into estrus one to nine days after giving birth. If she conceives, that conceptus develops to about the 100-cell stage before further development ceases. It remains as a dormant blastocyst until a few weeks before the preceding juvenile is permanently evicted from the pouch, or the pouch-young dies, then begins to develop again. It is born a day or so after the previous young quits the pouch. It is possible for a female to be suckling a young-at-foot (excluded from re-entering the pouch) and a pouch-young simultaneously (her mammary glands producing milk of different qualities and quantities for each of them), and to be carrying a dormant or developing fetus simultaneously.

During drought, females may lose pouch-young at a stage of high milk demand (middle to late pouch-life). However, that loss is quickly replaced by the implantation, development and birth of the quiescent blastocyst. Females can have a quick succession of young, none of which is weaned until the drought breaks, good conditions return, and a pouch-young is carried through to emergence and weaning. During prolonged drought some species (e.g., red kangaroo) may enter anestrus. For nearly all macropod females, this is the only break in a life of perpetual reproduction, often with two or three simultaneously dependent young.

Although many macropod species show seasonal peaks in births, only the Tammar wallaby (*Macropus eugenii*) and the island races of the banded hare-wallaby (*Lagostrophus fasciatus*, Prince, 1995) and red-necked wallaby (*M. rufogriseus rufogriseus* on Tasmania) are strictly seasonal breeders. In other species, females may enter estrus at any time of year. At a population level, males would have at best a weak ability to 'predict' a time of maximal frequency of estruses. No macropod males follow a strategy of restricting reproductive behavior to only part of the year, seeming to be constantly primed to respond to the unpredictable occurrence of estrous females.

At an individual level, the probability that a female might enter estrus relates to her cycle of development of existing dependent young. In most circumstances, she enters estrus soon after giving birth; thus, a mature female with a very small pouch-young is most likely to enter estrus. In contrast, a female with a medium-sized or large pouch-young is most unlikely to enter estrus. Eviction (or loss) of that pouch-young signals impending birth, soon to be followed by estrus. These stages of development of her young, readily read by human observers and presumably macropod males, predict the likelihood of forthcoming estrus.

Female sociality is also influenced by stage of development of the young. Many of the smaller and non-social macropods leave the unweaned young, once it has quit the pouch, 'lying out' while they forage (e.g., red-necked wallaby, Johnson, 1987). The larger macropods carry the young with them, or it follows 'at heel'. Among the social species, females about to deny their large pouch-young re-entry to the pouch avoid the company of other females with young at the same stage of development (Croft, 1981; Stuart-Dick, 1987). During this stressful period in the relationship between mother and offspring, the female is exceptionally often solitary.

Estrous female macropods have a few opportunities for mate choice (Walker, 1996). Females in homomorphic species are capable of vicious defense against males trying to inspect them, and could clearly withhold information about their reproductive condition by preventing inspection. A female in heteromorphic species may repel importunate smaller males trying to sniff at them. A female might determine for which male she will stand and urinate, another selective release of information. A female might accept or reject a male's attempts to establish a consortship, since a male cannot effectively stop her from moving away, although he may place himself between her and other kangaroos (personal observation).

At estrus, females may roam widely, even beyond their normal home-range (e.g., red-necked wallaby, Johnson, 1989; eastern gray kangaroo, Jarman & Southwell, 1986; red kangaroo, G. Moss, personal communication). Since all males are essentially free to seek out estrous females, a female in estrus may be attended by several males, even in normally solitary species (e.g., rufous bettong, bridled nail-tail wallaby, swamp wallaby, red-necked pademelon – personal observation). If harassed by importunate males, a female may flee, pursued by a train of males (Walker, 1995, 1996). Such flight may attract other males, and when it stops the most dominant may assert control. Eastern gray kangaroo females may flee to a structure (called a blocking structure by Walker, 1995) that allows her to 'get her back to the wall' and ward off males. Structures that have been noted include fallen tree branches, the roots and hollow base of a fallen tree, a wombat burrow, blackberry bushes, and the verandah of a house (J. Aldenhoven, personal communication; C. Johnson, personal communication; Jaremovic, 1984; Walker, 1995; personal observation). Again, once calm is restored, the most dominant male present associates with the female, and courts and mates. Wide ranging may assist a female to be detected by dominant males; and flight and use of blocking structures may attract attention and allow the male hierarchy to determine that she is courted and mated by the most dominant male.

Female macropods seem to control the duration of mounting and intromission. In homomorphic species, females break out of mounting and often attack the male (e.g., rufous hare-wallaby, McLean *et al.*, 1993; rufous bettong, personal observation). Even in the large kangaroos, a female can break out of the grasp of a mounted male (Walker, 1995; personal observation), often by jumping out of his arms, but does not directly attack the male.

Copulatory plugs have been observed in many macropods. They may fall out spontaneously, be removed by subsequent intromission, be removed by the female's grooming, or remain in place for hours or a day or two (e.g., McLean *et al.*, 1993). A vaginal plug, or its removal, may be considered part of female mate choice (Walker, 1996). Nothing is yet known about sperm competition in macropods, let alone its manipulation by females.

MALE REPRODUCTION

Male macropods attempt to inspect females by sniffing at the cloaca, pouch opening or body in general. Close inspection by some males may induce the female to urinate; the male tastes the urine, giving a flehmen display. Males can detect approaching estrus olfactorily several days in advance. Detection may induce some males to try to associate closely with the pro-estrous female; others increase the frequency of inspecting that female. More dominant males can displace less dominant males in the consort relationship. Estrus lasts about one day, during which the female may mate once or a few times, with one or more than one male. Courtship, mounting and copulation are brief, with some post-copulatory guarding by the male.

Among heteromorphic species, a size-related male hierarchy ensures that the locally most dominant male gets a high proportion of the available matings. However, he does not get all, and all other mature males show intense interest in estrous females. A male may be harassed, during courtship, mounting and copulation, by one or more other males even if he is normally dominant over them. He may break off courtship to chase them. Sub-dominant males have a chance to mate when the dominant male is either distracted (for example by chasing off others) or has failed to detect or relocate the estrous female.

A male's success depends upon a combination of achieving such dominance that he can displace any other male, and having good knowledge of the reproductive status of as high a proportion of females as possible in the local population. Male reproductive strategies reflect these two needs. Males invest time in growth and muscle development in the long term, and in male–male interactions in the short term. Dominant males also act to maximize their chances of contacting and inspecting females. Some spend time at localized resources where females will be (e.g., at water-holes or patches of fallen fruit, on access paths to female refuges, or even at females' nests); others move more widely and continuously than any other males, contacting a high proportion of the female population every day. This latter strategy leads to such males being alone more than other classes (Jarman & Coulson, 1989). Such males may not live long as dominants.

THE EXAMPLE OF THE EASTERN GRAY KANGAROO

The eastern gray kangaroo, the second largest extant macropod species, is highly heteromorphic; at first estrus, a female may be mated by a male up to five times her own weight. Eastern gray kangaroos eat almost exclusively grasses, selecting a high proportion of leaf and seedhead in their diet. They forage in grasslands, savannahs and light woodlands, avoiding forest.

The species conforms to the general pattern of sociality and mating systems outlined in the review above. This section illustrates some aspects of the socioecology of the species, mostly with data derived from a study, at Wallaby Creek, north-eastern New South Wales, in which all individuals in part of a population were individually known and monitored for over a decade (Jarman *et al.*, 1987, gives details of the study site and general methods).

Despite male-biased sex ratios at birth (Stuart-Dick & Higginbottom, 1989), the adult sex ratio was usually 0.3 to 0.6 males to each female. Females were philopatric, but males dispersed out of their natal range as late subadults; some dispersed again later in adult life. Home-ranges (95% isopleths) of males were greater than those of females (males 56.9 ha, SD 12.1, $n = 14$; females 35.9 ha, SD 5.7, $n = 15$, all individuals with >100 recorded locations). Home ranges of individuals overlapped extensively. No ranges were defended; no animals were excluded from any parts of the study area by the behavior of others.

Flexibility in grouping

Eastern gray kangaroo grouping is highly flexible, and this flexibility is fundamental to the mating strategies of males. Although these kangaroos usually forage and rest in open-membership groups (typically three to 12 animals), being in a group is not obligatory. Group sizes vary with habitat, being greater in more open and more productive habitats (Southwell, 1984; Heathcote, 1987; Jarman & Coulson, 1989).

The key feature of grouping in the Wallaby Creek population is flexibility: in size, rate of change of size, composition and internal spacing and orientation. Groups change membership frequently, a group of ten kangaroos having a 0.2 probability of changing composition in the next three minutes. Changes are particularly common around dusk and dawn, when small night-time groups amalgamate to form daytime ones twice as big (Clarke *et al.*, 1995). The presence of a large male in a group does not change the group's rate of flux, nor the composition of animals leaving or joining it.

Any class of kangaroo can occur in a group; none is excluded. Some groups contain females but no males, or *vice versa*. More groups than expected have several (>four) males. Some classes tend to aggregate with their own class within groups. Two classes are more likely to be solitary. Many females tend to separate themselves during the transition of their young from pouch life to following at foot. The largest class of males is generally more solitary than other males, and the local alpha male in particular may be alone in a quarter of his sightings.

Most females are rarely (4% of sightings) the only female with one or more males; females in or approaching estrus were in that situation in 31% of sightings, and were alone in only 4% of sightings (cf. 11% for all females). Within a group, those classes that aggregate also select their own class as nearest-neighbor significantly often. Large males are each other's nearest neighbors at random frequency. However, they are significantly more likely than by chance to have a female with no, or a very small, pouch-young as their nearest neighbor; that is the class of female most likely to enter estrus. That class of females and large males orient positively towards each other. Other adult males orient positively with their own class. No other class of females orients positively towards large males (Jarman, 1994).

Thus, eastern gray kangaroo group composition and nearest-neighbor association and orientation are by no means totally random. Kangaroos appear to respond to the relative rank of males and reproductive state of females that are in the group with them. Given the rate at which groups can change membership, individual kangaroos must be reassessing their relative status in the group very frequently.

Mating systems

In the studied population, most females first conceive when three years old and bear about 1.1 young a year. Females are philopatric, and a mother shares her range with adult daughters as well as unrelated females. Females with living mothers or sisters in the population are more successful than females without living female kin at rearing their first two offspring; thereafter, their advantage disappears (Jarman, 1994). Females tend to produce daughters early in life, maximizing the length of life over which they can benefit them. Sons come later in life; a female's last offspring is particularly likely to be a son (Stuart-Dick & Higginbottom, 1989). A mother may be unable to influence her son's reproductive success, since he must grow and develop for six to ten years after weaning before being dominant enough to contest matings.

There are no signs of females forming matrilineal or unrelated coalitions to repel the attentions of unwanted males. Females generally tolerate close approach and olfactory inspection by males larger than themselves, but they can push, growl and even rush at smaller males, driving them away. Larger males may induce a recumbent female to get to her feet by pulling at or gently hitting her. Once standing, she may consent to urinate for some larger (but not smaller) males who then taste her urine.

If her urine signals approaching estrus (detectable to males seven to ten days in advance), the male may test her readiness by making a clucking call while patting and pulling her tail, or he may simply accompany her closely. He may associate with her briefly, or guard her for several days. The consorting pair moves slowly and does not travel far. At intervals the male tests the female's receptivity by quiet clucking and tail pulling. The consorting pair may become separated from other females.

Some consortships break up for no obvious reason, and without mating. A more dominant male can displace a consorting male. Consortships established by the most dominant male are usually later in the pro-estrous phase and are likely to persist to estrus and lead to copulation. Often, as estrus approaches, a consortship breaks up and the female starts to move quite rapidly and widely, even outside her normal range. Her scent and the interest shown in her by males attract the attention of many males. She may be followed by a train of six or more males. A dominant male needs to slow her down and prevent her flight being triggered again by importunate males grabbing at her. At Wallaby Creek, estrous females sometimes used 'blocking structures' to ward off pursuing males. Such flights are obvious and allow a dominant male to catch up, drive off other males, and restore calm.

Mounting follows a bout of clucking and tail pulling to which the female responds by standing still. The male (two to five times her weight) keeps his hind feet on the ground, clasps the female below her ribs, and copulates with her. A female can escape before being clasped. She can, with some desperate struggling, break out of his grasp, but is most likely to do so when he is being harassed by other males. After copulation, the male may remain with the female, perhaps copulating again and keeping others away from her until the end of her estrus, or he may desert her. If he does, another male will attempt to court and mount her, which she may accept. Forced copulation was never recorded.

Females show few signs of mate selection; a female appears to accept the outcome of male–male competition, and usually mates with the most dominant male to find her. However, a female approaching and in estrus behaves in ways that make it most likely that she and her status will be noticed by the larger males. In some instances her behavior may induce male–male interactions.

Eastern gray kangaroo males grow throughout life, up to weights of 90 kg or slightly more (compared with 40 kg maximum weight for females). They establish a hierarchy through sparring and fighting, and display relative dominance and subordinate status in various ways. For reproductive success, an eastern gray kangaroo male depends in the longer term upon rising to the top of the dominance hierarchy, and in the short term upon finding as high a proportion of the estrous females as possible.

Almost all overt male–male interactions relate in some way to relative status. An unweaned male first spars with his mother; but after weaning, the subadult male tends to associate and spar with peers. A set of four or five subadult males may spar together, rather than the one-on-one sparring pairs seen among adults. Subadults use all the elements of fully escalated fighting, but most elements are poorly performed. As the male grows, he spends less time sparring. However, when two young adults begin sparring, others are likely to begin and a group may contain several pairs of sparring males for up to 15–30 minutes.

Sparring among large males is uncommon, escalated and prolonged, and might qualify as fighting. Two males may engage in such contests for prolonged periods (an hour or more) at intervals over several days. The outcome can reverse hierarchical position. To fight, they stand upright facing each other, grasp with the forepaws, wrestle with their arms around each other's heads, necks, shoulders and chests, throwing their heads back to keep their eyes out of the way of the opponent's paws, and, holding the opponent at arm's length, lean back on the stiffened and vertical tail and kick at the opponent's lower abdomen (which has particularly thick skin: Jarman, 1989b). The fights have a balanced exchange of blows and the animals generally remain on their feet. Occasionally, they break off and groom or even graze for a while before re-engaging. The fight ends when one animal turns away and avoids the other, showing signs of subordination.

The most highly escalated and damaging fighting is between the hitherto most dominant male in the local population and a challenger, and determines alpha-male status for a while (average tenure is one year, after taking eight to ten years to reach that status). Such challenges occur rarely; an alpha male may fight at this level only twice in his life: once when he gains alpha status and once when he loses it. Such fights sometimes last several hours, and can occur over one to three days. Although they may still be broken into separate bouts, they are more likely to be continuous. Combatants exchange kicks at a high rate (up to ten per minute during intense exchanges), and these are internally and externally damaging. In these intense fights, opponents may wrestle each other to the ground, and the victor jump on and kick the back, head or any other part of his downed opponent. If one male flees, the other pursues him, grabbing and kicking at him. Males are invariably damaged in such fights, and at least two of six males died soon after them.

Males display at each other as a prelude or alternative to fighting. Success in sparring or fighting relates to height, strength and reach, hence in part to age. Male display exhibits height (by the male drawing himself up to his full height, standing bipedally on the tips of his toes, supported by his stiffened tail, or on all fours but up on his hind toes), and forearm length and musculature.

Subordination is signaled by crouching low and giving an open-mouthed 'cough'. A dominant male when entering a group may advance toward a subordinate and display at him until the subordinate acknowledges his status by giving the subordinate cough. The dominant walks through the group, forcing males to acknowledge his status.

Dominant males spread scent from their throat and chest upon vegetation. They also stand tall, lean back, and spray urine in front of them – as much a visual as an olfactory display. Smaller males scent-mark, too, but less often and less vigorously. I have yet to see a kangaroo male respond to another male's scent mark, although the sight of one male marking may stimulate another to do the same.

Dominance, displays and scent marking are not used to exclude other males from a territory or a group. Relative rank allows a more dominant male to displace a subordinate from a resource such as food or shade, or from proximity to a female. Rank is rarely used to drive a male away by more than 10 m, except from the vicinity of an estrous female. Even so, a mating pair often has several other males within 30 m.

A male's best chance of finding an estrous female is to test olfactorily as many females as possible. Male eastern gray kangaroo males have larger ranges than females, and the largest males have the largest ranges of all (Jarman & Southwell, 1986). While most males content themselves with checking the females in several groups each day, the locally most dominant male searches over an exaggerated range, and contacts a very high proportion of the females. One dominant male, tracked for 24 hours, inspected 70% of the population of females during that time.

That 'roving male' strategy necessitates the males spending much time in solitary travel between groups, and they are alone significantly more often than other males. However, these largest males are able to fight off attacks by lone dingoes (Wright, 1993), although not without risk of injury. That reduces one cost of being so often solitary.

The rewards of this male strategy are relatively great. If a male can grow and live long enough to be the biggest and strongest in the local population, and if he can search efficiently enough to detect all the pro-estrous females in the population (and if their estruses are not too synchronized), then he can potentially father all the offspring of females in his range during his tenure as alpha male. Estimates of alpha male mating success range from 60% to 75% of available matings (Walker, 1995; personal observation).

The strategy is less than perfect because: the dominant male may simply fail to detect an estrus; he cannot be in two places at the same time, so will miss some synchronous estruses; while he is distracted by harassment by other males, the female may be mated by an interloper; or he may stop guarding a female (perhaps to search for another) before her fertile period has ended.

Subordinate males can gain matings in any of these circumstances, and behave so as to increase the probability that they do. For example, a male may reduce the chance of a pro-estrous female being detected by a more dominant male by forming a consort relationship with her and reducing her movement. Subordinate males may cause enough distraction, by harassing the dominant while he is courting or even mounting a female, to be able to sneak a mating.

However, because of the long wait to grow sufficiently large to be dominant enough to get any matings, a male faces years of potential mortality before having any hope of reproduction. Even most of the males that survive their first dispersal and enter the adult population do not live to reproduce. Males disperse from their natal range as subadults. Fully adult males may also disperse again, perhaps choosing to settle in a locality where their chances of becoming the locally most dominant male are higher.

Consequences of male numbers

In the local population, more or fewer males simply affect the numbers of males with whom an eastern gray kangaroo male must interact. The more males there are close to his rank, the more contests he will have to engage in. The more there are above him in the hierarchy, the lower are his chances of being the most dominant male to detect an estrous female, and the longer he must wait to reach alpha status (hence the greater chance that he will die without breeding), unless he emigrates to a range where he would face less competition. Numbers of males subordinate to him make little difference to a male's life.

In a group, having more or fewer males reflects the male's own choice, since all except the largest males tend to be in groups containing others of their class. Parties of same-class males move from group to group together; individuals thus appear not to be disadvantaged by each other's company. They may benefit by being with familiar males whose relative ranks are already established and known (as well as gaining antipredator benefits).

However, the largest males are neither attracted to nor repeled by peers. These males stand the best chance of being the most dominant male with an estrous female, and their strategy is to search independently for estrous females. Other large males in the group constitute immediate reproductive competition and the largest males assess and assert their dominance relative to other males upon entering a group.

The actual numbers of other males in a group around an estrous female will matter more to large but non-alpha males than to the alpha male himself. Any male can be challenged while courting, or even copulating, but the alpha male is least likely to suffer from this. For a non-alpha male, a quiet consortship with a pro-estrous female, keeping her away from groups and out of sight until she is ready to mate, is the ideal strategy. Discovery by other males leads to exposure, challenges, and mating chases, and will draw the attention of more dominant males. The more males there are, the less likely any male is to be able to keep his consort undetected.

ADAPTIVENESS OF KANGAROO MATING SYSTEMS

The mating system of large, social macropods, exemplified by the eastern gray kangaroo, derives, I argue, from the non-seasonal and spatially unpredictable occurrence of estrous females and their ephemeral grouping. Non-seasonal breeding selects for male strategies that will function year-round. Many forms of guarding of females by males (pair-bonding, harem-holding, or persistent group-living) could achieve that. However, kangaroo females occur in small, ephemeral, constantly changing groups; there is no such thing as a group home range. At any one time there will be several (two to ten) such groups within a male's home range. The better the resources in his range, the more groups there are likely to be. Groups split and join at a great rate. The male is faced with no permanent grouping of females that he might live with persistently or control access to, and too many transitory groups for him to be able to defend his range (or even a part of it) as an exclusive territory. Territoriality would not give him exclusivity because, as soon as he began courting a female in one group, other males would invade and court other females in other groups.

Given that kangaroos gain antipredator benefits by being in groups (however transitory those might be), the fundamental question becomes: why are females not in persistent groups? It may be because their individual resource demands vary with stage of development of their young, including the need to separate themselves from other females when their young is being evicted from the pouch. Without synchronized breeding, females at different stages of development of young, and consequently with different resource needs, may satisfy their resource requirements best by independent choice of activity schedule and foraging place, route and speed. The tendency for female eastern gray kangaroos to cluster with peers whose young are at the same stage of development supports this suggestion.

Moreover, loss of pouch-young or young-at-foot, through unpredictable resource depletion in drought, or loss to predators, or even through ejection of the young to aid the mother's escape from a predator, will switch a female's class of dependent young abruptly, without the delays inherent in a seasonally breeding species or one with prolonged gestation. Female kangaroos can fine-tune their social environment from minute to minute, if they need to.

Given that there is no invincible male strategy of long-term defense of females directly or of space that they might occupy, a strategy that is free of constraints of time or space would be adaptive. The 'roving male' strategy maximizes the alpha male's chances of detection of estrous females in a range that he can expand to the limits of his endurance of solitary travel. The relative success of the alpha male is great, fathering 60–75% of one year's cohort of young. Such genetic domination powerfully selects for competitive traits, leading to the huge dimorphism of the large kangaroos.

Males pay the price in prolonged growth and delay in the socially probable on-set of breeding. Evolutionarily it does not matter that very few males practicing this strategy are successful; all that matters is that those that are successful are practicing the strategy.

Under the roving male system, matings go (all else being equal) to the most dominant male to be with the female. However, all males have some chance of mating, so all males attempt to detect, associate with and court females. The openness of grouping allows them unrestricted access to groups containing females. Their pre-established hierarchy serves to rank their access in the presence of an estrous female. Males of different rank may follow slightly different tactics of seeking and associating with females, seen in their tendency to cluster with peers in groups. Those clusters of males make little difference to the most dominant local male. They will ensure that an estrous female is not mated by a male subordinate to any in the cluster, and will harass a less-than-alpha courting male. Female behavior reinforces the male hierarchy as a determinant of mating success.

The roving male system equally well serves many less social macropod species, too. However, use of spatially restricted resources may alter the males' searching tactics. Thus, in species that emerge from dense vegetation at night to feed in clearings, or that gather to feed on fallen fruit in the forest, males may wait at such sites to inspect arriving females. Where females regularly use fixed shelter sites, a dominant male may monopolize access to those sites, and thus the females they contain. These species are not as exaggeratedly heteromorphic as the largest kangaroos, perhaps because dominant males are sacrificing the potential for genetic swamping of the population's reproduction for more assured matings with a few females. In the solitary, homomorphic potoroines, males use known nests as a clue to female location, and check the nests early in the night's activities. However, a female uses a scatter of nests and builds new ones; so nest defense by the male would be at best a short-term tactic.

COMPARISONS BETWEEN MACROPODS AND PRIMATES

Primates do many things that macropods never do (Table 3.3), and *vice versa*. Some of the differences help us speculate upon the primary causes of sociality and social behavior in primates.

The most diurnal macropods still maintain groups, albeit smaller ones, at night, so grouping is not incompatible with nocturnal activity. However, convergence between primates and macropods suggests that small (<5 kg) species are too vulnerable to predators (such as eagles) to have adopted diurnal, terrestrial activity.

The greatest contrast between primates and macropods is in the forms of their grouping. Macropods do not form persistent groups, not even bonded pairs (*pace* Horsup, 1996). With rare exceptions, all reproductive pairs or at least some elements of the groups formed by primate species are persistent, often life-long. So strong is this difference that primatologists might not recognize the term 'group' as I have applied it to macropods (M. Pereira, personal communication), preferring a term such as 'party' or 'aggregation.'

The openness and fluidity of macropod groups are such that temporary membership is voluntary. No classes are excluded by the actions of others. Even their spatial relationships within groups are only lightly organized by inter-individual aggression. That is far from the case in the persistent groups formed by many primates, in which aggressive behavior of individuals often determines the group's composition and internal spacing.

Primate groups enjoy group identity, and often (not always) define and defend a group range. Just what is being defended may differ between classes within the group: females may defend food resources and their young, while males may defend access to females. Not having persistent groups, even social macropods do not have a group range to defend. Consequently, no class or individual gains any lifetime reproductive success through relative competence in defense. The size, strength and fighting ability gained by the greatly heteromorphic male are utilized entirely to promote his own reproduction; they contribute nothing to any other individual's survival and reproduction. There is no basis among macropods for male–female or male–male defensive coalitions.

Heteromorphism in primates is strongly correlated with sociality, although there are exceptions such as the largely solitary but highly heteromorphic orangutan (*Pongo pygmaeus*). Heteromorphism characterizes even many of the solitary macropods, serving to illustrate that selection for heteromorphism occurs whenever a larger male is able to out-compete a smaller male for access to estrous females. The roving male strategy, with a size-based hierarchy determining access to detected estrous females, can reward relatively greater male size even within species whose females forage solitarily, just as effectively as, say, persistent defense of a harem of females selects for it in social species. Indeed, the selection can be even stronger, since there is potentially

Table 3.3. *Comparisons between macropods and primates in aspects of their general socioecology and behavior*

Socioecology and behavior	Primates	Macropods
Diel activity	Mainly diurnal	Mainly nocturnal
Sociality	Mainly social, in closed-membership groups; some small and nocturnal species solitary	Mainly solitary, but a few of largest species forage in open-membership, ephemeral groupings
Group structure	Can be highly structured by composition and relatedness; structure partly enforced by individual or collective aggression	Groupings lightly structured by individual choice of association, usually not enforced by aggression
Defense of range	Group range commonly defended, by males or whole group	Individual range never defended; no such thing as group range
Size dimorphism	Heteromorphism in some social species, but exceptions in some lemurs	Smallest species homomorphic; larger, heteromorphic, even if solitary
Parental care	Prolonged in many species; male may carry and guard young	Prolonged carrying by mother; no paternal care
Infanticide	Reported in many taxa in which male guards monopolized female	Never reported
Mate guarding by males	Constant in many species, including bonded pairs	Only briefly before and during estrus
Soliciting by females	Common, occasionally even outside estrus	Never reported
Mate choice by females	Sometimes strongly exerted, including persistent pair-bonds	Reinforces mating determination by male rank

nothing to prevent the locally largest male from securing every mating in the population within his range, while the male of the group-forming species can never do more than mate with all females in his limited harem.

Having no persistent association with females, a macropod male has no opportunity to develop paternal care. Macropod society therefore lacks any of the complexities of investment by males in care of their offspring, the need to recognize probable relatedness, and the value to females of associating closely with care-giving males.

Infanticide is common among primate species, but has not been recorded in macropods, despite long 'carrying' of young. Infanticide would increase a male's reproductive potential only if it induced estrus when he could predictably control access to the female. If a male kangaroo killed a pouch-young, the next young into the pouch would be from an unimplanted blastocyst that he had probably not fathered. Without long-term control of access to a female or female groups, a macropod male would gain little from infanticide. Lack of persistent and therefore monopolizable grouping in macropod females makes infanticide by males pointless.

Without infanticide or potentially damaging forceful copulation, female macropods have nothing to fear from males. Thus, there is no need for female coalitions to ward off male coercion or male infanticide, reasons that have been advanced to explain the evolution of female bonding or persistent association in primates (van Schaik & Dunbar, 1990; Smuts & Smuts, 1993; Brereton, 1995; van Schaik & Kappeler, 1997).

Persistent grouping in primates makes possible reproductive strategies that involve mate guarding by males, including bonded pairs within or outwith larger groups. Macropod males show minimal mate guarding, mainly in defending access to a female in the few days leading up to estrus or during estrus. Primate mate guarding makes possible selection for secondary behaviors that increase the security of the guarding, or the quality (for the female or her offspring) of the guarding male. These include solicitation and several forms of mate choice by females. Mate choice by macropod females is weakly expressed, and (at least up to copulation) seems merely to reinforce the probability that the male hierarchy will ensure that the locally most dominant male detects a female's estrus and mates with her.

Without persistent grouping, macropod females can gain relatively little from coalitions in female–female competition, although evidence for matrilineal effects upon female reproductive success in eastern gray kangaroos (Jarman, 1994) shows that some gains are possible. However, such coalitions appear not to be directed against males, there being no male control of the group. Macropod males would gain nothing by persistent male coalitions, since females are not persistently monopolized and there is no monopolizing male to displace.

Despite these general differences between primates and macropods, there are some basic similarities, and occasional specific similarities. In both taxa, size equates vulnerability to predators and relates to forms of antipredator behavior, including crypsis or grouping, for example. There are similarities between the socioecology of the larger, social macropods and chimpanzees (*Pan troglodytes*). The small and constantly changing membership of chimpanzee 'parties' (Dunbar, 1988) is reminiscent of foraging groups of eastern gray kangaroos. However, the organization of male chimpanzees into coalitions that defend a communal range against other coalitions (Wrangham & Smuts, 1980) finds no parallel among macropods. The heteromorphy of male orangutans parallels that of some solitary macropods. Whether the adult male's mating system (Galdikas, 1985a) is the area-defense polygyny identified by Dunbar (1988) or not (as argued by van Schaik & van Hooff, 1996), it has strong similarities to a roving male strategy. However, the coercive mating by 'subadult' males (van Schaik & van Hooff, 1996) finds no parallels among macropods and its avoidance has played no part in the evolution of grouping in that family.

The major contrasts between the social systems of macropods and primates are revealed in the very different effects that varying male numbers would have, at the population or group level, in the two taxa. In very many primate societies, male behavior enforces intergroup and intragroup organization; thus, variation in numbers of males may crucially alter organization. In macropods, male behavior has no such effect, and variations in male numbers go virtually unnoticed. Yet these differences are merely the outcome of much more fundamental differences between the taxa in whether or not females associate persistently or temporarily.

ACKNOWLEDGMENTS

I would like to thank Dr Peter Kappeler, and the Directors and staff of the Deutsches Primatenzentrum, for inviting me to participate in the Göttinger Freilandtage. I am grateful to the organizations that sponsored the conference: Deutsche Forschungsgemeinschaft, Bonn; The Wenner-Gren Foundation for Anthropological Research, New York; the Niedersächsisches Ministerium für Wissenschaft und Kultur, Hannover; and Sparkasse Göttingen. I am also grateful to Michael Pereira, Peter Kappeler and an anonymous referee for helpful comments on drafts of this chapter.

4 • Social counterstrategies against infanticide by males in primates and other mammals

CAREL P. VAN SCHAIK

INTRODUCTION

While in many other mammals the sexes get together only during mating sessions that tend to last a few days at most (Hayssen *et al.*, 1993), primate males and females are often associated year-round and individual females may be sexually active for weeks at a time and often mate with multiple partners (van Schaik & Kappeler, 1997; van Schaik *et al.*, 1999). Catarrhine primates show these traits to the extreme. Year-round male–female association is nearly universal among anthropoids, multi-male groups are common, and long mating periods, including those covering the whole ovarian cycle, are the norm (cf. Hrdy & Whitten, 1987).

Long-tailed macaques (*Macaca fascicularis*) in a riverine rainforest on Sumatra show all these traits (van Noordwijk, 1985; van Noordwijk & van Schaik, 1985, 1988, 1999). Their large groups tend to contain up to ten non-natal adult and sub-adult males, but the number of males actually present in the group varies depending on the presence of sexually attractive females. Breeding is semi-seasonal, but even when not a single female is sexually attractive, one or more males always remain with the group containing adult females and immatures (Fig. 4.1). The most steadily present male is the male who was dominant during the previous mating season (Fig. 2 in van Noordwijk & van Schaik, 1988), and has sired the majority of the current infant cohort (de Ruiter *et al.*, 1994). Indeed, this male is often closely associated with these infants, who follow him like a shadow once they are spending most of their time off their mothers (Fig. 3 in van Noordwijk & van Schaik, 1988).

During the mating periods, female long-tailed macaques have variable degrees of morphological advertizing, from slight coloration of the perineal region in older females to conspicuous swellings near the base of their tails in young females (van Noordwijk, 1985). Any female may mate over a period of many weeks, without very clear patterns, but mating does tend to stop abruptly, probably following the occurrence of ovulation. Females tend to be guarded by the top-ranked male while they are most attractive, but in each mating period will mate with most, perhaps all, of the males present in the group. The consortships with the dominant male are maintained by both partners, but in years with uncertainty in the top-dominant position, females have been seen to terminate these consortships and actively switch between contending males. Mating behavior continues well into pregnancy, sometimes until a few weeks before parturition (van Noordwijk, 1985). During pregnancy, females tend to be less attractive and mate mainly with lower-ranking adult and subadult males. Each conception is accompanied by at least 500 and perhaps over 1000 matings.

Why are anthropoid males and females associated year-round, and why are their mating frequencies so high? Mating is fairly seasonal in long-tailed macaques (van Noordwijk & van Schaik, 1999) and highly seasonal in many other primate species. Nonetheless, in all of these species at least one adult male remains permanently attached to the group at all times, although males are often more closely associated with groups during the mating season (e.g., ring-tailed lemur, *Lemur catta*: Jolly, 1966) or some only enter the group during the mating season (*Erythrocebus*, *Cercopithecus*: Cords, 1988). Hence, the permanent presence of adult males in primate groups regardless of seasonality cannot be explained by continuous breeding activity (*pace* Zuckerman, 1932). For the same reason, it is unlikely that avoidance of harassment by strange males (Wrangham, 1979; cf. Rubenstein, 1986) is a major selective force for all primates. Recently, it has been proposed that the risk of infanticide by males unlikely to have fathered the females' infants has selected for both year-round intersexual association (van Schaik & Dunbar, 1990; van Schaik & Kappeler, 1993, 1997; see also Pusey & Packer, 1994; Mesnick, 1997) and for several aspects of female sexuality, including promiscuity, mating calls and sexual swellings (Hrdy, 1981; Hrdy & Whitten, 1987; O'Connell & Cowlishaw, 1994; van Schaik *et al.*, 1999).

In spite of detailed studies of lions (Pusey & Packer, 1994) and primates (e.g., Crockett & Sekulic, 1984; Watts, 1989; Sommer, 1994) suggesting that infanticide is adaptive

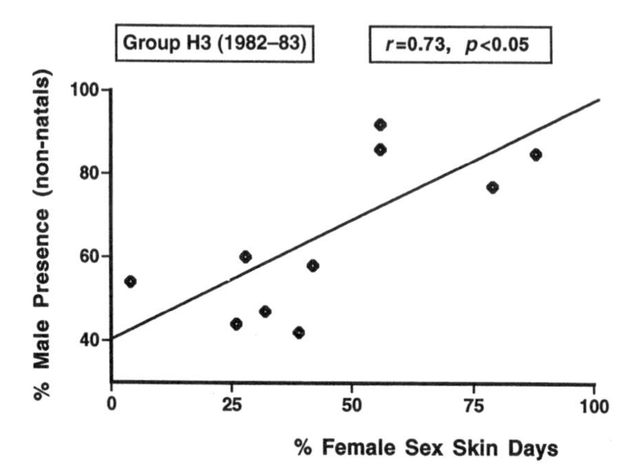

Fig. 4.1. The presence of immigrant adult and sub-adult males in relation to the number of sexually attractive females in group H of long-tailed macaques at Ketambe, Sumatra, over a ten-month period in 1982/3 (unpublished data; see also van Noordwijk, 1985). Presence is the percentage of hours an average male was recorded as present in the main party of the group in a monthly sample. Sex skin days are the percentage of days in a monthly sample in which the average female had some sex skin (reddening or swelling of the perineal region).

for males, there is nonetheless still widespread skepticism with respect to the adaptive significance of this behavior (Bartlett *et al.*, 1993; Sussman *et al.*, 1995; Rowell, 1996). Strictly speaking, the exploration of the consequences of infanticide by males for social and sexual behavior in primates does not depend on it being adaptive or pathological, as long as it occurs predictably in certain conditions. However, if infanticide is a result of targeted attacks by unlikely or impossible sires rather than general aggression, more finely tuned counterstrategies are expected to have evolved; hence, it is interesting to establish the extent to which infanticide by males is actually adaptive.

Infanticide by males is the expression of sexual conflict. It will be an adaptive male reproductive strategy if: (a) the probability that the male had sired the infant(s) was zero or close to zero; (b) the mother can be fertilized earlier than if the infant(s) had lived; and (c) the infanticidal male experiences a dramatic improvement in the chances of siring the next infant(s) relative to the current offspring (Hrdy, 1979; Hrdy *et al.*, 1995). All three conditions must apply simultaneously. Both a review of all published cases of infanticide in wild primates where infanticide was directly observed (Struhsaker & Leland, 1987) and detailed case studies of single species (especially Sommer's [1994] study of Hanuman langurs, *Presbytis entellus*) yield broad support for this hypothesis. First, infanticide by males often leads to a faster conception of the next offspring than without it (see also Crockett & Sekulic, 1984). Second, infants are usually killed at an age well below the average age at weaning, and killing of juveniles is very rare. Third, the male obtains a dramatic improvement in probability of paternity: he is not likely to kill an infant he has sired himself, and has improved mating access to the female during the mating period following infanticide. This has recently been confirmed for Hanuman langurs through paternity determination based on DNA analysis (Borries *et al.*, 1999). Finally, whereas females tend to get injured, males tend to sustain injuries only rarely (van Schaik, unpublished data), suggesting that infanticide is probably a low-cost strategy for males. This latter finding may explain why males commit infanticide even when their mating prospects are relatively slim (e.g., when meeting neighboring groups).

Clearly, the female suffers a major loss of reproductive output. Natural selection should therefore have favored counterstrategies involving reproduction (mating behavior, reproductive physiology: van Schaik *et al.*, 1999) and social behavior. These counterstrategies are sufficiently successful to make infanticide rare in most species most of the time, but obviously fail sometimes, in predictable circumstances, such as the absence of the protector. This chapter focuses on the main social counterstrategy: association with protective males.

The first aim of this chapter is to identify the life history that predisposes a female to the risk of infanticide by males. The relative duration of lactation and gestation is a good predictor of infanticide risk among primates and placental (eutherian) mammals in general. The second aim is to explore the social consequences of infanticide by males in the vulnerable taxa. This is done here in some detail for primates, and in a preliminary way for other mammals.

CAUSES OF INFANTICIDE BY MALES

Given the stringent conditions for the evolution of male infanticide laid out above, one can make several predictions as to where infanticide by males is expected in placental mammals. First, infanticide by males is less likely in species in which females can be fertilized soon after giving birth, and can therefore combine being pregnant and caring for the current dependent offspring. In such rapidly breeding species, post-partum fertilization should reduce the risk of infanticide unless early conception would seriously compromise the size or survival of the female's subsequent litter or

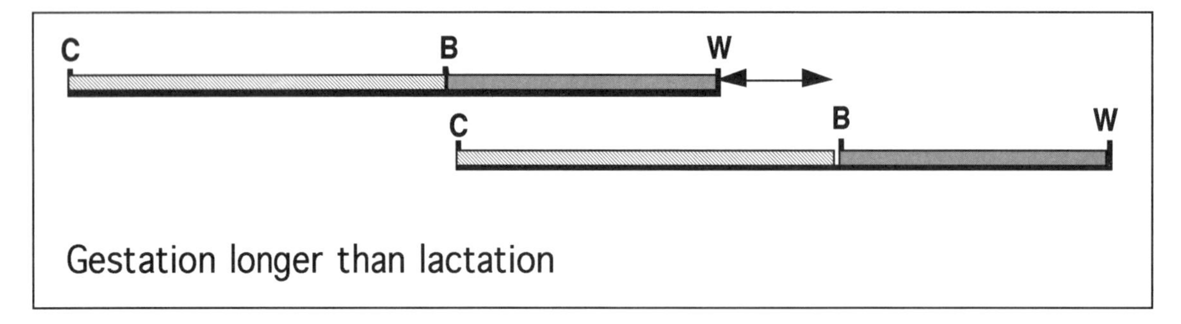

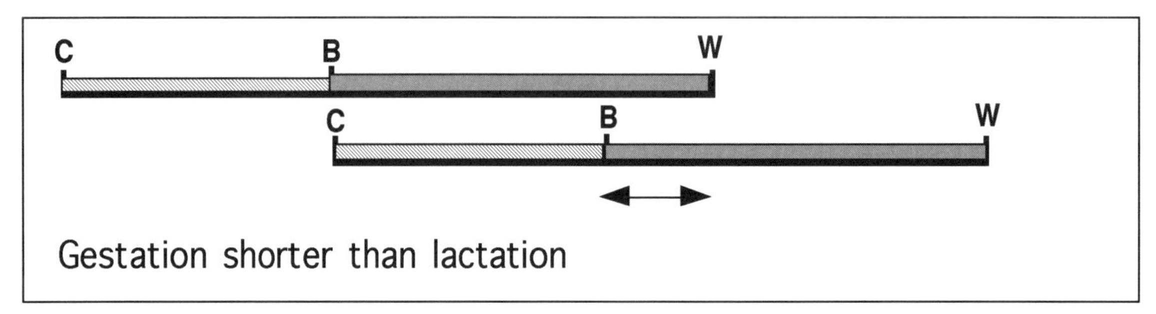

Fig. 4.2. The effect of the relative length of lactation and gestation on the occurrence of two sets of dependent offspring of different ages in non-seasonally breeding mammals. In case 1 (*top*), lactation is relatively short and conception can occur soon after the birth of one set of offspring. In case 2 (*bottom*), where lactation is relatively long, post-partum conception would lead to a period with two different sets of dependent offspring (indicated by dark bar). (C = conception, B = birth, W = weaning.)

the survival prospects of individual offspring. Second, infanticide by males is less likely in species in which infants are able to flee and hide independently, in other words in species in which infants are precocial, unless these infants are reliably associated with the mother. Third, infanticide by males is not expected when males cannot reliably mate with the females when they resume ovarian cycling, because the animals wander unpredictably over large areas, because they live in large anonymous herds, or because the male, the female, or both are unlikely to survive until that time. Finally, infanticide by males is less likely in strictly seasonal breeders that give birth only once a year, provided survival of the current offspring does not reduce the likelihood of giving birth in the next year. Where seasonality is severe and females undergo a period of reproductive quiescence in which they are neither pregnant nor lactating, reproduction in one year is less likely to constrain reproduction in the subsequent year.

The first condition is the most general one. For mammalian females which have long lactation periods, early post-partum conception is not a viable option. Whenever lactation length exceeds gestation length, such early post-partum conception would produce two dependent (sets of) offspring of very different ages (Fig. 4.2). The simultaneous presence of different-aged infants would create a range of problems for the mothers. Young of different size tend to require milk of different composition. They also drink at different intervals, creating a time budgeting problem for mothers leaving their young in a nest, whereas in species in which females carry their young it may be difficult to carry and look after two offspring of disparate ages. Finally, larger offspring should easily be able to displace newborns from the teats, and might even benefit from siblicide.

It is therefore predicted that females of species for which lactation lasts longer than gestation are forced to have lactational amenorrhea or anestrus, which forecloses the option of post-partum mating. Females with relatively long lactation would face an increased risk of infanticide by males unless they can evolve other counterstrategies. This argument assumes that when lactation is shorter than gestation, post-partum estrus, and hence overlap between gestation and lactation, is possible, which should act to reduce the risk of infanticide, even though it need not have evolved for this reason and need not eliminate the risk entirely. Thus, I predict an association across species between the relative length of lactation, the presence of post-partum mating activity and the incidence of infanticide by males.

TESTING THE PREDICTION

Methods

Which measure should one use to assess the feasibility of post-partum estrus, or conversely the need for lactational amenorrhea or anestrus? The most obvious measure may be the difference (d) between the interbirth interval (B) and the sum of lactation and gestation ($L+G$), i.e.

$$d=B-L-G.$$

If $d \geq 0$, then fertilization does not take place well before weaning, and the larger d is, the more a female should, all other things being equal, be at risk of infanticide. For negative values of d, as d gets more negative the risk should also decrease (provided that other aspects of the life history and lifestyle make infants vulnerable). Unfortunately, d is not a good measure because a long B can be long due to a period of 'time-out' (reproductive quiescence in which the female is neither lactating nor pregnant) imposed by the environment, due to strict breeding seasonality or to current local food scarcity. Such a long B would falsely suggest high infanticide risk: infanticide would be unlikely to incite the female to resume cycling in the 'time-out' period.

This problem is reduced if the L/G ratio is used, especially since L is often measured in situations with abundant food (the data frequently come from captive animals fed *ad libitum*). Because using L/G does not eliminate the problem of the seasonal breeders with a single reproductive event and a time-out period, they were excluded from the subsequent analyses. However, other seasonal breeders were included: those that have multiple breeding events in a year, those without clear time-out periods, and those that take more than a year to wean their young.

Gestation length is usually relatively easy to measure. In the following tests, only true gestation length is used (rather than the time between fertilization and birth) in order to exclude cases of delayed implantation. Most species with delayed implantation are strictly seasonal breeders with a single annual reproductive event and time-out (e.g., many carnivores, some bats: Daniel, 1970; Nowak, 1991; Hayssen *et al.*, 1993), and thus would have been eliminated anyway. Lactation is more difficult to estimate because weaning is often gradual rather than abrupt, but most researchers have developed clear, if somewhat arbitrary, criteria (cf. Lee *et al.*, 1991). While the duration of lactation can be quite variable, most data derive from situations with abundant food in which lactation tends to be uniform (and short). Data on lactation and gestation were taken from Eisenberg (1981), supplemented by Harvey *et al.* (1987) for primates, and from Gittleman (1986) and Moehlman & Hofer (1996) for fissiped (non-pinniped) carnivores. For the present purpose, I follow Nowak (1991) in distinguishing two rodent suborders. The data are presented in Appendix 4.1.

Post-partum estrus is usually defined as estrus within days or weeks after the birth of offspring (for haplorrhine primates, which lack estrus, the term post-partum mating activity is more appropriate). Here, a somewhat more liberal definition was adopted, considering mating within about one-third of the normal L for the species as post-partum estrus. Post-partum estrus does not invariably need to lead to conception, but it must do so at least occasionally to be effective in preventing infanticide by males, because the temporal proximity to birth makes it relatively easy for males to distinguish between consistently infertile periods and potentially fertile ones (unlike deceptive sexual attractivity during pregnancy, where clear markers of reproductive state are absent). When it does not lead to conception, infanticide need not be advantageous to the male, because it does not necessarily bring the female back into breeding condition (this also depends on food availability and seasonal triggers).

Data on post-partum mating were taken from Hayssen *et al.* (1993) (especially for the family level) and Nowak (1991) for mammals in general, and from van Schaik *et al.* (1999) for primates. The incidence estimates for post-partum mating are conservative because the absence of a condition is less confidently determined than its presence, and tends not to be reported even if it is known. Seasonal breeders with a single reproductive event per season and a time-out period were excluded (see above). The absence of post-partum estrus could sometimes be inferred from very long interbirth intervals even among favorable conditions. If species with such inferred post-partum estrus are included, the trends reported below become stronger. They will not be separately reported, however.

An association between two features across species can arise due to stabilizing selection, but it can also be an artifact of the inheritance of the same, selectively neutral features from a small number of common ancestors. Several methods have recently been developed to exclude the possibility of spurious association or correlation of non-adaptive traits. For continuous or mixed continuous–discrete variables, the independent contrasts method can be used (CAIC: Purvis & Rambaut, 1994), provided a credible phylogeny is available. Unfortunately, the phylogenetic relationships among mammalian orders remain disputed (cf. Graur, 1993), and those

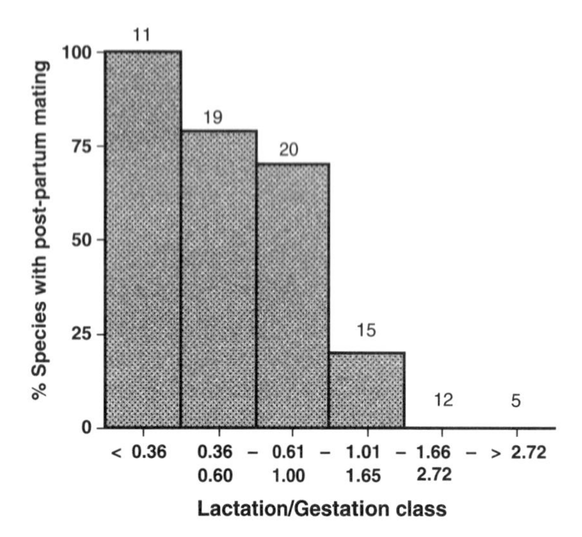

Fig. 4.3. The relationship between the lactation/gestation ratio, in classes, and the percentage of species with known data showing post-partum mating. Numbers of species in each class are indicated above each bar. Seasonally breeding species are excluded.

Table 4.1. *Correlates of variation in mode of infant care in primates: percentage of species (sample size in italics between brackets) with the listed properties*

	Parkers[1] (%)	Communal[1] carriers (%)	Non-communal[1] carriers (%)
Lactation > gestation	0 *(10)*	0 *(7)*	95 *(20)*
Post-partum mating	82 *(11)*	63 *(8)*	0 *(61)*
Male infanticide	0 *(14)*	0 *(8)*	49 *(61)*

Note:
[1] Parkers are species in which females leave young in nest or elsewhere while they forage; carriers are species in which females carry the young (almost exclusively because infants cling on the mothers' backs or bellies); communal carriers are infant-carrying species in which there is substantial infant carrying by non-mothers (males, juveniles).
Source: Based on data compiled in van Schaik *et al.*, 1999.

within orders are also still unsettled, making it difficult to make productive use of methods controling for phylogenetic inertia. For primates, I used Purvis's (1995) composite phylogeny even though this phylogeny too is disputed. Hence, all tests reported here are preliminary. They will have to be repeated with far more complete data sets and more refined methods.

Patterns at the species level

To evaluate the predictions, I examined patterns using only those species for which all the relevant information could be compiled. Figure 4.3 shows how, among placental mammals in general (data from Appendix 4.1), the occurrence of post-partum estrus is strongly dependent on the L/G ratio, with a sharp decrease in the tendency toward post-partum estrus as lactation becomes longer than gestation. The trend is repeated in all major mammalian groups. Among primates, the independent contrasts method gives four independent cases in which post-partum mating evolved; all four are associated with decreases in ln (L/G), as predicted, but not significantly so ($t=1.78$, $p=0.17$), probably due to small sample size.

Simpler tests can be done using association tests. L/G ratios are classified as high (≥ 1) or low (<1). As expected, among primates, post-partum mating is rather well predicted by the L/G ratio: none of the 17 species with $L/G>1$ shows post-partum mating, whereas 86% of the 14 species with $L/G<1$ do (Chi-square with continuity correction: $X^2_{[1]}=20.41, p<0.0001; n=32$). A similar, but less extreme difference is also found in the other placental mammals combined: 31% vs. 81%, respectively ($X^2=9.58$, $p<0.01$; $n=48$). These conclusions are tentative in the absence of corrections for phylogenetic non-independence. However, there is reason for confidence because the trend is repeated within each order with sufficient sample size, albeit with different pervasiveness.

A more detailed examination is possible for primates (cf. van Schaik *et al.*, 1999), because the L/G ratio among primates is tightly linked to their mode of infant care (Table 4.1). Lactation is relatively short in those taxa that give birth to litters of fairly altricial young which are kept in a nest (many Lemuroids, some Galagoids) or somewhat more precocial young that are parked (Lorisids). These taxa are referred to in Table 4.2 as infant parkers. Lactation is much longer in those taxa in which females carry their infants (some Lemuroids, all anthropoids) or are at least strongly associated with their dependent offspring throughout the active period (one tarsier) because infants develop slowly (cf. Charnov, 1993). However, it has, secondarily, been reduced again in several platyrrhines. In callitrichids and the pair-living cebids, infants are cared for communally, as evidenced

Table 4.2. *Life history and behavioral features of selected orders/suborders of placental mammals*

	Percentage of species with $L/G>1$[1]	Percentage of families with post-partum mating/estrus	Male infanticide common?	Stable M-F association[2]	
				Species (%)	Genera (%)
Carnivora (fissiped)	64	56	+	34	[32]
Sciurognath rodents	56	89	+	34	
Primates	51	30[3]	+	76	[73]
Insectivora	18	40	–	10	
Chiroptera	0?	(100)	–		[31]
Lagomorpha	0	(100)	–	20	
Hystricognath rodents	0	80	–	31	
Artiodactyla	0	88	–	22	[29]
Perissodactyla	?	100	(+)		[13]

Notes:

[1] L/G=Ratio of length of lactation/gestation.

[2] M–F = male–female.

[3] Post-partum mating in primates is reported in 30% (3/10) or 29% (4/14) of families (using the families distinguished by Jolly, 1985, and Richard, 1985, respectively).

Sources: female life history data from Appendix 4.1; post-partum estrus or mating periods from Hayssen *et al.* (1993); male infanticide from Hausfater & Hrdy (1984) and Parmigiani & vom Saal (1994); stable male-female association from Appendix 4.1 for species and from van Schaik & Kappeler (1997) for genera.

by heavy involvement of males and sometimes other group members in carrying and sometimes provisioning infants (cf. Lee, 1996; Ross, 1998). Both infant parkers and communally breeding infant carriers among primates tend to have post-partum mating, whereas none of the other infant-carrying species does (Table 4.1).

For primates, we can also examine how post-partum mating is related to infanticide by males. As predicted, infanticide by males is more common in species without post-partum mating (data from Appendix 4.1; 39% of 18 species vs. 0% of 12 species with post-partum mating; $X^2=4.11$, $p<0.05$). If we use mode of infant care rather than post-partum mating as our proxy for vulnerability, the number of species available for analysis is much greater. This analysis confirms the result (Table 4.1). Infanticide by males is reported for roughly half the species for which life history data are available (van Schaik, unpublished compilation) among the infant-carrying species without communal care (Hausfater & Hrdy, 1984; Struhsaker & Leland, 1987), whereas it is not reported in any of the species with the two other modes of infant care.

Is this pattern real? The incidence of infanticide is bound to be underestimated because infanticide must be rare and thus rarely observed, and not all species have been the subject of sufficiently intensive field studies. Unfortunately, this bias may be strongest in nocturnal, solitary species with absentee infant care, raising the possibility that the pattern is an artifact. On the other hand, the absence of infanticide by males in the communal carriers is real because it is not reported from the many field studies on these species, or from the numerous colonies where males are routinely placed into existing groups. This, incidentally, also suggests that the risk of infanticide by males is not a product of infant carrying *per se*, but rather due to the fact that infant carriers tend to have long lactation periods. Thus, the absence of post-partum mating is a good predictor of infanticide by males in primates.

Distribution among orders

Having established the relationship between L/G ratios and post-partum estrus at the species level, it is of some interest to see how these taxa are distributed over the eutherian mammals. Table 4.2 furnishes an overview of the relevant information for the better-known orders. Coverage is thin for most orders, so any pattern is currently only tentative, but three taxa stand out in having a large proportion of species with high L/G ratios: fissiped carnivores, sciurognath rodents and primates.

The incidence of post-partum estrus is estimated here as

the proportion of families in which it occurs in at least one species (due to the thin coverage of data for most groups). Post-partum estrus is common, being found in most families of placental mammals, except among fissiped carnivores and primates, as predicted by the L/G ratios. Post-partum estrus is surprisingly common among sciurognath rodents, even among species with $L/G>1$. This may appear impossible, but some species of rodents and insectivores are known to facultatively delay implantation (Daniel, 1970). Post-partum estrus accompanied by facultative delayed implantation may be a tactic to minimize lost time if the litter is lost, or may serve to prevent infanticide.

Infanticide by males is most often reported for primates, rodents and carnivores (Hausfater & Hrdy, 1984; Parmigiani & vom Saal, 1994), as expected by their high L/G ratios. Exceptions to this pattern may be due to inadequate information (e.g., for insectivores).

SOCIAL COUNTERSTRATEGIES AGAINST INFANTICIDE: PRIMATES

The conclusion from the previous section is that infanticide by males is a predictable threat to females with a life history that involves long infant dependence and thus lactational amenorrhea. We therefore expect that in taxa with the life history that predisposes them to infanticide risk, females and the likely sires have evolved counterstrategies to reduce this risk. For primate females, three major strategies of reducing the risk of infanticide have been suggested: (i) cutting losses through resorption, abortion or premature weaning; (ii) sexual strategies that manipulate the paternities and paternity assessments of potentially infanticidal and protective males; and (iii) social associations that may protect the infant, curiously mainly with males, because females are remarkably ineffective in preventing infanticide by males in primates (e.g. Hrdy, 1977; cf. Mesnick, 1997). The costly reproductive abandonment strategy is uncommon. Van Schaik *et al.* (1999) provide a detailed evaluation of the sexual counterstrategies. The focus here is on the social counterstrategies. I will first revisit them in primates, and then explore whether similar social patterns are also found in the mammals with the appropriate life history.

Contexts of risk

In order to explore the scope of the social counterstrategies, it is worth examining the contexts of infanticide and variation in rates as a function of social conditions. A review of 50 directly observed cases of infanticide in wild primates indicated that 86% of them happened in a context in which a reproductively capable male came into a position of top dominance, either by immigrating into the group and defeating or ousting the resident male, or by dramatically rising in rank within a group (van Schaik, unpublished data; cf. Steenbeek, 1996). Cases from captivity strongly tend toward the same pattern (Angst & Thommen, 1977; Böer & Sommer, 1992). Thus, infanticide risk is heightened when potential protector males are incapacitated or ousted.

The observed coincidence of infanticide with periods of male instability suggests that likely fathers play a particularly important role in protecting primate infants among infant-carrying species. There is, in fact, direct observational evidence for this (reviewed in van Schaik, 1996). In several species, likely fathers are reported to form protective associations with infants (Anderson, 1992; van Schaik & Paul, 1997; Palombit *et al.*, 1997) and come to their aid when they are threatened (Hauser, 1986), and females form close associations with these males too, sometimes following them out of their groups after the males are deposed (e.g., Sugiyama, 1966). In multi-male groups, such as those of long-tailed macaques, males also occupy top-dominance ranks only for a few years, but tend to emigrate only after the infants sired when they were dominant are weaned (van Noordwijk & van Schaik, 1988). It is most likely that these males protect infants against both infanticidal conspecific males and predators. *Thus, the first prediction is that species facing a serious risk of infanticide should show year-round male–female association.*

Rates of infanticide (expressed as the probability that an infant falls prey to infanticide) should vary predictably across kinds of social system, because the conditions in which infants are at risk (exposure to newly dominant male and loss of protector male in a situation where the new dominant can improve his paternity chances) should vary predictably with the social system. Although quantitative estimates are virtually absent, because long-term studies with semi-permanent presence of observers are few and a comprehensive analysis of variation in infanticide rates is as yet impossible, it is possible to examine the effect of some obvious risk factors.

One obvious risk factor is living in single-male groups. In these groups, immigration of a new male is accompanied by the defeat of the current resident and usually by his ouster. In multi-male groups, infanticide risk should be lower because immigrants tend not to oust the previous dominant male, who is therefore still present to protect the vulnerable offspring. Moreover, males who rise in rank within the group and take over top dominance may not benefit from infanticide for several reasons. First, in smaller groups, the former dominant may be willing to defend the offspring. In large groups, male dominance and siring rate are less strongly correlated, especially where breeding is seasonal (Oi, 1996; Paul, 1997). Hence, all males are potential fathers. Thus, second, males rising in rank from inside would risk killing their own offspring. And, third, where the male comes from outside, the benefit from infanticide is diluted because he will have to share mating access with others.

The primate data tend to support the expectation that infanticide risk is reduced in multi-male groups. First, Robbins (1995) showed that among mountain gorillas (*Gorilla gorilla beringei*) all known cases of infanticide occurred in single-male groups, and that infants in single-male groups were significantly more likely to die from infanticide than those in multi-male groups. Second, Newton (1986) noted that the great majority of reported cases of observed or strongly suspected infanticide among Hanuman langurs was from populations in which single-male groups predominated. Third, infanticide is rare or absent in some species with large multi-male groups which are among the most intensively studied species. Examples include rhesus, Japanese and barbary macaques (*Macaca mulatta, M. fuscata, M. sylvanus*). Fourth, some species form two-male groups in which male cooperation serves to keep potentially infanticidal males at bay during inter-group encounters (suggested for gorillas by Robbins, 1995) or reduces the probability of takeover (e.g. red howlers, *Alouatta seniculus*: Sekulic, 1983b; Pope, 1990; geladas: Dunbar, 1984). There is, however, one study going against this pattern. Borries' (1997) estimate of 31% infant mortality (8/26 infant deaths) due to infanticide by males in multi-male groups of Hanuman langurs is very similar to Sommer's (1994) estimate in single-male groups of the same species. There is no obvious explanation for this exception. *The second prediction, therefore, is that if infanticide risk has had a major impact on female social strategies, primate groups should tend to contain multiple males, all of whom will be sexually active.*

Another risk factor, especially likely to be serious in single-male groups, is the absence of female breeding dispersal. In groups subject to takeovers, females with vulnerable infants may be able to leave the group in the company of the former resident (Sugiyama, 1966). In a comparison of langur populations with single-male groups of several closely related species, Sterck (1998) showed that female dispersal and the presence or absence of male takeovers varied among populations. A crude estimate of infanticide risk (number of infants killed relative to number in vulnerable category) yields 7.1–11.5% (midpoint 9.3%) where females are philopatric (defined here as less than 0.1 migration per female per year; based on 484 infants, lower and upper estimate of infants killed) vs. 4.0% where they disperse (177 infants; a significant difference). Male takeovers have a very strong effect: where takeovers occur infanticide risk is higher (7.6%) than where they do not occur (1.0%). Interestingly, takeovers are much less common where females tend to migrate as adults (Sterck, 1998). These findings suggest that females are particularly at risk from infanticide when they live in single-male groups which they cannot easily leave to settle elsewhere. *The third prediction, therefore, is that if females live in single-male groups, they should be more likely to disperse as adults than if they live in multi-male groups.*

Association with protector males

The first prediction was that males and females should associate year-round where the risk of infanticide by males is serious. Using $\ln(L/G)$ as a proxy for infanticide risk, the four cases of evolution of year-round male–female association among primates are always associated with increases in relative length of lactation. The sample size is small, however, and the association does not reach significance ($t = 2.80$, $p = 0.07$). We can increase the resolution if we test the hypothesis in a more indirect way. Male association with one or more females can only be adaptation to reduce the risk of infanticide where mothers and infants are associated, i.e., where mothers carry their infants around with them (van Schaik & Kappeler, 1997). As noted above (Table 4.1), all of the non-communally breeding infant carriers have $L/G > 1$, and should therefore be vulnerable to infanticide by males. One should therefore see a strong correlation between spatial association between mothers and infants and year-round spatial association between males and females. The test focused on strepsirhines, because there is basically no variation in haplorrhines. Van Schaik & Kappeler (1997) found that the strepsirhine data strongly support the hypothesis: the correlation between male–female association and female–infant association is nearly perfect. It is also significant when the correlation is tested using Maddison's (1990) concentrated changes test to control for phylogenetic non-independence, using various different phylogeny reconstructions (*ibid.*).

Where females leave infants in a nest or park them while they forage, male–female association is not expected. The high incidence of post-partum estrus among this class of species should reduce their vulnerability, although one should still expect some infanticide by males in species with litters. Nonetheless, association with the female is not a viable strategy to reduce infanticide risk in these taxa; if anything, males are expected to guard the nests or parked infants when the females are away. This is what is found in ruffed lemurs (*Varecia variegata*), a lemur that leaves its young in a nest yet shows male–female bonding. White *et al.* (unpublished data) showed that males guard the nest and that the amount of nest guarding by the dominant male increased in the presence of potentially infanticidal males. Too little is known of the behavior in other solitary strepsirhines to assess whether male guarding of infants also occurs in other taxa.

The fit between male–female and female–infant association is so strong that it is impossible to determine the historical sequence of these changes. Most researchers assume that the ancestral state was one of solitary females that left their young in the nest (e.g., Martin, 1990). Females may have first begun to associate with males, and subsequently begun carrying their infants with them all the time. Or, alternatively, females may have begun carrying their infants and subsequently evolved protective associations with possible sires. Both scenarios are possible, although both assume a less plausible intermediate stage. In practice, the evolutionary scenario may have been more complex. As female life history changed toward slower reproduction, and thus longer lactation, males may have started to guard nests, much as in extant *Varecia*. As litters decline in size and infants become less altricial, females could carry the infant(s) and thus increase their foraging efficiency. So, in a way, male–infant association may have been the missing link between solitary life and constant intersexual association.

The communal carriers are an interesting exception to the anthropoid pattern. They tend to have post-partum mating and have relatively short lactation, yet show stable male–female association. The frequent incidence of post-partum mating among the communally rearing infant-carriers is clearly a derived condition. The ancestors of these platyrrhines were larger monkeys, which probably showed stable male–female association, which was the exaptation for the development of heavy male involvement in infant care. Dunbar (1995a) also concluded that intersexual association predated male infant care in this group.

Group composition and female breeding dispersal

At this stage, we cannot yet conduct critical tests of the other predicted social consequences of infanticide by males, for three reasons. First, some of the required basic data are still lacking; for instance, information on female breeding dispersal, as opposed to natal dispersal (Pusey, 1992), is incomplete. Second, the predictions assume that all other things are equal, i.e., they ignore social or ecological pressures that are incompatible with the social patterns purportedly selected for by infanticide avoidance. Third, there may be yet other alternative strategies to reduce the risk of infanticide, in particular mating tactics. I will therefore merely survey primate patterns here.

The second social prediction was that primate groups should tend to contain multiple males if infanticide by males is a serious threat. The multi-male groups are made possible by female behavior and reproductive physiology serving to break the monopoly of the dominant male. In the absence of concrete expected values, a good evaluation of this idea would be to compare the incidence of multi-male groups in primates with those in other orders. Lee (1994) compiled a data set on social organization in three major mammalian groups (primates, carnivores and African ungulates) that makes this comparison possible. Where females live in groups to which males are attached year-round, these groups are far more likely to contain multiple males in primates and carnivores than in African ungulates (Fig. 4.4; adding the unstable intersexual associations would not alter this pattern). Hence, compared with ungulates, primates and carnivores are likely to have multi-male groups (carnivores will be discussed below).

The third social prediction was that female breeding dispersal should be more common in populations with single-male groups than in those with multi-male groups. This prediction can be tested comparatively if reliable data on female breeding dispersal become available for enough species. Among primates, there are at least nine taxa with single-male groups (van Schaik, 1996). Female breeding dispersal is well established for gorillas (Watts, 1989), Thomas's langurs (*Presbytis thomasi*: Sterck, 1997), and for Tana River red colobus (*Colobus badius*: Marsh, 1979b), and is likely for proboscis monkeys (*Nasalis larvatus*: Rajanathan & Bennett, 1990). In gorillas and arboreal forest langurs, single-male groups form around a protector male, and the group ceases to exist after the male disappears or dies (Sterck *et al.*, 1997).

Some five taxa are left in which we find single-male

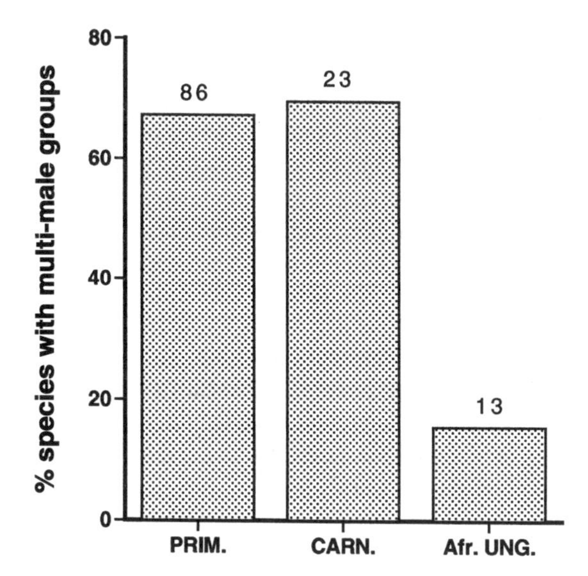

Fig. 4.4. Incidence of multi-male groups among primates, carnivores, and African ungulates. The measure is the percentage of species with female groups and year-round association between the sexes that contains multiple adult males. Numbers of species are indicated above columns. After data compiled by Lee (1994). Using the frequencies, the overall pattern is significant (X^2 [df = 2] = 13.67, $p = 0.001$); the two pair-wise comparisons involving African ungulates are also highly significant ($p < 0.005$).

groups without female breeding dispersal. The prediction was that these should be rare among primates because this raises the risk of infanticide by males. In some of them, an unusual multi-level structure may reduce the risk of infanticide. Although the majority of groups have only one male, a second male ('follower') resides in some groups. This may reduce infanticide risk because the follower is either the previously dominant male after having been deposed, or a younger subordinate male who is gradually attaining reproductive status and may eventually inherit the group. Examples include geladas (*Theropithecus gelada*: Dunbar, 1984) and hamadryas (*Papio hamadryas*: Kummer, 1968). It is possible that groups with males unlikely to be taken over are also less likely to have a follower. No data on extra-group matings are available in these species, nor in another species living in multi-level societies, snub-nosed monkeys (*Rhinopithecus bieti*: Kirkpatrick *et al.*, 1998).

Females in some of the taxa with single-male groups without female breeding dispersal are perhaps capable of reducing infanticide risk by yet another strategy, namely by actively seeking matings with extra-group males, and thus confusing paternity. Some species have groups that contain either one or a few adult males that often fuse to form larger groups when matings with multiple males are likely. This structure is found in pig-tailed macaques (*Macaca nemestrina*), drills and mandrills (*Mandrillus leucophaeus* and *M. sphinx*: Caldecott *et al.*, 1996), and perhaps others. In several African cercopithecines (*Cercopithecus* non-*aethiops*; *Erythrocebus*), groups contain a single male throughout most of the year, but experience an influx of non-group males during the relatively brief mating season (Cords, 1988). These males do mate. It can be argued that in this way females combine the benefits of having only a single male competitor during most of the year, and reducing the risk of infanticide by outside males.

Another way to test the idea that extra-group matings serve to reduce infanticide risk is to compare the proportion of extra-group matings between single-male groups with and without female breeding dispersal. This remains to be done.

Finally, some taxa in single-male groups do not seem to have any obvious protection. Sterck (1998) has shown that lack of female breeding dispersal in the Hanuman langur is a result of elimination of predators and habitat disturbance and fragmentation, with the resulting saturated local densities and lack of dispersal opportunities. Females in these groups suffer high rates of infanticide (see above). This leaves only some howler monkey populations as having at least some single-male groups without female breeding dispersal.

This survey indicates that several social arrangements as well as promiscuous mating behavior may all be effective in reducing the risk of infanticide by males. Of course this considerably weakens our ability to demonstrate that the social arrangements arose, at least in part, as a result of counterstrategies against male infanticide. At this stage, therefore, there is only tentative support for social consequences of infanticide by males beyond male–female association.

SOCIAL COUNTERSTRATEGIES AGAINST INFANTICIDE IN OTHER MAMMALS

Stable male–female association in primates can be ascribed to the risk of infanticide by males, which in turn is a result of long lactational amenorrhea and the absence of post-partum fertilization. Year-round male–female association is far from universal in mammals. I therefore hypothesize that, as in primates, other placental mammals form stable male–female associations where post-partum fertilization is impossible, probably as a result of lactational amenorrhea, and thus risk of infanticide by males is high.

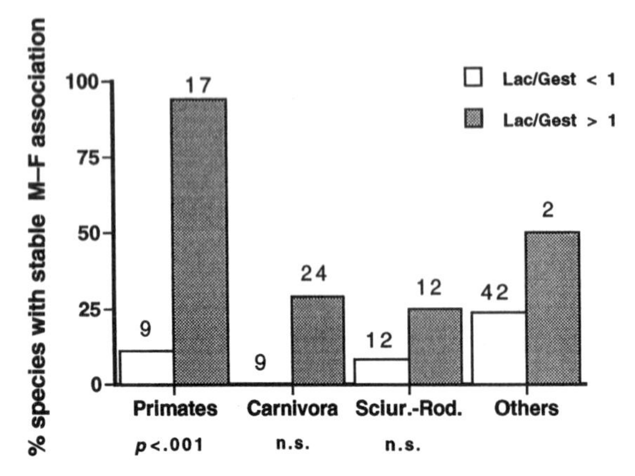

Fig. 4.5. Year-round male–female association and the lactation/gestation ratio in different mammalian taxa. Communally-rearing species and seasonal breeders with one reproductive event and reproductive quiescence are excluded. The number of species per category is indicated above each column.

The evidence from the general pattern across mammals is rather mixed (see Table 4.2). Year-round male–female association is much more widespread in primates than in any other mammalian group. It is only marginally higher among carnivores and sciurognath rodents, where a relatively large proportion of species has long lactation relative to gestation, than in the other taxa. On the other hand, the rather low incidence of male–female association in lagomorphs, artiodactyls and perissodactyls, as well as in other smaller orders not represented in Table 4.2 (e.g., Xenarthra [Edentata], Scandentia, Macroscelidea, Pinnipedia, Proboscidea) is consistent with the hypothesis. This pattern suggests that stable male–female association is a primate-specific response to high risk of infanticide by males.

We can also analyze this question at the species level, including only non-communally breeding species, which are not strictly seasonal (one breeding event per season and reproductive quiescence). The results (based on data presented in Appendix 4.1, but with the above restrictions) indicate that there is a trend in the predicted direction in carnivores and sciurognath rodents (Fig. 4.5). However, it is clear that the pattern is strongest among primates, suggesting that stable male–female association may be only one among several responses to risk of infanticide by males in non-primate mammals.

The taxa most similar to primates in lifestyle might be those most likely to show similar responses: groups of roaming females, always accompanied by their offspring. Equids have this lifestyle; among them, both permanently defended harems and loose and temporary male–female association can be found. Although post-partum estrus is found in perhaps all of them, infanticide by males in circumstances consistent with the sexual selection hypothesis is recorded for the species living in stable harems, not for two other species in which the sexes show only loose association (Table 4.3; see table for references). Thus, the equids show the predicted relationship between vulnerability to infanticide by males and year-round male–female association. (A possible alternative interpretation for infanticide in equids is the elimination of future rivals because new males may have a bias toward killing male foals [C. Feh, unpublished data], but there are insufficient data to test the critical prediction of whether females return to estrus sooner in the species with infanticide than would be expected otherwise.)

The reduced vulnerability to infanticide, and thus absence of permanent male–female bonding, in Grevy's zebra (*Equus grevy*) may be due to ecological peculiarities. Females with new foals concentrate near permanent water (Becker & Ginsberg, 1990), and Rubenstein (1986) suggests that infanticide is prevented here because females that lose their foals would immediately move away from the areas near water where males can no longer monopolize access to them. Life history factors may reduce vulnerability to infanticide as well. Foals of Grevy's zebra, and probably wild ass (*Equus asinus*), are feeding independently at a far younger age and disperse much earlier than those of the other equids (Becker & Ginsberg, 1990), suggesting earlier independence for the two species not forming stable harems.

In conclusion, non-primate mammals vulnerable to infanticide by males seem to have a variety of options to reduce infanticide risk, one of which is stable male–female association. Other social options, such as living in multi-male groups, remain to be explored. Figure 4.4 shows a high proportion of multi-male groups in carnivores, a taxon with many species vulnerable to infanticide. However, we do not know enough about the processes of social change to assess whether multi-maleness actually reduces infanticide risk in this taxon.

DISCUSSION

Infanticide avoidance in primates

Figure 4.6 provides an overview of the hypothesized socio-sexual consequences of a female reproductive style that

Table 4.3. *Social and sexual characteristics of the Equidae*

Species	Post-partum estrus	Year-round M-F association	Infanticide by males?	Source
Equus przewalski	+	+	+	1
E. caballos	+	+	+(+ feticide)	2
E. burchelli	+	+	?	
E. zebra	+?	+	+	3
E. grevy	+	–	–	
E. asinus	+	–	–	

Notes:

[1] Ryder & Massena (1988); C. Feh, peronal communication.

[2] Berger (1986); C. Feh, personal communication.

[3] Penzhorn (1984).

makes her vulnerable to infanticide by males. The first step claims that females in species with lactational amenorrhea or anestrus as a result of long lactation relative to gestation are unable to have post-partum fertilization, and are therefore also more likely to be subject to infanticidal attacks by males. This step has been found to be valid for primates. The species vulnerable to infanticide by males are therefore expected to have evolved counterstrategies.

Promiscuity and sex skins (Hrdy & Whitten, 1987) or female copulation vocalizations (O'Connell & Cowlishaw, 1994) among primates have been suggested to serve in reducing the risk of infanticide by males by confusing paternity. A review of primate sexuality (van Schaik *et al.*, 1999) compared the three classes of infant care distinguished here and found that the duration of mating within the cycle, the active pursuit of polyandrous mating, the length of the follicular phase, the incidence of situation-dependent receptivity (including during pregnancy), and the extent of additional advertizing through morphological or vocal signals were all related to the extent to which polyandrous mating would help to reduce the risk of infanticide by males. Among the infant-carrying species, similar trends with variation in infanticide risk were apparent. Although additional factors are likely to have played a role in molding primate sexuality, infanticide avoidance has been a major influence.

As to the social counterstrategies in primates evaluated in this chapter, there is good support for the idea that the need for protector males has favored year-round association between the sexes, the basis for all the elaborate male–female relationships we see in this order. It was also hypothesized that the risk of infanticide by males has selected for female behavior in primates that tends to lead to multi-male groups or, where this is not possible, to flexible female breeding dispersal. Rigorous tests of these predictions remain to be done, however. In particular, two clear predictions remain to be tested. First, female breeding dispersal should be more common among taxa living in single-male groups than those living in multi-male groups. Second, the proportion of extra-group matings should be higher in single-male group taxa where females are not able to disperse as adults than in those where they can.

Figure 4.6 also notes that where infanticide by males has selected for year-round male–female association, extensive male care for offspring can evolve, which sets the stage for infanticide by females (also to be discussed below).

Alternative explanations for primates

The infanticide-avoidance hypothesis for social patterns in primates examined here is part of a growing number of ideas that suggest that primate social systems (group size and composition, nature of social relationships) are not merely the result of ecological pressures but may also represent responses to social challenges. The infanticide-avoidance hypothesis makes predictions about intersexual association, the number of males in a group, and the conditions in which female breeding dispersal is expected. However, there are alternative explanations for each of these aspects of social organization.

Other possible explanations for year-round male–female association in primates are continuous receptivity and protection against harassment (reviewed in the introduction) but they do not have the same generality, although they probably provide additional pressures in specific cases.

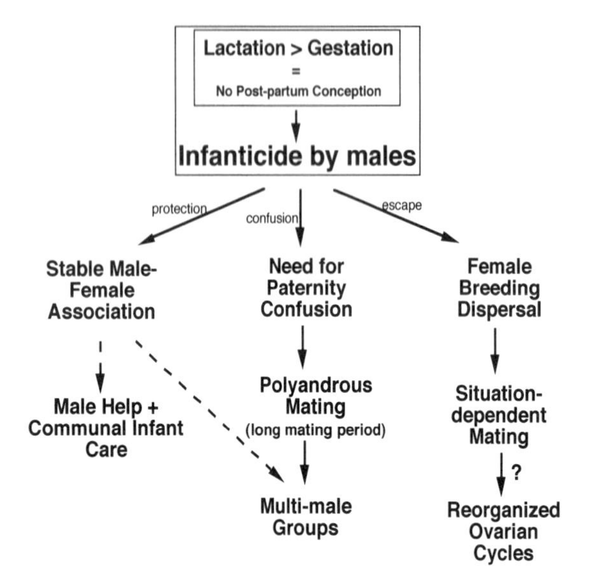

Fig. 4.6. Overview of the hypothesized causal and evolutionary relationships between life history, social and sexual features in species with inherent vulnerability to infanticide by males.

Adult male representation in groups is thought to be affected by the degree to which matings can be monopolized (Mitani *et al.*, 1996a; Nunn, 1999), by predation avoidance (Anderson, 1986; van Schaik & van Noordwijk, 1989; van Schaik & Hörstermann, 1994; Treves & Chapman, 1996), and by infanticide avoidance (as suggested here). Female breeding dispersal is predicted by infanticide avoidance but is also expected to be common where female within-group competition is absent or mainly through scramble (Sterck *et al.*, 1997).

In all of these cases, there is enough circumstantial evidence to take the hypotheses seriously. All the hypothesized processes can operate simultaneously, and they may interact. For instance, the little systematic work on female breeding dispersal in primates has focused on the role of infanticide avoidance (Watts, 1989; Sterck, 1997). However, where within-group contest competition for resources among females is strong, female breeding dispersal is very costly and thus not a very viable option (cf. Sterck *et al.*, 1997), forcing females in these situations to adopt other counterstrategies, which may or may not be equally effective. Thus, the challenge now is to develop tests, comparative or otherwise, that take these multiple pressures and their interactions into account. However, these tests require variation in these three aspects of social organization, as well as variation in the putative selective pressures. The latter, in particular, may not be easy to estimate in all cases.

Infanticide avoidance in non-primates

All the trends noted among primates are weaker or remain unexplored among other mammals. Species with high L/G ratios show a clear trend toward a lack of post-partum estrus. Facultative delayed implantation may be responsible for at least some of the deviations. However, in the absence of reliable comparative data on the incidence of infanticide in other placental mammals, it is impossible to test rigorously whether females of these species are more susceptible to infanticide by males. One way to proceed is to test a weaker version, which claims that species in which infanticide by males is reported should have an average $L/G > 1$ (an informative claim since the overall mammalian mean L/G is < 1).

As to the social consequences, stable male–female association is somewhat more common among mammals with L/G ratios > 1, but not nearly as common as in primates. At this stage, I can only speculate as to the reasons for this differential strength of the social effect. Two factors seem to be important. First, ecology may enforce solitary life; for instance, the species may use concealment as an anti-predation strategy, precluding pair-bonding. Second, where young are left in the nest or den for a relatively long proportion of lactation and mothers make long treks away from this central place, male–female association might not be an effective counterstrategy.

When these constraints on male–female association are serious, alternative counterstrategies would be needed. Thus, Wielgus & Bunnell (1995) showed that female grizzly bears (*Ursus arctos*) avoid the most productive habitats when they have dependent offspring because these are the habitats immigrant males are most likely to use and these males are most likely to commit infanticide. On the other hand, stable male–female association is found in some species with low L/G ratios, such as dik-diks (*Madoqua kirkii*: Brotherton & Manser, 1997), suggesting that factors other than the risk of infanticide by males may favor stable inter-sexual association among non-primate mammals (cf. Komers & Brotherton, 1997). However, in some of the species with relatively short lactation, the post-partum estrus is not always present, probably depending on food availability. Hence, even in these taxa, the risk of infanticide by males may still exist.

It is also not known to what extent variation in sexual behavior among non-primate mammals serves to reduce infanticide risk by males. Variation in the tendency toward promiscuous mating among different carnivores has been attributed to variation in risk of infanticide by males (Eaton, 1978). However, numerous additional factors may be involved (cf. Hoogland, 1998). In other orders, alternative, and perhaps quite sexually effective counterstrategies may exist. Thus, the post-partum estrus among rodents with relatively long lactation is probably the result of delayed implantation. It is conceivable that this effectively prevents infanticide by males.

Why long lactation?

The ideas pursued here raise a fundamental question. If long lactation increases the risk of infanticide by males, and in most cases predation as well, why did it evolve? At this stage, only some tentative answers can be proposed.

One factor predisposing toward relatively long lactation is altriciality. First, at similar post-natal growth rates, precocial young should require less time to reach independence than altricial young. Second, if precocial birth is an adaptation allowing newborns actively to escape from predators, one would expect precocial species to show faster post-natal growth rates, making it even more likely that lactation is relatively short. Third, precocial young are born after longer gestation (Martin & MacLarnon, 1985), thus automatically biasing toward a L/G ratio of <1. Among mammals in general, insectivores, carnivores and rodents have altricial young (Martin & MacLarnon, 1985). The latter two of these orders were found to have a relatively high incidence of infanticide by males. Although primate newborns are mostly rather precocial, they have unusually slow post-natal growth rates (Charnov, 1993), producing very long dependence on the mother.

Another factor may be body size. Among both primates and carnivores (Fig. 4.7), lactation increases relative to gestation length with female body weight. Reproductive rates show a negative allometry with body size (Harvey *et al.*, 1989), but it is not *a priori* obvious why lactation should slow down more with body size than gestation. Nonetheless, this trend suggests that the risk of infanticide by males should, all other things being equal, increase with body size in these orders. This seems to be generally true. There are not enough data to examine whether a similar relationship exists in other orders.

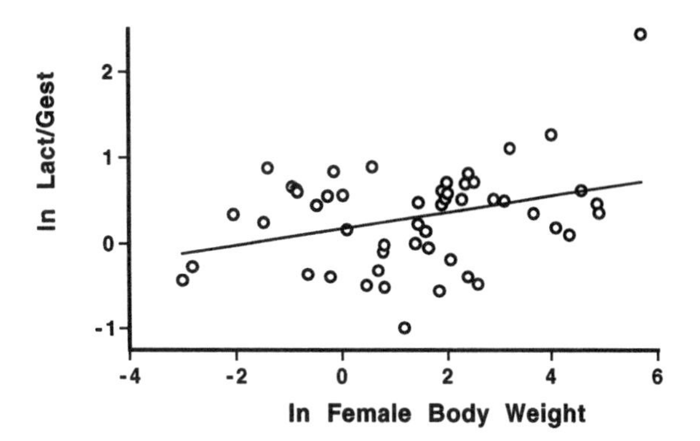

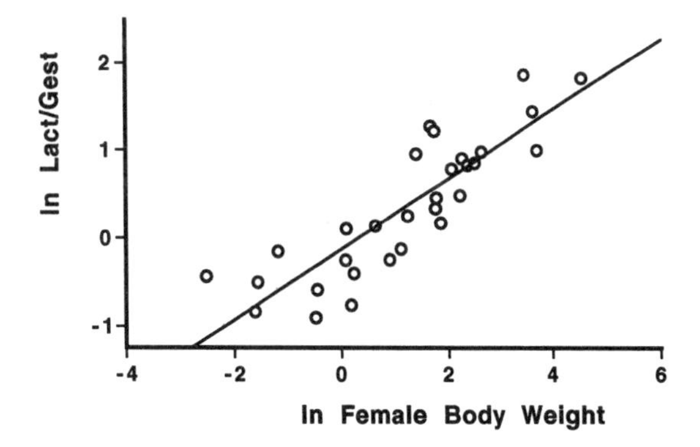

Fig. 4.7. The relationship between female body size and the lactation/gestation ratio for non-communally breeding species of carnivores (A) and primates (B). The slope is significantly positive for carnivores ($r=0.332$, $n=51$; $p=0.05$) and primates ($r=0.865$; n 30; $p=0.0001$).

Infanticide by males, male care, and infanticide by females

The hypothesis developed here for the relationship between life history, infanticide by males, and year-round male–female association can also help to clarify the distribution of male infant care and female infanticide in mammals. As noted earlier, the average mammalian male and female live largely separate lives. Due to the inherent female bias toward infant care as a result of internal gestation and female-only lactation, male mammals generally benefit more from pursuing additional matings than from caring for their offspring (Maynard Smith, 1977; Clutton-Brock & Parker,

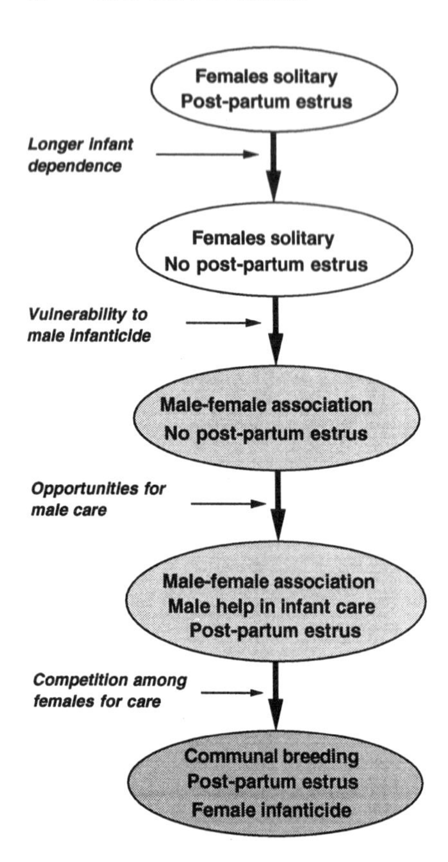

Fig. 4.8. Evolution of communal breeding and female infanticide in mammals, based on polarities inferred from logical preconditions.

1992; Hawkes *et al.*, 1995). Hence, male care for offspring is exceedingly rare among mammals, as compared to birds, the males of which can brood eggs and for which efficient male provisioning is possible right after hatching, or fishes with external fertilization, for which the potential for care is the same for both sexes (Clutton-Brock, 1991).

It can be argued that year-round male–female association is a necessary, albeit not sufficient, condition for the evolution of obligate male help in raising the offspring (see Fig. 4.8). Male care for offspring may have started as mating effort rather than parenting effort. Much of male care for infants in primates, even among the taxa with the most extensive male care, can be considered mating effort (Smuts & Gubernick, 1992; van Schaik & Paul, 1997). Inevitably, some of male care in carnivores is mating effort too, since males are known to care for offspring that cannot possibly be related to them. Although it is hard to see how one can test this historical scenario, patterns at the ordinal level support it strongly. Extensive male care, as witnessed by communal breeding, is only found in carnivores, rodents and primates (Solomon & French, 1996), exactly those orders in which the highest incidence of stable male–female association is found. Moreover, among the canids, communal breeding is found preferentially among the largest species (Moehlman & Hofer, 1996), those that historically were the most likely to have shown stable male–female association (cf. Fig. 4.7).

Once male care is established and females rely on male care for successful reproduction, and since most forms of male care such as carrying and provisioning cannot be shared with other females, females will compete for access to male helpers (cf. Fig. 4.8). In such species, females could evolve strategies to suppress the reproduction of other females within the social unit. One of these strategies would be female infanticide (L. Digby *et al.*, unpublished data). Thus, stable male–female association, itself the result of the threat of infanticide by males, may in the long run produce both male infant care and female infanticide.

ACKNOWLEDGMENTS

I thank the following for discussion or comments on the manuscript: Leslie Digby, Peter Kappeler, Charles Nunn, Karen Strier, Maria van Noordwijk, Richard Wrangham, and an anonymous reviewer. I thank Claudia Feh and John Vandenbergh for pointing to references on equids and mammalian reproduction, respectively. The work reported here was started while the author was at the Deutsches Primatenzentrum, supported by a Forschungspreis of the Alexander von Humboldt Foundation.

Appendix 4.1. *Data on reproduction, social behavior, and post-partum estrus in mammals*

Order	Species	Gestation (d)	Lactation (d)	Communal breeding	M–F association	Post-partum estrus
Artiodactyla	*Antilocapra americana*	235	60	N	N	
Artiodactyla	*Camelus dromedarius*	389	360	N	N	(Y)
Artiodactyla	*Capra ibex*	165	90	N	N	
Artiodactyla	*Choeropis liberiensis*	194	90	N	N	
Artiodactyla	*Dama dama*	240	107	N	N	
Artiodactyla	*Giraffa camelopardalis*	435	270	N	N	Y
Artiodactyla	*Kobus defassa*	280	210	N	N	Y
Artiodactyla	*Llama vicugna*	330	300	N	Y	N
Artiodactyla	*Madoqua kirkii*	174	42	N	Y	Y
Artiodactyla	*Odocoileus hemionus*	203	21	N	N	
Artiodactyla	*Odocoileus virginianus*	204	150	N	N	
Artiodactyla	*Ovibos moschatus*	240	139	N	Y	Y
Artiodactyla	*Ovis canadensis*	180	135	N	Y	
Artiodactyla	*Pudu pudu*	210	65	N	Y	Y
Artiodactyla	*Rangifer tarandus*	225	60	N	N	
Artiodactyla	*Tayassu tajacu*	123	50	N	Y	N
Artiodactyla	*Tragulus javanicus*	120	90	N	N	Y
Artiodactyla	*Tragulus napu*	162	21	N	N	Y
Carnivora	*Acinonyx jubatus*	91	109	N	N	N
Carnivora	*Alopex lagopus*	52	63	Y	Y	
Carnivora	*Arctictis binturong*	90	56	N	N	
Carnivora	*Bassariscus astutus*	52	120	N	N	
Carnivora	*Canis aureus*	63	63	Y	Y	
Carnivora	*Canis latrans*	63	57	Y	Y	
Carnivora	*Canis lupus*	63	56	Y	Y	
Carnivora	*Canis mesomelas*	60	56	Y	Y	
Carnivora	*Canis simensis*	60	70	Y	Y	
Carnivora	*Caracal caracal*	74	123	N	N	
Carnivora	*Cerdocyon thous*	56	90	N	Y	
Carnivora	*Chrysocyon brachyurus*	64	105	N	Y	
Carnivora	*Civettictis civetta*	68	140	N	N	
Carnivora	*Crocuta crocuta*	110	390	N	Y	
Carnivora	*Cuon alpinus*	62	58	Y	Y	
Carnivora	*Cynictus penicillata*	54	42	Y	Y	
Carnivora	*Enhydra lutris*	120	364	N	N	N
Carnivora	*Felis chaus*	65	102	N	N	
Carnivora	*Felis sylvestris*	67	84	N	N	
Carnivora	*Fossa fossa*	85	52	N	N	
Carnivora	*Galidia elegans*	83	56	N	N	
Carnivora	*Genetta genetta*	72	175	N	N	(N)
Carnivora	*Gulo gulo*	35	70	N	N	
Carnivora	*Herpestes auropunctatus*	46	32	N	N	
Carnivora	*Hyaena hyaena*	87	60	Y	Y	
Carnivora	*Ictonyx striatus*	36	56	N	N	N
Carnivora	*Leopardus geoffroyi*	70	63	N	N	
Carnivora	*Leopardus pardalis*	72	49	N	N	
Carnivora	*Lutra canadensis*	112	93	N	N	

Appendix 4.1 (*cont.*)

Order	Species	Gestation (d)	Lactation (d)	Communal breeding	M–F association	Post-partum estrus
Carnivora	*Lutra lutra*	66	112	N	N	
Carnivora	*Lutrogala perspicillata*	62	126	N	N	
Carnivora	*Lycaon pictus*	72	71	Y	Y	N
Carnivora	*Lynx lynx*	68	113	N	N	
Carnivora	*Lynx rufus*	63	60	N	N	(Y)
Carnivora	*Martes americana*	26	46	N	N	
Carnivora	*Martes pennanti*	77	46	N	N	Y
Carnivora	*Martes zibellina*	28	49	N	N	
Carnivora	*Meles meles*	42	95	N	Y	
Carnivora	*Mephitis mephitis*	63	46	N	N	
Carnivora	*Mustela altaica*	40	56	N	N	
Carnivora	*Mustela frenata*	24	30	N	N	
Carnivora	*Mustela lutreola*	38	70	N	N	
Carnivora	*Mustela nivalis*	42	32	N	N	(Y)
Carnivora	*Mustela rixosa*	37	24	N	N	
Carnivora	*Mustela sibirica*	29	56	N	N	
Carnivora	*Nyctereutes procyonoides*	61	70	N	Y	
Carnivora	*Otocyon megalotis*	60	105	Y	Y	
Carnivora	*Panthera leo*	106	150	N	Y	N
Carnivora	*Panthera onca*	104	115	N	N	
Carnivora	*Panthera pardus*	98	139	N	N	
Carnivora	*Panthera tigris*	104	165	N	N	N
Carnivora	*Poecilogale albonucha*	32	77	N	N	N
Carnivora	*Prionailurus bengalensis*	67	25	N	N	
Carnivora	*Prionailurus viverrinus*	92	53	N	N	
Carnivora	*Procyon lotor*	65	119	N	N	N
Carnivora	*Speothos venaticus*	67	120	N	Y	(N)
Carnivora	*Spilogale putorius*	30	56	N	N	
Carnivora	*Suricata suricatta*	77	56	Y	Y	
Carnivora	*Taxidea taxus*	42	42	N	N	
Carnivora	*Ursus americanus*	91	168	N	N	
Carnivora	*Ursus arctos*	63	730	N	N	N
Carnivora	*Vulpes velox*	50	49	N	Y	
Carnivora	*Vulpes vulpes*	53	56	Y	Y	
Carnivora	*Vulpes zerda*	52	61	N	Y	
Chiroptera	*Antrozous pallidus*	63	33	N	Y	
Chiroptera	*Carollia perspicillata*	110	42	N	Y	
Edentata	*Bradypus variagatus*	175	10	N	N	
Edentata	*Choloepus hofmanni*	332	20	N	N	
Edentata	*Euphractus sexcinctus*	63	32	N	N	
Insectivora	*Crocidura suaveolens*	28	20	N	N	Y
Insectivora	*Cryptotis parva*	22	22	N	Y	Y
Insectivora	*Echinops telfairi*	64	30	N	N	
Insectivora	*Erinaceus europaeus*	42	25	N	N	
Insectivora	*Hemicentetes nigriceps*	55	20	N	N	
Insectivora	*Hemicentetes semispinosus*	60	20	N	N	
Insectivora	*Microgale talazaci*	61	30	N	N	
Insectivora	*Neomys fodiens*	24	37	N	N	N

Appendix 4.1 (*cont.*)

Order	Species	Gestation (d)	Lactation (d)	Communal breeding	M–F association	Post-partum estrus
Insectivora	*Parascalops brewerii*	35	28	N	N	
Insectivora	*Suncus murinus*	30	19	N	N	
Insectivora	*Tenrec ecaudatus*	60	29	N	N	
Lagomorpha	*Lepus americanus*	37	15	N	N	
Lagomorpha	*Lepus californicus*	44	14	N	N	
Lagomorpha	*Lepus timidus*	49	28	N	N	
Lagomorpha	*Oryctolagus cuniculis*	34	20	N	Y	Y
Lagomorpha	*Sylvilagus aquaticus*	40	14	N	N	Y
Macroscelidea	*Elephantulus rufescens*	56	22	N	N	Y
Macroscelidea	*Rhynchocyon chrysopygus*	42	14	N	N	Y
Primates	*Alouatta palliata*	187	630	N	Y	N
Primates	*Aotus trivirgatus*	133	75	Y	Y	N
Primates	*Arctocebus calabarensis*	134	115	N	N	
Primates	*Ateles fuscipes*	226	365	N	Y	N
Primates	*Callimico goeldii*	154	65	Y	Y	Y
Primates	*Callithrix jacchus*	148	63	Y	Y	Y
Primates	*Cebuella pygmaea*	136	90	Y	Y	Y
Primates	*Cercocebus albigena*	177	210	N	Y	N
Primates	*Colobus satanus*	195	480	N	Y	N
Primates	*Eulemur fulvus*	118	135	N	Y	
Primates	*Galago crassicaudatus*	135	90	N	N	Y
Primates	*Galago demidovii*	111	45	N	N	Y
Primates	*Galago senegalensis*	124	75	N	N	Y
Primates	*Gorilla gorilla*	256	1583	N	Y	N
Primates	*Homo sapiens*	267	720	N	Y	N
Primates	*Hylobates klossii*	210	330	N	Y	N
Primates	*Hylobates lar*	205	730	N	Y	N
Primates	*Indri indri*	160	365	N	Y	N
Primates	*Lagothrix lagotricha*	225	315	N	Y	N
Primates	*Lemur catta*	135	105	N	Y	
Primates	*Leontopithecus rosalia*	129	90	N	Y	Y
Primates	*Lepilemur mustelinus*	135	75	N	N	
Primates	*Macaca fascicularis*	162	420	N	Y	N
Primates	*Macaca nemestrina*	167	365	N	Y	N
Primates	*Microcebus murinus*	62	40	N	N	Y
Primates	*Miopithecus talapoin*	162	180	N	Y	N
Primates	*Nycticebus coucang*	193	90	N	N	Y
Primates	*Pan troglodytes*	228	1460	N	Y	N
Primates	*Papio cynocephalus*	180	420	N	Y	N
Primates	*Perodicticus potto*	193	150	N	N	Y
Primates	*Pongo pygmaeus*	260	1095	N	N	N
Primates	*Propithecus verreauxi*	140	180	N	Y	
Primates	*Saguinus fuscicollis*	149	90	Y	Y	N
Primates	*Saguinus midas*	127	70	Y	Y	Y
Primates	*Tarsius spectrum*	157	68	N	Y	Y
Primates	*Theropithecus gelada*	170	450	N	Y	N
Primates	*Varecia variegata*	102	90	N	Y	
Rodentia-H	*Cavia porcellus*	68	21	N	Y	Y

Appendix 4.1 (*cont.*)

Order	Species	Gestation (d)	Lactation (d)	Communal breeding	M–F association	Post-partum estrus
Rodentia-H	*Chinchilla laniger*	111	49	N	N	Y
Rodentia-H	*Ctenomys tolarus*	130	35	N	N	Y
Rodentia-H	*Dasyprocta sp.*	120	20	N	Y	Y
Rodentia-H	*Erethizon dorsatum*	217	56	N	N	
Rodentia-H	*Galea musteloides*	52	21	N	Y	Y
Rodentia-H	*Hydrochaerus hydrochaeris*150	112	N	Y	N	
Rodentia-H	*Lagidium maximus*	153	30	N	N	
Rodentia-H	*Lagidium peruanum*	140	56	N	N	N
Rodentia-H	*Myocastor coypus*	132	56	N	Y	Y
Rodentia-H	*Octodon degus*	90	28	N	N	Y
Rodentia-H	*Proechimys guairae*	63	21	N	N	Y
Rodentia-H	*Proechimys semispinosus*	65	21	N	N	Y
Rodentia-H	*Thryonomys swinderianus*	155	30	N	N	
Rodentia-S	*Castor canadensis*	128	42	Y	Y	
Rodentia-S	*Citellus colombianus*	24	28	Y	Y	
Rodentia-S	*Citellus tridecemlineatus*	28	29	Y	Y	
Rodentia-S	*Clethrionomys glareolus*	21	18	Y	N	
Rodentia-S	*Cynomys ludovicianus*	30	49	Y	Y	
Rodentia-S	*Dipodomys merriami*	33	20	Y	N	
Rodentia-S	*Dipodomys nitratoides*	32	23	Y	N	Y
Rodentia-S	*Dipodomys panamintinus*	29	28	Y	N	
Rodentia-S	*Glaucomys sabrinus*	37	65	Y	N	
Rodentia-S	*Glaucomys volans*	39	65	N	N	
Rodentia-S	*Liomys pictus*	25	26	N	N	
Rodentia-S	*Meriones unguiculatus*	23	30	Y	Y	Y
Rodentia-S	*Mesocricetus auratus*	16	21	N	N	
Rodentia-S	*Microtus arvalis*	21	16	N	N	Y
Rodentia-S	*Microtus pennsilvanicus*	21	14	N	N	Y
Rodentia-S	*Neotoma albigula*	38	30	N	N	
Rodentia-S	*Neotoma cinerea*	30	21	N	N	
Rodentia-S	*Ochrotomys nutalli*	30	18	N	N	Y
Rodentia-S	*Ondatra zibethicus*	30	30	Y	Y	Y
Rodentia-S	*Onychomys leucogaster*	30	40	N	Y	Y
Rodentia-S	*Ototylomys phyllotis*	51	54	N	N	Y
Rodentia-S	*Perognathus californicus*	25	23	N	N	
Rodentia-S	*Peromyscus californicus*	24	21	Y	Y	
Rodentia-S	*Peromyscus crinitus*	24	28	N	Y	
Rodentia-S	*Peromyscus eremicus*	21	35	N	Y	
Rodentia-S	*Peromyscus maniculatus*	24	23	N	Y	
Rodentia-S	*Sciurus carolinensis*	44	60	N	N	
Rodentia-S	*Sciurus vulgaris*	38	57	N	N	
Rodentia-S	*Sigmodon hispidus*	27	20	N	N	
Rodentia-S	*Tamias striatus*	31	35	N	N	
Rodentia-S	*Tamiasciurus hudsonicus*	40	50	N	N	
Rodentia-S	*Tylomys nudicaudatus*	39	41	N	N	N

Note that no assessment for post-partum estrus is made for seasonal breeders with reproductive quiescence. See text for sources.

Part III
Variation in male numbers: taxon-level analyses

Chapters in the following section describe and analyze variation in the number of males across primate taxa. An attempt was made to provide a representative taxonomic coverage because it is not feasible to feature each taxon in the same detail. Colobines may appear over-represented at first, but because they exhibit some of the greatest variability in the traits of interest and because results of important recent field studies have not been included in previous reviews, it seemed appropriate to grant ample space to these taxa. So-called solitary primates, on the other hand, are not considered anywhere, even though the same problems related to sexual conflicts may present themselves to individuals of these species at a different level. Students of orangutans and prosimians should therefore discover inspiring ideas in many of these chapters. The same applies to a lesser extent to pair-living species. Some chapters consist of reviews, whereas others present illuminating case studies. Nevertheless, each chapter focuses on the most salient aspect of variation in the number of males in the respective taxon and goes on to explore its possible causes and consequences.

The independent radiation of Malagasy primates is examined by Peter Kappeler, who identifies the even adult sex ratios of group-living lemurs as the most unusual aspect of their social organization. Groups of less than five females can be monopolized by a single male in many other primates, whereas such small numbers of lemur females are always associated with several males. A review of existing studies indicates that these extra males provide females with no obvious extra benefits and that mechanisms of male reproductive competition do not differ fundamentally from those of anthropoids. The idea that these lemurs are in an evolutionary transitional state and that their demography lags behind their behavior could explain the observed patterns but requires additional studies.

Marmosets and tamarins are unusual among primates because many of them live in polyandrous groups in which reproductively active males outnumber females. Eckhard Heymann shows that this may be a response to the need for male assistance in carrying dependent young, which are typically born as litters. The nature of reproductive skew among sexually active males is not yet known because genetic data are lacking, but Heymann shows interesting variation among species in the degree of mutual tolerance among males which is accompanied by corresponding variation in testes size. He concludes that this tendency towards increasing reproductive skew among males may eventually lead to a monogamous mating system. Looking at another larger family of New World primates, Karen Strier discusses sex differences in natal dispersal as an important determinant of reproductive strategies and group composition. Using examples from the Cebids, she shows with long-term data that if males remain in their natal group, their numbers as adults are primarily determined by the birth sex ratio. Males in species that migrate into neighboring groups, on the other hand, appear to respond to variation in group composition and thus population sex ratios.

Guenons and their relatives are a group of Old World monkeys that exhibit variation between one-male and multi-male groups at several levels. Marina Cords summarizes this variation among and within species in her chapter. She examines to what extent female group size and breeding seasonality can explain the prevalence of single-male and multi-male groups, but these factors alone cannot explain all species differences. Guenons provide an example of added complexity because many of their groups are single-male for most of the year, but experience influxes of several males during the mating season. Because this phenomenon has only been studied in sufficient detail in one species, general explanations are not readily available. Baboons, which also show variation between one-male and multi-male groups, are examined by Robert Barton. He refines and further develops an earlier model to explain this variation. Central aspects of this model postulate a close link between ecological conditions, the nature of female feeding competition, and the types of social bonds among females. He argues that whenever female bonds are weak, the typically bigger and stronger males can isolate and monopolize small groups of females. This promising model should be tested in other taxa.

African and Asian colobine monkeys are obvious candidates for such further tests because they include taxa with single-male and multi-male groups, as well as female-bonded and non-female-bonded structures. Tom Struhsaker examines variation in group composition at several hierarchical levels in red colobus monkeys, a taxon in which most groups exhibit male philopatry. This provides an opportunity to turn the main question around and examine causes of variation in the number of females per group. Despite an impressive data base, there is so much variation within and between groups, population and over time that no clear answers emerge. Elisabeth Sterck and Jan van Hooff review data on Asian colobines that occur in single-male and multi-male groups. Because female group size does not differ much between these two groups, it is interesting to ask why so many groups include multiple males. A closer analysis reveals that multi-male groups are heterogeneous and often functionally single-male groups. Cooperative defense against infanticidal attacks by strange males appears to be an important benefit for many of these groups. Romy Steenbeek and her collaborators show in a case study of Thomas's langurs that many of these multi-male groups consist of father–son pairs, in which both may accrue benefits from this association. Results from this long-term study also emphasize the importance of the temporal component of variation in group composition by documenting a predictable sequence of single-male and multi-male phases. This aspect usually gets neglected in broad cross-sectional comparisons. In a final chapter on langurs, Carola Borries presents data on male dispersal from a Hanuman langur population that consists of mostly multi-male groups, but in which males also form all-male bands. Male dispersal is an important source of variation in the number of males that appears to be used strategically by these males because new immigrants have improved chances to attain high rank. Influxes of several males during the mating season into particular groups provide an interesting parallel to guenons, but their causes are also not yet understood.

Gibbons are the only monogamous primates discussed in detail in this section. Ulrich Reichard and Volker Sommer use a long-term data base of one well-studied population to show that sexual conflict is affecting male and female behavior as in other taxa. These gibbons may not be as flexible as dunnocks, but a surprisingly large proportion of groups does not exhibit the expected pair structure, and extra-pair copulations are also common. A similar theme is taken up by David Watts in his review of gorilla social systems. Also once considered a prime example for a species with single-male groups, Watts emphasizes the fact that a fair proportion of groups includes more than one adult male. Using the long-term records of mountain gorillas, Watts is able to compare expected reproductive payoffs of different male strategies. Because follower males do much better than bachelors in all-male groups and because females appear to prefer groups with more than one male, single silverbacks should experience strong resistance against their monopolization interests. As in gibbons, much of the reproductive behavior of gorilla females appears to be geared toward a reduction of infanticide risk. Chimpanzees are discussed in another context in the last section.

5 • Causes and consequences of unusual sex ratios among lemurs

PETER M. KAPPELER

INTRODUCTION

Over the last decade, several important pieces of the puzzle of lemur social evolution have been identified (Richard, 1987; Richard & Dewar, 1991; Goodmann *et al.*, 1993; van Schaik & Kappeler, 1993, 1996; Pereira, 1995; Jolly, 1998; Pereira *et al.*, 1999), but the more details about their social systems and life histories became available, the more questions about fundamental predictions of socioecological theory arose. These questions emerged because lemurs, as an evolutionary independent radiation, evolved ways and mechanisms of organizing themselves and their behavior that apparently differ in several respects from those characterizing most of their anthropoid cousins. In this chapter, I briefly review some of these idiosyncrasies and then focus on one of them, namely the unusually large number of males in their groups.

Primate socioecology

In order to highlight puzzling aspects of lemur socioecology, it is helpful to outline basic aspects of the current theoretical model which explains the observed association patterns of adult males and females as the outcome of sex-specific adaptations (N.B. Davies, 1991). Accordingly, the distribution of females is ultimately determined by the distribution of risks and resources in the environment, but females also respond to the presence of males (Emlen & Oring, 1977; Wrangham, 1987; van Schaik, 1996). Males, on the other hand, go where the females are, because their reproductive success is limited by access to mates (Trivers, 1972; Emlen & Oring, 1977; Ims, 1988; Altmann, 1990). The resulting association pattern between the sexes is ultimately governed by sex differences in potential reproductive rates, i.e. the maximum number of independent offspring that parents can produce per unit time (Clutton-Brock & Parker, 1992), which create conditions for sexual conflict (Kvarnemo & Ahnesjö, 1996).

The sex-specific integration of these ecological and social factors is reflected by the resulting social organization. At this level, each individual will find itself in one of five social situations (Table 5.1). Individual males and females can live alone, with one or more members of the same sex, with only one member of the opposite sex, with several members of the opposite sex, or in groups with one or more members of both sexes. These demographic categories broadly determine the operational sex ratio, defined as the ratio of males and females who are ready to mate, which in turn is an important predictor of sex roles and the intensity of mating competition (Kvarnemo & Ahnesjö, 1996).

In primates, as in other mammals, males are expected to compete among themselves for exclusive access, or at least for priority of access, to females. Such intrasexual competition has consequences at three levels. First, at the behavioral level, fighting, injuries, intolerance, dominance relationships and skewed mating and/or reproductive success in favor of dominant males are expected, and, indeed, often observed (e.g., Cowlishaw & Dunbar, 1991; de Ruiter & van Hooff, 1993; van Hooff, Chapter 16). Second, depending on the type of polygyny, intrasexual selection has predictable morphological consequences, resulting in prolonged male development, sexual dimorphism in body and canine size, as well as variation in relative testes size (Crook, 1972; Clutton-Brock *et al.*, 1977; Alexander *et al.*, 1979; Leigh, 1995; Harcourt, 1997). Third, male–male competition affects demographic parameters and may result in male-biased birth sex ratio and mortality rates, for example (Harvey & Zammuto, 1985; Clutton-Brock & Iason, 1986; van Schaik, 1992; Clinton & LeBeouf, 1993).

Lemur idiosyncrasies

Lemurs are unusual in this context because they exhibit most behavioral, but apparently not the morphological and demographic, consequences of intrasexual selection, which may suggest that access to several females cannot be monopolized. In this chapter, I focus on one fundamental demographic variable by examining adult sex ratios of group-living lemurs. This analysis is not only of interest to

Table 5.1. *Basic types of social organization*[1]

Type	Same sex	Opposite sex	Social organization
I	0	0	Solitary
II	≥1	0	All-male or all-female group
III	0	1	Pair
IV	0	>1	Harem or polyandrous group
V	≥1	≥1	Multi-male, multi-female group

Note:

[1] The social organization resulting from adding various numbers of individuals of the same or opposite sex to an adult male or female is shown.

students of lemur biology. Because the lemur radiation was separated on Madagascar more than 50 million years ago and 15 million years before the first anthropoids appeared (Purvis, 1995; Yoder *et al.*, 1996), it provides an opportunity to study convergent evolution within the framework of the socioecological model. Recent phylogenetic reconstructions indicated that group-living evolved twice independently among the lemur and indri families (Kappeler, 1999a; see also Ridley, 1986), but a lack of data presently precludes extending comparisons to that level.

As in most other group-living primates, male lemurs typically disperse from their natal groups and rarely provide direct paternal care (Wright, 1990; Kappeler, 1997a). In contrast, however, lemur males are generally neither bigger than females nor do they have larger canines, and they are often socially subordinate toward females (Richard, 1987, Kappeler, 1993a). In addition, regular nocturnal activity and extremely seasonal reproduction are common among group-living species (Richard & Dewar, 1991; van Schaik & Kappeler, 1993; Overdorff & Rasmussen, 1995). This combination of possibly interrelated traits is rare among other mammals and its adaptive significance remains poorly understood (van Schaik & Kappeler, 1993, 1996).

Lemurs also exhibit unusual traits related to the size and composition of their groups (van Schaik & Kappeler, 1993). First, lemur groups are smaller than groups of most anthropoids of the same body size (Kappeler & Heymann, 1996). The reasons for this difference are not yet clear, but as a result, lemur social units consist of relatively few adults. Second, lemurs exhibit great intraspecific variation in group composition (Table 5.2). Several species include groups of each basic type, namely one male and one female, one male and several females, one female and several males, and multi-male, multi-female groups, and this cross-sectional variability is probably accompanied by similar variation within groups over time. How can we explain this variability, which spans all traditional categories, and which is rarely found among anthropoids (Kappeler, 1999a)?

If, as the socioecological model suggests, at least population-specific adaptations of both sexes to local conditions are to be expected, these lemurs clearly deviate from this prediction and should be considered as living representatives of a 'socioecological null hypothesis'. Alternatively, this variation may indicate that lemurs respond in a very flexible manner to changes in local social and ecological conditions, and thus exhibit fine-grained adaptations at this level. How to distinguish between these two possibilities in primates, where experimental possibilities are limited (cf. Davies, 1992), remains a continuing challenge to field workers.

A third idiosyncrasy of lemur groups is that groups consisting of a single adult male and several females ('harem groups') are rare and never the modal category in those species in which they occur at all (Table 5.2). Instead, pairs are the most common category besides multi-male, multi-female groups. Apparently, most lemur females do not experience strong pressure toward associating with other females because roughly every third lemur female lives in permanent association, but without other females. In contrast to expectations of the socioecological model, the main reproductive options for males appear to be shared access to a few females and sole access to a single one, assuming that they cannot exclude rivals. This apparent lack of the ability to monopolize several females may reduce variance in average male reproductive success, compared to other primates, and thus contribute to some of the lemur idiosyncrasies, such as sexual monomorphism.

Finally, group-living lemurs have, on average, even adult sex ratios. Independent of group size, the average number of males more or less equals that of females across species (Fig. 5.1). Given the absolutely small average number of females per group, one obvious question is why the females are associated with relatively many males, or why single lemur males can apparently not monopolize groups of two to four females, as do many colobines, howlers, guenons or gorillas (Cords, 1988, Chapter 8; Pope, Chapter 19; Watts, Chapter 15). I will focus on this question for the remainder of this chapter.

Explanations of lemur sex ratios

I will begin by examining the even sex ratios from the perspective of both sexes by focusing on the operational sex

Table 5.2. *Lemur group composition*

Species	1M + 1F %	1M + MF %	MM + 1F %	MM + MF %	*n*
Eulemur mongoz	48.5	16.7	13.6	21.2	66
Eulemur fulvus	1.4	1.4	4.2	93.1	72
Eulemur rubriventer	54.8	9.7	9.7	25.8	31
Eulemur coronatus	10.3	17.2	20.7	51.7	29
Eulemur macaco	6.5	0	0	93.5	31
Varecia variegata	25.0	0	0	75.0	4
Lemur catta	0	3.6	0	96.4	28
Hapalemur griseus	80.0	0	0	20.0	5
**Hapalemur aureus*	100.0	0	0	0	1
Hapalemur simus	—	—	—	—	0
Propithecus diadema	0	20.0	20.0	60.0	5
Propithecus tattersalli	29.4	11.8	29.4	29.4	17
Propithecus verreauxi	12.0	9.8	18.7	59.6	225
**Indri indri*	100.0	0	0	0	13
Grand mean	24.4	8.2	10.5	56.9	

Note:
The proportion of censused groups consisting of one adult male and one adult female (1M + 1F), one male and several females (1M + MF), one female and several males (MM + 1F) and of multi-male, multi-female groups (MM + FF) is depicted for each species. Two pair-living species (*, not included in calculation of grand means) are included for comparison.
Source: Census data are from the following sources: *E. mongoz*: Curtis (1997); Harrington (1978); Tattersall (1978); J. Schmid (unpublished data); *E. fulvus*: Freed (1996); Harrington (1975); Overdorff (1996); Sussman (1974); Tattersall (1977); J. Ganzhorn (unpublished data); P. Kappeler (unpublished data); *E. rubriventer*: Dague & Petter (1988); Merenlender (1993); Overdorff (1988, 1996); J. Schmid (unpublished data); *E. coronatus*: Arbélot-Tracqui (1983); Freed (1996); Wilson *et al.* (1989); J. Ganzhorn (unpublished data); *E. macaco*: Andrews (1990); Colquhoun (1993); *V. variegata*: Morland (1991); Rigamonti (1993); White (1991); *L. catta*: Budnitz & Dainis (1975); Gould (1996a, 1996b); Hood & Jolly (1995); Jolly (1967; 1972); Koyama (1988, 1991); O'Connor (1987); Sauther (1991); Sussman (1974, 1992); A. Jolly & G. Wood (unpublished data); *H. griseus*: Mutschler *et al.* (1998); Overdorff *et al.* (1997); Wright (1989); *H. aureus*: Wright (1989); *P. diadema*: Wright (1995); *P. tattersalli*: Meyers (1993); *P. verreauxi*: Albignac *et al.* (1988); Brockman (1994); Jolly *et al.* (1982); Rakotoarisoa (1994); Richard (1974a); P. Kappeler (unpublished data); *I. indri*: Pollock (1975).

ratio. Because primate sex ratios are typically female biased as a result of higher male mortality (Clutton-Brock & Iason, 1986; van Schaik & de Visser, 1990; Owens & Thompson, 1994; Mitani *et al.*, 1996a), an obvious working hypothesis is that male mortality rates among lemurs are similar to those of females, and relatively lower than those of males in other species. However, females in several lemur taxa can get aggressively evicted from their natal groups, often by other females (Vick & Pereira, 1989; Pereira, 1993, 1995; Kappeler, 1997a). Additional observations indicated that a substantial proportion of lemur female dyads are characterized by a lack of alliances, rare grooming, frequent agonism and mating harassment (Pereira & Kappeler, 1997). Thus, the even sex ratio of lemurs could also be the result of relatively high female mortality, due to competition among females, instead of, or in addition to, low male mortality rate. Unfortunately, detailed demographic data on birth sex ratios and juvenile and adult mortality rates needed for a direct test are still unavailable for most taxa.

Some relevant data come from studies in captivity. Results of surveys of captive populations are to be interpreted with caution because they introduce several confounding factors to comparisons with wild populations, but at least they can suggest trends in the direction and magnitude of sex differences in birth sex ratios and mortality rates. Sex ratios at birth were not found to differ significantly from a 1:1 distribution in large samples of captive brown, black, ring-tailed and ruffed lemurs (*Eulemur fulvus*, *E. macaco*, *Lemur catta* and *Varecia variegata*: Watson *et al.*, 1996). Moreover, these same taxa exhibited no sex difference in pre-weaning mortality. Examination of another data base revealed no consistent sex differences in juvenile mortality

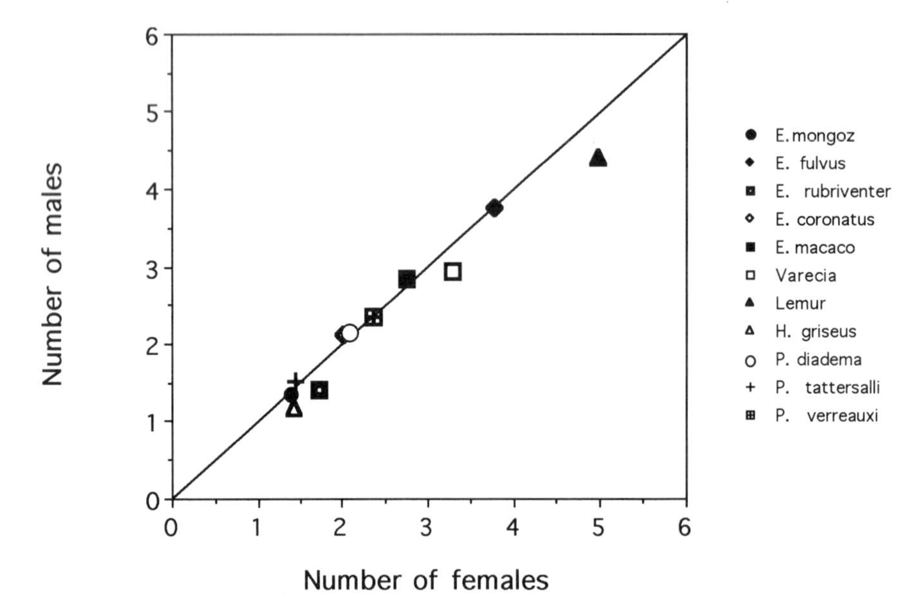

Fig. 5.1. Lemur sex ratios. The average numbers of adult females and males are plotted for 11 lemur taxa that regularly form groups of three or more adults. The slope of the linear regression ($Y = 0.893X + 0.226$) is not significantly different from 1.00 (solid line; lower and upper 95% confidence limits of the slope: 0.759 and 1.026, respectively). See Table 5.2 for references.

among three group-living species (Debyser, 1995). Data from a few well-studied wild lemur populations also indicate that sex differences in juvenile and adult mortality rates are negligible and not consistently biased in favor of one sex (Richard *et al.*, 1991; Sussman, 1991; Wright, 1995; Overdorff, 1998). Thus, the limited available evidence suggests that population sex ratios of group-living lemurs are on average even from birth on and are not modified by subsequent sex-specific mortality. Verification of these preliminary conclusions should be an important goal for future field studies of lemur demography, however.

To summarize, we are confronted with the following problem. At the population level, birth and adult sex ratios of group-living lemurs are even. Within populations, we find relatively small groups of highly variable composition. In only about 10% of groups is a single male associated with several females (see Table 5.2). Importantly, the average numbers of females in one-male and multi-male groups of the same species are not significantly different (Fig. 5.2). Thus, multi-female groups should in principle be monopolizable by a single male. In other words, these groups contain unusually many males. The same question arises for groups in which several males associate with a single female. If we assume for the moment that males and females are indeed born in equal proportions and that mortality rates do not differ between the sexes, group transfers become the main proximate determinants of group composition. At the ultimate level, the number of males may reflect the outcome of two processes: female choice and male–male competition. In the next sections, I examine the roles of males and females at these two levels.

Group transfers

The dynamics and behavioral circumstances of group transfers have been studied in only a few lemur populations (Jones, 1983; Vick & Pereira, 1989; Pereira & Weiss, 1991; Sussman, 1992; Richard *et al.*, 1993; Wright, 1995; Pereira & McGlynn, 1997). Male natal and secondary transfer is the rule among group-living lemurs. In most cases it appears to be voluntary. Female transfer, in contrast, is rare, and limited evidence suggests that it is often preceded by intense aggression, mostly from other resident females. Group sex ratios are therefore primarily regulated by male movements among groups. Unfortunately, next to nothing is known about the social histories of transferring males, the time they spend in transfer, and the risks they encounter during transfers (cf. Alberts & Altmann, 1995b; Borries, Chapter 13).

Potential immigrant males are often faced with aggression from resident males and females. This aggression has

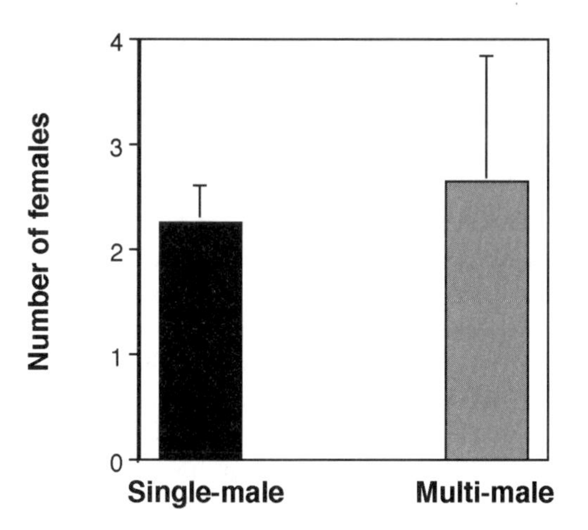

Fig. 5.2. Group structure and the number of females. The average number of adult females (±SD) in group-living lemurs is shown for single-male and multi-male groups of the same species (Wilcoxon test, T = 12, n = 8, n.s.).

been linked to the potential risk of infanticide these males may pose (Pereira & Weiss, 1991). After successful immigration, new immigrants may become favorite mates, however. Several strategies for transferring males appear to exist. Takeovers, consisting of intense fighting and expulsion of residents after a few days, have been observed in the brown lemur and Verreaux's sifaka (*Propithecus verreauxi*: Kraus, 1997; Pereira & McGlynn, 1997). Persistent shadowing of new groups over weeks, combined with visits of the old group, and followed by eventual integration, is known from ring-tailed and brown lemurs (Pereira & Kappeler, 1997). Male ring-tailed lemurs may also transfer in duos or trios (Jones, 1983; Sussmann, 1992). Finally, some Verreaux's sifaka males were regularly associated with two or three groups over several years (see below). Thus, many more field observations must accumulate before individual, and possible species-specific, tactics of transferring males as they relate to even sex ratios can be characterized and constructively analyzed.

Ultimate female interests

Females' interests may ultimately increase the number of males, if the presence of additional males provides them with benefits that offset the costs of increased within-group feeding competition. In species without paternal care, males could provide these services in several ways. First, they could be more vigilant than females and/or better at detecting or repeling predators, and females may promote their group membership to enjoy these benefits (van Schaik & Hörstermann, 1994). While there is some support for this hypothesis from studies of several anthropoids (Cowlishaw, 1994; Rose, 1994; van Schaik & van Noordwijk, 1989), studies of ring-tailed lemurs (Gould, 1996b) and of sifakas (Hussmann, 1996) revealed that males of these species are not more vigilant or more likely to detect predators than females (see also Baldellou & Henzi, 1992; Macedonia, 1993).

On the other hand, males decrease the per capita predation risk and females may prefer additional males over females for this reason, because they compete less with them for food. In fact, a recent study suggested that fossa (*Cryptoprocta ferox*), the main mammalian predator of lemurs, may prefer adult females (Wright *et al.*, 1997), even though males are often found in the most dangerous positions during travel and feeding (Jolly, 1966; Wright, 1998). Thus, males may buffer females from some predators, but, if predation risk is high, males may join groups primarily for selfish reasons. Their peripheral position within groups may then simply be a result of female feeding priority.

Additional studies of vigilance behavior that control for social vigilance and spatial position, experimental studies of predator detection and alarm calling, as well as quantification of intersexual feeding competition from a wider range of taxa will be necessary before the potential antipredator benefits of males for females can be evaluated with confidence. The presently available information indicates that male lemurs contribute to general vigilance levels and decrease per capita rates of predation, but apparently they do not provide females with any additional services that are costly to them and dramatically reduce female predation risk.

Second, males could benefit females by providing agonistic support in within-group and between-group conflicts (Cheney, 1987). Indeed, males in some *Eulemur*, *Propithecus* and *Hapalemur* species take an active role in between-group conflicts, but in ring-tailed lemurs and *Varecia*, females are much more active than males in such confrontations between groups, and males interact primarily with other males (Richard, 1985; Morland, 1991; Hörstermann, 1995; Pereira & Kappeler, 1997; Nievergelt *et al.*, 1998). In addition, male support of females in agonistic interactions with other resident females is either negligible or absent altogether (Kappeler, 1993b; Kraus, 1997; Pereira & Kappeler, 1997; Pereira & McGlynn, 1997). Thus, male agonistic support of females against other females appears to be an unlikely general male service, but additional studies of this aspect of male–female relationships in wild populations are required to test, and possibly differentiate, this preliminary conclusion.

Third, it has also been suggested that females establish friendships with particular males who protect them from sexual harassment and/or their infants from infanticidal attacks, and that these benefits tend to increase the number of males (Smuts, 1985; Smuts & Smuts, 1993). While this form of support may not be necessary in lemurs, because males are not larger than females and often subordinate to them, or because sexually selected infanticide is not *a priori* expected in seasonal breeders (Sussmann *et al.*, 1995), an increasing number of documented cases of attempted or successful infanticides (Hood *et al.*, 1994; Wright, 1995; Brockman *et al.*, 1998) suggests that female reproductive success is jeopardized by strange males and that females may indeed benefit from paternalistic aggression.

This idea about services by protective males receives general support in that, with the exception of *Varecia*, all lemur females that carry their infants are permanently associated with at least one male, whereas females that park their infants are not, assuming that carried infants are more vulnerable to infanticide than parked ones and that males can best protect them by permanently associating with the mother (van Schaik & Kappeler, 1997).

In addition, specific behavioral observations and paternity analyses of ring-tailed lemurs suggest that protective paternal aggression is a male reproductive tactic that coevolved with female choice (Pereira & Weiss, 1991). This idea has been extended to suggest that groups of cathemeral lemurs consist of multiple pairs, ultimately glued together by the same forces (van Schaik & Kappeler, 1993; Pereira & McGlynn, 1997). Recent reports of a lack of clear pair-bonds in some Eulemurs (Ostner, 1998; Overdorff, 1998) and attempts or actual cases of infanticide that were not challenged by resident sifaka males (Wright, 1995), on the other hand, do not support this hypothesis. Thus, there is limited preliminary support for the notion that lemur females prefer to live in groups with multiple males because their infants may be safer from infanticidal attacks (see also Steenbeek *et al.*, Chapter 12), but paternity analyses should be employed more widely to test the predicted relationship between male behavior and reproductive success directly. In addition, it is unclear why lemur males should be more susceptible than other primates to adapting to these female interests, and whether female choice is indeed the driving selective force (cf. Pereira & Weiss, 1991; Richard, 1992).

In conclusion, there are not enough data and consistent conclusions at present to support or reject the general hypothesis that lemur group sex ratios are primarily controled by female interests based on male services. One problem is that only a few studies have examined specific hypotheses explicitly with a focus on the problem of even sex ratios (e.g. Hussmann, 1996). It is also possible that the relative importance of different factors varies among species, and that, as a result, we find radically different social systems (e.g., Pereira & Kappeler, 1997), which would make the search for general explanations hopeless. To me, however, the similarities in group size and composition in two independent families indicate a general lemur-specific or Madagascar-specific (set of) cause(s) for these idiosyncrasies. Focused studies of female choice and other aspects of male–female relations should therefore provide important insights for such a general explanation.

The males' perspective

The ability of males to monopolize access to females in group-living species is thought to be primarily determined by the number and/or temporal distribution of receptive females (Andelman, 1986; Ridley, 1986; Altmann, 1990; Mitani *et al.*, 1996b; Paul, 1997). The highly seasonal reproduction of lemurs (Jolly, 1967, 1984; Sterling, 1994) may facilitate synchronization of female cycles, which in turn would reduce the monopolization potential of males. However, seasonality of reproduction is only a crude proxy for estrous synchrony because receptive periods clustered in time can nevertheless be asynchronized (Pereira, 1991). Only complete synchrony is apparently biologically relevant for males (Dunbar, Chapter 22). In addition, the small number of females in lemur groups should facilitate the ability of individual males to exclude rivals. However, demographic data summarized above indicate that lemur males cannot routinely monopolize access to small groups of females.

Why may lemur males be unable to exclude rivals from receptive males? Because multi-male lemur groups contain on average fewer adult females than multi-male anthropoid groups (cf. Andelman, 1986), and because the average number of females does not differ between one-male and multi-male lemur groups within species, the number of resident females *per se* cannot be responsible for the presence of supernumerary males. It is possible that groups of lemur females are less cohesive than groups of anthropoids with the same number of females (see, for example, Rigamonti, 1993), but quantitative data on group cohesion are not available for a direct comparison. It is also possible that males in many lemur species are unable to monopolize estrous females because they are regularly active at night and monitoring of several females may be difficult in the dark. For example, some cathemeral species increase their nocturnal activity during the mating season (Donati *et al.*, 1999), and normally diurnal species continue mating at night (e.g. Pereira & Weiss, 1991). These possibilities deserve further investigation in future field studies.

Preliminary data on the temporal distribution of receptive periods of lemur females do not support the possibility that socially mediated estrous synchrony within groups contributes to reduced male monopolization potential (Jolly, 1967; Richard, 1974b; Pereira, 1991; Brockman *et al.*, 1998). Thus, there is suggestive evidence that many lemur females could, in principle, be monopolized by a single male during their fertile periods. What has been observed, however, is that many lemur females mate with most or all resident males, and sometimes even with extra-group males (Jolly, 1967; Richard, 1974b, 1992; Pereira & Weiss, 1991; Sauther, 1991; Morland, 1993; Kraus, 1997; Brockman *et al.*, 1998; Ostner, 1998; Overdorff, 1998).

At the behavioral level, male competition among lemur males is not unusual. Fierce and bloody battles among males, including group takeovers, have been reported for several species, injuries are common, and mating success tends to be positively correlated with dominance status, even though dominance relations can break down or be reversed temporarily (Jolly, 1967; Richard, 1974b; 1992; Pereira & Weiss, 1991; Sauther, 1991; Morland, 1993; Kraus, 1997; Brockman *et al.*, 1998; Ostner, 1998; Overdorff, 1998). Genetic studies of the effects of dominance on male reproductive success revealed the expected skew in one captive group (Pereira & Weiss, 1991), but comparative studies of other populations are still lacking (but see Merenlender, 1993), and little is known about male tenure length (but see Richard *et al.*, 1993). In contrast, morphological and demographic consequences of male competition do not follow theoretical expectations (Kappeler, 1993a, 1997b).

A closer look at mechanisms and consequences of male–male competition may therefore generate hypotheses to explain the unusual sex ratios of lemurs without direct paternal care. To this end, it is useful to distinguish between two main competitive regimes: scramble and contest (van Hooff & van Schaik, 1994). Although many lemur females seem to mate with several or all resident males, as well as occasionally with extra-group males, we can exclude pure scramble as the predominant mode of competition in all species for three reasons. First, testes size among group-living lemurs is highly variable, indicating that the intensity of sperm competition varies across species (Kappeler, 1997b). Second, males of some species maintain dominance hierarchies year-round (Pereira & Kappeler, 1997); and, finally, vigorous fighting among males during the mating season is nearly ubiquitous (Jolly, 1967; Sussman & Richard, 1974; Richard, 1992). It remains to be determined by genetic studies, however, whether mating order, rather than frequency, is the most important proximate determinant of male reproductive success under these scramble conditions (cf. Pereira & Weiss, 1991).

Contest competition can have one of two consequences: males will either exclude rivals from reproductive opportunities altogether or dominate them so that priority of access to females is mediated by dominance. As noted earlier, exclusion of rivals is rarely achieved among lemurs because one-male groups are rare (see Table 5.1). In addition, we do find the behavioral, but not the morphological and demographic, manifestations of contest competition. Because the same patterns appear in two independent families inhabiting a wide range of habitats, there is no obvious ecological explanation for this discrepancy.

Four explanations

Four explanations for the observed combination of scramble and contest competition and the corresponding discrepancies concerning its evolutionary consequences are possible. First, the effects of male mate competition may be completely overridden by females. Several observations suggest that lemur females are free to chose subordinate or strange males (e.g., Pereira & Weiss, 1991; Richard, 1992). Female cooperation for successful matings is required in all primates, but a lack of power asymmetries or even female dominance puts lemur females in a very powerful position. Deference towards females or protector qualities have been suggested as male traits on which such female choice could operate (Pereira & Weiss, 1991; Richard, 1992; Pereira & McGlynn, 1997). Concealing ovulations from males has been suggested as another mechanism that females could employ in this context (Pereira & McGlynn, 1997). While it remains to be determined how and why such a system did arise, and why only lemurs among primates apparently evolved it, the study of reproductive tactics and their physiological underpinnings should become a major focus for future studies of lemur socioecology to examine this possibility in more detail.

The second possible explanation, poorly known male reproductive tactics, involving both competition and cooperation, may explain the observed discrepancies. Assuming that the relatively large number of lemur males is determined by some unknown factor, physiological suppression of reproductive functions in rivals could provide one mechanism that may be compatible with the lack of morphological and demographic consequences. Such suppressive effects are often mediated by olfactory signals, on which lemur communication relies heavily (Schilling, 1979). They have been demonstrated experimentally in a solitary species (gray mouse lemur, *Microcebus murinus*: Perret, 1992), and preliminary studies of scent-transmission, counter-marking and endocrinological profiles do not refute this possibility off hand for group-living taxa (Kappeler, 1990, 1998; Brockman *et al.*, 1998). This suppression may ultimately work because the underlying signals are honest, but additional investigations are necessary to substantiate the speculation that 'chemical castration' contributes to a discrepancy between demographic and operational sex ratios. In this case, evidence for male social benefits derived from the presence of other males would also be required (see below).

A second speculation in this context concerns alternative male strategies that could be employed to evade contest competition. In our sifaka population in Kirindy forest, for example, two young males have been regularly associated with two or three neighboring groups for more than two years. These roving males changed groups often on a daily basis, frequently during between-group encounters, and maintained grooming relationships with resident males and females in all groups (Kraus, 1997). They also obtained matings in more than one group, and paternity studies are underway to determine whether they were successful. Continuing observations will also reveal whether this tactic is age dependent, i.e., whether these males are making the best of a situation in which they cannot yet attain top rank.

Finally, in contrast to the impression that the focus on competitive behavior during the brief mating frenzy may have created, social relationships among lemur males throughout the rest of the year are noticeably friendly. In fact, in several species, affiliative behavior, such as grooming, occurs more frequently among males than among females or between the sexes (Gould, 1997; Kappeler, 1999b). Even in male ringtailed lemurs, which are known for their despotic dominance relationships, dyads of long-time adversaries (see Pereira & Kappeler, 1997, for a definition) started associating and grooming immediately prior to transferring and just as their group received new immigrants, respectively (M. Pereira, personal communication). This dimension of lemur male relationships is still poorly investigated, but closer examination of potential social benefits that males may provide for each other may shed light on the questions concerning their number and competitive relationships raised above.

A third, non-adaptive explanation postulates that the social systems of group-living lemurs are in a transitional state of evolutionary disequilibrium, following massive ecological changes in Madagascar over the last 2000 years, in particular the demise of large diurnal predators and competitors (van Schaik & Kappeler, 1996). Accordingly, group-living lemurs were nocturnal and pair-living until recently, and groups initially formed by the fusion of several pairs upon becoming diurnal. This hypothesis could explain many of the lemur idiosyncrasies, including the even sex ratios, as a by-product of this transition. Cathemeral species, in particular, were predicted to consist of multiple male–female pairs (van Schaik & Kappeler, 1993), but preliminary tests did not unequivocally support this prediction (Curtis, 1997; Ostner, 1998; Overdorff, 1998). Because it predicts deviations toward more typical anthropoid-like patterns, this hypothesis could nevertheless explain the observed discrepancies between behavioral and morphological traits shaped by sexual selection, if behavior does indeed change faster over evolutionary times than other anatomical and physiological traits (see Gittleman *et al.*, 1996). Additional detailed descriptions of lemur social systems under a wide range of ecological conditions should help to subject this hypothesis to stronger tests.

A fourth explanation posits that lemur social systems share few similarities, that each one is adaptive and equivalently affects female reproductive potential or reflects different sets of initial conditions (Pereira & McGlynn, 1997; Pereira *et al.*, 1999). It accounts for balanced sex ratios with species-specific explanations that build on a few shared underlying commonalities. So, for example, females of different species may get evicted from their groups due to female–female competition, but in each species females compete for different resources (see, for example, Vick & Pereira, 1989; Pereira, 1993). Similarly, supernumerary males may exist because they provide females with services that are beneficial to them, but these services may differ from species to species (see, for example, Pereira & McGlynn, 1997). Overall, male and female tactics are constrained by the overriding need to conserve energy during long, harsh, dry seasons (Pereira *et al.*, 1999). This elegant explanation integrates most behavioral and physiological information about the two best-known species (*L. catta* and *E. fulvus*), but only future studies of other taxa and their metabolic strategies, in particular, will show how far this approach can be carried.

In conclusion, either within the evolutionary disequilibrium hypothesis or an adaptive theoretical framework, questions about male behavior and numbers will most certainly continue to play a crucial role in further attempts to illuminate lemur social evolution. To gain a better understanding of male life histories, long-term studies of known individuals in as many taxa as possible must generate information about sex ratios, mortality rates, timing and circumstances of dispersal events, mating tactics, social relationships, reproductive success, nutritional ecology and metabolic strategies. Captive studies could provide important additional information about physiological correlates of social status and responses to experimental manipulations of group composition. Eventually, these pieces should help to complete the puzzle of lemur social evolution.

ACKNOWLEDGMENTS

I thank Alison Jolly, Jörg Ganzhorn and Jutta Schmid for sharing unpublished census data, and Alison Jolly, Julia Ostner, Carel van Schaik, and especially Michael Pereira for their helpful comments on an earlier draft. Thanks are also due to the German Primate Center and the German Science Foundation for their continuing support of my interests in lemur social systems.

6 • The number of adult males in callitrichine groups and its implications for callitrichine social evolution

ECKHARD W. HEYMANN

INTRODUCTION

The reproductive biology of callitrichines (marmosets and tamarins) except *Callimico* is characterized by the facultative monopolization of breeding by a single female in each group (reviews in Mittermeier *et al.*, 1988), a high amount of direct paternal care (Goldizen, 1988; Garber, 1997), and the birth of dizygotic twins (Hershkovitz, 1977). These traits are highly relevant to the number of adult males in callitrichine groups. Monopolization of reproduction by one female constrains the number of breeding opportunities for males (Garber, 1997). Paternal care influences potential reproductive rates and can lead to competition between females for males, with considerable implications for the operation of sexual selection (e.g., Clutton-Brock & Parker, 1992). Along with the lack of body size dimorphism, single female reproduction and paternal care have been considered as evidence for a monogamous social organization and mating system (Kleiman, 1977). In fact, direct parental investment by males in the form of infant carrying is only found in other monogamous primates, namely titi monkeys (*Callicebus*), night monkeys (*Aotus*), and siamang (*Hylobates syndactylus*) (Wright, 1984; Whitten, 1987).

In captivity, monogamous pairing has been the most successful way of keeping and breeding callitrichines (e.g., Hampton *et al.*, 1966). However, field studies revealed more flexible social and mating systems, including polyandry, polygyny, monogamy, and even polygynandry (Goldizen, 1988; Ferrari & Lopes Ferrari, 1989). While data from the wild and from the laboratory concur in the observation that there is usually only a single reproducing female per group (although multiple breeding females have been observed in some populations), the main controversy between field-based and laboratory-based workers is centered around the question of how many adult males are breeding in callitrichine groups. Thus, with regard to the number of adult males in callitrichine groups, two different questions have to be asked: (1) How many adult males are *living* in callitrichine groups and what determines their number? (2) How many adult males are *breeding* in callitrichine groups and what determines their number?

Although interrelated, the factors involved are different and the answers to these questions are not necessarily the same. Demographic data, which are increasingly becoming available, are required for the first question. Genetic data to answer the second question are largely unavailable, restricting any analysis to indirect measures, such as observations of matings, male competition and mate guarding.

To address the questions posed above, data were compiled from an extensive literature search. For interspecific comparisons, I calculated independent contrasts with the CAIC program (Purvis & Rambaut, 1994) using endpoints and all nodes of the callitrichine clade shown in Fig. 6.1.

THE NUMBER OF ADULT MALES IN CALLITRICHINE GROUPS

The number of adult males in callitrichine groups ranges from one to five, but the range differs between species (Table 6.1). All species have a modal number of one or two adult males per group. In the only species for which sufficient data are available from two different populations – saddle-back tamarins (*Saguinus fuscicollis*) – the two populations differ in the modal number and their range. Thus, variation in the number of adult males can be identified at three different levels: within populations, between populations of the same species, and between species. Due to the variability within the genus *Saguinus*, no intergeneric trend is apparent.

While in other primates a correlation between the number of adult males and females exists (Mitani *et al.*, 1996a), no correlation is expected in callitrichines, given the monopolization of reproduction by a single female and the constraint this imposes on male breeding opportunities. In fact, no correlation is found between the number of adult males and females ($r=0.10$, $n=12$, n.s.;

Fig. 6.1. Phylogenetic relationships of callitrichine species used in the present analyses. The arrangement of genera follows Schneider & Rosenberger (1996); the arrangement of species within the genus *Saguinus* follows Jacobs *et al.* (1995).

Fig. 6.2a), even if *Callimico*, with its low number of males and high number of females, is excluded from the analysis ($r = 0.51$, n.s.).

In all species and populations except Goeldi's monkey (*Callimico goeldii*), pygmy marmoset (*Cebuella pygmaea*), and black-mantle tamarin (*Saguinus nigricollis*), the number of adult males exceeds the number of females (Table 6.1). These three species are characterized by a modal number of one male per group. The fact that there are usually more males than females seems paradoxical because a single reproductive female should be easily monopolizable by a single adult male. This male should not tolerate the presence of other adult males, unless considerable benefits can be derived from their presence.

The number of adult males in relation to the costs of infant care

A major benefit of tolerating additional adult males can be derived from the help in raising offspring that these males may provide (Goldizen, 1987a, 1990). Litter mass in all callitrichines except *Callimico* is relatively high in relation to maternal body mass (Table 6.1), particularly when compared to other primates, where it is usually around 10% of maternal body mass (Ross, 1991). The high litter mass makes raising twins very costly. Goldizen (1987a) has shown for wild saddle-back tamarins that a pair without helpers is unable to raise twin infants successfully. Adult male cotton-top tamarins (*Saguinus oedipus*) – who are the

Table 6.1. *Proportion (%) of groups with different numbers of adult males, mean number of adult males and females, infant body mass and litter mass, and daily path length*

	Number of adult males					Mean number of adults		n[a]	Infant body mass[b]	Litter mass[b]	Daily path length (m)	Sources[c]
	1	2	3	4	5	Males	Females					
Callimico goeldii	66.7	33.3	—	—	—	1.3	2.7	3	8.1	8.1	2000	Encarnación & Heymann (1998); Masataka (1981); Pook & Pook (1981)
Callithrix jacchus	31.3	31.3	6.3	18.7	12.5	2.5	1.8	16[d]	8.5	17.0	1100	Digby & Barreto (1996); Koenig (1995)
Cebuella pygmaea	65.0	30.0	5.0	—	—	1.4	1.4	20	9.0	18.0	290	Soini & Soini (1990a)
Leontopithecus chrysomelas	20.0	60.0	20.0	—	—	2.0	1.4	5			1793[e]	Dietz *et al.* (1994); Rylands 1982
Leontopithecus chrysopygus						2.3	1.3	4			2290[e]	Keuroghlian (1990); Valladares-Padua (1993)
Leontopithecus rosalia	31.1	48.1	15.0	5.3	0.5	2.0	1.5	206[f]	7.6	15.2	1440[e]	Dietz & Baker (1993); Peres, quoted in Rylands (1993)
Saguinus fuscicollis (south-eastern Peru)	23.9	67.4	6.5	2.2	—	1.9	1.3	46[f]			1220	Goldizen *et al.* (1996); Terborgh (1983)
Saguinus fuscicollis (north-eastern Peru)	47.1	35.3	17.6	—	—	1.7	1.5	17			1720	Smith (1997); Soini (1990)
Saguinus geoffroyi						2.4	2.1	5			2061	Dawson (1977, 1979)
Saguinus imperator									9.0	18.0		Garber & Leigh (1997)
Saguinus labiatus						2.4	1.7	38			1487	Buchanan-Smith (1991); Puertas *et al.* (1995)
Saguinus mystax	20.5	43.2	29.5	6.8	-	2.1	1.6	44			1720	Smith (1997); Soini & Soini (1990b)
Saguinus nigricollis	77.8	22.2	—	—	—	1.3	1.3	18			1000	Izawa (1978); de la Torre *et al.* (1995)
Saguinus oedipus	7.7	84.6	7.7	—	—	2.0	1.6	13[f]			1700	Neyman (1977); Savage *et al.* (1996a)

Notes:

[a] Sample size for demographic data.

[b] As percentage of maternal body mass; source: Garber & Leigh (1997).

[c] Sources for demographic data and daily path length.

[d] May include repeated counts of some groups.

[e] Midpoint of ranges.

[f] Includes repeated counts of the same groups.

principal carriers of infants in this species – lose up to 5% of body mass during the first four weeks of infant development in captivity (Sánchez *et al.*, 1999). Tolerating other adult males as helpers could therefore represent a strategic option until related adult and subadult offspring become available as helpers (Rylands, 1982; Goldizen, 1987a).

In groups of moustached tamarins (*Saguinus mystax*) and common marmosets (*Callithrix jacchus*), the number of infants correlates with the number of adult males, but is unrelated to overall group size (Garber *et al.*, 1984; Koenig, 1995). This seems to indicate that adult males are particularly valuable helpers. Therefore, I tested the hypothesis that the costs of infant care are a determinant of the number of adult males.

Relative litter mass is very similar in all species except *Callimico goeldii*, the females of which give birth to a single offspring, and would not *a priori* predict different costs of infant carrying (Table 6.1). However, there are at least two additional factors that can influence the costs of carrying: infant growth rates and daily travel path length. Costs of infant carrying increase with increasing litter mass gain and increasing daily travel path length. One can therefore predict that the number of adult males should positively correlate with litter mass gain and with daily travel path length.

The number of adult males is indeed positively correlated with litter mass gain ($r=0.80$, $n=5$, $p=0.03$ [one–tailed]; Fig. 6.2b), but given the small sample size, I only tentatively accept this relationship. In contrast, I found no association between the number of adult males and the daily travel path length ($r=0.29$, $n=12$, n.s.; Fig. 6.2c). However, this lack of association is due to *Callimico*, which represents an outlier. If this outlier is removed from the analysis, the predicted positive correlation between the number of adult males and daily travel path length is found ($r=0.62$, $n=11$, $p<0.02$ [one–tailed]).

A tentative conclusion drawn from these analyses is that variation in the number of adult males in callitrichine groups is associated with the costs of infant carrying. The exception represented by *Callimico* is expected because litter size has been secondarily reduced, and more than one female seems to breed in a given group (Masataka, 1981; Encarnación & Heymann, 1998; A. Christen, personal communication). Although infant carrying by adult males has been observed in captive *Callimico* (Jurke & Pryce, 1994), I predict that they are unlikely to provide substantial care in the wild.

The number of males in relation to predation risk

The number of adult males in primate groups can vary in relation to predation risk (van Schaik & Hörstermann, 1994). Although it has been proposed that predation risk plays an important role in callitrichine socioecology (Caine, 1993), it is not clear whether it affects the number of adult males. This is apparent from a qualitative interspecific comparison. The number of adult males differs considerably between *Callimico*, *Cebuella* and black-mantle tamarins, on the one hand, and saddle-back tamarins, moustached tamarins and red-bellied tamarins (*Saguinus labiatus*), on the other hand (Table 6.1). These six species are distributed in western Amazonia (Rylands *et al.*, 1993) and exposed to the

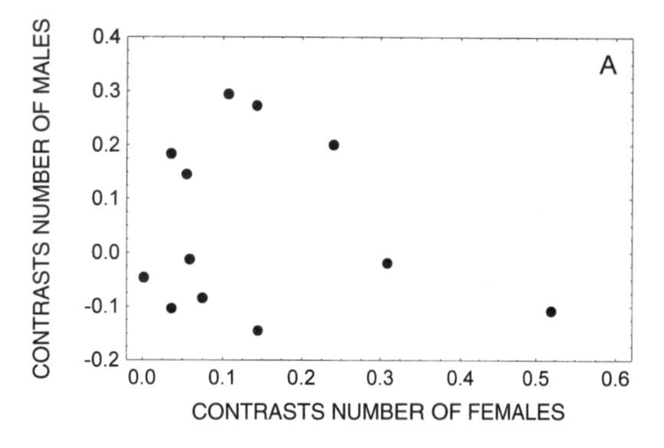

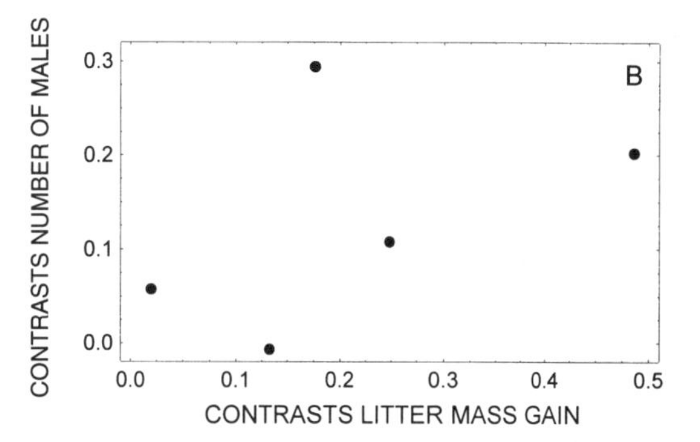

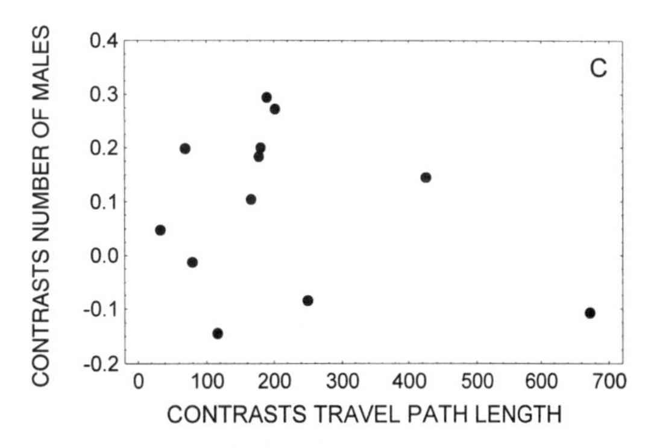

Fig. 6.2. (A) Phylogenetic contrasts in the number of males and females in callitrichine groups. (B) Phylogenetic contrasts in the number of males and litter growth rate (defined as litter size*infant mass gain during nursing per day per maternal body mass$^{0.75} \times 100$). Data on infant mass gain from Garber & Leigh (1997). Note: because no data are available on litter mass gain in *Saguinus mystax*, the value from the sister species *Saguinus imperator* was used. (C) Phylogenetic contrasts in the number of males and mean daily travel path length. Data used in the calculation of phylogenetic contrasts (except for litter mass gain) are provided in Table 6.1.

Table 6. 2. *Mechanisms of sexual competition between male callitrichines*

	Shared matings during fertile fertile periods of the female	Dominance relations	'Mate guarding'	Relative testes size (% of body mass)
Callithrix flaviceps	+/−	?	+	?
Callithrix intermedia	+	−?	?	?
Callithrix jacchus	−	+	+	0.30
Cebuella pygmaea	−	+	+	0.25
Leontopithecus rosalia	−	+	+	0.25
Saguinus fuscicollis, SE Peru	+	−	+	?
Saguinus fuscicollis, NE Peru	−	+?	+	0.36
Saguinus geoffroyi				0.13
Saguinus midas				0.36
Saguinus mystax	+	−	+	0.37
Saguinus nigricollis	−?	+?	+?	0.33
Saguinus oedipus				0.21

Note:
Data provided by Dawson & Dukelow (1976) on body mass and testes size for *S. geoffroyi* were not used because they render an unrealistically high value of 1.3%. (*Brachyteles* with very large testes have a value of 1.1%!) The value of 0.65% provided by Harcourt *et al.* (1981b), based on the data by Dawson & Dukelow, is also unrealistically high.
Sources for behavioral data: *C. flaviceps*: Ferrari (1992). *C. intermedia*: Rylands (1982). *C. jacchus*: Digby (1994). *C. pygmaea*: Soini (1987a). *L. rosalia*: Baker *et al.* (1993). *S. fuscicollis*: Goldizen (1987a); Goldizen *et al.* (1996); Soini (1987b). *S. mystax*: Garber *et al.* (1993b); Heymann (1996). *S. nigricollis*: Izawa (1978).
Sources for relative testes size: C. jacchus: Dixson *et al.* (1992). *C. pygmaea*: Harcourt *et al.* (1995). *L. rosalia*: Hershkovitz (1977). *S. fuscicollis*: Soini & Coppula (1981). *S. geoffroyi*: Hrdlicka (1925). *S. midas*: Hershkovitz (1977). *S. mystax*: Soini & de Soini (1990a). *S. nigricollis*: Hershkovitz (1977). *S. oedipus*: Hershkovitz (1977).

same set of predators (Hilty & Brown, 1986; Emmons & Feer, 1990). While one might argue that *Callimico* and *Cebuella* have strategies of predator avoidance that are different from those of the other species, black-mantle tamarins and saddle-back tamarins are ecologically and behaviorally very similar (Izawa, 1978; Soini, 1987b), suggesting that different antipredator strategies do not account for different numbers of adult males. However, for a final conclusion, actual predation risks must be known.

If the number of adult males were related to predation risk, one may also expect differences between species living in single-species and in mixed-species groups. Benefits of the formation of mixed-species groups between saddle-back tamarins and either moustached, red-bellied or Emperor tamarins (*S. imperator*) in terms of increased predator avoidance have been demonstrated (Peres, 1993; Hardie & Buchanan-Smith, 1997). Given that males are more vigilant than other group members (Savage *et al.*, 1996a), one might expect a higher number of adult males in single-species groups compared to groups of the same or related species in mixed-species groups. However, existing data do not confirm this prediction. The mean number of adult males in black-mantle tamarin groups (a species that does not form mixed-species groups) is lower than in groups of the closely related saddle-back tamarin, which forms mixed-species groups in both populations considered here (Table 6.1). Thus, the number of adult males seems not to be associated with the risk of predation, although this should be re-examined when more data become available from the wild.

The number of breeding males and mechanisms of sexual competition

Because genetic data on paternity are completely lacking, it is difficult to address the question of the number of males *breeding* in callitrichine groups directly. Therefore, mating patterns are used as a surrogate measure, i.e., the number of males copulating during fertile periods of the reproductive female. There are, however, two problems with this approach. First, callitrichines are not often seen copulating in the wild. Second, usually there are no external signs of estrus apparent to the observer (but see Sicchar &

Heymann, 1992; Heymann, 1996). Indirectly, fertile periods can be determined by calculating backwards from the time of birth of infants, but this is not always possible.

The number of breeding males is directly related to patterns of male competition for females, i.e., whether males compete directly through contest for access to females, or whether they scramble via sperm competition. Therefore, the occurrence or lack of dominance relations between males and of mate guarding, as well as relative testes size are examined here. In almost all groups of wild callitrichines that have been studied so far, a maximum of two males mated with the reproductive female. Only in Rylands' (1982) study on tassel-ear marmosets (*Callithrix intermedia*) were three of the four adult males of the study group seen copulating. Thus, the question of how many males are mating reduces to the question of whether or not matings are shared.

In the majority of species, matings are not shared (Table 6.2). Only in the saddle-back tamarin population in south-east Peru, in moustached tamarins and in tassel-ear marmosets do males share matings (Rylands, 1982; Goldizen, 1987a; Garber *et al.*, 1993b; Goldizen *et al.*, 1996; Heymann, 1996). Inversely related to this pattern is the occurrence of male dominance relations: in species with male dominance relations, matings are monopolized, whereas in species without such relations, matings are shared.

In most species, mate guarding has been observed, even when matings are shared (Table 6.2). This seems to be a contradiction, but need not be so. Any male would obtain the maximum share of fertilizations by strictly monopolizing access to the breeding female. However, mate guarding also incurs costs. It may interfere with prey foraging, and the male excluded from matings may reduce or refuse investment in future infant care. Furthermore, conflict with females will arise if polyandrous matings are in their interest. A female can maximize the amount of paternal care that her infants receive by partitioning matings (and paternity) between two (or more) males (Rylands, 1982; Harada & Iwasa, 1996). Relaxation of the reproductive skew towards one male may thus involve both concessions and limited control (Clutton-Brock, 1998).

Summarizing the available information on mechanisms of male sexual competition, it is apparent that there is behavioral evidence for the monopolization of matings by a single male and thus for a monogamous mating system in most species studied so far (Table 6.3). However, categorizations such as those in Table 6.3 can only reflect modal patterns and deny neither intraspecific flexibility in mating patterns observed in several callitrichine populations (e.g., Goldizen *et al.*, 1996), nor the role of extra-pair copulations (see below).

Table 6. 3. *Modal mating patterns of callitrichine primates*

Polyandrous	Monogamous	Polygynous
Callithrix intermedia?		
	Callithrix jacchus	
	Cebuella pygmaea	
	Leontopithecus rosalia	
Saguinus fuscicollis (south-east Peru)	*Saguinus fuscicollis* (north-east Peru)	
Saguinus mystax		
	Saguinus nigricollis	
		Callimico goeldii (?)

In a recent comparative analysis of testes size in primates, Harcourt *et al.* (1995) found that testes were larger than expected from a regression of testes size on body mass in *Saguinus* and smaller than expected in *Callithrix*, *Cebuella* and lion tamarins (*Leontopithecus*). This was considered as evidence for a polyandrous mating system in the former genus and a monogamous mating system in the latter genera. However, since the smallest testes are also observed in a member of the genus *Saguinus* (see Table 6.2), the situation is apparently more complex.

Examination of Table 6.2 reveals a trend toward larger relative testes in all species of *Saguinus* except Geoffroy's tamarins (*S. geoffroyi*) and cotton-top tamarins, and smaller testes in *Callithrix*, *Cebuella*, and *Leontopithecus*. Large testes certainly correlate with a polyandrous mating pattern, as, for example, in moustached tamarins, but they are not in line with a monogamous mating pattern, as, for example, in black-mantle tamarins (see Table 6.2). A further complicating factor that needs to be taken into consideration is the risk of extra-pair (extra-group) copulations. Extra-pair copulations have been observed in common marmosets (Digby, 1994) and in saddle-back tamarins (E. Tirado, personal communication). In black-mantle tamarins, the temporary merging of two or more groups (Izawa, 1978) may also provide ample opportunities for extra-pair copulations. For Geoffroy's tamarins and cotton-top tamarins, a monogamous mating pattern is predicted on the basis of relative testes size, with the presence of a dominance hierarchy and a lack of shared matings when multiple males are present in a group.

In summary, patterns of male sexual competition may

Table 6.4. *Sociocological contrasts between* Saguinus fuscicollis *and* Saguinus nigricollis

	Saguinus fuscicollis (south-east Peru)	*Saguinus fuscicollis* (north-east Peru)	*Saguinus nigricollis*
Group size	5.1	5.5	5.2
Number of males : females	1.5	1.1	1.0
Interbirth interval (months)	? (Long)	8	? (Short)
Annual distribution of births	Unimodal	Unimodal	Bimodal
Modal mating system	Polyandry	Monogamy ?	Monogamy
Number of months with < 100 mm rainfall	3	0	0 ?

Sources: S. fuscicollis (south-east Peru): Goldizen *et al.* (1988, 1996); Terborgh (1983); *S. fuscicollis*, (north-east Peru): Castro Coronado (1991); Garber (1993); Soini (1987b, 1990); *S. nigricollis*: Izawa (1978); de la Torre *et al.* (1995).

vary from largely tolerant to strong monopolization, but behavioral evidence suggests a prevalence of a monogamous mating system in most species.

Determinants of the number of breeding males

As the number of mating males was used as a surrogate measure to determine the number of *breeding* males in callitrichine groups, the question of what determines this number cannot be addressed in the same quantitative way as the number of males *living* in callitrichine groups. Rather, conclusions will be drawn from qualitative comparisons between species.

It is hypothesized here that the costs of infant care also play a role in determining the number of breeding males. When costs of infant care are reduced, males should try to monopolize matings more strongly. Costs of infant care to adult males can be reduced by shortened daily travel distances and the availability of non-reproductive helpers. This is demonstrated by the comparison between common marmosets and *Cebuella*.

In common marmosets, high productivity resulting from two annual births results in large groups (mean group size [without infants]: 7.3; Koenig, 1995) and the availability of many related adult and sub-adult helpers. Adult males should therefore be less inclined to share matings with other males. In contrast, *Cebuella* has the smallest group size of all callitrichines (mean group size: 4.6; calculated from data in Soini & Soini, 1990b, excluding infants), despite high productivity due to two annual births. Upon maturity, offspring are peripheralized and emigrate from their natal group (Soini, 1988). The costs of infant care are substantially reduced due to infant parking. Infant parking is possible because of the short daily travel path length (see Table 6.1) which results from the activity of a group usually being concentrated around a single exudate source (Soini, 1988). Infant parking reduces the need for helpers and thus allows males to monopolize breeding. This conclusion is supported by the fact that no correlation exists in *Cebuella* between the number of infants and the number of adult males (Heymann, 1997; Heymann & Soini, 1999).

In black-mantle tamarins, costs of infant care are also reduced due to a daily path length that is shorter than in any other *Saguinus* species (see Table 6.1; see also Garber, 1993). This is in line with the monogamous mating pattern observed in this species. However, it is also apparent from Table 6.4 that other factors must play a role. The daily path length is also relatively short in south-eastern Peruvian and relatively long in north-eastern Peruvian saddle-back tamarins; nevertheless, the modal mating patterns are polyandry and monogamy, respectively. Habitat productivity (approximately measured by patterns of rainfall) and its effect on the reproductive output (as measured by the interbirth interval and the number of birth peaks per year) may also affect the number of breeding males. Where high reproductive output is permitted by favorable environmental conditions, males may invest less in offspring and more in the monopolization of matings. In a more seasonal habitat as in south-eastern Peru, a higher investment may be required to guarantee offspring survival, resulting in the need for help, and thus in the acceptance of other males mating with the reproductive female.

Thus, qualitative evidence suggests that the number of breeding males also varies with the costs of infant care. However, factors such as habitat seasonality and productivity, through their influence on reproductive output and infant survival, may influence patterns of male sexual competition and the degree of reproductive monopolization by a single male.

TRENDS IN CALLITRICHINE SOCIAL EVOLUTION

It has been hypothesized that a monogamous mating pattern represents the ancestral condition for callitrichines (Goldizen, 1990; Dunbar, 1995a). This view has been opposed by Garber (1994), who reconstructed the ancestral callitrichine social organization and mating system as small multi-male, multi-female groups with polygynous matings. Garber's argument was based on the consideration of *Callimico* as an ancestral callitrichine. However, molecular evidence suggests that the most likely position of *Callimico* is within the callitrichine clade as a sister taxon to *Callithrix*/*Cebuella* (Schneider & Rosenberger, 1996). Thus, a different scenario should be considered for the evolution of patterns of social organization and mating in callitrichines.

In many New World primates, strong affiliative bonds exist between adult males (Boinski, 1994; Strier, 1994b, Chapter 7). Strong affiliative bonds are also observed between adult males in all *Saguinus* species studied in the wild so far (Goldizen, 1989; Heymann, 1996). Such male bonding, particularly if it is between relatives, provides a firm base for the evolution of cooperative male care for infants, as observed in the genus *Saguinus* in particular (for an opposing view see Rylands, 1982). Once a system of male care evolved, it may have created competition between females for male helpers, leading to reproductive monopolization by a single female per group and a polyandrous mating pattern. The evolution of dizygotic twinning may have stabilized the system of male care: dizygotic twinning increases the probability of paternity for each male because the two ova can be fertilized by different males. Accordingly, it is conceivable that twinning even evolved as a female strategy to recruit more help from males (see also Rylands, 1982).

Increased offspring production through twinning also means that there is more paternity to steal (Hawkes *et al.*, 1995). This should select for male strategies to monopolize matings. Depending on the balance between the relative costs and benefits of infant care and of sharing or monopolizing matings, as well as female strategies, different modal mating patterns will arise. Overall, a trend for an increasing prevalence of a monogamous mating pattern emerges which is thus not the starting point, but rather the modal state to which callitrichine social evolution is tending. Under specific conditions (immigration of a new male, habitat saturation), polygyny may occasionally occur (Digby & Ferrari, 1994; Coutinho & Corrêa, 1995; Dietz & Baker, 1993; Savage *et al.*, 1996b). Notably, offspring survival can be strongly reduced when multiple females are breeding (Coutinho & Corrêa, 1995; Dietz & Baker, 1993; Digby, 1995), indicating that polygyny is not a stable system. Only in *Callimico* has a stable polygynous mating system evolved, which is associated with the secondary reduction of twinning.

Recent theoretical analyses of bird mating strategies by Iwasa & Harada (1998) may also shed some light on callitrichine social evolution. Iwasa and Harada concluded that monogamy is the evolutionary stable mating system when parental ability does not vary between members of a sex (a situation that should apply when costs of infant care are not too high). However, polyandry is the evolutionary stable pattern if the cost function for male care is strongly accelerating and if differences in female fecundity are very large. The monopolization of reproduction by a single female per group and the lack of reproduction in other females is the largest possible difference in female fecundity. The more often reproductive monopolization is breached, the smaller becomes the difference in female fecundity, a condition in which polyandry is not stable (Iwasa & Harada, 1998). This model could explain why polygyny is apparently more often observed in species which otherwise are considered here as having monogamy as their modal mating pattern.

ACKNOWLEDGMENTS

I am grateful to Peter Kappeler for inviting me to speak at the first Göttinger Freilandtage, and to Hannah Buchanan-Smith, Andreas Koenig, Christoph Knogge, Anthony B. Rylands and two anonymous referees for critical and helpful comments on a previous draft of this chapter.

7 • From binding brotherhoods to short-term sovereignty: the dilemma of male Cebidae

KAREN B. STRIER

INTRODUCTION

The New World Cebidae differ from many Old World monkeys, and more closely resemble the apes, in at least two important respects that affect the number of males in their groups. The first is that most of the Cebidae have slow, ape-like reproductive rates relative to their body sizes due, in part, to their comparatively long interbirth intervals (Ross, 1991). The second is that most of the Cebidae, along with other platyrrhines, lack the strong female kin bonds so prevalent among cercopithecines (Strier, 1994a). Even among capuchin monkeys (*Cebus*), in which, typically, females remain in their natal groups and males disperse, the strength and dynamics of female relationships with the alpha males in their groups distinguish their kin bonds from those characteristic among other female-bonded primates (Strier, 1999a).

Both life history strategies (Ross, 1991; Kappeler, 1996) and dispersal regimes are phylogenetically conservative traits among primates (Strier, 1990, 1994a, 1999a; DiFiore & Rendall, 1994). These traits appear to be linked to one another and to divergent patterns in male reproductive strategies in both Malagasy prosimians (Kappeler, 1996) and New World platyrrhines (Strier, 1996a). Dispersal regimes may also account for the different mechanisms by which the relationship between male numbers and female group sizes is achieved across primates (Mitani *et al.*, 1996a). For example, whereas dispersing males may improve their reproductive options by seeking membership in female groups with the most favorable sex ratios, philopatric males are constrained by the number of males born and surviving in their natal groups and the ability of these brotherhoods to attract dispersing females. Thus, dispersal regimes determine whether it is males or females that bear the primary responsibility for adjusting their respective numbers and the breeding sex ratios in primate groups.

This chapter evaluates the interacting effects of demography and dispersal in an effort to extend traditional explanations for the variation in the number of males in primate groups. Long-term demographic data are examined from five Cebidae taxa representing different dispersal regimes: (i) male philopatry (muriquis, *Brachyteles*, and woolly monkeys, *Lagothrix*); (ii) male-biased dispersal (white-throated capuchin monkeys, *Cebus capucinus*); and (iii) dispersal by both sexes (red howler monkeys, *Alouatta seniculus*, and mantled howler monkeys, *A. palliata*). These Cebidae also encompass a diversity of male reproductive strategies, ranging from those in which males compete for reproductive sovereignty by monopolizing access to female groups or estrous females within multi-male groups (e.g., some *Alouatta* and *Cebus*), to those in which male kin display tolerance toward one another's sexual activities in exchange for cooperation in competitive contests with other male kin groups over access to females (*Brachyteles* and possibly *Lagothrix*).

Some of these differences in male reproductive strategies can be explained by the constraints that female reproductive rates, particularly when coupled with reproductive seasonality, impose on male dispersal options (Strier, 1996a). Others, however, appear to be a product of demographic processes that influence the variance in level and intensities of within-group and between-group male competition under each dispersal regime. Together, dispersal and demography play critical roles in determining the number of males in these Cebidae groups, and in defining the dilemma between competing for reproductive sovereignty and cooperating with brothers or unrelated males that male Cebidae – and some other primates – face.

DISPERSAL REGIMES AND DEMOGRAPHIC EFFECTS

The levels at which demographic events influence the number of males in primate groups under different dispersal regimes are summarized in Table 7.1. Under conditions of male philopatry and female-biased dispersal, the number of sexually mature males in a group is a direct result of the number of males born in the group that survive to sexual

Table 7.1. *Primary demographic effects on the number of males in primate groups under different dispersal regimes*

	Female residence	Female dispersal
Male residence	XXXXXXXXX	Group infant sex ratios and sex-specific survivorship
Male dispersal	Group infant sex ratios and population sex-specific survivorship	Population infant sex ratios and sex-specific survivorship

maturity. Larger male kin groups may be more attractive to dispersing females than smaller male kin groups, but under this dispersal regime, it is the behavioral adjustments by females, rather than males, that determine group size and the ratio of breeding females to males in their groups. When males are philopatric, their numbers, and the corresponding consequences of male numbers on their competitive strategies, can therefore be reduced to the demographic factors that affect infant sex ratios and survival within their natal groups.

Patterns of manipulating infant sex ratios vary with local conditions under different dispersal regimes (reviewed in Strier, 1999b). The effects of these demographic events on the number of males in groups are not restricted to primates practicing male philopatry, but they operate beyond the natal group under other dispersal regimes. When dispersal is male biased, for example, it is the number of males in the population that influences male competition for membership in female groups. In cases where both sexes disperse, population-level demography also dictates whether males have the additional option of forming new groups with dispersing females instead of being limited to joining extant groups, as is the case when dispersal is male biased and females are found only in established matrilineal groups.

This dichotomy, which distinguishes between the primary roles of natal group demography under conditions of male philopatry and of population demography under conditions of male dispersal, is clearly an oversimplification. For example, the number of sexually mature males in a population will influence the relative competitive ability of male kin groups to defend the females that join their groups from other groups of related males. Similarly, the number of males born and surviving to dispersal age within their natal groups will influence the overall levels of competition that males face when they set out in search of new female groups to join or females with which to form new groups. Dispersing male Cebidae, like other dispersing male primates, may ultimately end up in multi-male groups because they disperse preferentially with kin (e.g., *Saimiri*: Mitchell, 1994) or because they disperse into the same groups that older male kin have previously joined (e.g., brown capuchin monkeys, *Cebus apella*: Izawa, 1994a, 1994b).

EFFECTS OF DEMOGRAPHY UNDER MALE PHILOPATRY

The prevalence of male philopatry and female-biased dispersal regimes among the three atelin genera (*Brachyteles*; spider monkeys, *Ateles*; and *Lagothrix*) has been attributed to their common phylogenetic ancestry (Rosenberger & Strier, 1989; Strier, 1992a, 1999a). All live in variably sized, multi-male, multi-female groups, which differ in their degree of cohesiveness across populations in response to the proportion of patchy fruits versus other foods, such as leaves (e.g., *Brachyteles*) or insects (e.g., *Lagothrix*), in their diets (Moraes *et al.*, 1988; Chapman, 1990b; Symington, 1990; Strier, 1992a, 1999a, 1999b; Stevenson *et al.*, 1994; Peres, 1994, 1996).

Despite the benefits that larger groups of male kin may gain in competition with other groups of related males over access to females, philopatric males have few overt mechanisms for increasing their numbers. Instead, the competitive ability of male atelin kin groups (and other primates with female-biased dispersal regimes) in between-group competition is determined by the survival and maturation of their sons in their groups. Philopatric males born into groups with male-biased infant sex ratios should have competitive advantages in direct contests against groups with fewer males. Indeed, tolerance among philopatric males may increase the survivorship of male kin, thereby increasing the number of male allies and the ability of their kin groups to win in contests against other groups of related males.

Brachyteles

Demographic data obtained from August 1982 through December 1987 during an ongoing study of one group of *Brachyteles* inhabiting a small (860 ha) forest at the Estação

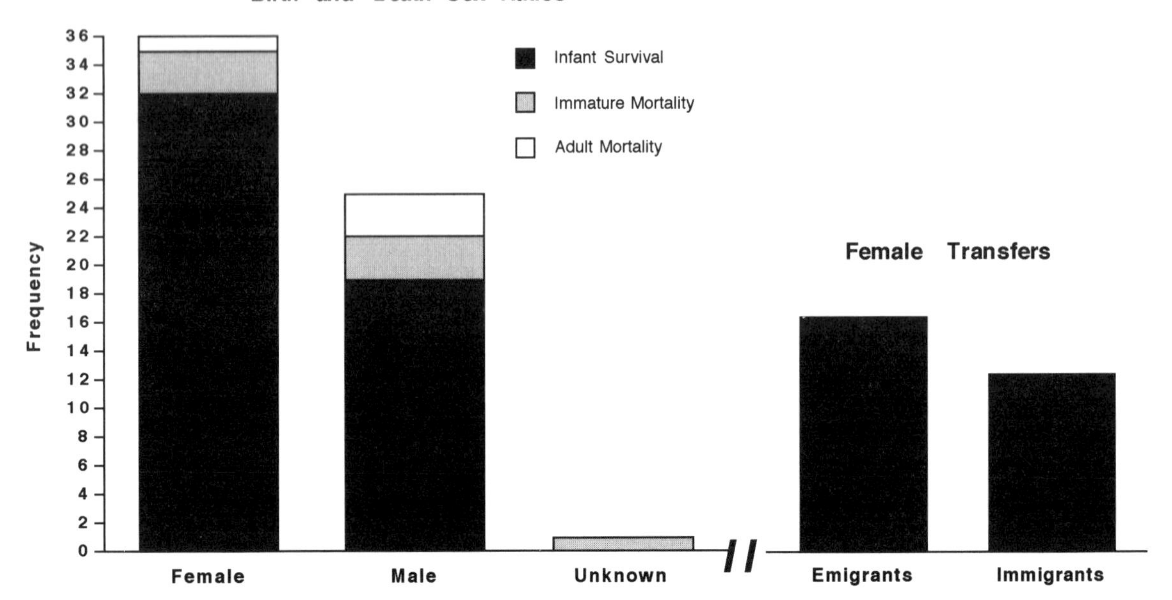

Fig. 7.1. Demographic events in one group of *Brachyteles*, 1982–97. Data from Strier (1996a, 1996c, 1999b, 1999c). Comparative data on group sizes from other muriqui populations include fewer than 35 individuals (references in Strier, 1997a, 1999b, 1999c; Moraes *et al.*, 1998).

Biologica de Caratinga, situated on Fazenda Montes Claros in Minas Gerais, Brazil (Strier, 1992b), illustrate an increase in group size from 22 to 63 members due to the low mortality rates documented for all age–sex classes (Fig. 7.1). All but one of the natal females born in this group since the study began dispersed between five and eight years of age, whereas all natal males surviving to this age, when they also become sexually active, have remained in the group (Strier, 1996b).

The group has increased steadily in size since the onset of the study (Fig. 7.2A), but changes in the number of sexually active males and the ratio of sexually active females to males have been less consistent. Maturing natal males replaced adult males that disappeared and are presumed to have died, resulting in a total of six to eight sexually active males (hereafter, males) for the first ten years of the study (Fig. 7.2B). During this same decade of minimal fluctuations in male numbers, female emigrants ($n=8$) were replaced by comparable numbers of immigrants ($n=9$). The ratio of sexually active females to males nearly doubled from its original value (Fig. 7.2C) because of the female-biased infant sex ratio that has characterized this group since the onset of the study (see Fig. 7.1; Strier, 1999b).

More recently (1993–1997), the number of males has increased with group size (Fig. 7.2 B), and the breeding sex ratio has dropped to its original value (Fig. 7.2C). These shifts can be attributed to the maturation of natal males without corresponding adult male mortality and the disproportionate number of female emigrations ($n=8$) relative to immigrations ($n=3$) during these last few years. Male numbers in this patrilineal society reflect natal group infant sex ratios and male maturation rates relative to adult mortality. Similarly, the breeding sex ratio, and its implications for levels of male–male competition, is a consequence of natal male demography and female migration rates.

Lagothrix

The demography of one *Lagothrix* group monitored over a seven-year period at La Macarena, Colombia, tells a similar story to that of the *Brachyteles* group, as expected from their similarly female-biased dispersal regimes (Nishimura, 1992, 1994). This *Lagothrix* group, like the *Brachyteles* group, has increased in size since the onset of the study period due to low mortality rates among all age–sex classes (Fig. 7.3A). One adult male lost from the group was replaced by the only natal male that has reached sexual maturity during this study period (Fig. 7.3B). This has resulted in relative stability in the number of males in the group, similar to the stability that characterized male numbers during the first ten years of the *Brachyteles* study. By contrast to the *Brachyteles*

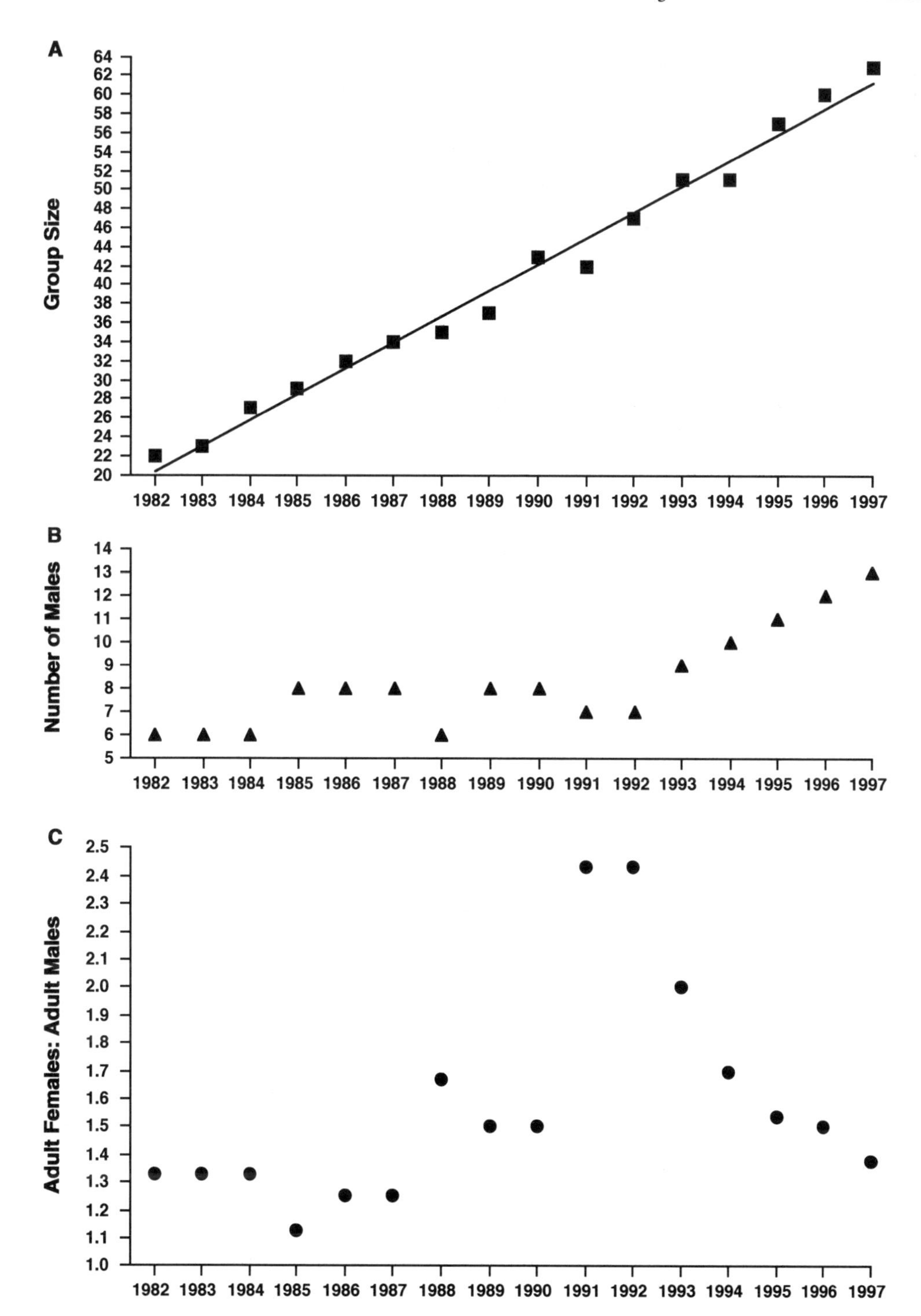

Fig. 7.2. Changes in *Brachyteles* group demography, 1982–97. Data from Strier (1996a, 1999b): (A) group size, (B) number of sexually active males, and (C) ratio of sexually active females to males.

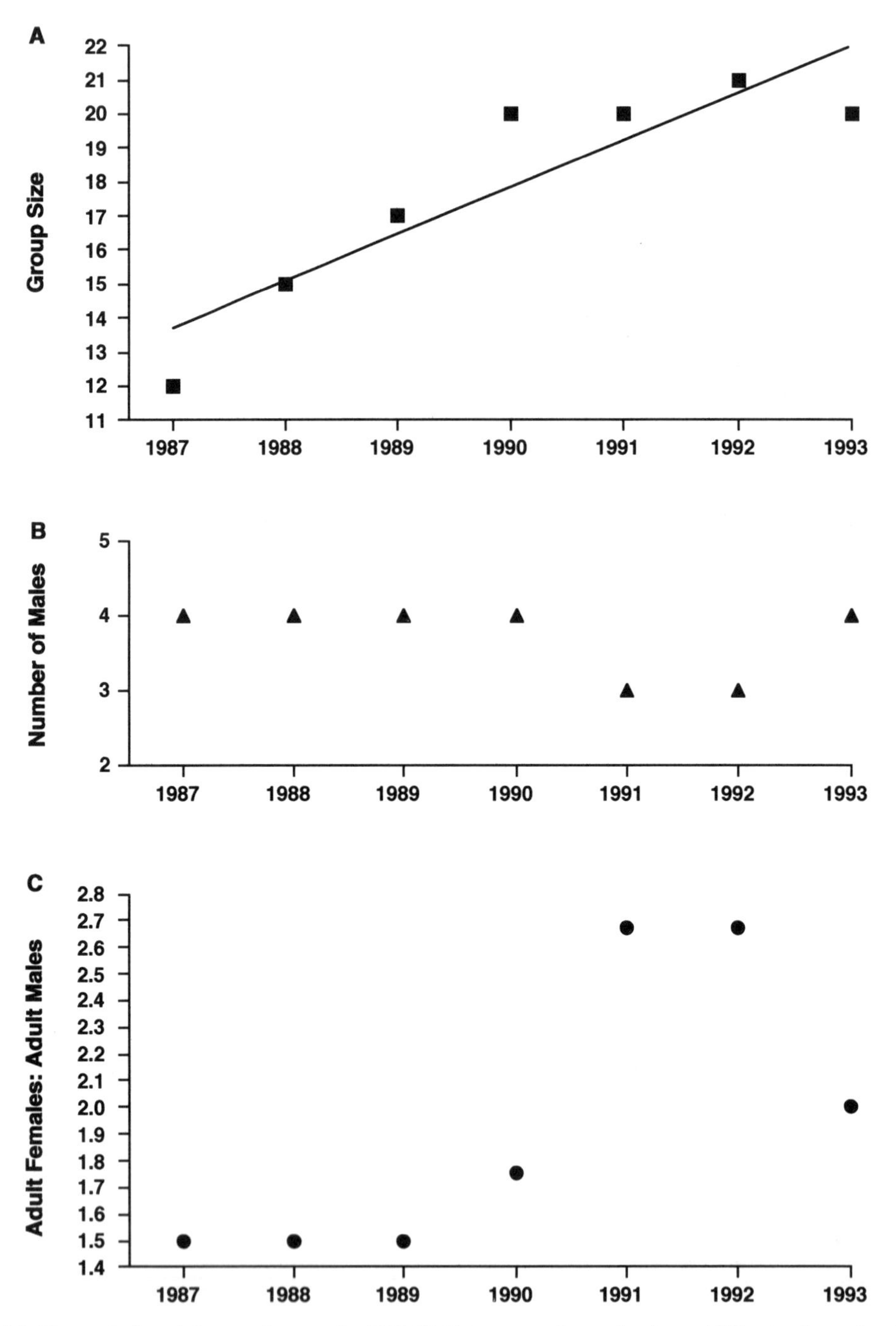

Fig. 7.3. Changes in *Lagothrix* group demography, 1897–93. Data from Nishimura (1994): (A) group size, (B) number of sexually active aged males, and (C) ratio of sexually active females to males.

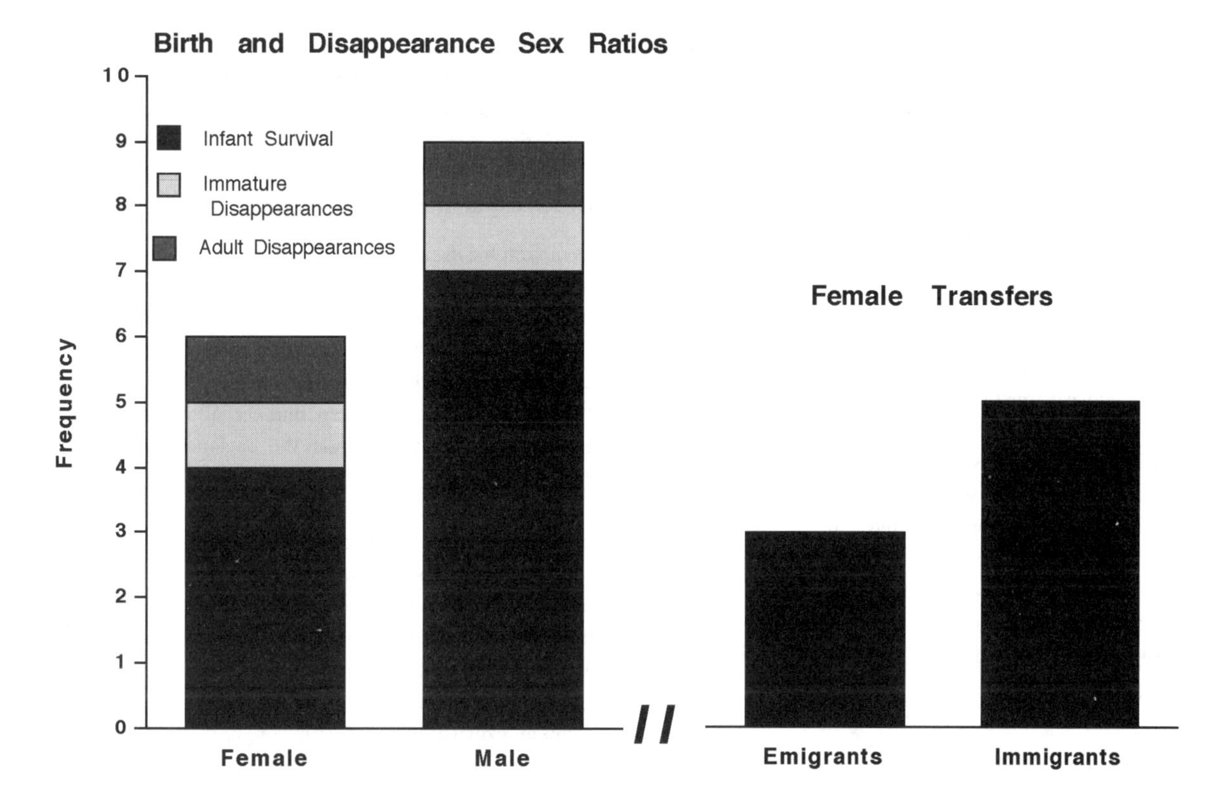

Fig. 7.4. Demographic events in one group of *Lagothrix*, 1987–93. Data from Nishimura (1994).

group, however, female immigrations have outnumbered emigrations, and infant sex ratios have been skewed in favor of males instead of females in this *Lagothrix* group (Fig. 7.4). Yet, as in the case of *Brachyteles*, male numbers reflect natal group demography, and breeding sex ratios (Fig. 7.3C) reflect the interaction between natal male demography and female migration rates.

Behavioral corollaries of changing demography

Changes in mating patterns accompanied the increases in the ratio of sexually active females relative to males in both species. Despite the persistence of tolerant, egalitarian mating opportunities among *Brachyteles* males (Strier, 1997b), there has been an increase in the proportion of copulations involving a subset of males from a neighboring group. By 1992, the increase in both group size and the relative number of females in the group and the greater tendency of females to split up into temporary feeding associations may have decreased the ability of natal male kin groups to monitor and monopolize access to group females (Strier *et al.*, 1993; Strier, 1994b). The fact that mothers of one or more sexually mature sons in this group have accounted for a majority of the extra-group copulations raises the possibility that this apparent shift in *Brachyteles* mating patterns reflects female strategies to avoid inbreeding (Strier, 1997b). Nonetheless, these and other females still prefer unrelated males over extra-group males as mates.

The lack of observations of extra-group copulations in the *Lagothrix* study (Nishimura, 1992) may be a consequence of the duration of the study relative to the natal male maturation rates or of lower densities of *Lagothrix* at this site compared to *Brachyteles* (Strier, unpublished data). Nonetheless, demographic changes in this *Lagothrix* group may also be implicated in what appears to be a shift in their mating patterns. In an early report, Nishimura (1990) described a despotic mating system in which the dominant male monopolized the majority of copulations. In subsequent reports based on larger sample sizes, however, copulations were more equitably distributed among *Lagothrix* males, and as in many other groups of primates, more than one *Lagothrix* male was observed to copulate with a particular female on the same day (Nishimura, 1992).

Although the possibility of sampling biases due to the small number of copulations observed during the early study period cannot be discounted, it may be significant that a shift to multi-male mating practices observed in this *Lagothrix* group also coincided with the first steep increase in its socionomic sex ratio. Such a shift to multi-male mating might be expected if the ability of the alpha male to monopolize reproductive opportunities declined with the increase in the number of estrous females in his group. Indeed, if reproductive competition among philopatric male primates is linked to demographic variables in their groups, then it is possible that the hierarchical relationships described among these *Lagothrix* males will relax into more egalitarian relationships, such as those exhibited by *Brachyteles*, with increases in the number of both males (through natal recruitment of maturing males) and females (through migration). This is a hypothesis that can only be evaluated with further data from this *Lagothrix* group as its size and composition fluctuate.

EFFECTS OF DEMOGRAPHY UNDER MALE-BIASED DISPERSAL

Male-biased dispersal regimes are rare among New World platyrrhines. Among the Cebidae, male-biased dispersal is only known to occur in Peruvian squirrel monkeys (*Saimiri sciureus*) and various species of *Cebus* (reviewed in Strier, 1999a). Male *Saimiri* in Peru disperse as cohorts from their natal groups, and form long-lasting alliances that may improve immigration and subsequent mating success (Mitchell, 1994). Some *Cebus* males may also disperse with natal kin, leading Izawa (1994a, 1994b) to suggest that brown capuchin males in Colombia maintain community networks that encompass both their natal groups and the adjacent groups that they join at maturity.

The life-long residence of *Cebus* females in their natal groups poses a unique problem for male *Cebus* compared to other platyrrhines (Strier, 1999a). Dispersing males must venture into their populations to find groups of these females as potential mates. The range of variation in female group sizes they encounter and the number of male competitors that have previously dispersed into these groups will reflect the demography of groups within the population. Under such male-biased dispersal regimes, female-biased infant sex ratios will mean fewer males competing for positions within non-natal groups, whereas male-biased infant sex ratios will mean stronger competition unless high male mortality reduces the number of dispersing males in the population.

Cebus capucinus

Demographic data are available from population-wide censuses of white-throated capuchin monkeys conducted over an eight-year period at Santa Rosa National Park, Costa Rica (Fedigan *et al.*, 1996). Mean group sizes during this period increased from 11.5 individuals in 1983 to 18 individuals in 1992 (Fig. 7.5A). The mean number of males in these groups also increased, while the socionomic sex ratio declined (Fig. 7.5B–C).

Under male-biased dispersal regimes, increases in group sizes may be achieved through the maturation of females, which remain in their natal groups, and through the maturation and subsequent immigration of males dispersing from other groups in the population. Extrapolating from the number of females, which remained fairly constant (4–5.6), and the male-biased infant sex ratios observed in three capuchin monkey study groups (Fig. 7.6) to the censused population suggests that dispersing males were responsible for the documented increase in mean group sizes (Fedigan *et al.*, 1996). Indeed, male-biased infant sex ratios, despite higher male mortality, appear to have overridden the effects of natal female maturation in these *Cebus* groups (Fig. 7.6).

Behavioral corollaries

The implications of population demography on the mating patterns of these *Cebus* groups may be as important as the implications of demographic changes within groups of philopatric male atelins. For example, the multi-male mating patterns and shorter tenures of alpha male white-throated capuchins compared to other species of *Cebus* may be a consequence of the relatively large number of dispersing males in this population competing for membership in female groups and the unfavorable socionomic sex ratios within these groups (Fedigan *et al.*, 1996). Traditional explanations for the reproductive sovereignty that alpha males achieve through female choice in some brown capuchin and wedge-capped capuchin (*C. olivaceus*) groups have emphasized ecological variables that affect female dependency on alpha males for access to food resources (e.g., Janson, 1984, 1986; O'Brien, 1991; Fedigan, 1993; O'Brien & Robinson, 1993). Demographic data from other *Cebus*

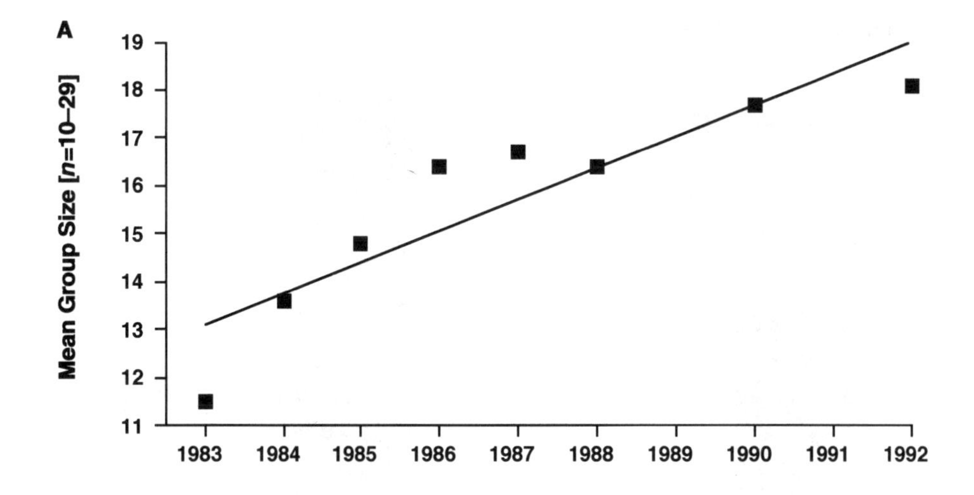

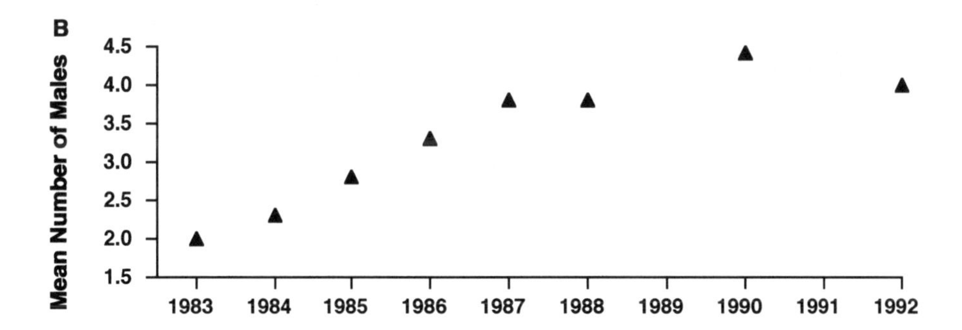

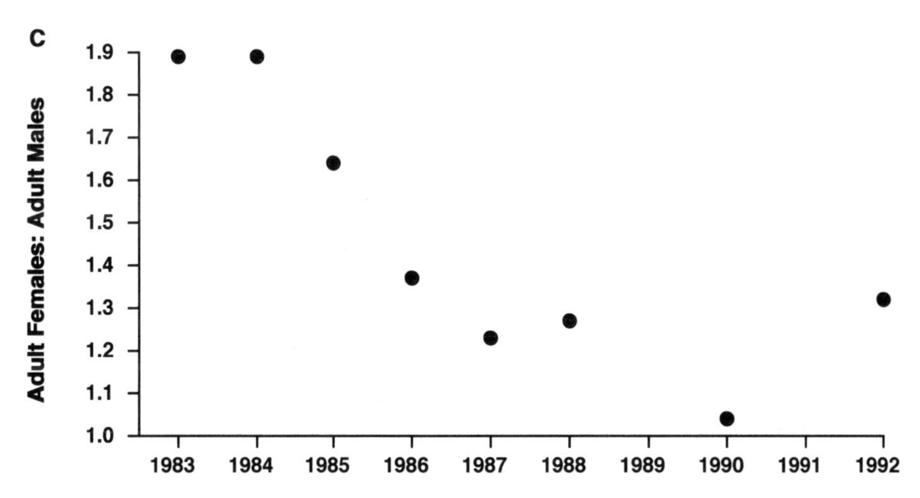

Fig. 7.5 Changes in mean group demography from census population of *Cebus capucinus*, 1983–92. Data from Fedigan *et al.* (1996): (A) group size, (B) number of sexually active-aged males, and (C) ratio of sexually active females to males.

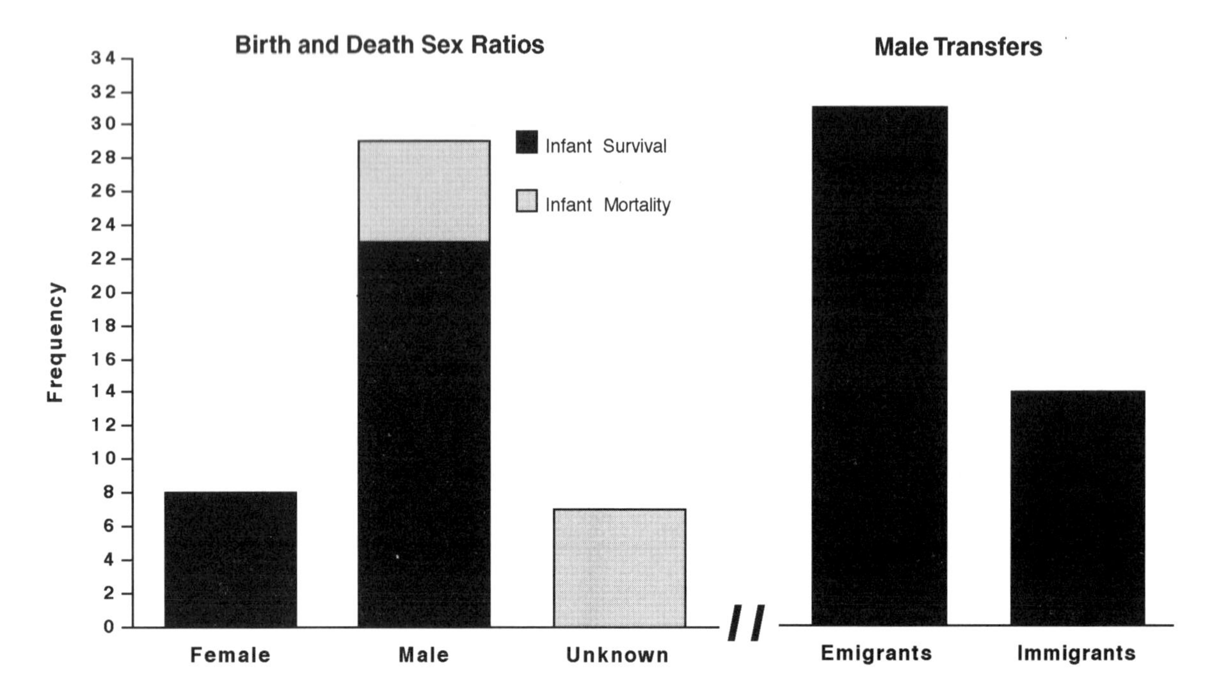

Fig. 7.6. Demographic events in three study groups of *Cebus capucinus,* 1985–93. Data from Fedigan *et al.* (1996). Male transfers exclude nine transfers among three study groups.

populations comparable to those from this white-throated capuchin population are needed to evaluate the possibility that variation in *Cebus* male reproductive patterns can be attributed, at least in part, to the effects of population-wide demographic processes.

EFFECTS OF DEMOGRAPHY WHEN BOTH SEXES DISPERSE

The effects of demography on male reproductive strategies are more complicated when both sexes disperse (as occurs among various species of *Alouatta*) than they are when dispersal is sex biased. Whenever one sex is philopatric, opportunities to establish new groups with one or more members of the opposite sex are limited or non-existent. But, when both sexes disperse, the possibility of forming a new breeding group may be a preferable alternative to joining groups with unfavorable sex ratios or remaining in their natal groups as non-breeding adults. Group demography, including infant sex ratios and survivorship, affects competition both for recruitment into breeding positions within the natal group and for successful dispersal into extant groups in the population. Additionally, population-wide densities and sex ratios affect whether disenfranchised members of both sexes are available to establish new groups.

Natal group dispersal should occur at younger ages when opportunities to inherit breeding positions within the natal group are limited by competition from other dispersing individuals when populations approach carrying capacities and opportunities to form new groups become limited. Conversely, offspring should be tolerated longer in their natal groups in growing populations as they wait for either inherited breeding positions within their natal groups or individuals of the opposite sex from other groups to reach dispersal age and join them in establishing new groups. In the latter case, secondary dispersal should also be common among dispersing individuals seeking membership in groups with more favorable sex ratios.

Alouatta comparisons

Long-term censuses of populations of red howler monkeys (*Alouatta seniculus*) at Hato Masaguaral, in Venezula (Crockett & Pope, 1993; Crockett, 1996), and mantled howler monkeys at La Pacifica, in Costa Rica (Glander, 1980, 1992; Clarke & Zucker, 1994), illustrate how population demography affects male options of forming new troops versus joining established troops. Mantled howler monkeys appear to maintain larger average troop sizes, with

Table 7.2. *Population demography of* Alouatta seniculus *and* A. palliata

	A. seniculus[1]				*A. palliata*[2]	
	Gallery		Woodland		1984	1991
	1981	1987	1981	1987		
Adult females in troops (mean *n*)	2.5		2.9		8.9	6.8
Adult males in troops (mean *n*)	1.2		1.6		2.1	2.5
Adult females: males (mean)	2.1		1.8		3.0	2.8
Troop size (mean)	7.6	~10.5	10.5	~10.5	15.7	12.6
Density (individuals/km^2)	36	<70	112	>130	77	103
Number of troops	17		26		16	27

Notes:

[1] Crockett & Eisenberg (1987: for 1981 data); Crockett & Pope (1993); Crockett (1996: 1987 values estimated from Figs. 1 and 4 in Crockett [1996] using ecological density for gallery forest and established troop sizes for both forests).

[2] Glander (1992); Clarke & Zucker (1994).

a higher proportion of males than red howler monkeys, but in both populations there appears to be a limit to increases in group sizes that is independent of population density (Table 7.2).

Although males and females in both populations disperse, over 20% of red howler monkey females remained in their natal troops (Crockett & Pope, 1993). Natal female recruitment may function only until troops achieve what appears to be a maximum of four adult females, after which daughters are expelled (Crockett, 1996). Because extant troops recruit breeding females from among their daughters, there are few, if any, opportunities available to dispersing female red howler monkeys to join extant troops. Whereas dispersing red howler females must find mates to form new troops with them, in mantled howler monkeys, of which only 3% of natal females remain in their troops, dispersing females have greater opportunities to join extant troops as well (Glander, 1992).

Behavioral correlates and life history comparisons

Such differences in the dispersal options of females affect those of males. Red howler males are not only tolerated much longer (four to six years) in non-breeding positions in their natal troops, but they also practice secondary dispersal, in contrast to mantled howler males, which disperse at much younger ages (one to two years) when they are evicted by males other than their fathers. These young mantled howler males may live alone for years until they attain the competitive size and ability they need to gain membership in extant troops (Glander, 1992; Crockett & Pope, 1993).

Perhaps not unrelated to these differences in male dispersal ages is the disproportionately high incidence of infant mortality among mantled howler males (60%) compared to mantled howler females (7%) and both male (17%) and female (21%) red howler monkeys, despite comparably even numbers of males and females born in both populations (Glander, 1992; Crockett & Pope, 1993). A higher proportion of surviving red howler males in the population may increase the pressure of takeover attempts, and give fathers that tolerate their maturing sons in their natal troops a competitive advantage in defending their troops against these incursions. The genetic consequences of such a strategy on the genetic composition of *Alouatta* troops are discussed in Chapter 19 by Pope.

LIFE HISTORY CORRELATES OF DISPERSAL REGIMES

The age at which females disperse, along with female age at first reproduction and interbirth intervals, are among the most critical life history variables affecting primate reproductive rates. Yet, although *Alouatta* and *Lagothrix* females have similar adult body weights (Ford, 1994), these life history traits in *Lagothrix* are more comparable to their closer relatives, the other, larger-bodied, female atelins (Strier, 1999c; Table 7.3).

Table 7.3. *Comparative life history variables among five Cebidae taxa*[1]

	Cebus[2]	*Alouatta*[3]	*Lagothrix*[4]	*Ateles*[5]	*Brachyteles*[6]
Female body weight (kg)[7]	2.67	5.35, 5.60	5.75	7.46	9.45
Age males disperse (years)[8]	>5	1–2, 4–6			
Age females disperse (years)		2–4	4–5	5–7	5–7
Female first birth (years)	5	4–7	6–10	7	7.5–14
Interbirth interval (months)	26.4	20–22.5	33.8	34.7	36.8

Notes:

[1] Table modified from Strier (1999c).

[2] Fedigan & Rose (1995); Fedigan *et al.* (1996: for *Cebus capucinus*).

[3] Glander (1980); Crockett & Pope (1993); Fedigan & Rose (1995).

[4] Nishimura (1992, 1994).

[5] Chapman & Chapman (1990); Fedigan & Rose (1995).

[6] Strier (1996a, 1999c).

[7] Ford (1994: for *Cebus capucinus, Alouatta palliata* and *A. seniculus*, respectively, *Lagothrix lagotricha, Ateles geoffroyi, and Brachyteles arachnoides*).

[8] Glander (1992: for *Alouatta palliata*); Crockett & Pope (1993: for *A. seniculus*).

The fact that *Lagothrix* also resemble the other atelins in their male philopatry and female-biased dispersal regimes is consistent with suggestions that female reproductive rates influence male reproductive strategies (Strier, 1996a), and that their life history strategies and dispersal regimes are phylogenetically conservative and linked with one another (Strier, 1999a). Male *Alouatta* may practice more competitive reproductive strategies than male atelins because of the faster reproductive rates of their mates. Conversely, male philopatry in *Lagothrix*, as well as the other atelins, may be bound by the benefits of tolerance among brothers for the competitive advantages that they gain in competition with other groups of male kin, particularly when reproductive opportunities within their groups are constrained by delayed maturation of females and long interbirth intervals (Strier, 1996a).

Life history variables may also be mediated by the effects of demography on male reproductive strategies. For example, *Cebus* females have slow life histories even though they are much smaller in body size than any of the four atelin genera (Ross, 1991). Both male dispersal age and female age at first reproduction are comparable to those among red howler monkeys (Table 7.3), and the proportion of their interbirth intervals attributed to lactation and cycling (82%) resembles that of *Ateles* (78%) more closely than *Alouatta* (69%; Fedigan & Rose, 1995). These life history traits appear to be similar across *Cebus* species (Robinson & Janson, 1987), so other factors must account for the variation in male reproductive strategies reported for the genus. The female-biased infant sex ratios reported for wedge-capped capuchins (Robinson & O'Brien, 1991), compared to the male-biased infant sex ratios of white-throated capuchins (Fedigan *et al.*, 1996), may represent one such factor affecting the ability of males to monopolize matings and the length of male tenure in their groups.

FUTURE DIRECTIONS

The interaction between primate life histories, dispersal regimes, and group and population demography exerts a powerful, but often neglected, influence on the level and intensity of competition among males. This influence begins with the mechanisms by which males end up in groups with females, and persists through the dynamics of male reproductive strategies as demographic variables within groups in a population fluctuate.

The possibility that distinctions between hierarchical or egalitarian male relationships may be a consequence of demographic events has provocative implications, and strong cautionary indications, for how seemingly apparent differences in primate behavioral patterns are interpreted (Strier, 1997a). Fluctuations in group and population sizes and sex ratios were associated with behavioral shifts in *Brachyteles* and *Lagothrix*, and were consistent with differences in male reproductive strategies among closely related species of both *Cebus* and *Alouatta*. While ecological variables may be responsible for stimulating changes in group and population sizes and skewed infant sex ratios, dispersal regimes and life history variables influence the direction of these demographic responses and their consequences for the number of males in primate groups (Strier, 1999b).

Demographic influences on male primate reproductive strategies may take multiple generations to manifest themselves, and may shift direction over time as a result of both adaptive strategies, including sex ratio manipulations, and stochastic processes (Strier, 1996c, 1999b). Incorporating the influence of these demographic variables into comparative models will be critical to understanding the mechanisms underlying the behavioral diversity observed among Cebidae and other primate males.

ACKNOWLEDGMENTS

I am grateful to Dr Peter Kappeler for inviting me to participate in the first Göttinger Freilandtage he organized on male primates, the funding agencies that supported the conference, the other participants at the conference for their stimulating comments on the version of this chapter I presented there, and to Dr Kappeler and two anonymous reviewers for their comments on a previous version of this manuscript. Fieldwork on *Brachyteles* was made possible by permission from CNPq, sponsored by C. Valle, C. Ades, and G.A.B. de Fonseca. Fundação Biodiversitas provided infrastructure at the Estação Biologica de Caratinga. NSF grants BNS 8305322, BNS 8619442 and BNS 8959298, the Fulbright Foundation, Grant 213 from the Joseph Henry Fund of NAS, Sigma Xi, the L.S.B. Leakey Foundation, the World Wildlife Fund, the National Geographic Society, the Seacon Fund of the Chicago Zoological Society, the Liz Claiborne and Art Ortenberg Foundation, the Scott Neotropical Fund of the Lincoln Park Zoo, and the Graduate School of the University of Wisconsin–Madison provided support. N. Bejar, A. Carvalho, D. Carvalho, C.G. Costa, P. Coutinho, L.T. Dib, J. Gomes, M.A. Maciel, F.D.C. Mendes, F. Neri, S. Neto, C.P. Nogueira, A. Odalia Rímoli, A. Oliva, L. Oliveira, R. Printes, J. Rímoli, W. Teixeira, and E.M. Veado contributed to the long-term demographic data.

8 • The number of males in guenon groups

MARINA CORDS

INTRODUCTION

The African guenons (the genera *Cercopithecus*, *Miopithecus*, *Erythrocebus*, and *Allenopithecus*) are a diverse group of primates in many respects, including their pelage, geographic distribution, and habitat preferences. In terms of social structure, there are still many species about which we know little (especially those living in forests), but even limited information reveals that this group of monkeys also exhibits considerable diversity in terms of the number of adult males that reside in social groups.

In some species, like vervets (*Cercopithecus aethiops*), talapoins (*Miopithecus talapoin*) and probably swamp monkeys (*Allenopithecus nigroviridis*), groups almost always contain multiple adult males. In other species, like the forest *Cercopithecus* and patas monkeys (*Erythrocebus patas*), however, individuals are most often found in one-male groups, though not necessarily throughout the year. Variation in the numbers of males in groups thus occurs among species, among populations of a given species, and over time within populations as well.

This chapter focuses on the causes of this variation, by examining factors that correlate with variable male numbers. Considering both the quantity and quality of comparable data, this is a daring enterprise. In general, relevant data on guenons are scarce, and comparable data are even harder to find. Most of the data addressing intraspecific variation come from just a few species. Therefore, the conclusions drawn here must be viewed as preliminary, but they will have served their purpose if they stimulate further research.

WITHIN-SPECIES VARIATION IN MALE NUMBERS

Multi-male phases in predominantly one-male groups

DESCRIPTION

In many species of forest *Cercopithecus*, as well as in the woodland/savannah-dwelling patas monkey, numerous reports indicate that while a uni-male social structure is the modal social unit over long periods, more than one male may be present in the group on more than a passing basis (review by Cords, 1988, for forest-dwelling species; for more recent reports, see Mitani, 1991; Hill, 1994; Beeson *et al.*, 1996; Glenn, 1997). Many of these reports come from census studies in which group composition is studied through enumeration of members. For three species, however – *C. mitis* (blue monkeys and samangos), *C. ascanius* (redtail monkeys), and patas monkeys – more detailed accounts are available of the transitions between uni-male and multi-male status, the behavior of males and females in both types of groups, and how changes in social structure are related to various environmental factors (for references more recent than the Cords, 1988, review, and/or including patas monkeys, see Chism & Rowell, 1986; Harding & Olson, 1986; Henzi & Lawes, 1988; Jones & Bush, 1988; Butynski, 1990; Ohsawa *et al.*, 1993; Rowell, 1994; Chism & Rogers, 1997).

These reports describe how a one-male group can be invaded by multiple males for periods that typically last several months and that coincide with the mating season. During these multi-male influxes, the number of males in a group on a given day is typically three to six, and often varies from one day to the next. Former long-term residents may persist during such influxes, or may disappear from the group. Most of the males in the group during the influx are relative newcomers, though some may have had limited social contact with females previously. A few of the intruding males are long-term residents from neighboring groups, although these males are typically very short-term visitors during the influx, since they spend most of their time with their own groups. I know of one case (in *C. mitis*) in which a former resident returned to his group for an extended time (two months) during an influx, before disappearing entirely. (Henzi & Lawes, 1988, also report a case of a former resident who returned to his group as a peripheral visitor for two days during a non-influx breeding season.)

The coincidence of these intrusions by previously non-resident males and sexual activity of females suggests that males move into bisexual groups for mating opportunities. Females do mate with the newcomers, and at least in *C. mitis* have been observed to go out of their way (quite literally) to do so, bee-lining to new males and soliciting them for copulations even when the newcomer is on the group's periphery, or the female is being followed persistently by the resident. Some of these females show little or no sexual interest in the long-term resident if he is still in the group.

These multi-male phases in predominantly one-male groups resemble only superficially the regularly multi-male groups of other Cercopithecine monkeys (such as vervets, talapoins, baboons, and mangabeys; Henzi, 1988). Aside from their passing nature, a big difference is that the identity of the males present changes often, frequently from day to day. While three to six may be a typical number of males in the group on a particular day, over the mating season as many as 25 males may have visited or resided in a group for variable periods. Some of these males are more regularly present than others, and day-to-day fluctuations in numbers depend on which particular individuals are present.

Given the fluidity of their membership, it is perhaps not surprising that co-resident males do not appear to develop friendly relations with each other. Although we have seen adult male blue monkeys groom each other, and maintain spatial associations with particular other males, such overt affiliation occurs only when they are not part of a heterosexual group. Similarly, in patas monkeys, males can live together peacefully as long as females are absent (Gartlan & Gartlan, 1973; Rowell & Chism, 1986). When males are together with females, male–male relations in these species are typically antagonistic, and at best grudgingly tolerant.

Agonism among males that reside together in heterosexual groups often occurs in predictable directions so that dominance relationships and even hierarchically organized sets of relationships can be recognized by observers (Cords, 1984; Tsingalia & Rowell, 1984; Cords *et al.*, 1986; Harding & Olson, 1986; for a partial exception, see Chism & Rogers, 1997). If a former long-term resident remains in the group along with newly arrived intruders, he usually takes the top position in such a hierarchy, although that does not guarantee him priority of access to mates (Tsingalia & Rowell, 1984; Cords *et al.*, 1986; Henzi & Lawes, 1988, Ohsawa *et al.*, 1993; Rowell, 1994).

INTER-ANNUAL VARIATION IN THE NUMBER OF MALES IN GROUPS

A population of blue monkeys in the Kakamega Forest, Kenya (see Cords, 1987a, for a description of the site), has been monitored since 1979, and provides the most extensive data set with which to evaluate correlates of year-to-year variation in male numbers. Figure 8.1 shows which males were present during the breeding season in various years from 1979 to 1997 in six blue monkey groups. The most complete records are available for groups T and G, which have been most intensively monitored. The diagram simplifies a more complex situation by dichotomizing breeding seasons as influx or non-influx years. Even during the non-influx years, a long-term resident male was occasionally joined by other males who lurked on the edge of the group or even entered it, and sometimes mated with females. On the whole, however, influx years could be distinguished by the greater proportion of days during the breeding season on which more than one male was in or near the group (typically 50–70%, versus 7–28% in non-influx years, data from T_w and G troops in 1995–7 inclusive); by the maximum number of males present on a single day (eight during influxes, four during non-influx years, 1995–7); by the average number of males present on a single day (1.9–2.4 in influx years, 1.1–1.2 in non-influx years, 1995–7), by the duration of visits by non-residents (who often stayed for many hours or a whole day during influxes, but typically less than two hours in non-influx years; see also Henzi & Lawes, 1988); and by the way in which residents regularly chased non-residents in non-influx years, but seemed to relax this intolerance during influxes.

These data make several noteworthy points.

1. Overall, 21% of breeding seasons ($n = 80$) were characterized by multi-male influxes. The proportion in different groups varied from 0% to 33%, with the two extreme values coming from the groups that were monitored for the smallest number of years.
2. Every group monitored for more than five years has experienced at least two years in which a multi-male influx occurred.
3. No males have been the sole residents in more than one of the groups we monitored. These groups are localized in space: each one shares boundaries with two to three others. So, if a male takes up residence more than once in his lifetime, he probably moves at least one home range diameter away in his second tenure as resident. There is no evidence that males do have more than one tenure as resident, although we also cannot rule out this possibility.

Struhsaker (1988) provides some comparable data for redtail monkeys in the Kibale Forest, Uganda. Studying seven groups over several years, he similarly found variation in the number of males present, with multiple males being found in groups about 30% of the time. These data included non-mating periods, however, so the proportion of time

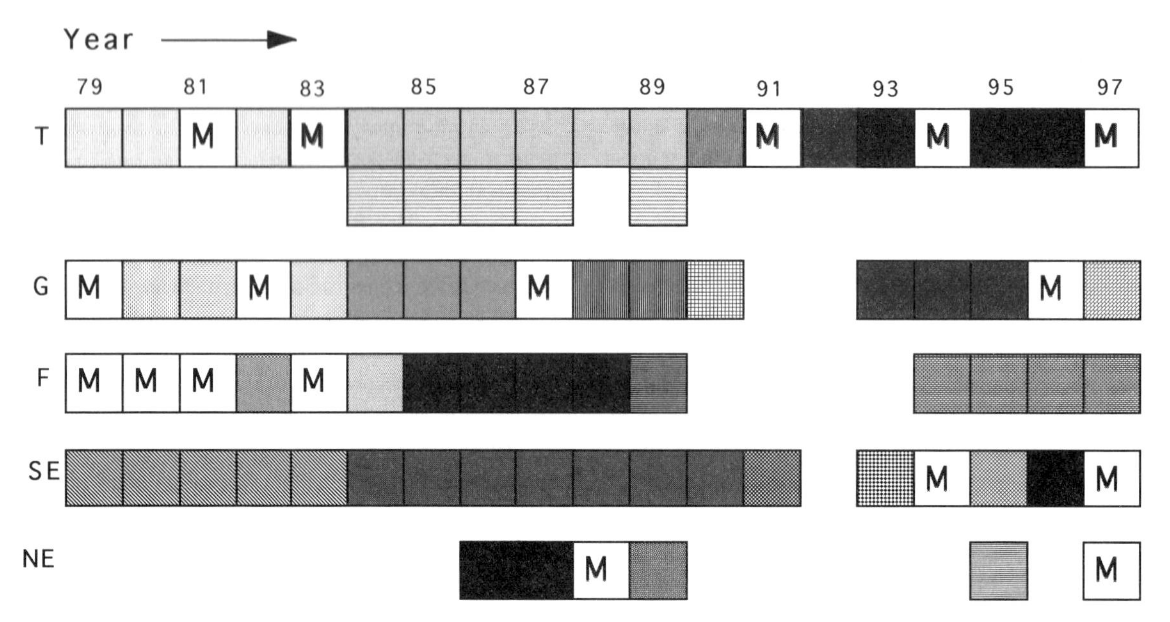

Fig. 8.1. Males present in five Kakamega blue monkey groups during the 1979–97 breeding seasons. Each row of boxes represents one group. T group fissioned in 1984 into T_w (top row) and T_e (adjoining lower row). Each box represents the breeding season for a particular year. Each shading pattern represents a particular male who was present in that year; boxes with an M represent breeding seasons in which there was a multi-male influx (see text).

with several males might be higher if analysis were restricted to the mating period, as was done with the Kakamega data.

Whether these figures for the proportion of time that groups contain multiple males are typical for redtails, blue monkeys, or any guenon species, is debatable. We shall return to this point in discussing interpopulational variability in male numbers (below). In any case, however, the existing variation can be used to explore factors that might explain it.

WHY DO INFLUXES OCCUR IN SOME YEARS BUT NOT OTHERS?

The fact that influxes occur during the mating season, and that intruding males mate with females, strongly suggests that the presence of supernumerary males is related to the presence of estrous females. (The word *supernumerary* is used in the sense of exceeding the usual number, and not in the sense of exceeding a desired or necessary number. The usual number of males is one.) The behavior of the former resident, who (if he persists) chases intruding males whenever he sees them, indicates that he would prefer those males to be absent. Together, these observations suggest that defendability of females may be a key factor influencing the number of males in a group, and specifically that multi-male influxes occur when resident males cannot exclude intruders.

One aspect of defendability is the number of females that need to be defended (Altmann, 1990). Influxes should be more likely in breeding seasons in which the number of fertile females is high. To test this prediction, the first step is operationally to define 'fertile' females. I evaluated fertility in blue monkeys in two ways. First, using demographic data on female reproductive histories, and recognizing that interbirth intervals in this population are minimally two years when the first infant survives, I assumed that adult females with infants less than 12 months old would not be fertile. The remaining parous females were considered fertile, and I also included in this class any nulliparous females that were observed to mate. Second, I used conception as an indicator of fertility, and counted as fertile only those females that actually gave birth after each breeding season. Because observations were not continuous, some conceptions may not have been registered, including those that ended in pre-term abortion, still-birth, or death of the infant at a young age. Defined in either of these ways, however, the number of fertile females did not differ in influx versus non-influx years (females without young infants: Mann-Whitney U Test, two–tailed, $p=0.21$, $n=6$ influx, 18 non-influx years in T_w and G groups; females that conceived: Mann-Whitney U Test, two–tailed, $p=0.87$, $n=$ 5 influx, 15 non-influx years in T_w and G groups). Jones & Bush (1988) carried out a similar analysis for the Kibale

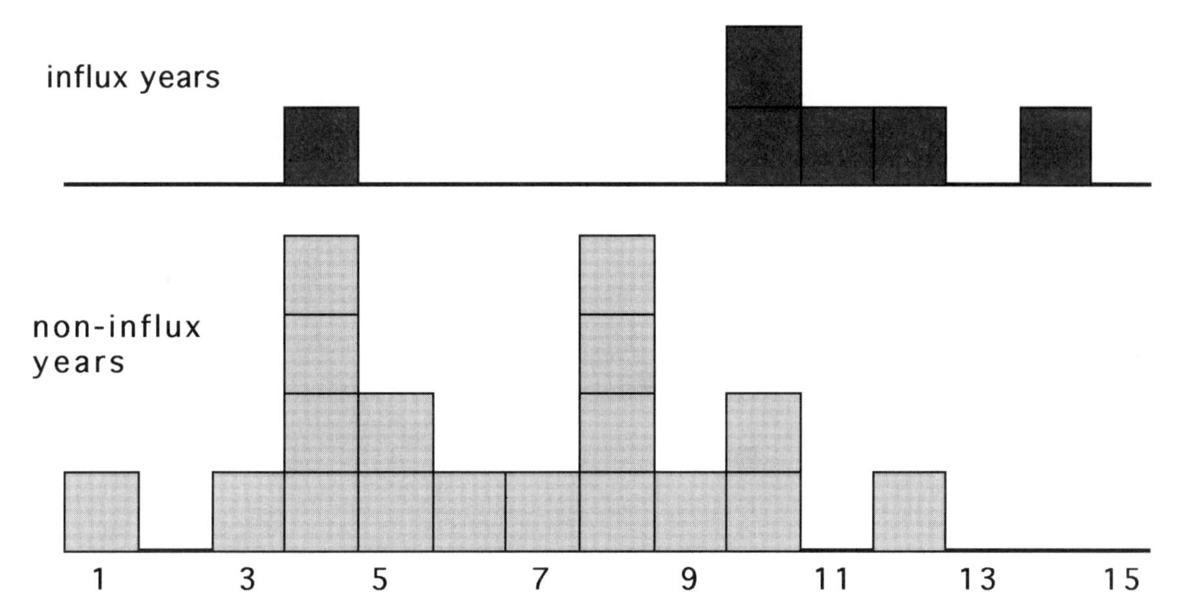

Fig. 8.2. The number of females that mated in influx and non-influx breeding seasons of Kakamega blue monkeys. Data from T_w group (1979–97) and G group (1993–7). The low outlying value of four females for the influx years is almost certainly an underestimate, as data were collected early in the study (1981) by colleagues who were not yet able to identify individual females reliably.

redtail monkeys, relying on the absence of infants to indicate fertility, and also found no relationship between the number of non-resident male sightings and the number of fertile females per group. In contrast, Henzi & Lawes (1988) reported that in South African samangos, more females bore young after an influx than in a subsequent non-influx year. This study, however, compared only two breeding seasons in one group.

From a proximate perspective, we might expect male blue monkeys to use behavioral cues to assess female fertility, especially since morphological signals of reproductive cycling, at least those detectable over a distance, do not occur in this (or any congeneric) species. Therefore, I compared influx and non-influx years in terms of the number of estrous females, namely those observed to mate or to engage in the conspicuous proceptive behavior of lip-puckering. (Puckering is frequently observed in blue monkey females that mate, and is not known to occur in any other context. Puckering occurs more often than copulation, since females may follow males with puckering for some time before copulating. Thus, including puckering along with actual mating probably makes a behavioral assay of sexually active females more sensitive.) The number of estrous females judged according to these criteria tended to be smaller than the number of potentially fertile females as assessed with data on reproductive histories (above), perhaps because the period of lactational amenorrhea varied in length across individuals. The number of estrous females was higher in years with multi-male influxes than in non-influx years (Fig. 8.2, Mann-Whitney U Test, two-tailed, $p = 0.02$).

However, this analysis says nothing about the time course of female mating within a given mating season. From a male's point of view, it is not the number of females that will mate over an entire season that determines female defendability, but rather the number of females that are sexually active at any one time (Ridley, 1986; Altmann, 1990). Thus, we might predict that supernumerary males will be more likely to attend a group of females in years when there were many days with multiple estrous females available. To examine this issue, I used data from our two focal blue monkey groups (T_w and G) from 1995–7 inclusive, over periods of about four months per year during the breeding season. The two troops were very similar in size (each containing 15 adult females), and were observed by similar-sized teams of observers using standardized methods. The data set includes six breeding seasons in all, two of which were characterized by multi-male influxes. Figure 8.3 shows that in years when there were many days with at least two females in estrus, there were also many days on which more than one male was in the group ($r_s = 0.9$, one–tailed, $p < 0.025$). The same positive correlation was found if the analysis was expanded to include days with just one estrous female in the

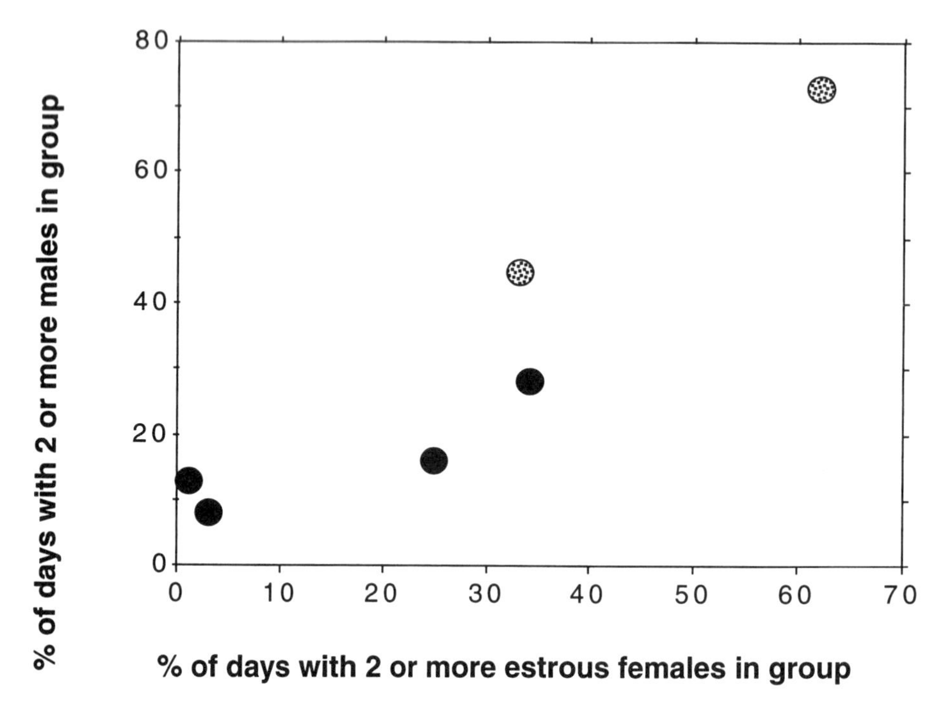

Fig. 8.3. The proportion of days in a given breeding season in which there were multiple males in the group as a function of the proportion of days in which there were at least two estrous females in the group. Each point represents one breeding season in one group of Kakamega blue monkeys. Data are from G and T_w groups, 1995–7 inclusive. The two stippled points represent breeding seasons with multi-male influxes.

group: years with a high proportion of days with at least one female in estrus were likely also to include a large proportion of days with more than one male in the group ($r_s = 0.9$, one-tailed, $p < 0.025$). These results thus confirm the relationship between sexual activity in females and presence of more than one male.

WHAT DETERMINES DAY-TO-DAY VARIATION IN THE NUMBER OF MALES DURING THE BREEDING SEASON?

In every year, whether there is a multi-male influx or not, the number of males in a group during the breeding season can vary from day to day. To test the hypothesis that such variation could also be related to variation in the number of estrous females, I again used data from two focal groups of blue monkeys monitored over three breeding seasons in Kakamega. For each day, the number of estrous females and the number of males in the group were tallied. As Fig. 8.4 shows, on days with more estrous females, the mean number of males in the group was higher (Kruskal–Wallis Test, $p = 0.0001$ for each of the two groups). This pattern was very similar for both study groups.

While these data on day-to-day variation agree with those on year-to-year variation in showing that the number of males in a group is related to the number of estrous females, these analyses do not pin down the direction of the causal arrow. Does the presence of many estrous females bring males to their group? Or might it be that the presence of unfamiliar males stimulates estrus in the females?

This second view is not unreasonable, considering behavioral interactions between females and unfamiliar males. These males often appear to be more attractive to females, some of whom solicit matings and actually mate exclusively, or much more often, with the newcomers than with long-term residents still present. Mating with newcomers appears to require more effort on the part of females, who engage in various sneaky tactics apparently to avoid detection by the ever-vigilant resident. Some females even leave their group for several hours, or an entire day, to go on safari with a relatively unfamiliar male. These pairs have been found hundreds of meters from the female's group, completely out of visual contact, and out of auditory range as well, except perhaps for the loudest of adult male calls.

We have suggested previously (Cords *et al.*, 1986) that

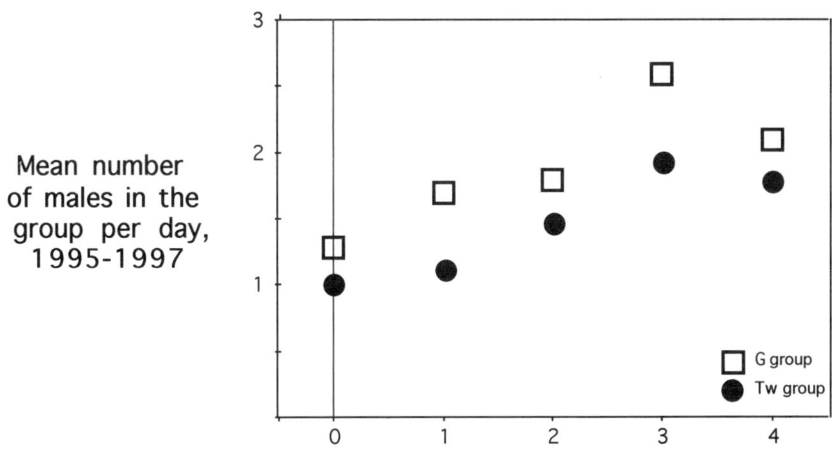

Fig. 8.4. The number of males in a blue monkey group per day as a function of the number of estrous females in the group on that day. Data are presented for two study groups, but for each of them come from three breeding seasons, 1995–7 inclusive.

both of these mechanisms – estrus bringing new males to a group, new males bringing females into estrus – may be at work simultaneously. To understand the determinants of variation in male numbers within a population over time, however, it is important to evaluate their relative magnitude.

Two kinds of information would be useful for this task. First, it would be helpful to know more about the factors that determine the spatial distribution and number of non-resident males. In Kakamega, these males appear to be non-randomly distributed in the forest, being found especially often near the interstices of heterosexual group territories, and quite often near one another (see Henzi & Lawes, 1988; T. Rowell, unpublished data). Furthermore, there appears to be variation from year to year, and perhaps on shorter time scales, in the number of males detected (see also Henzi & Lawes, 1988, for discussion of 'wandering' males in another population). Because the behavior of these animals when away from heterosexual groups is so cryptic, however, it is difficult to assess their numbers, and thus to know whether changes in their proximity to females precede (and so could have caused) the advent of estrus.

A second kind of information that might help in evaluating the cause and effect relationship of male presence and female estrus is a detailed chronological record of these states at the beginning of the breeding season. Does female estrus or supernumerary male presence typically come first? In the Kakamega blue monkey population, there are seven breeding seasons, including three influx years, for which we have relevant data. One year is uninformative, as supernumerary males and estrous females were both present from the first day of observations. In four of the remaining six breeding seasons (and two of the three influx years), there were multiple days of female estrus before supernumerary males were ever observed in the group. While these data suggest that the presence of estrous females at least sometimes may bring males to the group, it seems that there are also seasons in which the presence of new males precedes female estrus (see Fig. 1 in Cords *et al.*, 1986). Furthermore, although our observation periods begin in June, which is one month before conceptions typically begin to occur in this population, we may not have picked up the very first estrous females or male intrusions of the season. In general, then, these analyses fall short of demonstrating clearly that male numbers change as a result of the presence of estrous females, although this remains a viable, if perhaps incomplete, explanation.

WHAT DETERMINES VARIATION AMONG POPULATIONS?

If defendability of estrous females determines the number of resident males, multi-male breeding periods should occur more often in those populations in which a resident male's ability to monopolize females is relatively low. Several factors that vary across populations might influence a resident's ability to monopolize estrous females. Ecological factors include the degree to which female estrus is synchronized within groups, the dispersion of females, and the visual opaqueness of the particular habitat. A demographic factor is the ratio of extra-group males to groups

Table 8.1. *Interpopulational comparisons of the ratio of males to heterosexual groups and the frequency of multi-male influxes during the breeding season in* C. mitis

Population	Cape Vidal (South Africa)	Kakamega (Kenya)	Ngogo (Uganda)	Kanyawara (Uganda)	Ngoye (South Africa)
non-resident male density (per km^2)	6.9	5.2–10.4	0.5	0.08	2.3
group density (per km^2)	6.9	5.2	0.38	2.28	3.0
ratio of males: groups	1.0	1.0–2.0	1.3	0.35	0.76
influxes (proportion of breeding seasons)	~50%	21%	Rare	Never	Never

Note:
Values for non-resident male density in the Uganda sites differ from Henzi and Lawes' original compilation of data: whereas they took the lowest of a range of values originally reported by Butynski (1990), I have taken the midpoint of this range. The group density figure for Ngoye is from Lawes (1990).

of females in the population, which is a measure of intruder pressure.

C. mitis is the only guenon species for which comparative data are available from several populations, but these data concern only the males-to-groups ratio. Table 8.1 expands on an analysis by Henzi & Lawes (1988) by adding data from Kakamega. While there is undoubtedly error in all of these measurements, especially in estimates of the density of extra-group males, a general trend is still apparent: groups are more likely to include multiple males during the breeding season when the population ratio of extra-group males to female groups is greater than one. This result is consistent with the hypothesis that defendability of females is an important explanatory variable. Clearly, however, data on additional parameters, from additional populations, and from other species will be needed to see how general this result may be.

Variation in male numbers in species whose groups are typically multi-male

Vervets, talapoins and swamp monkeys are the three guenon species whose groups typically include several adult males. Within each of these species, however, the number of males per group is variable, and there is some evidence that this number is a function of the number of adult females in the group. Cheney & Seyfarth (1987, Table 1) present 80 annual group censuses of vervets in Amboseli, with figures coming from 14 groups monitored for one to ten years. A regression analysis of these data shows that variation in the number of females explains a significant though small part of the variation in the number of males ($r=0.4$, $r^2=0.2$, $F_{1,78}=17.6$, $p=0.0001$). Similar results come from Hall and Gartlan's cross-sectional data on 18 vervet groups in Uganda ($r=0.7$, $r^2=0.5$, $f_{1,16}=16.1$, $p=0.001$). With fewer groups to compare, Kavanagh (1981) also found that male numbers increased with the number of adult females across four groups in Cameroon, as did Struhsaker (1967), studying six vervet groups in Amboseli more than a decade earlier than Cheney and Seyfarth. Gautier-Hion (1971, Table 2) provides information on four talapoin groups in Gabon, for which an increase in male numbers as a function of female numbers is also apparent. Data on *Allenopithecus* are not available: although Gautier (1985) and McGraw (1994) report groups quite different in size, McGraw (personal communication) was not able to determine the number of males in the parties he observed. Taken together, however, these data on intraspecific variation in predominantly multi-male species suggest quite strongly that male numbers within groups are related to the number of females, and that when female numbers become very small, there may be only one adult male. Given that males in these species are not known to live outside of groups for any extended periods, a concomitant increase in male and female numbers would seem to be a demographic necessity. Other factors, however, like those mentioned in the previous section, may also be sources of variation in male numbers, but data to evaluate their importance are not available.

BETWEEN-SPECIES VARIATION IN MALE NUMBERS: PREDOMINANTLY ONE-MALE VERSUS TYPICALLY MULTI-MALE GROUPS

So far, we have considered temporal and cross-population variation in the number of males residing with females within single species. Still to be addressed is the variation in male numbers among guenon species. In particular, why do most forest species and patas monkeys exemplify a modally one-male social structure with occasional multi-male phases, while vervets, talapoins and swamp monkeys live in essentially permanently multi-male groups? I will consider ecological explanations for these different social systems, but do not mean to imply that all interspecific variation is driven directly by ecological factors.

There seems to be an evolutionary history of male social coexistence in the multi-male species that is lacking in the one-male species. Vervet males, for example, have a specialized communicative repertoire that may function to mediate the coexistence of competitors (Henzi, 1985); this communicative repertoire is lacking in the closely related one-male guenon species. Furthermore, animals removed from a natural environment and held in captivity show differences in social tolerance, as expected from their natural groupings: multi-male species can be held in captivity in groups including several males, whereas the one-male species cannot (Rowell, 1988b; Fairbanks, 1993). Here, we will investigate the proposition that these different social tendencies evolved to deal with ecological differences between species, and we will assume that these ecological differences can be assessed by examining extant populations.

Talapoins and swamp monkeys

For talapoins and swamp monkeys, there is evidence supporting the hypothesis that defendability of females is a critical factor influencing the number of males in the group. Groups of these species have been reported to be large, especially relative to the forest-dwelling *Cercopithecus*. Talapoin groups living in the proximity of human villages, where they had access to crop-plants as food, and where they could be most closely studied, had at least 69 individuals in Cameroon (Rowell, 1972, 1973), and averaged 112 animals in Gabon (Gautier-Hion, 1970, 1971). Non-commensal groups in Gabon were estimated to have average group sizes of 66 (Gautier-Hion, 1971). Adult females accounted for at least 15–25% of the total. Swamp monkeys have been very little studied, but based on partial group counts and vocalizations heard, Gautier (1985) estimated that their groups contain more than 40 individuals, although much smaller groups of about four to eight individuals were reported by McGraw (1994; personal communication) at another site in Zaire.

While large group size alone may make male defense of females difficult, reproductive seasonality can exacerbate the problems of monitoring several estrous females at once. Data on reproductive seasonality are available only for talapoins, in which births, and presumably conceptions, occur during a limited period of about two months per year in any one group (Gautier-Hion, 1971; Rowell, 1972, 1982). This is relatively tight seasonality for guenons, and so may well contribute to the difficulty a male would have in defending access to reproductive females in this species.

Other explanations for the regular presence of multiple males in talapoin and swamp monkey groups may also be relevant, however, especially in light of the fact that males remain in groups even outside of the breeding season. Field studies have not been sufficiently detailed to examine carefully the behavior of males in groups, and thus to evaluate thoroughly either a male or a female perspective on the costs or benefits of year-round co-residence with more than one male. In the case of talapoins, however, it seems that males have little to offer females, at least during the day. In fact, Rowell (1973) reported that males are socially segregated from females outside of the breeding season, and tentatively attributed this segregation to the behavior of females, who have been observed to be extremely aggressive to males (even harassing them to death) in captivity. Gautier-Hion (1970) described how adult males can act as leaders and at least sometimes direct group movements with particular loud calls; however, since only one male in a group behaves in this way, this service performed by males cannot explain why several males should be present. At night, however, the situation may be different. Gautier-Hion (1970) described how adult males form a protective shield at group sleeping sites, by taking positions peripheral to adult females and juveniles, and thus guarding the access route of potential predators on these small and vulnerable monkeys. The anti-predator advantages of group living may also, of course, benefit the males themselves.

Vervets

Vervets have been studied much more extensively than talapoins and swamp monkeys, but the multi-male nature of their groups remains puzzling as well. Unlike talapoins and swamp monkeys, vervets do not live in particularly large groups. The range of values for the number of adult females per group in vervets is essentially identical to that for other guenons (Table 8.2). Female group size is only a rough measure of the number of breeding females, however, because it does not take into account variation in interbirth

Table 8.2. *Number of adult/reproductive females per group in vervets versus other guenons with modally uni-male groups*

Number of adult females		Site	Reference
Vervets			
7		South Africa	Baldellou & Henzi (1992)
5		Kenya, Amboseli	Struhsaker (1967)
5		Kenya, Amboseli	Cheney & Seyfarth (1987)
8–9		Kenya, Samburu/Isiolo	Whitten (1982)
4		Uganda, Lolui Island	Hall & Gartlan (1965)
≥3		Cameroon, Bakossi	Kavanagh (1983)
6		Cameroon, Bakossi	Kavanagh (1983)
6		Cameroon, Buffle Noir	Kavanagh (1983)
18		Cameroon, Kalamloue	Kavanagh (1983)
7–9		Senegal, Mt Asserik	Harrison (1983)
1–4		Senegal, Niokolo-Koba	Dunbar (1974); Galat & Galat-Luong (1978)
12		Senegal, River	Galat & Galat-Luong (1978)
4–10		Senegal, Siné-Saloum	Galat & Galat-Luong (1978)
2–3		Senegal, Bandia	Galat & Galat-Luong (1978)
2		Senegal, Casamance	Galat & Galat-Luong (1978)
4		St Kitts	Poirier (1972); Horrocks & Hunte (1983)
Non-vervet guenons whose groups typically contain just one adult male			
6	*C. ascanius*	C.A.R., near Bangui	Galat-Luong (1975)
10	*C. ascanius*	Kenya, Kakamega	Cords (unpublished data)
7–10	*C. ascanius*	Uganda, Kibale	Struhsaker (1977); Struhsaker & Leland (1979); Jones & Bush (1988)
3	*C. campbelli*	Ivory Coast, Adiopodoume	Hunkeler *et al.* (1972)
2–3	*C. campbelli*	Ivory Coast, Tai	Noë (unpublished), Bourlière *et al.* (1970); Galat & Galat-Luong (1985)
4	*C. cephus*	Gabon, Makokou	Quris *et al.* (1981)
2.8	*C. cephus*	Cameroon, Campo Reserve	Mitani (1991)
6–10	*C. diana*	Sierra Leone, Tiwai	Whitesides (1989; Hill (1994)
5–7	*C. diana*	Ivory Coast, Tai	Zuberbühler *et al.* (1997); Galat & Galat-Luong (1985)
~5–7	*C. erythrotis*	Cameroon, Douala Edea	Whitesides (1981)
8–10	*C. mitis*	Uganda, Kibale (Kanyawara)	Butynski (1990); Struhsaker & Leland (1979)
7–12	*C. mitis*	Uganda, Kibale (Ngogo)	Butynski (1990)
4	*C. mitis*	Uganda, Budongo	Fairgrieve (1993, 1995)
4	*C. mitis*	DRC, Kahuzi-Biega	Schlichte (1978)
~4	*C. mitis*	Malawi, Zomba	Beeson (1985)
15	*C. mitis*	Kenya, Kakamega	Cords (unpublished data)
6–8	*C. mitis*	South Africa, Ngoye	Lawes *et al.* (1990)
9–10	*C. mitis*	South Africa, Cape Vidal	Lawes & Piper (1992)
3	*C. mona*	Grenada, West Indies	Glenn (1997)
1	*C. neglectus*	Gabon, Makokou 1	Gautier-Hion & Gautier (1978)
1–2	*C. neglectus*	Gabon, Makokou 2	Quris (1976)
3	*C. neglectus*	Kenya, western	Wahome *et al.* (1993)
2–3	*C. nictitans*	Gabon, Makokou	Gautier-Hion & Gautier (1974)
~5–7	*C. nictitans*	Cameroon, Douala Edea	Whitesides (1981)
3.8	*C. nictitans*	Cameroon, Campo Reserve	Mitani (1991)
3–7	*C. petaurista*	Ivory Coast, Tai	Noë (unpublished); Galat & Galat-Luong (1985)
3+ (≤6)	*C. pogonias*	Gabon, Makokou	Gautier-Hion & Gautier (1974)
~5–7	*C. pogonias*	Cameroon, Douala Edea	Whitesides (1981)
3.4	*C. pogonias*	Cameroon, Campo Reserve	Mitani (1991)

Table 8.2 (*cont.*)

Number of adult females		Site	Reference
~3	*C. preussi*	Cameroon, Kilum Mountain	Beeson *et al.* (1996)
1–22 ($\bar{X}=11$)	*E. patas*	Kenya, Laikipia	Chism & Rowell (1988)
4–12	*E. patas*	Uganda, Murchison	Hall (1965)
3–17	*E. patas*	Cameroon, Waza	Struhsaker & Gartlan (1970)
6–13	*E. patas*	Cameroon, Kala Maloue	Nakagawa (1989, 1995); Muroyama (1994)

interval, which may influence the number of breeding-age females that actually do mate and produce offspring in a given year. Data from wild populations are quite rare, but suggest that vervets have relatively short interbirth intervals (less than two years) compared at least to forest-dwelling species with modally uni-male social structures (often more than two years; Table 8.3). Patas monkeys are exceptional, in that they also have short interbirth intervals, despite a modally uni-male social structure (Cords, 1987b).

Vervets also are more tightly seasonal breeders than most other guenons. Butynski (1988, his Table 16.1) summarized birth seasonality across species and populations of guenons. Although there is considerable unevenness in the quality of the available data, it is clear that vervet births typically occur over a period of two to three months, whereas in other guenon species the birth season is typically three to seven months. A few populations with exceptionally long birth seasons stand out from these general patterns, but even here the difference between vervets and other guenons is apparent: the exceptionally long birth seasons among vervets last four to six months, whereas in the other species they last 8–12 months (at which point 'seasonality' becomes something of a misnomer). The tighter reproductive synchrony in vervets corresponds with the typically more seasonal habitats in which they occur. Non-vervets with the shortest birth seasons (patas, *C. mitis* in South Africa) also live in relatively seasonal habitats. However, these latter species are not routinely multi-male as are the vervets, even though their birth seasons are equally short.

Assuming that estrus periods last equally long in vervets as in those guenon species that usually live in uni-male groups, and given comparable group sizes, the shorter interbirth intervals and tighter breeding synchrony should mean that the number of females in estrus per day is higher in groups of vervets than in groups of these other species. (Data that would allow direct comparisons are not available.) As discussed earlier in this chapter, this variable is closely related to temporal variation in the number of males in groups of guenons that usually contain just one male. Based on the available evidence, it may also relate to the difference in male numbers between vervets and most of the modally uni-male guenon species.

In this context, the case of patas monkeys merits special attention. Patas groups are like those of most forest guenons in containing just one adult male outside the breeding season, and variable numbers during the breeding season. In terms of breeding seasonality and interbirth interval, however, patas monkeys are more like vervets (with whom they are broadly sympatric) than they are like forest guenons. Group sizes in patas are relatively large compared to vervets (see Table 8.2). Thus, one might expect patas to show multi-male groups like those of vervets, and yet they do not. It is therefore difficult to understand the differences in social structure between patas and vervets in terms of female defendability: other factors must be involved.

The idea that other factors may be involved in explaining interspecific differences in the number of males per group of females is further suggested by the fact that in seasonally breeding animals, issues of defendability relate to variation in male numbers only during the breeding season. For vervets, as for talapoins and swamp monkeys, we still need to explain why males that are present in heterosexual groups during the breeding season remain there for the rest of the year, or why they do not leave. We may need to focus attention on times other than the breeding season, and consider the interests of the various members of the social system – a dominant male who might exclude others, subordinate males who might have been excluded, and adult females.

Taking this approach, Baldellou & Henzi (1992) have argued that the enduring presence of multiple males in vervet groups is related to predation pressure, which they take to be more significant in the savannah–woodland environment typical of vervets than in the more continuously

Table 8.3. *Typical interbirth intervals (in months) of wild, unprovisioned guenons*[1]

Interbirth interval		Site	Reference
Vervets			
20.1		Samburu, Kenya	Whitten (1982)
13.8–21.3		Amboseli, Kenya	Cheney *et al.* (1988)[2]
Forest guenons with uni-male social groups			
52	*C. ascanius*	Kakamega, Kenya	Cords (1988)
18.2	*C. ascanius*	Kibale, Uganda	Struhsaker & Pope (1991)
~12	*C. campbelli*	Adiopodoume, Cote d'Ivoire	Bourlière *et al.* (1970)
28.5	*C. mitis*	Kakamega, Kenya	Cords (1988, unpublished)
28–50	*C. mitis*	Kibale, Uganda	Butynski (1990)[3]

Notes:

[1] Intervals have not been differentiated according to whether the first of two infants survived. Values are means as reported in the cited references, except for Kakamega *C. ascanius* (median) and Kibale *C. mitis* (see note 3 below).

[2] Mean values differed across groups; reported here is a range of group-mean values.

[3] These figures have been derived as the inverses of birth rates reported in Butynski's Table 3.

forested environments in which most uni-male guenons live. According to their argument, subordinate males prefer to be in groups, even during the non-breeding season, because of the safety of numbers. They considered the possibility that a dominant male and the females might have an interest in keeping such males around because of the services they might offer, particularly in terms of predator defense. But they rejected this hypothesis after finding that subordinate males were not more efficient than females in detecting predators, and less likely than the dominant male to alarm call or to challenge predators directly. Instead, they concluded that subordinate males were parasitic on the rest of the group, benefiting in terms of safety, while being either a neutral presence or an unavoidable disadvantage for the females and the dominant male.

If one wished to challenge Baldellou and Henzi's view, one might do so on at least two grounds. First, it rests on the assumption that predation pressure is more intense in savannah–woodland than in forested environments, at least for individuals who do not reside in groups. This is why vervet males should be better off remaining in a group, and why males of forest species are not as pressured to remain in heterosexual groups beyond the breeding season: in fact, they may do better being on their own, leading an intensely cryptic existence. The belief that savannah–woodland (coupled with a more terrestrial way of life) is a more dangerous place than forest (coupled with a more arboreal way of life) is commonly held and supported by circumstantial evidence (Dunbar, 1988), although there are few direct data that support it (Cheney & Wrangham, 1987; Isbell, 1994). Data on the vulnerability of extra-group individuals in these different habitats, which would be at the heart of Baldellou and Henzi's argument, are non-existent.

Second, patas monkeys may pose a problem for Baldellou and Henzi's hypothesis. Patas live in open, vervet-like habitats (in fact, both species co-occur in some sites) and should be vulnerable to the same predators; but patas groups do not regularly contain multiple males. Perhaps there are differences in the antipredator strategies of patas and vervets that can account for the difference in male social behavior in these species. More than vervets, patas appear to rely on being cryptic, using early detection of predators to initiate early flight so that they remain undetected by the predator (Chism & Rowell, 1988). Because their general antipredator strategy emphasizes early detection and being cryptic, lone patas males may be less vulnerable than lone vervet males, in the same way that lone males of the forest species are supposedly less vulnerable than lone vervet males. Data relevant to this speculation would be welcome.

Another service that male vervets might perform is range defense. In vervets, males participate conspicuously in intergroup encounters, and may participate more often than females, especially when encounters include displays and aggression (as opposed to vocalizations only; Cheney, 1981; Kavanagh, 1981; Harrison, 1983; Whitten, 1984). Data from the Amboseli vervets suggest that success at intergroup competition influences home range size and quality, and thus has significant consequences for the reproductive success of females (Cheney & Seyfarth, 1987) and probably of males as well (Isbell *et al.*, 1991). If the presence of subordinate males improves a group's competitive power *vis à vis* other groups by adding additional manpower, it may be to everyone's advantage to have such males in the group.

Comparisons among guenon species partially support this idea. Although patas monkey groups, like vervets, are often hostile to neighboring groups when they are encountered, patas groups typically occur at lower densities than vervet groups, and encounters between groups are comparatively rare. Summarizing information from two populations, Harrison (1983) gives minimum figures for vervets of one intergroup encounter (involving aggression or display) every three to four days, whereas long-term studies of patas have each reported rates of encounter substantially lower (two encounters in 638 observation hours spread over ten months: Hall, 1965; one encounter every two weeks: Chism *et al.*, 1984; five encounters in four years: L.A. Isbell, personal communication). In some special cases, intergroup encounters can occur more frequently in patas, as when water is limited to very few water holes at the peak of the dry season (Struhsaker & Gartlan, 1970). When aggressive intergroup encounters occur in patas, females participate more vigorously and more often than males, except during the mating season when males also join in (Struhsaker & Gartlan, 1970; Chism *et al.*, 1984; Chism & Rowell, 1988). If intergroup aggression is relatively rare in patas, and males are only seldom involved, the range-defense benefits of having multiple males should not be as substantial for this species as it is for vervets.

The uni-male, forest-dwelling species appear to be more problematical, however, as many populations have been reported to be territorial in the sense of engaging in site-specific intergroup aggression (Cords, 1987b; see Lawes & Henzi, 1995, for a discussion of within-species flexibility in territorial behavior), and yet they contain only one adult male for most or all of the year. It is consistent with their residence patterns, however, that those males that do live in groups are conspicuously uninvolved in aggressive defense of range boundaries, at least in the two species (*C. ascanius* and *C. mitis*) whose intergroup behavior has been most thoroughly studied. It seems to be primarily females that fight the neighbors, so that adult group members, whether male or female, should gain little from adding additional males in terms of improving the effectiveness of their range defense. Still, it would remain to be explained why forest guenons can aggressively defend their ranges without much male support, while such support is so critical in vervets that it explains their multi-male nature.

There are other complications with the idea that supernumerary males provide an important service in terms of range defense. First, vervet populations vary in terms of how much territoriality they exhibit (Kavanagh, 1981; Harrison, 1983; Fedigan & Fedigan, 1988). The degree of home range overlap between groups may vary, and while some groups engage in frequent hostile intergroup encounters, others do so very infrequently; and yet, essentially, all vervet groups are multi-male. Numerous factors appear to influence the frequency of hostile intergroup encounters, including characteristics of the food supply and predation pressure (Kavanagh, 1981; Harrison, 1983), and social factors (Cheney, 1981). Whether intense intergroup competition is or was characteristic of the species remains an open question. Second, in those populations in which intergroup aggression occurs, it seems to be uncertain whether males are defending a food supplying area, which would benefit everyone in their group, or whether they are defending reproductive females, which would benefit primarily themselves. Baldellou & Henzi (1992) point out that males mostly fight other males (also Cheney, 1981), and that when there are fights between the sexes, females typically initiate them. Herding of females during intergroup encounters has been reported from several populations, and suggests that males are defending mates (Harrison, 1983).

CONCLUSION

In summary, available evidence suggests that several factors must be considered to understand variation in male numbers in groups of guenons. Within-species variation in one group over time, across groups, and among populations is related to mating opportunities for males, so that there are more males when there are more estrous females available. Other factors, both ecological and social, may also be important, but available information does not allow an evaluation of their impact.

Between-species differences in male numbers – especially why some species routinely include multiple males in their groups while others form mostly uni-male groups – are more difficult to explain. While breeding opportunities may be relevant, we also need to consider, especially in light of the seasonal breeding that characterizes this group of monkeys, why males remain in female groups during the non-breeding season, and what the consequences of non-gregarious life would be for males. It would also be useful to know the degree to which males and/or females can control male numbers, and the degree to which their interests coincide or conflict.

While the above discussion rather assumes that characteristic ecological conditions have determined an optimal solution, we might also consider the possibility that phylogeny has constrained evolutionary responses in at least some of the guenon taxa. At present there is still no consensus on the phylogenetic history of these monkeys (Disotell, 1996), but it is noteworthy that those species whose groups routinely include multiple males (especially *Allenopithecus* and *Miopithecus*) are generally viewed as having branched off earliest from more papionine-like stock. Confident placement of both vervets and patas will be needed to evaluate thoroughly the possibility that the number of males in a species-typical group depends at least in part on whether that species diverged early or late in the evolutionary history of the guenons.

ACKNOWLEDGMENTS

For discussions and advice during the preparation of this manuscript, I thank Janice Chism, Lynne Isbell, Thelma Rowell, Pat Whitten, and one anonymous reviewer. I am grateful to the Office of the President, Government of Kenya, for permission to study blue monkeys in the Kakamega Forest; to the Zoology Department, University of Nairobi, for local sponsorship; and to the Kakamega Forest Station staff for their cooperation. My field work has been supported by the National Science Foundation (SBR 95-23623), the L.S.B. Leakey Foundation, the Wenner-Gren Foundation, and Columbia University. I am grateful to many Kenyan and American student-interns for contributing to the Kakamega data, to Jeff Hatcher for help searching the literature, and to the reviewers and Lynne Isbell for comments relevant to the preparation of the manuscript. Finally, thanks to Peter Kappeler for inviting me to participate, and for his assistance throughout.

9 • Socioecology of baboons: the interaction of male and female strategies

ROBERT A. BARTON

INTRODUCTION

Baboons (genus *Papio*) present an interesting model system for primate socioecology, because their occupancy of a range of habitats across the African continent is associated with subtle yet striking variation in social organization. Traits that vary include the number of males in breeding groups, the number of females, female coalition formation and social cohesiveness, and sex-specific patterns of philopatry. The aim of this chapter is to describe and explain such variation. A model is presented here which suggests that these social traits are linked to one another and covary. The model is an elaboration of one presented elsewhere (Barton *et al.*, 1996), incorporating additional factors, building in each factor step by step, more explicitly than previously, and extending the principles to the explanation of broader phylogenetic variation. It is based on the approach pioneered by Wrangham (1979, 1980) and van Schaik (1989), which derives primate social organization from the interaction between the divergent reproductive strategies of males and females (Bateman, 1948; Trivers, 1972). It is argued that the behavior of males both influences and is influenced by the way in which females organize themselves. This dynamic interaction between male and female strategies is held to explain several features of social organization, including the number of males in social groups.

The phylogeny of the genus *Papio* is still somewhat controversial (Disotell, 1994), but it is thought that hamadryas baboons (*Papio hamadryas*) were an early offshoot from the rest of the genus – olive baboons (*P. anubis*), yellow baboons (*P. cynocephalus*), chacma baboons (*P. ursinus*), and Guinea baboons (*P. papio*). A recent estimate of the phylogenetic relationships of these species and other cercopithecines is presented in Fig. 9.1. The main focus of this chapter is variation among *Papio* species, but this variation will be placed in wider phylogenetic and socioecological context by discussing contrasts and parallels with other Old World monkeys. Proposed general influences on baboon social organization, such as terrestriality, are tested using comparative analysis across the whole primate order. This is used to assess whether social and ecological traits exhibit statistically significant correlated evolution (see Harvey & Pagel, 1991). If they do not, there is little justification for assuming that such traits are linked in any specific case, such as that of baboons.

Social organization of baboons

The following is a brief summary of the main features of social organization in baboons. This will be expanded when ecological influences are discussed, and further details can also be found in Kummer (1984), Melnick & Pearl (1987), Stammbach (1987), and Dunbar (1988).

Many populations of baboons exhibit what is thought of as the 'classic' savannah baboon social system,[1] consisting of large and relatively cohesive, multi-male groups in which females form the social core: males usually transfer between groups in search of breeding opportunities, whereas females usually remain in their natal groups to breed, and form stable, linear dominance hierarchies (e.g., Hausfater *et al.*, 1982). Adult females and males form so-called 'special relationships' or 'friendships' (Smuts, 1985; Strum, 1987), which persist beyond the periods when the female is sexually receptive. Females have highly differentiated grooming relationships and alliances with one another, forming social networks of as many as 20 or 30 adult animals and their offspring. Female–female alliances are predominantly formed with close kin (see Gouzoules & Gouzoules, 1987; Walters & Seyfarth, 1987; Harcourt & de Waal, 1992), but can also be formed with more distantly related, or even completely unrelated, females, as on the rare occasions when a female transfers to a new group (Strum, personal communication; personal observation). Females seem to transfer only when they have no effective allies in their natal group. Hence, alliances among females may be a cause rather than a consequence of philopatry and high coefficients of relatedness (see Goldberg & Wrangham, 1997, for similar conclusions about chimpanzees (*Pan troglodytes*); cf Gouzoules & Gouzoules, 1987).

The classic savannah baboon system described above

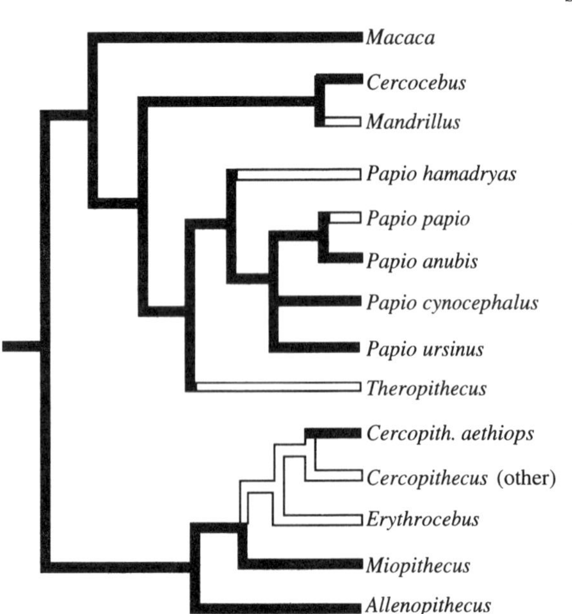

	System	Group size	Sexual swellings	Testis size	Sexual dimorphism	Substrate
Macaca	MM	27.7	✔	1.7	1.5	2
Cercocebus	MM	23.9	✔	0.7	1.7	2
Mandrillus	S/F-F	17.9/?	✔	1.0	2.1	3
Papio hamadryas	S/F-F	7.3/68	✔	0.3	1.8	3
Papio papio	S/F-F	?/?	✔	1.6	--	3
Papio anubis	MM	44.7	✔	1.5	1.9	3
Papio cynocephalus	MM	61.5	✔	1.0	1.9	3
Papio ursinus	MM	63	✔	1.1	1.8	3
Theropithecus	S/F-F	10/145	*	-0.3	1.7	3
Cercopith. aethiops	MM	24.2	✖	0.5	1.4	2
Cercopithecus (other)	SM	14.4	✖	-0.9	1.5	1
Erythrocebus	SM	28.1	✖	-0.5	1.9	3
Miopithecus	MM	112	✔	0.1	1.2	1
Allenopithecus	MM	40	✔	0.8	1.8	1

Fig. 9.1. Evolution of mating systems and associated traits in cercopithecine primates. Assuming that the ancestral condition was a multi-male group structure, the system of one-male groups within fission–fusion bands must have evolved four times within the papionines. Phylogeny based on Purvis (1995). Mating system: MM = multi-male, SM = single-male, S/F–F = single-male units within fission–fusion bands. Presence/absence of sexual swellings in females from Hrdy & Whitten (1987). Social group sizes were taken from Smuts *et al.* (1987) and Kappeler & Heymann (1996); the two figures given for fission–fusion species are one-male unit size (the smaller figures) and band size. Testis size measurements were taken from Harcourt *et al.* (1995) and corrected for body size by regressing testis volumes of haplorhine primates on female body weight, and taking residuals. The computed residuals were very similar, whether derived from an interspecific regression slope or from a slope based on phylogenetically independent contrasts (Purvis & Rambaut, 1994), and the former are presented here. Sexual dimorphism is the ratio of male to female body weight (data from Plavcan & van Schaik, 1997; Smith & Jungers, 1997). While this measure is positively correlated with mean body weight across primates (larger primates tend to be more dimorphic), it has been argued that this allometry reflects increased male–male competition among large-bodied species, rather than an allometric constraint (Mitani *et al.*, 1996b; Plavcan & van Schaik, 1997). In fact, it makes little difference to the overall picture here if ratios are replaced by allometric residuals. Substrate is ranked on a scale from 1 (most arboreal) to 3 (most terrestrial).

contrasts with that for the hamadryas baboon, which inhabits generally more arid areas of North Africa. Hamadryas social structure consists of small, single-male units or harems which comprise the building blocks of a multi-level, fission–fusion social system. One-male units aggregate within bands that have stable membership. In turn, bands aggregate to form large 'troops', but these do not have stable membership, and may be essentially the same phenomenon as the aggregations of savannah baboon groups found at sleeping sites. An intermediate level of organization in hamadryas baboons, between the one-male unit and the and, is the clan, consisting of a group of units within which les have social bonds. Hamadryas baboons are thought of as unusual particularly in the fact that females transfer between one-male units, and, unlike savannah baboons, do not have extended grooming and alliance networks. They do, however, groom the adult male in their unit, and may compete for social access to him.

The social system of the Guinea baboon is relatively little known, but what evidence exists points to a multi-level, fission–fusion organization reminiscent of hamadryas baboons. Very large groups have been observed (up to 250), apparently containing small units led by a single adult male (Byrne, personal communication). In captivity, Guinea baboon females seem to show a similar social focus on adult males as seen in hamadryas females (Boese, 1975).

Small, usually single-male, units that forage alone and do not aggregate into large bands have been reported from certain populations of hamadryas baboons (in Saudi Arabia: Kummer *et al.*, 1985; but see also Biquand *et al.*, 1992) and of chacma baboons (in the Drakensberg of South Africa: Whiten *et al.*, 1987).

It has been suggested that Old World monkeys display 'a marked uniformity in patterns of social organization,' based on a phylogenetic analysis of variation in key social traits in primates (DiFiore & Rendall, 1994). While this may be true at a relatively crude level of analysis, the brief review above suggests it neglects more subtle variation. The analysis of DiFiore and Rendall was carried out at the genus level, consequently classifying all *Papio* species together. Hence, the conclusions of broad brush phylogenetic analyses are not necessarily in conflict with the assumption that social organization has diversified in response to ecological factors.

Homologies in the savannah and hamadryas baboon social systems

In the past, emphasis was placed on the differences between the social systems of savannah and hamadryas baboons (see Stammbach, 1987). Dunbar (1988), however, has pointed out that it is easy to derive the one-male units of hamadryas baboons from the special relationships found in savannah baboons, and that hamadryas 'bands' are probably the demographic and ecological equivalents of multi-male groups. These are crucial insights for interpreting the socioecology of baboons. The identification of homology at each level allows the assembly of a modular theory dealing separately with each component of the social system, instead of seeking a unitary and over-simplistic explanation for the system as a whole. Hence, instead of asking why savannah baboons live in large multi-male groups whereas hamadryas live in small one-male groups within bands, we can ask:

(a) Are hamadryas bands and savannah groups demographic and ecological equivalents?
(b) If so, why are hamadryas bands less cohesive, and segregated into discrete one-male units?
(c) Why do female hamadryas not form strong bonds and extended alliance networks as in multi-male groups?
(d) Are the bonds formed between some male hamadryas within a band simply more stable versions of the coalitions formed by males within multi-male groups?
(e) More generally, how do male and female strategies influence variation at each social level?

Dunbar (1988) suggested that hamadryas bands are just the size expected for a conventional *Papio* multi-male group in a habitat with the same rainfall. Following this, Barton *et al.* (1996) found that hamadryas bands and savannah multi-male groups do not differ significantly in size. Furthermore, while female hamadryas transfer between units, they generally remain within the same band (Sigg *et al.*, 1982; Dunbar, 1988). Close reading of the literature reveals a property of hamadryas social organization often neglected: subtle relationships among band females, which seem to play a role in clan structure and the composition of one-male units (Kummer, 1984; Colmenares, 1992; Zinner, personal communication). There is thus some reason to believe that bands are effectively multi-male groups, with less cohesion among females, and perhaps more cohesion between females and males, than is the case for savannah baboons. The presumed correspondence between male–female friendships and one-male units on one hand, and between multi-male groups and bands on the other, is highlighted in Fig. 9.2. Assuming the ancestral condition to have been multi-male groups (see Fig. 9.1 and discussion below), the shift to a hamadryas-type system could occur simply by a weakening of female–female bonds. Such a shift would be likely to have feedback effects on the evolution of male behavior.

Parallel social evolution in papionines?

Social structures which have been described as 'analogous' to those of the hamadryas baboons (Kawai *et al.*, 1983; Stammbach, 1987) are found in the gelada (*Theropithecus gelada*) and in mandrills (*Mandrillus sphynx*) and drills (*Mandrillus leucophaeus*). Here again we see small, one-male units assembled within larger bands. The differences are that female–female bonds within units are stronger in these species than in hamadryas baboons, females do not (in gelada at least) routinely transfer between units, male aggression against females plays a minor role in establishing and maintaining units, and bands appear to be less cohesive (Kawai *et al.*, 1983; Dunbar, 1984, 1988; Stammbach, 1987).

The reconstruction in Fig. 9.1 of the evolution of group structure suggests that multi-male grouping represents the ancestral condition, for cercopithecines generally, and for papionines and macaques in particular. For the whole clade, this assumption is slightly more parsimonious than assuming that one-male groups were ancestral (five versus seven evolutionary transitions in group structure required). For the papionine/macaque clade alone, parsimony does not clearly distinguish between the two possibilities, and uncertainty

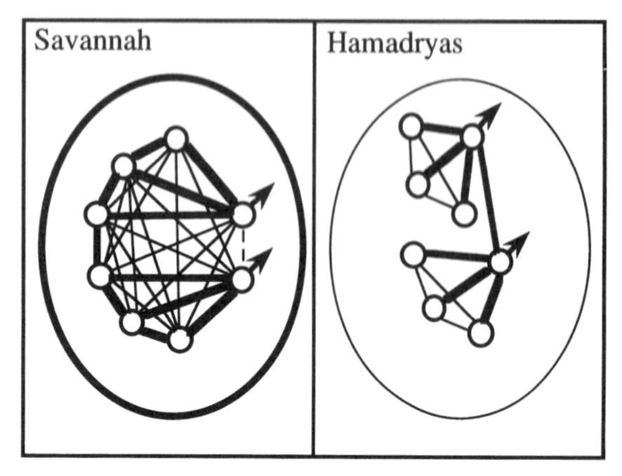

Fig. 9.2. Diagram of social structure in hamadryas and savannah baboons. Females are represented by circles, males by the standard symbol. The thickness of lines connecting individuals indicates the relative strength of social bonds, but the use of similar lines for inter-sexual and intra-sexual bonds is not meant to imply that bonds are qualitatively similar. The thick boundary around the savannah group indicates greater cohesiveness, particularly of females, than is the case for hamadryas bands. Male savannah baboons have transient coalitionary relationships, whereas some male hamadryas baboons within a band have more stable and more affiliative bonds, perhaps partly based on kinship. Following Dunbar (1988), it is suggested that the following social groupings in savannah and hamadryas baboons, respectively, are homologous: multi-male groups and hamadryas bands; male–female special relationships and one-male units; male–male coalitions and 'clans' (Kummer, 1984).

about how to resolve the *Papio* tree should in any case preclude firm conclusions. However, evidence from reproductive morphology suggests that *all* of these species, with the probable exception of the gelada, either have an essentially multi-male mating system, or did so in the recent evolutionary past. All exhibit female sexual swellings, a characteristic closely linked to a multi-male system across primates generally (Clutton-Brock & Harvey, 1976; Hrdy & Whitten, 1987; Sillén-Tullberg & Møller, 1993). Also like other multi-male species (Harcourt *et al.*, 1995), relative testis size is comparatively large. It is true that available estimates indicate smaller testes in hamadryas baboons than in other members of the genus, but hamadryas testes are substantially larger than in gelada and in the one-male guenons (*Cercopithecus* and *Erythrocebus*), indicating perhaps moderate levels of sperm competition. Almost nothing is known about paternity in hamadryas baboons, but it is known that one-male units often have follower males (either young males that may later take over the unit, or deposed leaders). Although the dominant male generally prevents other males from gaining access to the females, there is little justification for the assumption that his control is absolute.

If the reconstruction in Fig. 9.1 is correct, then the fission–fusion system comprised of stable one-male units within larger bands has evolved several times independently among the papionines. Consequently, its multiple appearance in these taxa can be regarded as parallelism. In that sense, the hamadryas, gelada and mandrill systems are indeed, as previously suggested, analogous. It may nevertheless be true that the behaviors giving rise to these parallels are essentially homologous. That is, species of this clade have inherited from their common ancestor behavioral tendencies predisposing them to develop such systems, and ecological conditions dictate whether or not they appear. The only other taxon in which a similar system is known to occur is the snub-nosed leaf monkeys (genus *Rhinopithecus*) of Asia. Both papionines and snub-nosed leaf monkeys exhibit terrestrial habits and extreme sexual dimorphism. The possible role of these factors is explored below.

A SOCIOECOLOGICAL MODEL

Identifying homology and parallelism is the starting point for developing a socioecological model. The next step is to explain how ecology acts on homologous behavioral tendencies to produce divergence and parallelism at each level of social structure. I start with group size and its possible relation to mating system.

Group size in baboons

A consensus is emerging that much group size variation among and within primate species is at least partly related to the extent and type of predation risk (e.g., Alexander, 1974; van Schaik & van Hooff, 1983; Dunbar, 1988; Isbell, 1994; but see Cheney & Wrangham, 1987; Sterck *et al.*, 1997). Large groups provide protection by reducing the individual's probability of being taken during any single predation event (Hamilton, 1971), through mobbing of predators (Cheney & Wrangham, 1987; Cowlishaw, 1994), and through greater or more efficient vigilance (Powell, 1974; Caraco, 1979; van Schaik *et al.*, 1983; de Ruiter, 1986). In many parts of Africa, the terrestrial habits of baboons make them vulnerable to attack by large carnivores such as leopards, lions and hyenas (Cowlishaw, 1994), and baboons use refuges (trees or cliffs) both at night and during the day to

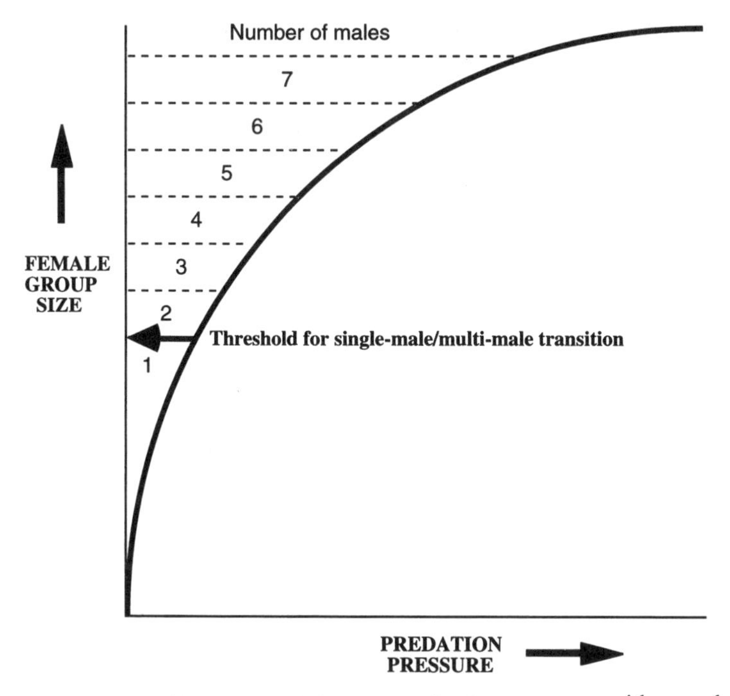

Fig. 9.3. Relationships among predation pressure, female group size, and the number of males.

reduce the risk of predation (Cowlishaw, 1997b). Phylogenetic comparative analysis reveals an evolutionary correlation between large group size and terrestriality in primates.[2] Where hamadryas baboons are not subject to high predation risk, their one-male units tend to forage alone instead of congregating within bands (Kummer *et al.*, 1985). Thus, large group size in baboons, whether in the form of cohesive multi-male groups or fission–fusion bands, may be a response to the increased risk and/or different nature of predation on the ground. (See Wrangham, 1980, for an alternative view, and Barton & Whiten, 1993, and Cowlishaw, 1995, for arguments against that alternative view as applied to baboons, and Sterck *et al.*, 1997, for more general objections.)

Relationship between group size and mating system

The number of males in primate groups may be simply determined by the number of females (Andelman, 1986; Altmann, 1990). Because a single adult male can monopolize only a few females, where females form large groups (because of predation, for example), they will be associated with several adult males. A transition between single-male and multi-male mating systems will occur at the threshold number of females monopolizable by a single male (Fig. 9.3). Note that, in the case of between-species variation, this transition is not simply a trivial demographic issue, because single-male and multi-male mating systems are associated with different morphological, physiological and behavioral specializations (see above).

In principle, therefore, one way to explain variation in the number of males is to invoke the effect depicted in Fig. 9.3: predation, in combination with other factors, dictates group size, which in turn dictates the number of males. In practice, however, the group size effect does not explain all the variation in the number of males. It may explain much intraspecific variation (e.g., Crockett & Eisenberg, 1987), and some of the interspecific variation (Altmann, 1990; Mitani *et al.*, 1996a). Other factors are also at work, however. Among primates generally, mating system is not correlated with terrestrial habits as group size is,[3] indicating that mating system and group size are not closely linked. Among colobines, at least some of the variation in the number of breeding males is directly related to predation risk, rather than indirectly through differences in group size (van Schaik & Hörstermann, 1994). Cords (Chapter 8) shows instances of variation in mating systems among guenons that are not explained by differences in the number of females. More relevant to the present chapter, additional factors would seem to be required to explain the details of social variation in baboons. In particular, why are the bands of hamadryas (and other species) substructured into one-male units, and what prevents this happening in the more cohesive multi-male groups?

The role of male–female bonds and sexual dimorphism

A number of authors have drawn attention to the unusual importance of bonds between males and females in the social organization of baboons (Smuts, 1985; Strum, 1987; Dunbar, 1988; Byrne *et al.*, 1990). These bonds are probably based on two behavioral tendencies: the tendency of males, in competition with one another, to form exclusive mating relationships with females, and the tendency of females to seek protection from harassment and infanticide by other males (Wrangham, 1979; Smuts, 1985; Dunbar, 1988; van Schaik & Dunbar, 1990; van Hooff & van Schaik, 1992; Barton *et al.*, 1996). The importance of males to females may be related to sexual dimorphism, which is extreme among papionines (see Fig. 9.1). Adult males are up to twice the size of females, and are thus both a particular threat, and particularly useful as allies and protectors. Females often use males as agonistic buffers (Smuts, 1985; Strum, 1987), or line up in their 'attack shadow' (Stammbach, 1987). Both observations and experiments indicate that friendships benefit females by protecting against infanticide (Palombit *et al.*, 1997). Males use a mixture of coercion and affiliation to persuade females to associate with them, with the balance differing between species; coercion, including 'herding' behavior, is more characteristic of hamadryas baboons than of savannah baboons (although herding is in the behavioral repertoire of all male baboons), and hamadryas males may 'abduct' females from other units (Kummer, 1984; Stammbach, 1987). It is important to note, however, that female decisions probably play a crucial role in determining the outcome of a male's strategies (Kummer, 1984; Colmenares, 1992).

In hamadryas and perhaps Guinea baboons, male–female bonds are manifested in discrete one-male units, whereas in savannah baboons they are not. Why?

The role and ecology of female–female bonds

Barton *et al.* (1996) argued that the critical difference in the social organization of hamadryas and savannah baboons is the strength of bonds between females. It is noted above that female decisions are likely to be important in determining the effectiveness of males' attempts to form exclusive mating associations, and this may apply to between-population comparisons as well as to the reproductive careers of individuals within a population. Several studies have documented extended alliance networks and grooming relationships among female savannah baboons (e.g., Seyfarth, 1976; Melnick & Pearl, 1987). These correlate with the presence of linear dominance hierarchies that have considerable long-term stability (Hausfater *et al.*, 1982; Barton & Whiten, 1993). Barton *et al.* (1996) suggested that female sociality acts as a constraint on the options for males: where females form extended alliances and grooming networks, males cannot partition groups into independent one-male units. Essentially, there are two forces at play: the tendency of females to form bonds that make groups cohesive, and the tendency of males to partition these groups into exclusive breeding units. The relative strength of the two forces, one centripetal and one centrifugal, determines the outcome in terms of social structure.

This raises the question of what determines female sociality. The characteristics of female sociality described above for savannah baboons are found in many Old World monkeys (as well as some other primates) and have been attributed to contest competition for food, either intergroup (Wrangham, 1980) or intragroup (van Schaik, 1989). The despotic behavior and steep linear dominance hierarchies commonly found among female cercopithecines suggest that intragroup contest competition is certainly one factor in female sociality, and a number of studies of female-bonded primates, including savannah baboons, have now shown that high rank among females confers foraging advantages (e.g., Janson, 1985; van Noordwijk & van Schaik, 1987; Janson & van Schaik, 1988; Barton & Whiten, 1993). Two comparative tests, one on squirrel monkeys (Mitchell *et al.*, 1991), and one on baboons (Barton *et al.*, 1996), support a link between intragroup contest for food and female bonding.

In the baboon study (Barton *et al.*, 1996; see also Byrne *et al.*, 1987, 1990; Whiten *et al.*, 1987; Barton & Whiten, 1993; Henzi *et al.*, 1997a, 1997b), we compared a large group of olive baboons, in a broadly typical savannah baboon habitat, with two small groups of chacma baboons living in a subalpine habitat in the Drakensberg of South Africa. The olive baboons conformed to the label 'female-bonded' (Wrangham, 1980; van Schaik, 1989): adult females were philopatric, showed extensive affiliative behavior toward one another, devoted most of their grooming to each other, formed coalitions relatively frequently, and had a linear dominance hierarchy that correlated with foraging success. High-ranking females monopolized food such as flowering *Acacias*, and the three top-ranking females had food intakes 30% greater than the three bottom-ranking females (Barton & Whiten, 1993). The single case of female transfer observed over a ten-year period involved the lowest-ranking female transferring into a smaller group where, although still bottom-ranking, there were fewer females dominant to her. Group fissions appeared to be linked to female–female aggression during times of food stress.

In contrast, contest competition for food was virtually absent from the South African groups. Even allowing for the difference in group size, feeding supplant rates in the savannah group were five times greater than the pooled rates for *all* types of supplant among the South African females. The South African females exhibited very low frequencies and less diversity of affiliative interactions, and did not form coalitions. Although they, like the Kenyan females, do not regularly transfer between groups (Henzi, personal communication), they do frequently leave with an adult male to form a new one-male group, as in hamadryas baboons (Henzi & Lycett, 1993; Henzi *et al.*, 1997a). In contrast to the situation at Laikipia, fission in mountain baboons is apparently unrelated to feeding competition (Henzi & Lycett, 1993; Henzi *et al.*, 1997b). Instead, fission is actively initiated by adult males, resulting in eight out of 11 cases in the formation of one-male units (Henzi & Lycett, 1993). During encounters between one-male units, males attacked and herded their own females (Byrne *et al.*, 1987), suggesting that male reproductive strategies are an important force maintaining these groups. Thus, the small, frequently one-male groups in the South African population may be maintained by the same general mechanisms maintaining hamadryas units. The lack of cohesion among the females in both cases shifts the social balance in favor of the centrifugal force (male-initiated breakaways). This does not deny that there are some differences. Not all groups in the South African population are single male (Henzi & Lycett, 1995), for example, and hamadryas males appear to be behaviorally more specialized to maintain such units (Kummer, 1984; Colmenares, 1992).

The interaction between female–female and female–male bonds can thus be summarized as follows. All baboon groups are to some extent 'cross-sex bonded' (Byrne *et al.*, 1990), in that strong male–female bonds are observed. Only some, however, are in addition female-bonded, where this term refers to groups in which females are long-term residents and form strongly differentiated agonistic, affiliative and coalitionary relationships with one another (Wrangham, 1980; van Schaik, 1989). Where such female–female bonds are weak or lacking, individual males can exploit the situation by hiving off their own group of females into small units. An important assumption is therefore that the categories 'female-bonded' and 'cross-sex bonded' are not mutually exclusive.

Predation revisited

An obvious difference between the South African baboons just described and the typical hamadryas system is that the small groups of the former did not coalesce into bands. Barton *et al.* (1996) attributed this difference to a lack of predation at the South African site. Elsewhere in the Drakensberg, the cohesion of bands correlates with the presence of leopards (Anderson, 1981, 1992). As noted above, where hamadryas baboons are not subject to predation, the band structure tends to disintegrate, and one-male units more often forage alone (Kummer *et al.*, 1985). Most savannah baboon populations have some tendency for temporary subgrouping during foraging (Hamilton & Bulger, 1992), and the extent of this may be related to predation pressure. It might therefore be argued that the hamadryas system occurs simply under moderate predation pressure, leading to reduced cohesion, exploitable by males, regardless of levels of feeding competition among females. This is possible, but there is no evidence for it. In fact, Guinea baboon populations with a similar organization appear to be under intense predation pressure (Byrne, 1981; Dunbar, 1988). The idea also fails to explain populations of multimale groups that are intermediate in size rather than in cohesion (Melnick & Pearl, 1987; Dunbar, 1992).

Food availability and maximum tolerable group size

Predation pressure may set a lower limit on group size, but something else must set an upper limit (Dunbar, 1992). In principle, increased food availability should permit larger group size and higher population density. Dunbar (1992) introduced the idea of a maximum ecologically tolerable group size in baboons, determined by food availability (assayed by rainfall), above which the required increases in foraging time could not be absorbed by time budgets. Foraging travel distance is a potential confounding variable: the response to greater food availability might be reduced travel distance rather than increased group size, hence saving time and energy. Indeed, populations in low biomass habitats tend to range further (Whiten *et al.*, 1987; Dunbar, 1992; Barton *et al.*, 1992). Nevertheless, there must be an upper limit to daily travel distance, dictated by attainable average speed, and hence a limit to the extent to which increased ranging can compensate for lower food availability. No population of baboons has a recorded mean daily travel distance exceeding 9.0 km (Barton *et al.*, 1992; Dunbar, 1992), and interpopulation variance in group size is positively correlated with rainfall,[4] a proxy for food availability. Intrapopulation comparisons of baboons also indicate a positive relationship between habitat quality and group size (Dunbar, 1988, p. 135). Finally, provisioned populations tend to have larger group sizes (for example, the rhesus macaques, *Macaca mulatta*, of Cayo Santiago).

Synthesis

In summary, the following factors, two universal (a and b) and three variable (c–e) appear to influence social structure and the number of males in baboon groups.

(a) *Sexual dimorphism*: the relatively large size and strength of male papionines render them socially important to females.
(b) *Mating strategies of males*: males have a universal tendency to form relationships with females that exclude other males.
(c) *The cohesion and size of female alliances*, which are related to intragroup contest competition, restrict male strategies. Large networks based on strong bonds among females make segregation into discrete one-male units impossible or uneconomic for males.
(d) *Predation pressure* dictates how large groups or bands must be.
(e) *Food availability* dictates how large groups or bands can be.

The interaction of these factors is illustrated in Fig. 9.4. Female group size increases with increasing predation risk, and the number of males tracks the number of females. The way these larger groups are organized depends on the potential for within-group contest competition (after van Schaik, 1989). Where such competition occurs, females are cohesive and form alliance networks, resulting in the classic female bonded, multi-male group. Where such competition does not occur, females do not form extensive alliance networks, and males are consequently able to segregate small units within the group. Populations may occupy the space above the curved plane in Fig. 9.4 (i.e., groups can be larger, consequently with more males, than the minimum required for effective defense against predators), but only to the extent that food biomass allows.

Although this model is apparently consistent both internally and with what information is currently available, further data from the field will be required to test it properly. These tests will have to take into account the fact that some variables, such as group size and sex ratio, vary markedly within populations, from group to group and over time (e.g., Bronikowski & Altmann, 1996). The significance of population differences in mean values for such variables will not be apparent until large samples have been collected for representative populations, and analyzed appropriately (for example, using analysis of variance to determine whether population means truly differ). Very little is known about the distribution or monopolizability of food patches in the various populations. In addition, detailed study of intrapopulation variability may shed more light on the way that ecological conditions set the stage for larger-scale, persistent differences between populations. Intra-population variability *per se* should not therefore be regarded as contrary to an ecological model.

Extension to other species

As in hamadryas baboons, gelada social structure is based on one-male units within larger bands. Unlike hamadryas baboons, gelada are said to be female bonded: they groom and form coalitions together within units, and do not generally transfer between units. Furthermore, there is very little coercion by males, with the formation and maintenance of units based on affiliation among females and between females and males.

For these reasons, it has often been stated that the similarities between the social systems of hamadryas baboons and gelada are 'superficial' (e.g., Kawai *et al.*, 1983; Stammbach, 1987). One could, however, assert the opposite: that the structural similarities are fundamental and the behavioral differences are superficial. Part of the problem here is the label 'female-bonded'. Although gelada females have grooming and coalitionary bonds within units, their networks are small, and based largely on kin ties (Dunbar, 1984). This is because, as with hamadryas baboons – and chacma baboons in the Drakensberg – females are willing to leave with males to form new units. Hence, in both hamadryas and gelada, the

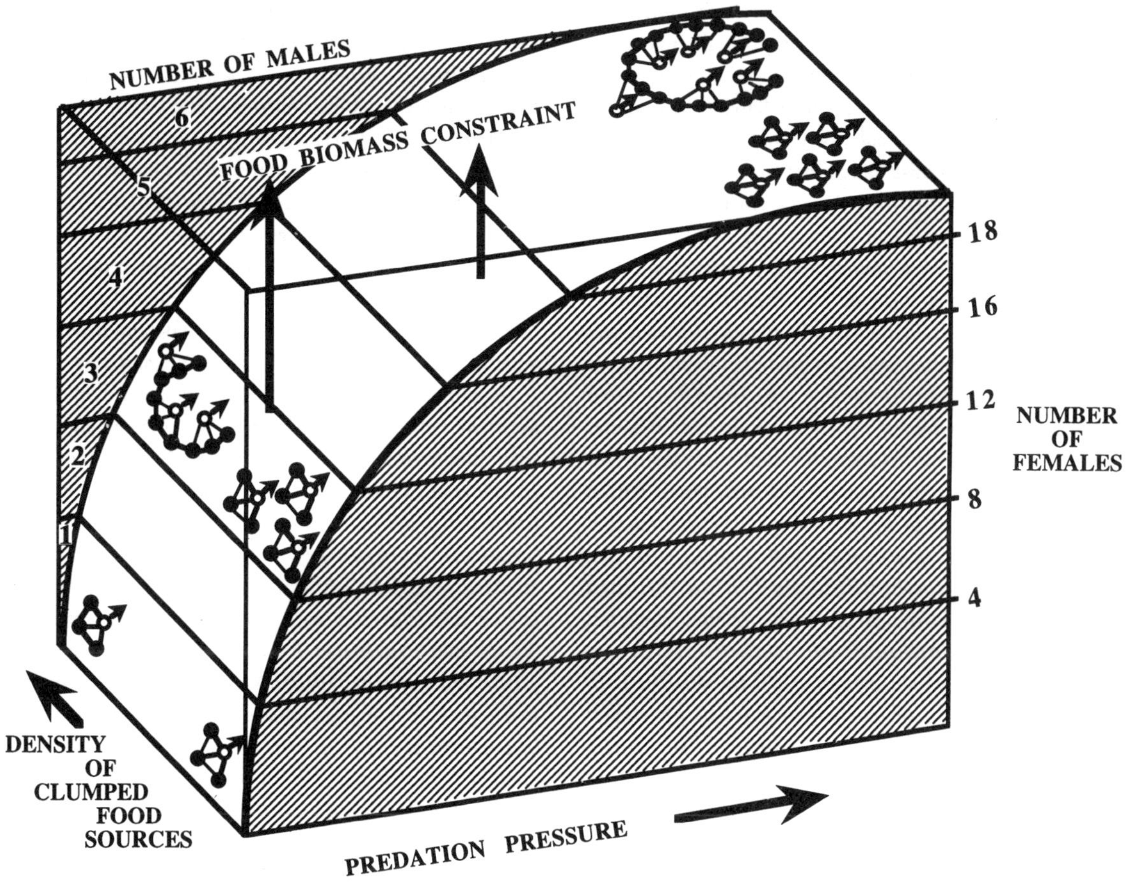

Fig. 9.4. Diagrammatic socioecological model of baboons. This is an elaboration of the model presented in Barton *et al.* (1996), differing in treating variables as continuous rather than categorical, and in recognizing a possible constraint on group size exerted by food availability. As in Fig. 9.3, female group size tends to increase with predation pressure, in turn determining the number of males. Another dimension is introduced by considering the effects of the ecological potential for within-group contest competition for food (here assayed by the density of clumped food sources). This determines the structure of the larger groups found at higher predation pressure; where females compete for food, they form linear dominance hierarchies and alliance networks, thereby constraining the tendency of males to form exclusive breeding units, and resulting in the classic female-bonded multimale groups of savannah baboons (and macaques). Where females do not compete for food, their bonds with each other are weak, and males are able to partition groups into stable units (the classic hamadryas system). The curved plane represents the minimum tolerable group or band size (determined by predation pressure), taking into account the number of males that accompany the females. Group size can exceed this minimum (and often should, given the additional antipredator benefits) to the extent that this is allowed by food availability. Hence, the model assumes that food biomass puts a ceiling on group size (similar to Dunbar's, 1992, 'maximum ecologically tolerable group size'), while predation sets the floor. It is therefore predicted that once range size is controlled for, the size of groups and bands is determined jointly by predation pressure and food availability. At very low predation pressure, the food availability constraint is unlikely to be an issue. As predation pressure increases, the floor on group size it sets becomes closer to the ceiling set by food availability, leaving less room for group size variability. This suggests that intra-population group size variability should be more constrained in high predation/low biomass habitats.

lack of interest among females in forming the extended networks characteristic of multi-male groups facilitates the formation of small units, whether by coercion, affiliation, or a mixture of the two. Conversely, female–female relationships are not non-existent in hamadryas. To some extent, bands in both species contain 'distributed matrilines' (Dunbar, personal communication), with female–female kinship and affiliation playing a role in determining what units are formed (Dunbar, 1984; Kummer, 1984; Colmenares, 1992). These considerations, together with the reconstruction of mating systems in Fig. 9.1, support Dunbar's (1988) suggestion that gelada one-male units evolved by condensing out of multi-male groups. Small testis size, however, indicates that the gelada one-male mating system is stricter than in hamadryas and mandrills, or has a longer evolutionary history, or both. This still leaves the problem of female bonds within gelada units, but this problem is solved if we assume that gelada females contest access to small feeding sites that cannot accommodate many individuals. This would create a pressure for the establishment of dominance hierarchies, but severely limit the size of coalitions contesting access. A candidate for a small but valuable resource is the rhizome-containing holes dug by individuals at considerable effort, and from which subordinates are frequently supplanted (Wrangham, 1977). In effect, gelada may fall somewhere between the two extremes of contest competition represented at the top of Fig. 9.4.

The other papionines that form bands containing one-male units, the mandrills, drills and Guinea baboons, are presumed to fit the same model, although little is known about the details of their social organization and behavioral ecology. There is evidence for mandrills that their version of the system may be gelada-like, rather than hamadryas-like: in a study in captivity, females devoted most of their grooming to each other, and relatively little to the male (Stammbach, 1987). The same has been found for one of the snub-nosed monkey species (*Rhinopithecus bieti*), which have also evolved a one-male unit structure within bands (Kirkpatrick *et al.*, 1998). Like the papionines, *Rhinopithecus* species exhibit extreme sexual dimorphism, whereas their close relative the douc langur (*Pygathrix nemaeus*) is much less dimorphic and lives in multi-male groups (Kirkpatrick, 1997). The latter comparison hints again at a link between dimorphism, male–female relationships and a predisposition to form one-male units. Wrangham (1979) previously made a similar argument for the evolution of one-male groups in gorillas, another highly dimorphic species. A prediction is that a lack of dimorphism should lead to a more peripheral role for males; in squirrel monkeys and talapoins (*Miopithecus*), phylogenetically distant taxa that both live in large multi-male groups and lack dimorphism, males and females interact very little outside of the mating season, and tend to travel in distinct subgroups (Melnick & Pearl, 1987). In talapoins, females are known to be dominant to males (Rowell & Dixson, 1975), and in squirrel monkeys, female coalitions often defeat males (Robinson & Janson, 1987). Lack of dimorphism and the consequently reduced influence of male aggression on the partitioning of groups may help to explain squirrel monkey and talapoin group sizes, which are unusually large for arboreal primates.

Is dimorphism a cause or a consequence of mating systems? Polygynous species tend to be more dimorphic than monogamous species (Clutton-Brock *et al.*, 1977; Mitani *et al.*, 1996b), and this is usually interpreted as evidence that male–male competition in polygynous systems has selected for large male size. Perhaps the particularly high levels of sexual dimorphism in the one-male groups discussed reflect greater male–male aggressive competition. This would be arguable if the association between dimorphism and one-male groups were consistent, but the evidence presented suggests that dimorphism only *predisposes* to one-male groups: the multi-male savannah baboons are slightly more dimorphic than hamadryas and gelada. Among primates generally, there is no significant association between weight dimorphism and multi-male versus one-male groups.[5] According to the model, this is because it is only where a lack of female sociality allows that dimorphism and male influence lead to one-male groups. A general prediction is that male–female bonds and female–female competition for access to males are correlated with dimorphism.

Differences in male behavior as responses to social constraints

The behavioral tendencies of males that lead to the formation of one-male units appear to be present or latent in males of all baboon populations, and savannah males herd females and form units opportunistically under a variety of circumstances (Hamilton & Bulger, 1992). Such units, however, tend to be temporary. For example, in the Kenyan savannah baboons, we have seen small subgroups form when a male breaks away from the main group with an estrous female, under pressure from follower males. These subgroups can become detached from the main group for minutes, hours or even days, but they invariably return to the main group, apparently under the impetus of female–female bonds.

Subgrouping seems to be more frequent in an area where predators are rare, supporting the proposed role of predation discussed above. Herding is not seen frequently in typical savannah groups, but appears under appropriate conditions, when opportunities to form one-male units arise (Byrne *et al.*, 1987; Hamilton & Bulger, 1992).

These 'facultative responses to social circumstances' (Hamilton & Bulger, 1992) serve to emphasize that male baboons have the behavioral flexibility to optimize reproductive strategies under changing conditions; they will herd females and prevent other males from joining the group wherever possible, either to maximize mating opportunities or to protect against infanticide. While this sort of facultative behavioral response is undoubtedly the proximate cause of variation in social structure and can occur at least transiently in a variety of circumstances, it is ecological conditions that set the stage for persistent differences between populations.

It seems that where female bonds are strong, coercion is an ineffective strategy. The presence of bonds between certain male hamadryas within the band may similarly be interpreted as the development of a behavioral pattern common to males of all *Papio* species. Coalitions are formed between males in multi-male groups, but these are transient phenomena related to competition for single estrous females (e.g., Noë, 1992). For hamadryas males, however, the long-term maintenance of a unit must place a premium on stable coalitions.

The suggestion that the strength of female bonds determines the efficacy, and hence occurrence, of male coercion reverses the causal link envisaged by Brereton (1995). Brereton argued that it is male coercion that causes females to bond and form defensive alliances against males. This idea, however, fails to explain the variation in female bonds amongst baboons, since sexual dimorphism (and hence the potential for male coercion) is high in all populations. Furthermore, the populations in which males are most coercive, such as hamadryas baboons, are those in which female bonds are weakest, not strongest. (See Sterck *et al.*, 1997, for more general objections.)

CONCLUSION

Once, it was hoped that simple correlations would permit simple inferences about the relationship between social structure and ecology (Crook & Gartlan, 1966; Clutton-Brock & Harvey, 1976). The lack of such simple correlations has led to more detailed attempts to model socioecology, based largely on the insight that male and female strategies have different goals and must be modeled as separate but co-dependent factors (Wrangham, 1979, 1980). Because different sets of factors appear to be involved in different taxonomic groups, our ability to construct general models is still seriously limited. A way forward is to use relatively narrow taxonomic groupings, such as the papionines, as model systems, for which the interactions of each component of the system can be worked out in detail. If successful, attempts can then be made to build from these specific models more general models of primate socioecology.

ACKNOWLEDGMENTS

I thank John Crook, both for effectively inventing socioecology (Crook, 1964; Crook & Gartlan, 1966), and for inspiring me through his undergraduate teaching to study it. Many discussions with Robin Dunbar, Carel van Schaik, Andy Whiten, and Dick Byrne have been crucial to the development of the ideas presented here. Dietmar Zinner, Charlie Nunn, and Peter Kappeler suggested improvements to the manuscript.

NOTES

1. Species for which this system has been reported are *Papio anubis*, *P. cynocephalus*, and *P. ursinus*.
2. The computer program CAIC (Purvis & Rambaut, 1994) was used to generate independent contrasts in group size where transitions in substrate use (terrestrial versus arboreal) have occurred. Of eight contrasts, seven showed the predicted larger group size in terrestrial compared with arboreal lineages ($t = 2.2$, one-tailed $p = 0.03$). The exception was the contrast between *Erythrocebus* and *Miopithecus*. The analysis used data from Kappeler and Heymann (1996) and Smuts *et al.* (1987), and the primate phylogeny in Purvis (1995).
3. I used the method of Read and Nee (1995) to determine whether evolutionary transitions in substrate use are associated with consistent changes in breeding structure (more males in terrestrial taxa). Out of six changes in substrate use, three were associated with more males in the terrestrial taxa, and three with fewer males in the terrestrial taxa.
4. Group size is positively correlated with rainfall across baboon populations ($t = 4.2$, df $= 13$, $p = 0.002$), using the data (log-transformed for linear regression) mentioned in Barton *et al.* (1992), and controling for day range length and home range size by multiple regression.
5. Independent contrasts analysis using CAIC as above: seven of nine contrasts indicate greater dimorphism in one-male than in multi-male taxa, $t = 0.93$, $p = 0.39$.

10 • Variation in adult sex ratios of red colobus monkey social groups: implications for interspecific comparisons

THOMAS T. STRUHSAKER

INTRODUCTION

Red colobus monkeys (*Procolobus/Colobus badius*) have one of the most variable and complex social systems of all non-human primates. Their patrilineal social organization differs from that of most primate species, but is similar to that of chimpanzees and many traditional human societies. The adult sex ratio of red colobus is female biased in their social groups and total populations, as with the majority of primates. A better understanding of red colobus social systems depends on identifying the variables that most strongly influence the number of adult females in social groups. This is because it is the female red colobus that disperse and transfer between groups. Parameters affecting group size and the size of male coalitions will, in turn, influence patterns of female dispersal.

An analysis of the variability in red colobus social systems has important implications for interspecific comparisons and our attempts to understand the relative contribution of ecological and phylogenetic variables in shaping social organization. Unless the variation within populations, species and superspecies is taken into account, interspecific comparisons based on mean values are of limited utility, if not counterproductive, in understanding evolution and behavioral plasticity.

This chapter describes the high degree of variability in adult sex ratios within specific social groups over time, between social groups of the same population, within populations over time, and between different subspecies and/or species of red colobus. It examines a number of variables that probably influence adult sex ratios in social groups and offers hypotheses of causal relationships.

Background information on social organization

Red colobus in Kibale and, apparently, most other sites live in patrilineal social groups. All females leave their natal groups to join other groups. Furthermore, parous females are readily able to transfer between groups and this can occur several times in an individual female's lifetime. Even females with young juveniles (one to two years old) occasionally transfer between social groups. These transfers appear to be immediate, without any aggression from the resident group members (Marsh, 1979b; Struhsaker & Leland, 1979, 1985; Struhsaker & Pope, 1991; T.T. Struhsaker, unpublished data).

In contrast, although about half of the males leave their natal groups, they are rarely able to join other groups. Males remaining in their natal groups form a coalition among themselves. There is a dominance hierarchy within these male coalitions which is expressed through displays and priority of access to food and space. The highest ranking male does most, but not all, of the copulations (Struhsaker, 1975; Struhsaker & Pope, 1991). These coalitions include males of all ages, and dominance ranks and mating success of individuals change over time. No male has been dominant and the chief copulator for more than about three to four years, but some males were reproductively active for nearly 12 years (Struhsaker & Pope, 1991). Coalitions of males fight one another and the outcome of these encounters may be important in determining patterns of female dispersal and intergroup transfer. In other words, the fights may serve to attract mates and in some situations may also involve competition for food. Females do not usually participate in these intergroup aggressive encounters. The Tana River red colobus (*P. b. rufomitratus*) are exceptional in that their social groups contain only one or two adult males and adult males are able to immigrate and take over social groups from incumbent males (Marsh, 1979b). All-male groups (uncommon and unstable) have been seen only in the Tana River population (Marsh, 1979b) and solitary males are rare at all sites.

Territoriality is absent in the populations of red colobus living in the Kibale Forest of Uganda (Struhsaker, 1975), the Jozani Forest and adjacent agricultural areas of Zanzibar (Siex & Struhsaker, 1999), Gombe in Tanzania (Stanford, 1999), and along the Tana River in Kenya (Marsh, 1979b). Groups often have extensive, if not complete, overlap in

Table 10.1. *Red colobus group size, sex ratio, predation pressure, and gross habitat quality*

Taxa	Location (source)	*n*	Mean group size	Mean adult females	Mean adult males	Crowned hawk-eagle	Chimps	Habitat
temminckii	Eastern Gambia (1)	2	16.5	6.5	3.0	R	A	2
temminckii	Senegal (2)	12	32.1	13.3	6.7	R	A	2
temminckii	Gambia (3)	2	26.0	10.5	2.5	R	A	2
badius	Tiwai (4)	1	33.0	13.0	7.0	R	C	1
badius	Tai (5)	1	32.0	13.0	3.0	C	C	1
badius	Tai (6)	1	72.5	28.0	17.0	C	C	1
badius	Tai (7)	1	26.5	12.0	4.0	C	C	1
tephrosceles	Kibale (8)	3	40.0	14.2	5.6	C	C	1
tephrosceles	Kibale (9)	2	61.0	26.5	11.5	C	C	1
tephrosceles	Gombe (10)	1	82.0	28.0	11.0	C	C	1
tephrosceles	Gombe (11)	2	24.0	7.8	4.5	C	C	1
gordonorurm	Magombera (12)	6	31.3	14.0	1.8	C	A	2
gordonorurm	Kalunga (13)	1	24.0	11.5	1.5	C	A	2
gordonorurm	Sonjo (14)	1	32.0	12.5	1.5	C	A	2
rufomitratus	Tana (15)	13	18.1	9.8	1.5	A	A	2
rufomitratus	Tana (16)	17	11.6	5.9	1.1	A	A	2
kirkii	Zanzibar (17)	15	26.9	10.1	2.1	A	A	2
kirkii	Zanzibar (18)	3	37.9	16.8	3.1	A	A	2

Notes:
Predators: A = absent, R = rare, C = common (see text).
Habitat: 1 = large forest blocks (> 15 km^2) and relatively aseasonal; 2 = small, fragmented blocks and/or highly seasonal, semi-deciduous, and often lower plant diversity. These latter forest blocks were often heavily disturbed by human activities.
Sources: 1 Struhsaker (1975). 2 Gatinot (1975). 3 Starin, cited in Oates (1994). 4 Davies, cited in Oates (1994). 5 Galat and Galet-Luong (1985). 6 Honer *et al.* (1997). 7 R. Noë and B. Beerlage personal communication. 8 Struhsaker (unpublished). 9 & 10 Clutton-Brock (1972). 11 Stanford (1999). 12 Decker (1994b). 13 & 14 Struhsaker (unpublished). 15 Marsh (1979b). 16 Decker (1994a). 17 Siex and Struhsaker (1999). 18 Mturi (1991).

home ranges. Exceptions exist in the heavily logged areas of Kibale and the woodland habitat of Senegal, where there is little overlap in home ranges and intergroup encounters are rare (J.P. Skorupa, personal communication; Gatinot, 1975). Temporary division of red colobus social groups into smaller foraging parties that vary in size and composition from day to day (fusion–fission) has been reported for a heavily logged area of Kibale (Skorupa, 1988) and the Jozani Forest (Siex & Struhsaker, 1999). Both of these areas have relatively low diversity of food plants. It is not known how new groups of red colobus are formed.

Among the 15–18 forms (species and/or subspecies, Kingdon, 1971; Colyn, 1991; Oates, 1996) of red colobus that are spread across Africa from Senegal to Zanzibar, social groups range in size from four to 85. Even within one forest (Kibale), group size is extremely variable (8–85). Specific social groups also change dramatically in size over time, e.g., the CW group of Kibale ranged from eight to 37 between 1970 and 1987. Group size tends to be smaller in the drier and more seasonal habitats, such as in Gambia, Senegal, and the Tana River. Larger groups predominate in rainforests and wetter woodlands, e.g., Tai, Kibale, and Gombe (Struhsaker, 1975; Struhsaker & Leland, 1979; Struhsaker, 1999; Table 10.1).

Adult sex ratios in social groups are also highly variable, ranging from 0.2 females per male to 15 females per male across Africa ($n = 84$ social groups, Table 10.1). Sex ratios differ between groups of the same population (e.g., an extreme example from Kibale: 10:1 vs. 0.2:1) and even between groups occupying completely overlapping home ranges (Ngogo, Kibale: 2.57:1 vs. 0.2:1). The sex ratio of the CW group in Kibale varied from 1.5 to ten adult females per male during the period of 1970 to 1987. In most groups, however, there were usually at least two adult females per adult male.

Determinants of group size

A recent review of this subject for red colobus (Struhsaker, 1999) concluded that group size is influenced by a complex interplay of predation pressure and habitat quality (diversity and dispersion of food). The extent of intolerance among adult males can also influence group size and this, in turn, often closely parallels phylogeny among Old World monkeys (Struhsaker & Leland, 1979). In situations where crowned hawk-eagles (*Stephanoaetus coronatus*) and chimpanzees (*Pan troglodytes*) – the two most important non-human predators of red colobus – are common, red colobus groups are often larger than where these predators are rare or absent. However, where foods are in low density, in clumped distributions, and/or species diversity is low, red colobus groups split into smaller foraging parties in spite of high predation pressure from eagles (e.g., in heavily logged parts of Kibale a social group was divided into smaller units 33% of the time: Skorupa, 1988; Struhsaker, 1999). In other words, in some habitats the advantages of large social groups to predator avoidance are apparently outweighed by foraging costs, leading to a fusion–fission society. Hunting by humans does not generally appear to select for smaller groups, as evidenced by the large red colobus groups in Tai, Cote d'Ivoire and Korup, Cameroun (Struhsaker, 1975) where some hunting of colobus by humans occurred. Once hunting by humans becomes too intense, however, it can result in smaller groups and the eventual elimination of red colobus, as it apparently did in Bia, Ghana (Struhsaker, 1997).

Determinants of male coalition size

The male coalitions in red colobus social groups appear to be composed of individuals that are more closely related to one another than they are to the males in coalitions of other groups. This high degree of tolerance among closely related males or at least those born in the same group seems to be a function of evolutionary history, i.e., phylogeny.

There are significantly fewer adult males than adult females in the social groups and total population of red colobus and this can only be attributed to differential mortality and/or sex ratio at birth. We do not know whether there was a biased sex ratio at birth because it was difficult to determine the sex of very young infants at Kibale. There was, however, a weak indication of a biased sex ratio by the age of two years. Combining 17 years of data for the CW group at Kibale gave an offspring sex ratio of 1.44 females per male at 24 months of age ($n = 42$, not statistically significant) (Struhsaker & Pope, 1991). Adult males are killed less than expected by crowned hawk-eagles (Struhsaker & Leakey, 1990) and chimpanzees (Busse, 1977; Stanford, 1999), but die from other causes more than expected by chance (Struhsaker & Leakey, 1990). These other causes of mortality are poorly documented. Adult males certainly fall much more than adult females, and this often occurs during fights between males (intragroup and intergroup) and while performing leaping-about aggressive displays. Broken fingers, as evidenced by distortion (stiff or permanently bent), are indicators of falls and these are very common among adult males, but rare, if not absent, in adult females. A herpes-like disease (inflamed mouth, lower jaw, and genital area, followed by loss of hair, and later by slow and arthritic-like locomotion, and eventually by disappearance – presumed death) in the Kibale population of red colobus was lethal to adult males, but not to adult females or juveniles. In one of the study groups (RUL at Ngogo) five of the ten adult males apparently died from this disease over a 2.5-year period.

A less well-documented variable influencing intragroup sex ratios and the size of male coalitions is the number of young males of the same age cohort that are growing up together in the same group. Young males, meaning large-sized juveniles and subadults about three to four years old, seemed to be less tolerant of one another than were adult males. This impression is based on nearly 17 years of observation of one group. When there were two to three males growing up together, they seemed to fight one another as subadults much more than they did with adult males. A single subadult male was often harassed by adult males and gave stylized presents to them, but did not engage in physical aggression (Struhsaker, 1975). When there were two to three young males growing up together, it was not uncommon for one or two of them to leave the group, apparently because of the fighting between them. In one case, the smaller of three subadult males left the group (CW) for nine months, during which time he was solitary. He returned after the other two subadult males had fought and severely wounded one another around their mouths (presumably from canine slashes). The smaller male became dominant to the other two after his return, largely, it appeared, because they had temporarily lost their ability or inclination to fight due to their wounds.

In summary, the determinants of the size of male coalitions in red colobus groups are poorly understood. Chance

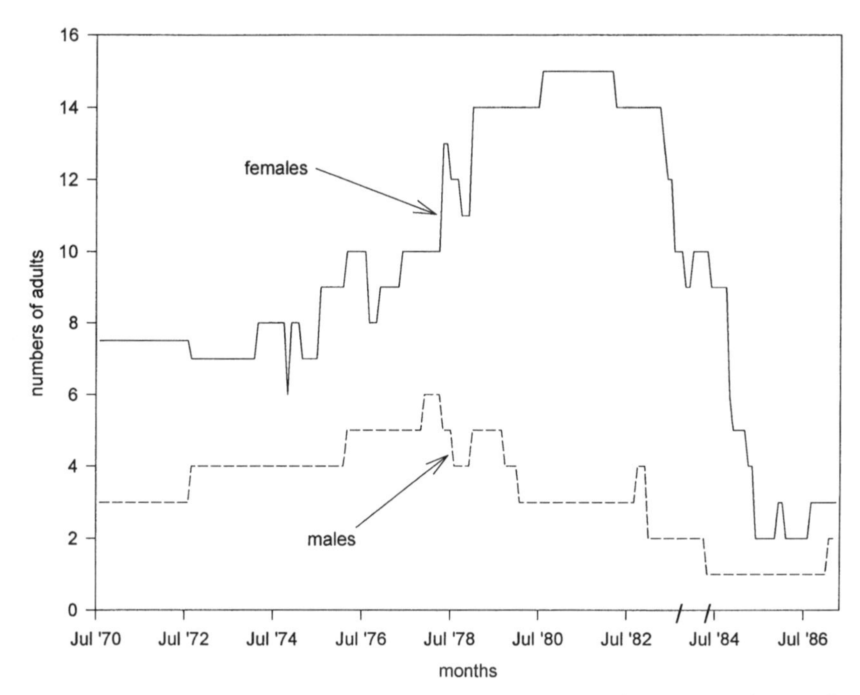

Fig. 10.1. Changes in numbers of adult females and males in the CW group of red colobus over time. Kanyawara study site, Kibale Forest, Uganda. The majority of changes in female numbers are due to dispersal, and in males to mortality.

events seem to play a major role. Fatal falls, lethal wounds, and disease are some of these events. In contrast to the Kibale and Zanzibar populations, the male red colobus on the Tana River are generally intolerant of one another in heterosexual social groups, which have only one or two adult males.

DETERMINANTS OF THE NUMBER OF ADULT FEMALES IN SOCIAL GROUPS

Variables that probably influence the number of adult females and sex ratio in red colobus groups include:

1. the number of adult males in the group;
2. predation pressure;
3. group size;
4. habitat quality; and
5. quality of adult males in the group.

Number of adult males in groups

Because adult male red colobus are generally larger, have longer canines, and more massive head musculature than adult females, one might expect them to be effective deterrents against predators. Consequently, where predators are common one might expect selection to have favored greater tolerance amongst males and therefore for there to be more males per female in social groups, i.e., lower adult sex ratios (females:males), e.g,. see van Schaik & Hörstermann (1994).

VARIATION WITHIN SPECIFIC GROUPS OVER TIME

Three social groups were studied in Kibale for long periods, ranging from four to 17 years. During this period all three groups showed dramatic changes in the number of adult males, females, and sex ratios (Figs. 10.1 to 10.4). The CW group living in the Kanyawara area of Kibale ranged in size from eight to 37, with the adult sex ratio varying between 1.5 and ten females per male. Some 10 km to the south in the Ngogo study area of Kibale, two other groups of red colobus were studied for a shorter period, but they too showed pronounced changes in size and composition. These two groups occupied the same area of forest, with completely overlapping home-ranges. The smaller group (HTL) followed the larger group (RUL), often within 50 m or less. This occurred so frequently that we referred to the HTL group as a satellite of the RUL group. RUL ranged in size from about 50 to 25 individuals until after about 5.3 years of study, when it no longer existed because of dispersal. During this time the ratio of adult females to males ranged from 1.33 to 2.57. The satellite HTL group was much more dynamic in size and composition and not typical of most red colobus

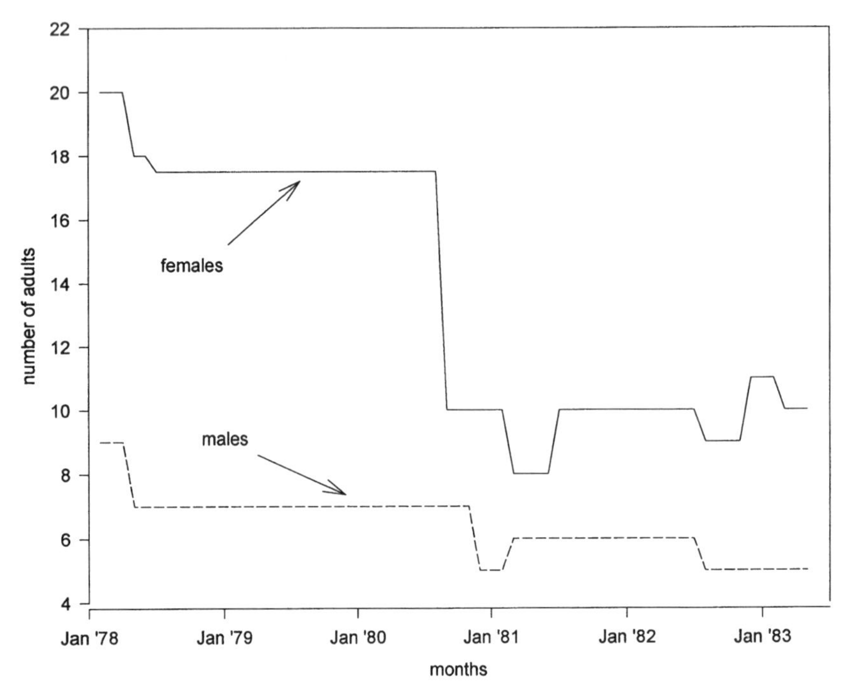

Fig. 10.2. Changes in numbers of adult females and males in the RUL group of red colobus over time. Ngogo study site, Kibale Forest, Uganda. The majority of changes in female numbers are due to dispersal, and in males to mortality.

groups in Kibale. Its mean monthly size ranged from seven to 17 individuals, and the adult sex ratio varied from 0.2 to 1.33 females per male. This is the only group I know of where adult males often outnumbered females. It broke up after four years of study.

In the CW group there was an indication that as the number of adult males increased, more females joined the group (Fig. 10.1). This was not always the case, however, because in 1979–1982 the numbers of adult females remained high and actually increased slightly in spite of a decline in the number of adult males (Fig. 10.1). The most dramatic changes occurred in 1984 and 1985, when the number of males declined to two and, eventually, only one. Emigration of adult females soon followed on a large scale and the group size decreased accordingly. The sex ratio in the CW group also declined, but there were more adult females per males when the group was small than when it was large (Fig. 10.4). The CW group broke up some time between 1988 and 1991.

The RUL group showed trends similar to those of the CW group. As numbers of adult males declined due to disease, adult females left the group and were often accompanied by their juvenile offspring. In other words, the numbers of adult females in a group tended to track the numbers of males.

In contrast, the HTL group had a very dynamic membership. The only consistent members during the four-year study were three adult males, one adult female, and one old juvenile. Adult, subadult, and old juvenile females moved in and out of the group on a regular basis. Most of these dispersing females stayed in the HTL group for only one month or less, and none for more than a year, in spite of a coalition of three to five apparently healthy adult males (Fig. 10.3). Contrary to the CW and RUL groups, the numbers of adult females in the HTL group were not dependent on the numbers of adult males.

Regression analysis of the data in Figs. 10.1 to 10.3 revealed strikingly different patterns in the relation between numbers of adult females (dependent variable) and numbers of males (independent variable because they form the stable core of the group and do not transfer) (Fig. 10.4). Even though there was a significant correlation between the numbers of adult females and adult males for all three groups ($p<0.01$ for CW and RUL groups and $p<0.05$ for HTL), only in the RUL group did the number of adult males account for an appreciable amount (65%) of the variation in numbers of adult females ($r^2=0.65$). In the CW group the number of males accounted for 17% of the variation in numbers of females and only 11% in the HTL group. Note that in Fig. 10.4 and most of the other regression

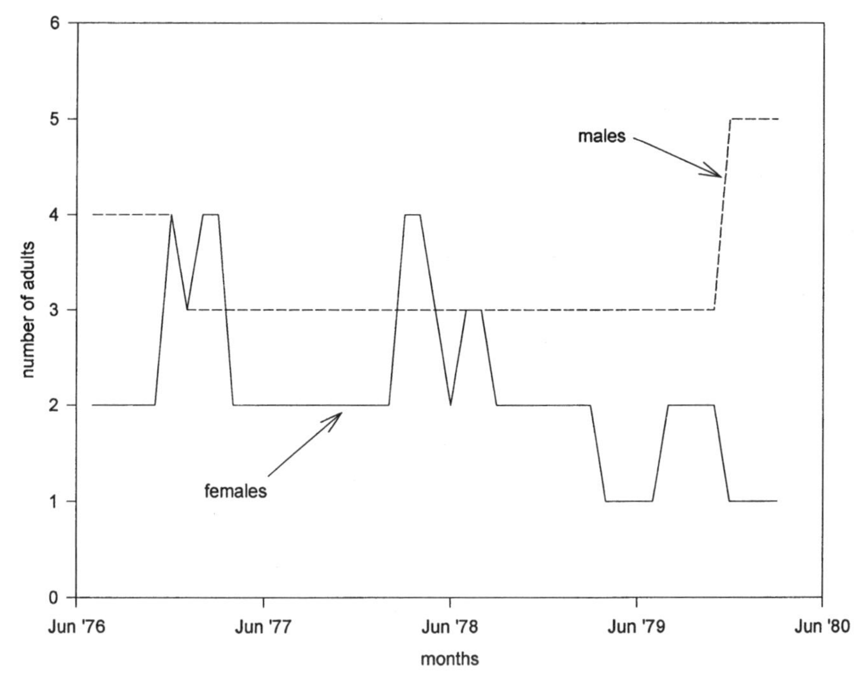

Fig. 10.3. Changes in numbers of adult females and males in the HTL group of red colobus over time. Ngogo study site, Kibale Forest, Uganda. The majority, if not all, changes in female numbers are due to dispersal, and in males to mortality and dispersal. Lysa Leland (unpublished data).

figures, the line does not pass through the zero point of the *y*-axis. In positive regressions where the line intersects the *y*-axis above zero, there are always more females per male at the lower end of the slope than at the upper end. For example, in the CW group when there were five males in the group, the ratio of females to males was on average approximately 2:1, but when there was only one adult male, the ratio averaged approximately 6:1. This shift in sex ratio was independent of predation pressure, which remained the same.

VARIATION WITHIN POPULATIONS OVER TIME

The Tana River population of red colobus (*Procolobus badius rufomitratus*) was studied during two different periods, separated by 14 to 19 years (Marsh, 1979b; Decker, 1994a). The first study was done between 1973 and 1974 at a time when many of the forest blocks had only recently been reduced in size by shifting cultivators. Population densities of red colobus at that time were high (approximately 165 individuals/km^2 at the Mchelelo site), and it is believed that these high densities were the consequence of habitat destruction followed by population compression. Fourteen to 19 years later, group size, population density (56 individuals/km^2), and adult sex ratio (females:males) had declined significantly (Decker, 1994a; Struhsaker, 1999). As the population crashed, it appears that the mortality rate of adult females increased relative to that of males. The number of adult males in each group was still only one or two, but the number of adult females per group declined significantly (Struhsaker, 1999). The regression slope of numbers of adult females against males changed dramatically (Fig. 10.5). The number of adult males in the group accounted for only 0.4% of the variation in numbers of adult females in 1973–4, while in 1987–92 they accounted for 50.6% of the variation. There was a significant positive correlation between the numbers of adult females and males in 1987–92 ($p = 0.01$), but not in 1973–4. Persistent habitat loss and senescence appeared to be the most important variables influencing these demographic changes. Neither of the two major non-human predators (crowned hawk-eagles and chimps) of red colobus occur in the Tana study site, so these significant differences cannot be attributed to changes in predation pressure.

VARIATION BETWEEN POPULATIONS/TAXA (SUBSPECIES AND/OR SPECIES)

Sex ratios and regression slopes of adult females (dependent variable) against adult males (independent variable) were highly variable among the six different taxa of red colobus

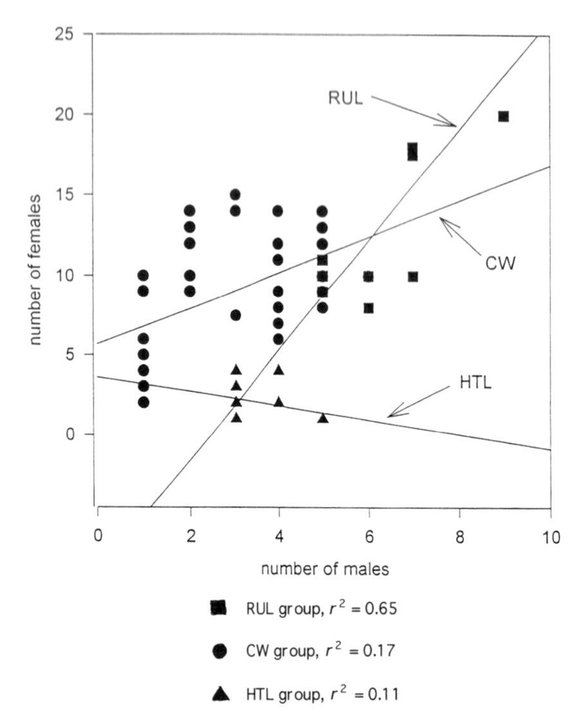

Fig. 10.4. Regression of numbers of adult females against adult males over time, presented as monthly values. Sample size in months: CW = 200; RUL = 64; HTL = 45. (Same data as in Figs. 10.1–10.3). In this and all subsequent figures, the Y intercept = b and the slope = m in $Y = mX + b$. Here, $Y = 1.12X + 5.68$ for CW; $Y = 3.42X + [-8.37]$ for RUL; and $Y = -0.45X + 3.61$ for HTL

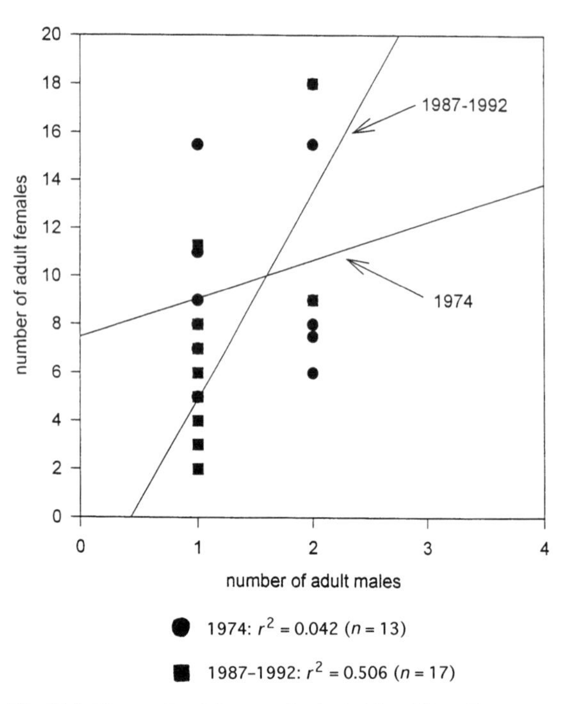

Fig. 10.5. Changes in adult sex ratios in the Tana River, Kenya population (*rufomitratus*) over time. Mean values for 13 groups in 1973–4 (Marsh, 1979) and 17 groups from 1987 to 1992 (Decker, 1994a). $Y = 1.60X + 7.48$ for 1973–4, and $Y = 8.61X + [-3.73]$ for 1987–92.

compared (Table 10.1, Fig. 10.6). These comparisons are based on counts of 84 social groups from 12 sites collected by 13 scientists during 18 different studies (Table 10.1). The data have been grouped according to relative predation pressure. The relative abundance of crowned hawk-eagles was classified into three categories: (1) common (seen at least once each week; often seen daily); (2) rare (seen approximately once every two to three years or no records, but within the broad geographic range of this species); and (3) absent (no sightings and outside known range) (Struhsaker, 1999). The same general criteria apply to chimps.

Except for the three sites where eagles are common and chimps are absent (Uhehe red colobus, *P. gordonorum*, in Magombera, Kalunga, and Udzungwa in Tanzania and all close together), there were significant correlations between the numbers of adult females and males (Fig. 10.6, $p < 0.01$). The numbers of adult males accounted for 50–84% of the variation in numbers of adult females at all but the former three sites (Fig. 10.6). Only at the sites where eagles were rare or absent and chimps absent (AA and RA in Fig. 10.6), did the regression slopes approach the zero point, i.e., the sex ratio remained approximately the same regardless of the number of males. At sites where eagles were common and chimps common or absent (CA and CC in Fig. 10.6), there was an inverse relationship between the numbers of adult males and the ratio of females to males. So, in groups with relatively few females (usually small groups, Fig. 10.7), the sex ratio was higher than in groups with more females. This relationship held in spite of major differences in predation pressure, e.g., compare sites where eagles were rare and chimps absent with sites where both predators were common (RA and CC in Fig. 10.6). These results do not support the hypothesis that predation pressure exerts a strong influence on adult sex ratios in red colobus (see also Struhsaker, 1999).

The fact that there were fewer females per male in larger groups (groups with more males, Fig. 10.6) at sites where chimps were common may reflect the fact that chimps attack larger groups of red colobus more frequently than smaller groups ($n = 5$, $r^2 = 0.85$) and selectively kill more adult

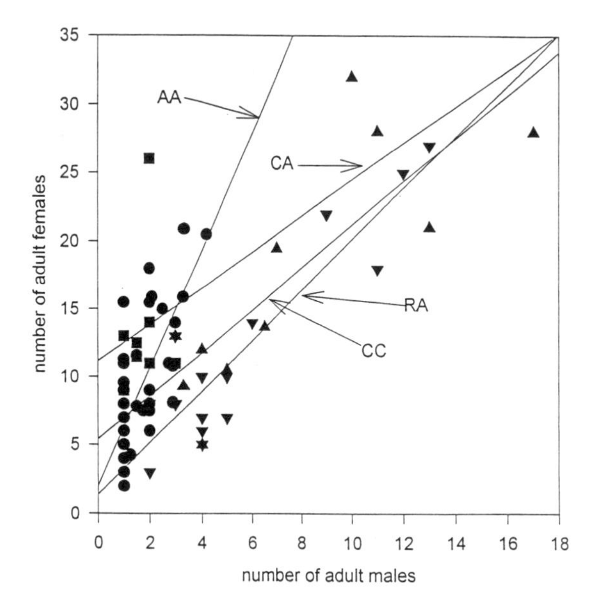

Fig. 10.6. Differences in adult sex ratios between taxa of red colobus under different predation pressure and gross habitat quality (also see Table 10.1). Sample size: AA (48 groups; *rufomitratus* and *kirkii*); CA (8; *gordonorum*); RA (16; *temminckii*); CC (11; *badius* and *tephrosceles*). $Y = 4.29X + 2.00$ for AA; $Y = 1.33X + 11.17$ for CA; $Y = 1.58X + 5.41$ for CC; $Y = 1.88X + 1.39$ for RA.

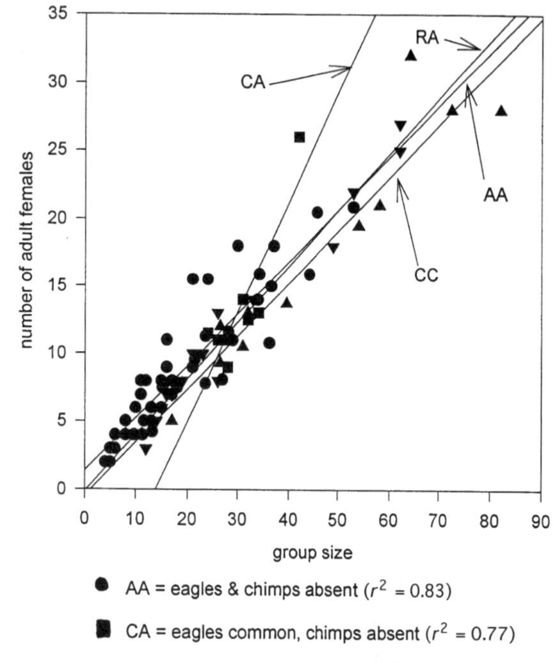

Fig. 10.7. Regression of numbers of adult females against group size. Samples as in Fig. 10.6. $Y = 0.39X + 1.40$ for AA; $Y = 0.81X + [-11.25]$ for CA; $Y = 0.39X + [-0.44]$ for CC; $Y = 0.41X + [-0.13]$ for RA.

females than males (Stanford, 1999). In support of this hypothesis, the likelihood and success of an attack by chimps on a red colobus group were independent of the number of adult male (potential defenders) red colobus in the group. Contrary to intuition, there was a positive, but non-significant, correlation between the likelihood and success of chimps attacking a red colobus group and the number of adult males in the group (Stanford, 1999). In other words, the number of adult male red colobus in a group was not sufficient to deter the success of an attack by chimps.

VARIATION BETWEEN TAXA WITH SIMILAR PREDATION PRESSURE AND GROSS HABITAT

A comparison of the relation between the numbers of adult females (dependent variable) and adult males (independent variable) in the Tana River population (*P. b. rufomitratus*) with the Zanzibar population (*P. kirkii*) reveals interesting differences. In both populations there is a significant correlation between the numbers of females and males ($p < 0.01$), but on Zanzibar the number of adult males accounts for much more of the variation in the numbers of adult females than on the Tana (63.6% vs. 27.5%, Fig. 10.8). Neither population was exposed to crowned hawk-eagles or chimps, and both lived in highly seasonal habitats with relatively low diversity of plant food species. The most striking difference between these two taxa is that the Tana groups usually had no more than two adult males (only one group with three; Decker, 1994a), whereas those on Zanzibar had up to six. Male mortality relative to females was clearly higher in the Tana population than on Zanzibar.

Predation pressure

As indicated in the preceding section and by Struhsaker (1999), sex ratios are not closely linked to predation pressure. They are not necessarily lower where eagles are

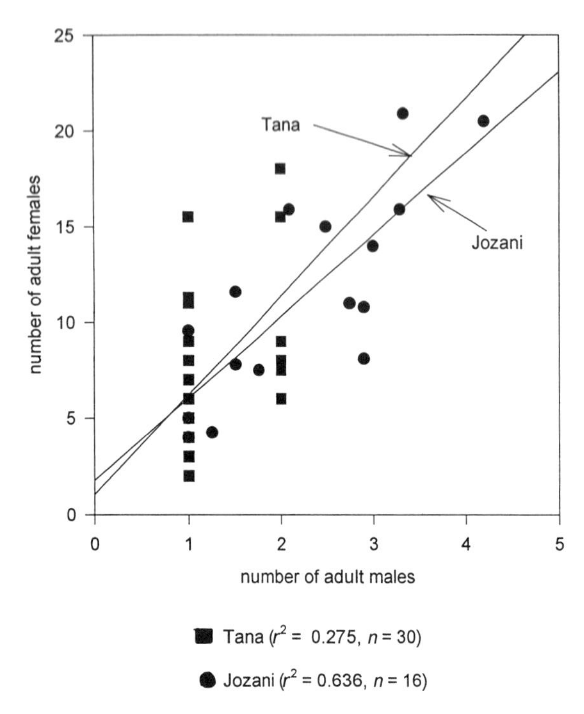

Fig. 10.8. Differences in adult sex ratios between taxa under similar conditions of predation pressure and gross habitat quality. Although the slopes are somewhat similar, the contribution of numbers of adult males to the variation in adult female numbers differs between populations; Tana (*rufomitratus*) and Jozani (*kirkii*). Mean values for each group. $Y = 4.26X + 1.78$ for Jozani, $Y = 5.16X + 1.06$ for Tana.

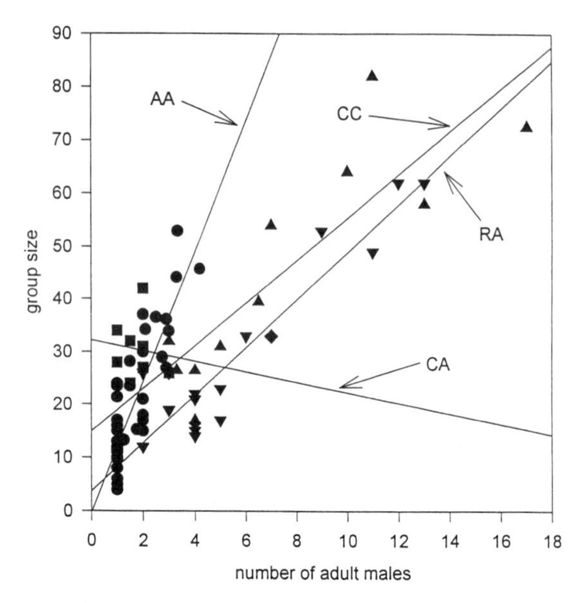

Fig. 10.9. Regression of group size against numbers of adult males. Same samples as in Figs. 10.6 and 10.7. $Y = 12.25X + [-0.22]$ for AA; $Y = -1X + 32.25$ for CA; $Y = 4.04X + 14.93$ for CC; $Y = 4.51X + 3.74$ for RA.

common than where they are rare (see Fig. 10.6). It is possible that predation by chimps may actually lower the sex ratio in larger groups by selectively preying on adult female red colobus compared to males, an explanation that is contrary to the generally accepted hypothesis, namely that sex ratios decrease as a defense response against predators.

Understanding the relative importance of predation pressure to sex ratios is compounded by the fact that red colobus groups tend to be larger in rainforests than in more seasonal and less species-rich woodlands and forests. It is in rainforest habitats that both eagles and chimps tend to be most abundant. In other words, high-quality habitat often covaries with predation pressure. Bioko Island (Equatorial Guinea) could prove to be an important test case. Red colobus on Bioko live in rainforest where there are no predators other than humans. Preliminary observations there indicate that red colobus group or foraging party size is small (approximately 14 or less), but no complete counts are available from Bioko to evaluate adult sex ratios (Struhsaker, 1999). So, in the absence of non-human predators, group/party size is small, as might be predicted if large group size is a defense mechanism. Small group size on Bioko is not likely to be the result of hunting by humans because there was very little hunting in the study area and it had only recently begun there. Furthermore, hunting pressure by humans was less on Bioko than at either Tai, Cote d'Ivoire, or Korup, Cameroun, where red colobus groups were large. If predation pressure is important in shaping sex ratios, one would expect that, in the absence of predators, there would be a high ratio of adult females to males on Bioko. But, recall that even where both eagles and chimps are common, groups of red colobus with relatively few males have higher sex ratios than those with more males (CC in Fig. 10.6) and that group size and numbers of adult males are often correlated (Fig. 10.9).

The great variation in group size and sex ratios in populations under high predation pressure also indicates that predation pressure *per se* is a weak predictor of group size and sex ratio in red colobus. In contrast, selective predation on adult males by eagles in Kibale appears to play a major role in shaping adult sex ratios among black and white colobus (*Colobus guereza*) and mangabeys (*Cercocebus albigena*; Struhsaker & Leakey, 1990).

One of the strongest examples arguing against the role of crowned hawk-eagles in lowering adult sex ratios comes from the Uhehe red colobus (*P. gordonorum*). In spite of eagles being common (seen daily), groups contain relatively few adult males (one to three), and adult sex ratios are high (3.6:1 to 13:1, Fig. 10.6). These sex ratios are not significantly different from those on the Tana River where eagles are absent (Struhsaker, 1999). In both of these populations (*P. gordonorum* and *P. rufomitratus)* male mortality appears to be unusually high relative to females.

Group size

The significant regressions ($p<0.01$) between adult females (dependent variable) and group size (Fig. 10.7) indicate that when dispersing, female red colobus may be selecting for larger groups rather than small groups. This may be to reduce the probability that they as individuals will be preyed upon by chance alone, even though larger groups are preyed upon by chimps more than smaller groups. In other words, they may be selecting for safety in numbers rather than greater defense by more adult males (i.e., lower sex ratios). Although the regression and correlation between numbers of adult females and group size are pronounced regardless of predation pressure (Fig. 10.7; Struhsaker, 1999), groups are often larger where chimps are common and the habitat is good (see Table 10.1). In contrast, Stanford (1995) suggests that the very intense predation by chimps at Gombe may reduce red colobus group size either as the direct result of killing immature monkeys or because the colobus modify their grouping tendencies and form smaller groups to reduce detection and, thereby, predation by the chimps.

Habitat quality

Group size is often linked to habitat quality (Struhsaker, 1999). Groups tend to be larger in rainforests than drier, more seasonal and less species-rich habitats, and this often covaries with chimp abundance (see Table 10.1; Struhsaker, 1975, 1999). Heavy logging of rainforest can result in fusion–fission groups of red colobus (Skorupa, 1988; Struhsaker, 1997, 1999). The total absence of non-human predators may also account for the smaller groups in the rainforests of Bioko Island (Struhsaker, 1999).

Male quality, mate choice, and reproductive success

Although breeding among red colobus is not panmictic (Struhsaker, 1975; Struhsaker & Pope, 1991), female options for mate choice are to some extent determined by the number of adult males in the group. The number of potential mates for females depends on the size of the male coalition. In addition, larger coalitions of adult males have the potential to dominate smaller coalitions over food resources, as well as the potential to deter predators more effectively. The effect of male coalition size on female reproductive success might be reflected in conception rates. Survivorship of offspring and life-time reproductive success of females as a function of male coalition size are much more difficult to evaluate because of the dynamics in group size and membership.

The only data available to evaluate the reproductive advantages to females of male coalition size come from the long-term study of the CW group in Kibale. The number of conceptions per adult-female month was determined by back-dating six to seven months from birth dates. Conception dates were then compared with the size of the male coalition at that time. Eleven time periods were evaluated in this manner between September 1970 and May 1988. The results demonstrate a variable relationship, but are suggestive of a positive trend (Fig. 10.10). When all 11 periods are considered, the correlation is not significant, and only 11.6% of the variation in conception rates is accounted for by the number of adult males in the group ($r^2=0.116$, $Y=0.37$, $X+2.37$). Observations during the last of these 11 sample periods (June 1987–May 1988) were somewhat atypical in that the CW group was unusually small (nine to ten individuals) and had only one to two adult males and two to three adult females. Two of the three females conceived during this period and only after an unusually long interbirth interval (34 months vs. the mean of 24.4 months: Struhsaker & Pope, 1991). If this sample is excluded (open square, uppermost point in Fig. 10.10), the correlation between conception rate and number of males is significant ($r=0.64, p<0.05$), with the number of males accounting for 41% of the variation in conception rates (Fig. 10.10).

The relation between reproductive success for males and the number of males in the group is less clear. Larger male coalitions tended to attract more females (potential mates),

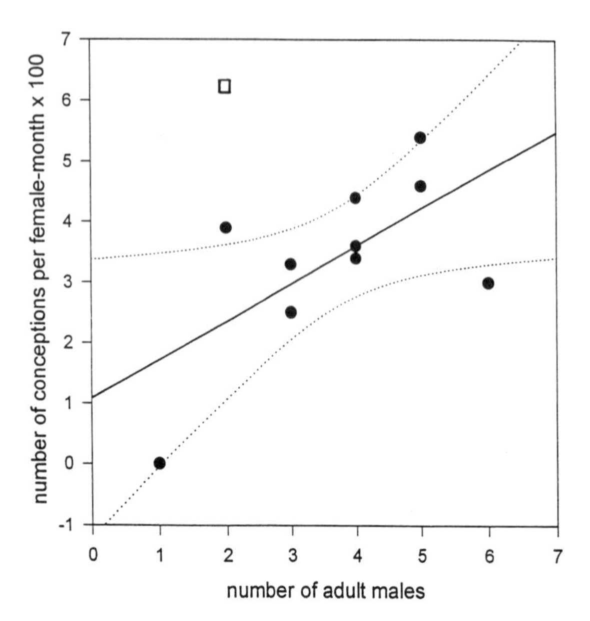

Fig. 10.10. Regression of numbers of conceptions against numbers of adult males in the CW group. The last sample period (June 1987–May 1988, open square) was excluded from the statistical analysis (see text). $r^2 = 0.41$, $Y = 0.63X + 1.09$. Dotted lines are 95% confidence limits.

but not invariably (see above). Counteracting this possible advantage, larger male coalitions result in potentially more male–male competition for mates. Lifetime reproductive success among males was more dependent on variation in the frequency of successful copulation than variation in reproductive lifespan or the number of females in the group (Struhsaker & Pope, 1991). Exceptions to these trends in male reproductive success have been described (Struhsaker & Pope, 1991) and emphasize the importance of inter-individual differences.

Data from specific groups in Kibale (Figs. 10.1 to 10.4, and 10.10) and other study sites (Figs. 10.5 to 10.8) clearly demonstrate that there is more to attracting females to groups than numbers of adult males alone. Interindividual differences in male attributes or quality may well be the explanation for much of the variation in sex ratios and group size.

Marsh (1979b) presents data from his main study group of Tana River red colobus that support this idea. Adult female numbers had steadily decreased through emigration from 14 to eight during the five-month tenure of the resident male until he was replaced. When this male was replaced with a new incumbent male, the number of adult females increased from eight to 19 over the next 22 months. With only one exception, this increase was due to immigration. I suggest that this is a case in which differences in male quality had a profound effect on group size and sex ratio.

Stanford (1999) also presents data supporting the idea that male quality is important in shaping group size and sex ratios. In Gombe, the success of chimp attacks on red colobus groups is not affected by the numbers of adult male red colobus *per se*, but, within limits, it is negatively affected by the number of male colobus that counterattack the chimps. Male quality may well explain why some individuals counterattack chimps more readily than others.

Although 'quality' of individuals is a nebulous concept, an operational definition of which runs the risk of circularity, it is something that the majority of behavioral primatologists are acutely aware of. In addition to the examples given here, it may explain, for example, why the dominant male is not necessarily the largest male and why a coalition of two or three males may be more attractive to immigrating females than five or six males.

CONCLUSION

Adult sex ratios in social groups of red colobus are highly variable within specific groups over time, within populations at any one time and over time, and between populations and taxa. An understanding of this variation depends on clarifying the relative contribution of numerous parameters that can be broadly categorized as habitat quality, predation pressure, differential mortality between the sexes, male quality, and phylogeny. With the possible exception of phylogeny and its consequences for male–male tolerance, the importance of any one of these parameters in determining sex ratios and group size in red colobus is likely to vary between sites and even within a site over time. A working hypothesis of the interplay of these variables is schematically summarized in Fig. 10.11. Exceptions to this diagram occur at all levels of comparison and are emphasized throughout this chapter. For example, high densities of crowned hawk-eagles do not necessarily lead to large groups of red colobus or to low sex ratios. Chance events are particularly important in shaping sex ratios and group size within specific social groups. The value of this diagram is to highlight the most important variables and how they are likely to interact with one another.

The magnitude of variation described here emphasizes

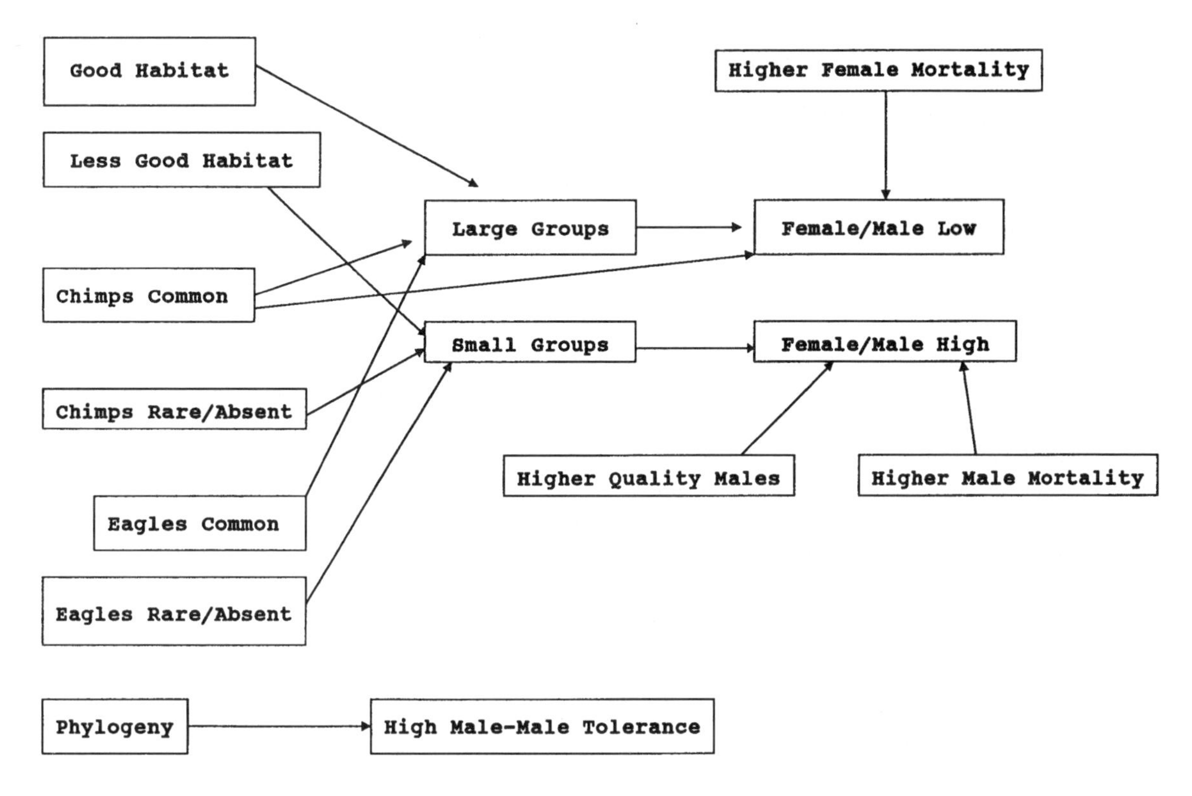

Fig. 10.11. Schematic diagram summarizing the main variables influencing red colobus group size and adult sex ratio (see text and Table 10.1 for definitions and discussion).

the need to better understand intrapopulational and intraspecific variation if we are to advance our comprehension of the plasticity and evolution of primate social organization and behavior. Until this level of variation is understood and taken into consideration, broad interspecific comparisons and generalizations are misleading, if not counterproductive, in furthering the field of behavioral ecology.

ACKNOWLEDGMENTS

I thank Lysa Leland who collected most of the information on the HTL group and helped with the research on the other red colobus study groups in Kibale. Dr Ronald Noë and Mr Bart Beerlage kindly permitted me to use their unpublished count of a red colobus group in Tai. Dr Peter Kappeler and the German Primate Center are thanked for organizing the conference and funding for me to attend, thereby providing the motivation for this chapter. Dr Thelma Rowell and an anonymous reviewer are thanked for useful comments on the manuscript.

11 • The number of males in langur groups: monopolizability of females or demographic processes?

ELISABETH H.M. STERCK &
JAN A.R.A.M. VAN HOOFF

INTRODUCTION

The number of males in primate groups differs widely between species. As a result, the group structure of a species is often categorized as single male or multi male. Some species, however, do not fit this distinction because both single-male and multi-male groups are found, sometimes even within one population (e.g., howler monkeys, *Alouatta* spp.: Crockett & Eisenberg, 1987; mountain gorillas, *Gorilla gorilla beringei*: Robbins, 1995) and the percentage of one-male groups in different populations can differ (e.g. Eisenberg *et al.*, 1972; howler monkeys: Crockett & Eisenberg, 1987; Hanuman langurs, *Presbytis entellus*: Newton, 1988; Srivastava & Dunbar, 1996). This variation in the number of males in a group may be due to different causes and be achieved through different processes.

First, the presence of more males in a group could be due to a low monopolizability of females (e.g., Emlen & Oring, 1977; Mitani *et al.*, 1996a), i.e., to an inability of a dominant male to exclude other (nearly) adult males. In Hanuman langurs, for example, Srivastava & Dunbar (1996) could, indeed, relate the number of males per group to the monopolizability of females and to the costs for males of searching other groups.

Second, multi-male groups sometimes exist, even though the respective groups of females could be monopolized by one male. An adult male may tolerate additional males when his own fitness is not impeded, but when these additional males yield advantages, such as improved anti-predator protection (van Schaik & Hörstermann, 1994) or support against infanticidal assaults by strange males on the infants in the group (van Schaik, 1996). Additional males should give such support more easily if they have a fitness interest in the protection of other group members, i.e., because they might have fathered the group's infants or are otherwise related. Additional adult males may come to a group through immigration, but they can also mature in their natal group. This gives rise to an age-graded group with a high degree of between-male relatedness (Eisenberg *et al.*, 1972). In such a group, the original male must have been resident long enough for his sons to mature to adulthood. For these sons, an extended stay in the parental group could be a mating strategy, aimed at succeeding their father (e.g., mountain gorillas: Robbins, 1995) or improving their own prospects to obtain a breeding position elsewhere (e.g., red howler monkeys, *Alouatta seniculus*: Pope, 1990; Thomas's langur, *Presbytis thomasi*: Steenbeek *et al.*, Chapter 12).

When several males reside in a group, they can form coalitions. In principle, the formation of coalitions between males to obtain sexual access to females is hampered by the fact that fertilization is an all-or-none matter that cannot easily be divided between coalition partners (e.g., van Hooff & van Schaik, 1992, 1994). Yet such coalitions do occur, especially in between-group interactions (van Schaik, 1996), but in some rare cases also within groups (e.g., savannah baboons, *Papio cynocephalus*: Noë & Sluijter, 1995).

As noted earlier, langurs exhibit intra-specific variation in the number of adult males in a group. Males can either reside in bisexual groups, or outside such groups, namely as solitary males or in all-male bands. When we regard the number of males most often found as the typical group composition (e.g., Struhsaker & Leland, 1987), we may neglect the fact that the number of males reflects alternative adaptive strategies.

So far, the most widely studied langur species is the Hanuman langur. Srivastava & Dunbar (1996) identified various factors that influence the number of males in a group, namely: (1) birth seasonality; (2) number of females in a bisexual group; (3) home range size or population density. These were most plausible determinants of the monopolizability of females and of the search time of males for a new group. (4) One additional factor has been suggested by Treves & Chapman (1996), namely the extent to which the predator community in an area is intact, measured by the disturbance index of Bishop *et al.* (1981). This influence, however, was not significant for Hanuman langurs. This last variable may also affect other processes that influence the number of males in a group. The rate of male takeovers, where an extra-group male ousts a resident male of a bisexual group and becomes the new resident male, is thought to be higher in disturbed areas (Rudran, 1973a). Here, age-graded groups will not have time to develop.

Furthermore, demographics and their social consequences are thought to be affected by habitat disturbance (Sterck, 1998): Female split-merger, the process in which dispersing females start new bisexual groups with a male and other females emigrating from an existing bisexual group, has been observed more frequently in undisturbed areas, whereas male takeover of bisexual groups has been found more often in disturbed areas. Female split-merger has been seen in all studied langur species, although not in all studied populations, whereas male take-overs have been found in all Hanuman langur populations for which there are relevant data. By contrast, takeovers have been seen much less often in the other langur subgenera.

This chapter examines the determinants of the number of males in the other langur species and compares the results with those for the Hanuman langurs (Srivastava & Dunbar, 1996; Treves & Chapman, 1996). The chapter then examines whether the same explanations hold by identifying the differences and similarities of male mating strategies among these species and relating these to the ecology and to female behavior. Our conclusion is that, whereas differences in monopolization of females is the essential determining variable for the Hanuman langur, this is not the case in the other species because all groups are monopolizable. When these groups are age graded, benefits accrue to both the natal males and their resident father. However, multi-male stages may also represent gradual male takeovers, and these are costly.

METHODS

Data set

The available literature was searched for data on group composition for all langur populations, except Hanuman langurs (Table 11.1). Data concerning the Hanuman langur used in the comparisons were taken from Srivastava & Dunbar (1996).

Five populations were represented by two data sets taken at different points in time. Including both of these data sets in the analyses would lead to an unjustified inflation. Therefore, the samples from the most extensive studies (Kuala Selangor; Madhupur N.P.; Tanjung Puting) or with the most typical group composition were used – Kuala Lompat: the 1981 data of the banded langur, *Presbytis melalophos*; in the opinion of Chivers & Raemaekers (1980), the high proportion of multi-male groups in 1970/1 was atypical; Ketambe: one-male groups are typical in this population of Thomas's langurs, because less than half the groups contain several adult males, and only for a minor part of the time: (Sterck, 1997; Steenbeek *et al.*, Chapter 12; see also Table 11.4). The analyses concerning the percentage of one-male groups were also conducted using these data.

Taxonomy

Phylogenetic differences may affect the results of a comparative analysis (Harvey & Pagel, 1991). The phylogeny of the langurs is still debated (e.g., Thorington & Groves, 1970; Weitzel & Groves, 1985; Gupta, 1998). Several authors, however, distinguish four subgenera (e.g., Wolf & Fleagle, 1977). We followed them in order to determine possible phylogenetic effects. The subgenera are *Semnopithecus* (Hanuman langur, *S. entellus*); *Kasi* (Nilgiri, *Presbytis johnii*, and purple-faced langur, *P. senex*); *Trachypithecus* (silver, *P. cristata*, golden, *P. geei*, dusky, *P. obscura*, and capped langur, *P. pileata*); and *Presbytis* proper (Sunda Island langur, *P. aygula*, banded, *P. femoralis*, maroon, *P. rubicunda*, and Thomas's langur).

Variables

DEMOGRAPHIC VARIABLES

The average group size (N) and the average number of females in a group (F) in a population were calculated for those groups for which complete counts have been reported. The percentage of one-male groups (OMG) was calculated when the number of adult males was reported with certainty. The birth rate (B) is the average number of infants born per female per year. The birth index (S) of Srivastava & Dunbar (1996) is a measure of birth seasonality. The criteria for calculating this index were not precise and it could not be replicated. Therefore, we used a more objective value of variability, namely the Shannon Index (Bronikowski & Webb, 1996). We included studies in which 15 or more births were recorded and data were collected for at least one year. We recalculated S for those populations of Hanuman

Table 11.1. *Demographic, reproductive and environmental data for populations of the langur subgenera* Kasi, Trachypithecus *and* Presbytis *proper*

Reference	Location	Langur species	Sub-genus	*G*[a]	*N*[a]	*F*[a]	OMG[a]	*D*	*B*	*S*(*n*)	*P*	*R*	*T*	*L*	PD	HRS
1	Ootacamund	Nilgiri	k	8	9.5	2.8	75	2.50	—	—	1510	3[f]	14.9[g]	11°4′N	—	—
2	Periyar	Nilgiri	k	4	15.4	8.0	100	2.00	—	—	—	—	—	9°5′N	—	—
3	Horton Plains	Purple-faced	k	20	8.9	3.3	100	1.00	0.34	0.89 (88)	2640	0	15	8°N	93	6.8
3	Polonnaruwa	Purple-faced	k	29	8.4	4.4	93	2.50	0.31	0.94 (47)	1727	2	26	8°4′N	215	2.5
4	Cagar Alam	Silver	t	6	10.3	8[d]	100[d]	1.50	—	—	4557	0	26.3	7°4′S	175	6.5[d]
5	Kuala Selangor (1977)	Silver	t	5[b]	30.6[b]	14.0[b]	60[b]	3.25	0.53	0.94 (76)	—	—	—	3°N	550	—
4	Taman Wisata	Silver	t	6	13.5	6.8	83	1.75	—	—	4557	0	26.3	7°4′S	360	4.5[d]
6	Jamduar	Golden	t	7	12.7	5.1	86	1.25	—	—	5641	—	—	26°2′N	—	—
6	Raimona	Golden	t	2	15.5	6.5	100	1.25	—	—	5015	—	—	26°2′N	—	—
7	Kuala Lompat	Dusky	t	5	14.2	5.0	60	1.00	—	—	2120	0	27.7	3°4′N	31	17
8	Madhupur N.P. (1986–8)	Capped	t	9	8.2	3.6	100	2.00	—	0.67 (15)	2885	4	24.9	24°5′N	52	18
9	Kamojang	Sunda Island	p	4	8.0	3.0	100	1.75	—	—	3000	—	16[h]	7°1′S	11	38
9	Patenggang	Sunda Island	p	5	5.8	1.4	80	2.25	—	—	3000	—	16[h]	7°1′S	35	14
10	Perawang (1984)	Banded	p	8[c]	11.0[c]	5.5[c]	100[c]	2.50	0.38	0.86 (40)	2379	0	26.5[g]	0°4′N	42	23.9
11	Kuala Lompat (1981)	Banded	p	3	15.0	7.7	100	1.00	—	—	2120	0	27.7	3°4′N	108	29.5[i]
12	Sungai Tekam	Banded	p	3	14.0	5.7	67	1.00[e]	—	—	2207	0	—	4°1′N	48	14[i]
13	Sepilok	Maroon	p	1	6	2	100	1.00	—	—	2977	1	27.2	5°5′N	19	84
14	Tanjung Puting (1975)	Maroon	p	9	6.1	2.6	100	1.25	—	—	2970	0	27.6	2°5′S	—	70.8[j]
15	Bohorok	Thomas's	p	23	8.0	3.6	85	2.25	—	—	4575	0	26.7[h]	3°3′N	21	14
16	Ketambe (1989)	Thomas's	p	7	8.9	3.6	100	1.00	0.49	0.98 (94)	3229	0	24.9	3°4′N	28	37.7[i]

Notes:

a. Based on the number of groups and group composition at the start of the study.
b. Based on the number of groups and group composition at the middle of the study.
c. Based on data of 1983 or, when not available, of 1984.
d. Based on one group.
e. Low, therefore we took 1.00 (situation before logging).
f. 'Three months virtually dry' (Poirier, 1970, p. 260), therefore we took 3.
g. We averaged the minimum range and the maximum range temperature and, subsequently, averaged these for an estimate of the average temperature.
h. Average temperature as given by the author.
i. Based on two groups.
j. Based on five groups.

References: 1. Poirier (1969, 1970). 2. Tanaka (1965). 3. Rudran (1973a, 1973b). 4. Kool (1989, 1993). 5. Wolf & Fleagle (1977); Wolf (1984). 6. Mukherjee & Saha (1974). 7. Curtin & Chivers (1978); Chivers & Raemaekers (1980); Curtin (1980); van Schaik *et al.* (1992) (for PD *P. obscura*). 8. Stanford (1991a, 1991b). 9. Ruhiyat (1983). 10. Megantara (1989); van Schaik *et al.* (1992). 11. Bennett (1983); Chivers & Raemaekers (1980); Raemaekers *et al.* (1980); Bennett personal communication. as cited in Sterck (1998). 12. Johns (1983, 1985); Chivers & Raemaekers (1980). 13. A.G. Davies (1984, 1987, 1991); van Schaik *et al.* (1992). 14. Supriatna *et al.* (1986); Galdikas (1988). 15. Gurmaya (1986). 16. Assink & van Dijk (1990); van Schaik (1986); van Schaik *et al.* (1992); Sterck (1995, 1997); Sterck, Steenbeek & Wich, unpublished data.

Subgenus: k, *Kasi*; t, *Trachypithecus*; p, *Presbytis* proper.

G: number of groups; *N*: average group size; *F*: average number of females in a bisexual group; OMG: percentage of one-male groups; *D*: human disturbance index (cf. Bishop *et al.* (1981); *B*: birth rate (infants per female per year); *S*: birth index calculated with the Shannon Index, *n*: number of births; *P*: annual rainfall (mm); *R*: number of months with less than 50 mm rain; *T*: temperature (°C); *L*: latitude; PD: population density (individuals/km^2); HRS: home-range size (ha).

Table 11.2. *Average values of the demographic, environmental, and ecological variables per subgenus*

Subgenus		N	F	OMG	B^a	S^b	P	R	T	L	PD	HRS
Kasi	average	10.6	4.6	92	0.33	0.92	1959	1.7	18.6	9.3	154	4.7
	sd	3.3	2.3	12	0.02	0.04	600	1.5	6.4	1.5	86	3.0
Trachypithecus	average	15.0	7.0	84	0.53	0.81	4129	1.0	26.3	14.0	234	11.5
	sd	7.3	3.4	18	–	0.19	1343	2.0	1.1	11.0	220	7.0
Presbytis proper	average	9.2	3.9	92	0.44	0.92	2940	0.1	24.1	4.1	39	36
	sd	3.4	2.0	12	0.08	0.09	732	0.4	5.1	2.2	30	25
Semnopithecus	average	29.2	12.5	53	0.45	0.73	1342	6.5	22.5	23.2	41	170
	sd	12.6	5.8	34	0.19	0.15	622	1.8	5.1	5.8	39	197

Notes:

a. The values for *Kasi*, *Trachypithecus*, and *Presbytis* proper combined are 0.41 (0.10).

b. The values for *Kasi*, *Trachypithecus*, and *Presbytis* proper combined are 0.88 (0.11); values of *Semnopithecus* are from Table 11.3.

The data of *Semnopithecus* come from Srivastava & Dunbar (1996, $n = 23$ populations).

For explanation of variables, see Table 11.1.

langurs in which these were available and for two other Hanuman langur populations (Mt Abu and Wilpattu) in which the proportion of infants born per two months was published. The Shannon Index is calculated as:

$$S = -\text{SUM}\,(p_i/n \ln (p_i/n)) \,/\, \ln (m)$$

where p_i is the number of births per month, n is the total number of births, and m the number of periods (months) = 12; for Mt Abu and Wilpattu, p_i is the proportion of infants born in a two-month period, $n = 1$ and $m = 6$; and when $p_i = 0$ we used in the calculations $(p_i/n \ln (p_i/n)) = 0$.

ENVIRONMENTAL VARIABLES

The annual rainfall (P) has been measured in mm and the average annual temperature (T) in °C. The temperature was calculated as the average of monthly average minimum and maximum, averaged over the year (cf. Dunbar, 1992). The number of months with less than 50 mm rainfall (R) was counted. The latitude (L) of the study area has been reported for most sites. The values for Periyar and Kuala Selangor were taken from an atlas; the value for Horton Plains is the average latitude for Sri Lanka.

ECOLOGICAL VARIABLES

The population density (PD), measured in individuals/km^2, and home range size (HRS), measured in ha, were taken from the literature. The human disturbance index (D) that Bishop *et al.* (1981) developed is the average of four indices: (i) classification of home range; (ii) harassment of animals; (iii) habituation to humans; and (iv) presence of predators. These indices are rated from 1, no human disturbance, to 4, highest degree of human disturbance.

Statistics

The data set consists of 20 populations from 11 different langur species grouped into the three different subgenera *Kasi*, *Trachypithecus* and *Presbytis* (see Table 11.1). The data of the fourth subgenus, *Semnopithecus*, concern 23 populations (Srivastava & Dunbar, 1996). One outlier was detected by inspecting box plots of the different variables: the silver langur of Kuala Selangor has a larger group size and larger number of females and a lower percentage of one-male groups than the other populations. The analyses were therefore conducted with and without this population.

For each demographic variable, we first performed a one-way ANOVA to test for differences between the four subgenera. In case of overall significance, this was followed by various *post hoc* comparisons, using the Scheffé Test. Subsequently, the effect of the different factors that may influence the percentage of one-male groups was tested for the langur populations excluding the Hanuman langur. This was done with single regression analyses. All tests are two-sided and considered significant at $p < 0.05$.

RESULTS

Comparison of subgenera

Table 11.2 presents a comparison of the demographic variables of the *Semnopithecus* subgenus (the Hanuman langur: Srivastava & Dunbar, 1996) with those of the other langur subgenera. The average group sizes of subgenera were significantly different (ANOVA: $F_{3,39} = 11.61$, $p < 0.001$), as were the average number of adult females (ANOVA: $F_{3,39} = 9.56$, $p < 0.001$) and the percentage of one-male groups

Table 11.3. *The birth index (calculated with the Shannon Index) of Hanuman langur populations (*Semnopithecus*)*

Reference	Location	n	S
17	Wilpattu	71	0.74
18	Dharwar	26	0.67
19	Mt Abu	56	0.79
20	Jodhpur	398	0.99
21	Ramnagar	51	0.56
22	Rajaji	27	0.62

References: 17. Muckenhirn (1972). 18. Sugiyama *et al.* (1965). 19. Moore (1985). 20. Sommer & Rajpurohit (1989). 21. Koenig *et al.* (1997). 22. Laws & Vonder Haar Laws (1984).
Abbreviations as in Table 11.1.

(ANOVA: $F_{3,39}=6{,}29$, $p<0.001$). *Post hoc* tests revealed that for these variables the subgenera of *Kasi*, *Trachypithecus* and *Presbytis* proper did not differ significantly from each other. However, the average group size of *Semnopithecus* was significantly larger than those of the other langur subgenera. The average number of adult females of *Semnopithecus* was significantly larger than in *Kasi* and *Presbytis* proper groups. The percentage of one-male groups of the *Semnopithecus* was significantly smaller than that of *Presbytis* proper.

The differences between the environmental variables were also investigated. Significant variation was found in average yearly rainfall (ANOVA: $F_{3,36}=23.60$, $p<0.001$), number of dry months (ANOVA: $F_{3,32}=35.76$, $p<0.001$) and latitude (ANOVA: $F_{3,39}=22.82$, $p<0.001$). Average temperatures did not differ significantly (ANOVA: $F_{3,33}=1.55$, $p=0.22$). *Post hoc* comparisons indicated that *Trachypithecus* populations live in significantly wetter areas than *Semnopithecus* and *Kasi*, and that *Presbytis* proper lives in significantly wetter areas than *Semnopithecus*. The number of dry months was higher for *Semnopithecus* than for the other three subgenera. *Semnopithecus* was found at significantly higher latitudes than the other three subgenera, and *Trachypithecus* occurred significantly higher than *Presbytis* proper.

Few data on the birth rate and birth index of *Trachypithecus*, *Kasi* and *Presbytis* proper were available. Therefore, these data were treated as one subgeneric category and compared with *Semnopithecus*. Birth rate did not differ significantly between these two subgeneric categories (t-test: df=23, $t=0.41$, $p=0.69$). The recalculated birth index data of *Semnopithecus* are presented in Table 11.3. The birth index tended to be shorter in *Semnopithecus* than in the other three subgenera (t-test: df=10, $t=1.97$, $p=0.08$). When the provisioned *Semnopithecus* populations, i.e., those with artificially long birth seasons (i.e., Jodhpur and Mt Abu: Srivastava & Dunbar, 1996) were omitted, this difference was highly significant (t-test: df=8, $t=3.62$, $p=0.007$).

The population densities of the subgenera differed significantly (ANOVA: $F_{3,26}=6.24$, $p=0.002$). Population density was significantly larger for *Trachypithecus* than for *Semnopithecus* and *Presbytis* proper. The home-range sizes tended to differ (ANOVA: $F_{3,25}=2.52$, $p=0.08$).

In summary, the Hanuman langurs differ in many features from the other three subgenera. They live in larger groups with more females and more often with multiple males. Their population densities are of intermediate value. They live at higher latitudes, the climate in their habitat has more dry months, and the yearly rainfall is lower. These factors are reflected in seasonality in births: Hanuman langur females more often give birth in a specific period of the year than the combined other species.

Number of males in *Kasi*, *Trachypithecus* and *Presbytis* proper

GROUP SIZE

On average, larger groups contained more adult females ($F_{1,18}=110.86$, $p<0.0001$; adj. $r^2=0.853$). These two independent variables have each been entered in a single regression analysis with birth rate, birth index, environmental and ecological data as dependent variables. Both average group size and the average number of adult females appeared to be determined by population density:

$N=7.66+0.03\times \mathrm{PD}$ ($F_{1,13}=20.90$, $p=0.0005$, adj. $r^2=0.587$)

$F=3.16+0.02\times \mathrm{PD}$ ($F_{1,13}=28.60$, $p=0.0001$, adj. $r^2=0.663$)

However, when these analyses were repeated without the silver langur from Kuala Selangor, which was an outlier (see Methods), only the regression analysis for the number of females reached significance:

$F=3.54+0.01\times \mathrm{PD}$ ($F_{1,12}=5.49$, $p=0.04$, adj. $r^2=0.257$)

Because much less variation is explained, the previous result must have depended mainly on this one extreme data point. None of the measures that represent habitat quality reached significance. This is in contrast to the findings of Srivastava & Dunbar (1996).

Table 11.4. *Demographic data from study sites with two data points*

Reference	Location	Langur species	*G*	*N*	*F*	OMG	HRS
23	Kuala Selangor (1966)	Silver	4	—	15.6	75	—
5	Kuala Selangor (1977)	Silver	5	30.6	14.0	60	—
24	Madhupur Forest (1976)	Capped	13	9.0	3.0	100	—
8	Madhupur N.P. (1986–8)	Capped	9	8.2	3.6	100	18.0
7	Kuala Lompat (1970/1)	Banded	4	15.0	5.8	25	17.3
11	Kuala Lompat (1981)	Banded	3	15.0	7.7	100	29.5
14	Tanjung Puting (1975)	Maroon	9	6.1	2.6	100	70.8
25	Tanjung Puting (1987)	Maroon	1	9.0	4.0	100	—
16	Ketambe (1989)	Thomas's	7	8.9	3.6	100	37.7
26	Ketambe (1994)	Thomas's	7	8.4	2.9	43	—

References: 23. Bernstein (1968). 24. K.M. Green (1981). 25. Salafsky (1988). 26. Romy Steenbeek, personal communication.
Abbreviations as in Table 11.1.

BIRTH RATE AND BIRTH INDEX

Birth rate was significantly related to home range size:

$B = 0.3 - 0.005 \times \text{HRS}$ ($F_{1,2} = 28.16$, $p = 0.03$; adj. $r^2 = 0.901$).

As in *Semnopithecus*, the birth index (S) was significantly lower at higher latitudes:

$S = 0.97 - 0.01 \times L$ ($F_{1,4} = 9.56$, $p = 0.04$; adj. $r^2 = 0.63$).

RANGE SIZE

In the Hanuman langur the relation between the natural logarithm of home range size and population density shows a 'compression curve' (Srivastava & Dunbar, 1996). Therefore, log-transformed values of the variables were also used for the other subgenera. A single regression indicated that the natural log of population density was significantly related to the natural log of home range size:

$\ln(\text{HRS}) = 5.7 - 0.7 \times \ln(\text{PD})$ ($F_{1,12} = 19.84$, $p = 0.0008$; adj. $r^2 = 0.592$)

The relation between home range size and population density shows a 'compression curve' in the other subgenera as well.

MATING SYSTEM: BETWEEN-POPULATION VARIATION IN MONOPOLIZABILITY

The relationship between the percentage of one-male groups and average group size, average number of females per group, environmental and ecological variables was calculated. The relationship between the percentage of one-male groups and average group size was significant:

$\text{OMG} = 104.1 - 1.3 \times \text{N}$ ($F_{1,18} = 6.01$, $p = 0.02$; adj. $r^2 = 0.209$).

This is contrary to what the monopolization hypothesis predicts. The relationship ceased to be significant, however, after the data for the silver langur of Kuala Selangor (the outlier) were omitted. The other factors did not result in significant relationships.

Some populations were measured twice (Table 11.4). The data not used in Table 11.1 on two occasions presented an atypical (see Methods) low percentage of one-male groups. When these data substituted those in Table 11.1, not one of the above regression analyses was significant. These results therefore indicate that variation in monopolizability does not explain the number of males in groups in populations of the subgenera *Trachypithecus, Kasi* and *Presbytis* proper. This is in contrast with the results for *Semnopithecus*.

MATING SYSTEM: WITHIN-POPULATION VARIATION IN MONOPOLIZABILITY

Some groups of *Kasi*, *Trachypithecus* and *Presbytis* proper contain more than one male. The question is why. Multi-male groups of these subgenera may differ from the one-male groups in the same population. Especially the number of females in such groups may differ. A comparison of populations that were measured at two different moments revealed that the percentage of one-male groups differed in three populations (Table 11.4). Again, the result is the opposite from what the monopolization hypothesis predicts: there were fewer one-male groups when the average number of females in groups was smaller. Comparisons of

Table 11.5. *Comparing the number of adult females in single-male and multi-male groups*

			Average		Single-male group(s)		Multi-male group(s)		
Reference	Location	Langur species	*F*	OMG	*G*	*F*	*G*	Number of males	*F*
1	Ootacamund	Nilgiri	2.8	75	6	2.8	2	3.0	2.5
3	Polonnaruwa	Purple-faced	4.4	93	27	—	2	2.0	—
27	Horton Plains and Polonnaruwa	Purple-faced	3.6	91	43	3.6	4	2.0	4.3
17	Kuala Selangor (1966)	Silver	15.6	75	3	16.5 (14.7–18.3)	1	2.0	12–14
5	Kuala Selangor (1977)	Silver	14.0	60	3	12.0	2	3.0	17.0
4	Taman Wisata	Silver	6.8	83	5	6.4	1	2.0	9
6	Jamduar	Golden	5.1	86	6	4.8	1	2.0	7
7	Kuala Lompat	Dusky	5.0	60	3	4.0	2	1.5	6.5
9	Patenggang	Sunda Island	1.4	80	4	1.0	1	2.0	3
7	Kuala Lompat (1970/1)	Banded	5.8	25	1	5–6	3	2.7	5.7–6
12	Sungai Tekam	Banded	5.7	67	2	6.0	1	2.0	5
15	Bohorok	Thomas's	3.6	83	19	3.0	4	2.0	6.3
16	Ketambe (1994)	Thomas's	2.9	43	3	3.3	4	2.25	2.5

References as in Tables 11.1 and 11.4; 27. Manley (1986). Abbreviations as in Table 11.1.

other relevant variables, such as birth index and home range size, were not possible due to a lack of data.

Alternatively, groups with multiple males may contain more females than groups with single males of the same population (Table 11.5). Multi-male groups in eight out of 12 populations contained more females and in four contained fewer females (Wilcoxon Matched-Pairs Signed-Ranks Test: $n=12$, $x=4$, $Z=-1.412$, $p=0.16$). Thus, monopolization of females cannot explain why some groups of *Kasi*, *Trachypithecus* and *Presbytis* proper contain several males and others do not.

MALES IN GROUPS WITH MULTIPLE MALES: A DETAILED DESCRIPTION

A more detailed review of the characteristics of males in multi-male groups may explain their existence.

Nilgiri langur: Poirier (1969) described how a group became multi-male after three males who formed an all-male band joined a bisexual group. The group was unstable and at the end was often seen to split in three subgroups, each with one male of the original all-male band. The original resident male tried to keep these subgroups together, but did not succeed. The study ended before a stable situation was reached. Male–male aggression was common.

Purple-faced langur: Rudran (1973*a*) suggested that multi-male groups are age graded, i.e., that the additional adult males matured in their natal group. Groups became single-male after a new male took over. Rudran did not have many detailed observations but deduced that male takeovers lead to eviction of all natal males. He noted the disappearance of immature animals and additional adult males from groups after a male take-over. One such group contained two males before one evicted the other.

Silver langur: Bernstein (1968) noted that one of two adult males in his group W seemed of very advanced age. Some 11 years later, Wolf (1984) studied the same population and found maturing males in all five study groups. In three cases a maturing male became the dominant male. In one case two males from another group entered the group. First one and then the other obtained the dominant position in the new group. In all five groups the former dominant male stayed in the group after loosing his dominant position. In four cases other maturing males were either evicted ($n=9$) or left voluntarily ($n=6$). Males had tolerant or aggressive relationships, and coalitions were not observed. In another population of silver langurs, at Pangandaran, two males were present in one group, one clearly being dominant and the other one staying in the periphery of the group (Kool, 1989).

Golden langur: Mukherjee & Saha (1974) observed much aggression between adult males residing in the same bisexual group, but not when the group contained subadult males and only one adult male.

Dusky langur: Curtin (1980) reported one group with two adult males. She did not provide any details about these males.

Sunda Island langur: Ruhiyat (1983) also reported that amongst his groups there was one with two adult males. Their coexistence was not always peaceful: during a between-group encounter the two males became very excited and attacked each other. This was the only physical attack observed during the study.

Banded langur: Curtin (1980) reported the occurrence of two males in some groups, but not in all. Some of these males differed in their tendency to emit loud calls, to display, and to confront neighbors, suggesting that one of the two may have become just adult. Johns (1983) reported on a group with two adult males, one of which eventually left the group and became solitary.

Thomas's langur: Multi-male groups of Thomas's langurs (Steenbeek *et al.*, Chapter 12) usually consist of natal males that have become adult and an older resident male, so that they are age-graded groups *sensu* Eisenberg *et al.* (1972). In one case, a (solitary) adult male regularly associated with a bisexual group. No aggression was observed between this male and the resident male. All these groups had in common that they contained multiple males at the end of the resident male's tenure.

DISCUSSION

Monopolization of females

The percentage of one-male groups varies among populations of all langur species. In Hanuman langurs (subgenus *Semnopithecus)* there is considerable intraspecific variation. This is best explained in terms of the monopolizability of females, i.e., the number of females in a group and birth seasonality, whilst the costs for males of searching and becoming a member of another group (measured as home range size or population density) may be important as well (Srivastava & Dunbar, 1996). An effect of the completeness of the predator community on the number of males, indicating a response to predation pressure, was only visible as a trend (Treves & Chapman, 1996).

In a similar analysis for populations of the other langur subgenera, we could establish no such effects. In particular, monopolizability could not be identified as a variable influencing the presence of additional males. When we compared the same population at different points in time, we found that the percentage of one-male groups was not related to the number of females. Similarly, single-male groups did not contain fewer females than multi-male groups of the same population. Therefore, differences in monopolizability do not determine the number of males in groups of these species.

The groups of these subgenera should be easier to monopolize than those of *Semnopithecus*, as the number of females in a group is on average smaller and birth seasonality is less pronounced, whereas the home range size or population density of *Semnopithecus* gave intermediate values. These differences may be caused by habitat differences. Whereas Hanuman langurs are found in a wide array of habitats, ranging from desert-like habitats to forests, the other langur species live in relatively lush (rain)forests. For the Hanuman langur, the number of females in a group was largest in the driest and, therefore, poorest habitats (Srivastava & Dunbar, 1996). These are also the habitats that will promote seasonality in reproduction.

The humid and relatively rich areas coincide with smaller groups and reduced seasonality of births, both in Hanuman langurs and in the other three subgenera. Srivastava & Dunbar (1996) give three possible explanations for the difference in the average number of females in a group: in drier habitats the rare but sufficiently large resources may be more patchily distributed, forcing females into larger groups; females may defend clumped resources against other groups; refuges to escape from predation, such as trees, are evidently scarcer, which may compel females to group. The factual reasons remain to be elucidated. Notwithstanding this, the ease with which groups of *Trachypithecus*, *Kasi* and *Presbytis* proper can be monopolized by one male can be linked to features of their habitat.

Tolerance of additional males?

Thus, in the subgenera *Trachypithecus*, *Kasi* and *Presbytis* proper, the monopolizability of groups of females is always high and one-male groups would be expected. Yet, multi-male groups have been observed regularly. Multi-male groups may develop because these males provide improved protection against predators. This hypothesis, however, is not corroborated by the data: the percentage of one-male groups is not related to the index of environmental distur-

Table 11.6. *Characteristics of males in langur groups with multiple males*

	Silver langur	Nilgiri langur	Purple-faced langur	Thomas's langur
Previous dominant male	+	–	–	–
(Presumed) brothers	+	–	–	–
Male joins at end tenure	–	–	–	+
During male takeover	+	+	–	–
Age-graded (end tenure)	+	–	+	+

bance. Thus, predation risk does not seem to affect the number of males in a group. In addition, two other processes that may be measures of the same factor, i.e., the rate of take-overs (Rudran, 1973*a*) and the nature of demographic changes (Sterck, 1998), did not determine the percentage of one-male groups either.

Our survey suggests the alternative, namely that multi-male groups in langur species other than *Semnopithecus* are age graded. They usually contain one prime breeding male and, near the end of a male's tenure, one or more young adult males (Table 11.6). These young males are often natal males, making the group an age-graded group. These groups contain multiple males at the end of the resident male's tenure, and, therefore, are temporary phases in a demographic development. They are not based on an inability of the dominant male to exclude the others, but must be based on tolerance on his part and an interest in their staying.

A group can also temporarily contain multiple newly immigrated males, each of whom tries to obtain the dominant position. This occurred in Nilgiri langurs, but seemed to constitute an unstable condition (Poirier, 1969). Poirier's observations ended before a stable situation was reached, but probably the process will have led to separate one-male groups. Such gradual male takeovers were also reported for Hanuman langurs. These multi-male phases can exist for several months (e.g., Vogel & Loch, 1984; Agoramoorthy 1994a).

The situation of the silver langur of Kuala Selangor was somewhat different. Wolf (1984) found multi-male groups containing an old, formerly dominant male, who stayed after a rank change. This seemed to be a comparatively stable situation. Moreover, in one case two males immigrated and remained in a group. This was also the population with the largest groups and with the most females in the groups. Therefore, an increased problem of the dominant resident male with the monopolization of the females could well be the cause of the demographics in this population. This would come closest to the multi-male situation found for the Hanuman langur. In the other species and populations, little is known of the males in multi-male groups. Often these males were aggressive to one another or their presence was temporary.

Benefits and costs of additional males

Thus, resident males often tolerated mature natal males, or, in one population, a former resident male. The presence of these additional males may have been beneficial for the resident male. However, aggression suggests that in other populations additional males are not always welcome, especially not when they seem to be in the process of ousting the resident male (cf. Nilgiri langur).

One of the benefits may be increased protection of infants against infanticide. This is especially expected when the additional males are related to the infants, as was found, for instance, in Thomas's langurs (Steenbeek *et al.*, Chapter 12). As the natal males matured they took part increasingly in loud calling and in confrontations with neighbors. Even if the help of young, inexperienced males may be modest, it might make a difference by lengthening their father's tenure and promoting the survival of their infant brethren. The data, however, are not convincing for the silver langur of Kuala Selangor. Cooperation of several natal males in group defense has not been reported (Wolf, 1984). Alternatively, if a former dominant male manages to stay, this could allow him to protect his own offspring against infanticide. It is not known whether this does indeed help. At least this strategy does not seem very effective in the described population, as infanticide after male takeover has been reported (Wolf & Fleagle, 1977).

A further benefit may be the increased direct fitness of the additional males and indirect fitness of the resident male, their presumed father. In Thomas's langurs, the natal males seem to profit directly because they postpone their exposition to the evident hardships of roving alone or in an all-male band (cf. Rajpurohit & Sommer, 1991; Rajpurohit *et al.*, 1995), and may better be able to start their own bisexual group (Steenbeek *et al.*, Chapter 12). Inheritance of the breeding position in a natal group was found in the silver langur population of Kuala Selangor. Silver langur males also dispersed together and entered a new bisexual group together. Thus, direct benefits for the natal males and indirect benefits for the father accrue from age-graded groups. Yet it remains unclear why some natal males are evicted by their presumed father in silver and Thomas's langurs, whereas others are not. Similarly, it remains unclear why in some populations or species, natal males can inherit their father's breeding position, and not in others.

A last benefit may be the formation of coalitions against extra-group males. These have been reported in langurs, be it rarely. For example, Steenbeek *et al.* (Chapter 12) mentioned how adult sons may participate in between-group conflict in Thomas's langurs. Hohmann (1989) reported one case of coordinated aggression by two males of one group against a neighbor for the Nilgiri langur. This coordination, however, was alternated with aggression between these two males and eventually resulted in the splitting of their group. In general, it is unknown whether deliberate coordination does take place, since males seem neither to check the actions of other males nor to coordinate their behavior with other males actively. They may well act in parallel because similar behavior is triggered by the same stimulus (e.g., Curtin, 1980: banded langurs). In silver langurs, the participation of natal males in between-group conflicts has not been reported (Wolf, 1984).

In contrast, the presence of additional males may have costs for the resident male. This becomes evident when they evict him. Such a gradual male takeover was probably the outcome of the multi-male phase reported for the Nilgiri langurs (Poirier, 1969). The aggression described among adult males in some less intensively studied populations, i.e., the golden and the Sunda Island langurs, points in the same direction. The multi-male phase then represents a period in which neither the resident male nor the newly immigrated male is able to monopolize the group, probably because of a stalemate.

In conclusion: amongst langur species one finds different types of multi-male groups. In Hanuman langurs these arise where dominant males cannot prevent immigrants and sexual rivals from entering and staying. In other multi-male groups, where effective monopolization of access to fertile females is possible, the resident male may tolerate sons when they grow up. The resident male, the females, and the grown-up sons will profit in their own way from the lasting association with their relatives.

ACKNOWLEDGMENTS

We thank Carola Borries and Volker Sommer for their constructive comments on an earlier version of this chapter, and them, Jim Moore and Han de Vries for their help in revising the index of birth seasonality. The research on the Ketambe Thomas's langurs was financially supported by WOTRO (the Netherlands Foundation for the Advancement of Tropical Research).

12 • Costs and benefits of the one-male, age-graded, and all-male phases in wild Thomas's langur groups

ROMY STEENBEEK, ELISABETH H.M. STERCK, HAN DE VRIES & JAN A.R.A.M. VAN HOOFF

INTRODUCTION

In non-human primates there is great variation in the number of adult males in bisexual groups (van Hooff & van Schaik, 1994). Species, or populations, are usually classified as typically having one-male or multi-male bisexual groups as basic social units. Nevertheless, there is appreciable intra-specific variation. In many species classified as having one-male social systems, groups with more than one male are occasionally reported (Pusey & Packer, 1987). Eisenberg *et al.* (1972) first described for several species that one-male groups can grow into age-graded groups with more than one adult male when the dominant male is tolerant of his offspring and allows them to stay after maturation. One-male, age-graded male, and multi-male groups were seen as gradations within social structures, with increasing complexity based on increased tolerance amongst adult males. Eisenberg *et al.* also suggested a strict differentiation between age-graded and true multi-male species. The concept of tolerance provides no testable predictions to study this variation empirically. Since then, cost/benefit analyses have been introduced to study such questions.

The costs and benefits of having more than one male in a group can be studied most profitably in those species in which males regularly occur in more than one social setting: one-male groups, age-graded groups, and all-male bands (and/or solitary males). This is the case in the Thomas's langur (*Presbytis thomasi*). The life-span of a bisexual group in this species is restricted to the tenure of its reproductive male. This chapter presents evidence that variation in male group membership is not random, but follows a pattern that develops in a predictable way. In this chapter, we document this development and examine the costs and benefits of males' life history phases corresponding to their successive membership of different types of groups.

Possible benefits of age-graded groups, compared with one-male groups

The degree to which access to females can be monopolized has been proposed as a major factor determining the number of adult males in a group (Emlen & Oring, 1977; Hanuman langurs, *Presbytis entellus*: Newton, 1988; Srivastava & Dunbar, 1996; black and white colobus, *Colobus guereza*: Dunbar, 1987; primates in general: Struhsaker & Leland, 1979; van Hooff & van Schaik, 1992, 1994; Mitani *et al.*, 1996a). Whereas females mainly compete for food, which can, in principle, be partitioned and shared, males compete for fertilizations, which cannot be shared (Trivers, 1972; Alexander, 1974; Wrangham, 1980). If there are no costs to the parents in terms of survival and fecundity, sons can be allowed to wait for breeding opportunities in the natal group (Waser, 1988). However, maturing sons could become potential competitors, for food as well as for mates. A male can deal with mature or maturing sons in two ways: he can either expel these rival males from the group or tolerate them. The potential benefits of tolerating the rivals may offset the costs of expelling them. In Thomas's langurs, both expulsion and tolerance have been observed. This raises the questions of under which conditions tolerance occurs and what benefits might accrue from such tolerance.

In general, males in groups with multiple males can benefit by sharing the vigilance task for predators (Elgar, 1989; Roberts, 1996) and/or against hostile extra-group males (Rose & Fedigan, 1995), although there may be a collective action problem (Nunn, Chapter 17). Van Schaik & van Noordwijk (1989) suggested that males may have a superior ability in detecting predators. However, Steenbeek *et al.* (1999) could not show that adult breeding male Thomas's langurs were more vigilant than females in predator-sensitive positions. Therefore, not only maturing sons, but any additional individual could be beneficial for the adult breeding male, which makes the 'shared vigilance for predators' hypothesis an unlikely explanation for the tolerance of maturing sons. Steenbeek *et al.* (1999) also showed that adult breeding males with infants in their group, as well as females with an infant, were vigilant in those locations and positions where they were prone to attacks by extra-group males. Therefore, 'shared vigilance for extra-group males' can be an advantage of age-graded groups in Thomas's langurs.

In mountain gorillas (*Gorilla gorilla beringei*), groups most often have one adult male (60 %), but age-graded and all-male groups also occur (Robbins, 1995). Watts (1989) and Sicotte (1993) reported for mountain gorillas that males in groups with more than one adult male can benefit by collaborating in excluding extra-group males. Thus, they can prevent contact between these extra-group males and their females, and they can lower the risk of infanticide by these males (see also Watts, Chapter 15). Because of these advantages, Robbins (1995) suggested that reproductive success can be higher for dominant males in multi-male groups than in one-male groups. Sekulic (1983a) found that subordinate male red howler monkeys (*Alouatta seniculus*) may prevent the entry of outside males into the group and thus deter infanticide. In addition, Pope (1990) found that coalitions of males had a superior competitive ability over single males in establishing and maintaining tenure, which increased the reproductive success of dominant males. Byrne *et al.* (1987) report for mountain baboons (*Papio ursinus*) that males in two-male groups collaborate during between-group encounters. However, Dunbar (1987) found for black and white colobus that multi-male groups had lower reproductive rates than one-male groups, but this study took place when there was a multi-male phase during a male takeover. So, two more possible advantages of developing into age-graded groups are 'collaboration in keeping extra-group males away from the females and infants', and, as a consequence of this, 'an increase in reproductive success'.

To achieve tolerance and co-operation between males, the conflict over direct reproductive interests has to become subordinate to other interests. One of these other interests can be inclusive fitness. Kinship appears to contribute to the development of male bonds (Gouzoules & Gouzoules, 1987), although it is not a prerequisite (Hill & van Hooff, 1994). Because males in age-graded groups are related, 'an increase in inclusive fitness' could also be one of the benefits.

Therefore, we will discuss these possible benefits of the development into age-graded groups; namely, 'shared vigilance for extra-group males', 'collaboration in keeping extra-group males away from the females and infants', 'increased reproductive success', and 'increased inclusive fitness'.

All-male bands, compared with bisexual groups

When a male is not tolerated by a dominant male in a group, he has the choice of remaining solitary or of joining other 'lone' males, thus forming an all-male band. These males have to make the best of a bad job, but there could be some advantages for males in all-male bands, compared to males in bisexual groups. Because group sizes of all-male bands are often smaller than those of bisexual groups, this should result in a lower level of food competition (Alexander, 1974) – in Thomas's langurs, at least reduced scramble competition (van Schaik, 1989), although food does not seem to be a limiting factor in this species (Sterck, 1995; Steenbeek, 1999). Van Schaik & van Noordwijk (1986) suggested that adult males in bisexual groups have to adjust their diet to that of the females, whereas adult males without females in the group can feed on a more preferred diet, consisting of relatively more energy-rich foods, such as fruits and flowers, and fewer leaves. Another advantage for males in all-male bands could be a lower level of aggression, and/or a higher level of affiliative behavior. In a number of species in which all-male bands occur, interactions among these males are conspicuously peaceful and friendly. This is in marked contrast to the hostility between these males when they are in bisexual groups (Pusey & Packer, 1987). We will investigate, therefore, whether males in all-male bands are better off than males in bisexual groups.

All-male bands, compared with living solitarily

When males aggregate in an all-male band, group size is larger than when males stay solitary and, consequently, food competition will increase (Alexander, 1974). This cost should be outweighed by benefits. The following benefits have been proposed.

1. Advantages in collective foraging or information sharing, these being especially important for juveniles (e.g., Wrangham, 1980), although others have argued that these benefits must be considered secondary (van Schaik & van Hooff, 1983; Dunbar, 1988).

2. Increased protection against predators (Struhsaker, 1969; Pusey & Packer, 1987, but data are not consistent; Cheney & Wrangham, 1987).
3. Increased protection against harassment by extra-group males (mountain gorillas: Yamagiwa, 1987a).
4. The formation of a male alliance that increases the chance of entering a bisexual group (lions, *Panthera leo*, and red howlers: Pusey & Packer, 1987; Hanuman langurs: Sommer, 1988; Rajpurohit *et al.*, 1995), although in most primate species co-operation between all-male band males is usually only slight, and it is rare for more than one male to stay in the bisexual group after a takeover (Pusey & Packer, 1987).

Therefore, we will investigate why males aggregate in all-male bands, instead of staying solitary.

Thomas's langurs

This chapter investigates the dynamics of changes in male group membership in wild Thomas's langurs. Thomas's langurs live in comparatively small groups, with one to six females per group and typically one breeding male. All-male bands and solitary males are also observed. Both males and females disperse from their natal groups, female secondary dispersal is also common (Sterck, 1997; Steenbeek, 1999), and the timing of female dispersal is influenced by infanticide risk (Sterck, 1997). Mortally wounded infants and infanticidal attacks by extra-group males have been observed (Steenbeek, 1996, 1999; Sterck, 1997). Males have larger canines than females but there is no sexual dimorphism in body size (Sterck, 1997). Dominance relationships are weakly expressed (inside food patches) or not apparent (outside food patches), but the adult breeding male is always dominant over all other group members (Sterck & Steenbeek, 1997).

Home ranges of neighboring groups partly overlap and in this overlap area groups may interact. Two types of interactions between groups have been distinguished (Steenbeek, 1999). (1) Groups approach each other to within 50 m; this is defined as a between-group encounter. These can be neutral, but males can also chase individuals of other groups. (2) Males can silently approach a group and suddenly attack the individuals (often females); this is defined as a male attack. In Thomas's langurs, infanticidal attempts are only observed during male attacks. Both between-group encounters and male attacks only take place in areas where the home range overlaps with that of other langur groups – both bisexual groups and all-male bands (Steenbeek, 1999).

METHODS

Study area and subjects

This study was conducted at the Ketambe Research Station (3°41′ N, 97°39′ E), Gunung Leuser National Park, situated in northern Sumatra, Indonesia. The study area, approximately 200 ha, mainly consists of undisturbed primary rainforest, as was described by Rijksen (1978) and van Schaik & Mirmanto (1985).

Data were collected on 11 bisexual groups, seven all-male bands, and two solitary males. The composition of the study groups is shown in Table 12.1. All individuals were recognized individually. We distinguished infants (younger than two years and still spending time in the ventro-ventral position with the mother), juveniles (from weaning until sexual maturity), young adult females (natal, nulliparous, and older than four years and two months – the age of the youngest female that ever conceived), young adult males (natal or in all-male bands/solitary; descended testes were observed, in combination with attempts to produce loud calls or sexual behavior; over five years of age), adult females (sexually mature females in non-natal group), adult breeding males (resident males in non-natal bisexual groups), adult males (sexually mature males; age not known; in all-male bands/solitary), and post-tenure males (in all-male bands/solitary; males that already had tenure). Details about potential predators for this species have been described elsewhere (Steenbeek *et al.*, 1999)

Data collection

Data on demographic changes were collected under the responsibility of EHMS (1988–91), RS (1992–5), and Amanda H. Korstjens (1996). Detailed data on male behavior were collected from March 1993 until July 1996 by seven different observers. Observations were only made after observers had been trained to reach inter-observer agreement percentages over 90% with RS.

The age-graded groups for which detailed data were collected were groups K1, N, L, and J1. In addition to the adult breeding male, the following adult males were present in the groups: in group K1 there were two young adult males, estimated to be around six and seven years of age; in groups N and L there was one young adult male who was estimated to be around seven years of age; and in group J1 there was one young adult male known to be 5.5 years of age.

The analysis of all-male bands concerns ten individual

Table 12.1. *Group composition of the study groups before male tenure ended and during the subsequent AMB phase*

Group	Phase	Adult males	Young adult males	Juvenile males	Adult females	Young adult females	Juvenile females	Total	Duration of tenure (months)
U	Bisexual	1(2)[a]	0	2	2	0	2	8	?
	AMB-U	2	0	2(4)[b]	0	0	2	6(8)	
B1	Bisexual	1	0	5[b]	4	0	1	11	?
	AMB-B1	1	0	3	0	0	0	4	
B2	Bisexual	1	0	2	3	2	2	10	81
	AMB-B2	1	0	2	0	0	1	4[c]	
B3	Bisexual	1	0	0	6	0	0	7	4
	AMB-B3	1[d]	0	0	0	0	0	1	
J1*	Bisexual	1	1	2	3	2	2	11	75
	AMB-J1	1	1	2	0	0	2	6	
M1	Bisexual	1	0	2	3	1	3	10	69
	AMB-M1	0[e]	0	2	0	0	1	3	
M2*	Bisexual	1	0	2(1)[f]	4	1	0	8	31
	AMB-M2	1	0	1	0	0	0	2	
N	Bisexual	1	1	1	1	0	0	4	?
	AMB-N	1[g]	1[h]	1	0	0	0	2	
K1	Bisexual	1	2	4	4	2	0	13	92
	AMB-K1	1(0)[i]	1[j]	4	0	0	0	6(5)	
L*	Bisexual	1	1	2	1	1	0	6	?
	AMB-L	1	1	2	0	0	0	4	
R	bisexual	1	0	1	4	1	0	7	?
	AMB-LR	1[k]	0	1	0	0	0	6[l]	
?	AMB-X	0–3	2–3	3–4	0	0	0	5–11	
L,R,K1*	AMB-LRK	3	2	6	0	0	0	5–11[m]	

Notes:

Phase: bisexual = group composition before male tenure ended in a bisexual group. Directly below is the group composition of the remaining all-male band (AMB) after all adult females had left. The classes 'Juvenile males' and 'Juvenile females' include infants which were weaned because the mother had left the group. Duration tenure: the duration of male tenure of the sum of the bisexual group phases.

* These groups and all-male bands were used in the detailed analysis.

[a] Six months before group U became an all-male band (there was still one adult female in the group), an adult male joined the group. Nothing is known about his relationship with the group members or his age (young adult or past breeding).

[b] Two juvenile males from group B1 joined AMB-U. It was not known whether they had been expeled by their father or had left their group voluntarily.

[c] Because the individuals of AMB-B2 could no longer be found, we do not know whether the individuals stayed together after the last females had left.

[d] The tenure of the male of B3 ended because of an aggressive takeover. After the takeover, the male disappeared.

[e] The tenure of the male of group M1 ended because the male had died (Steenbeek, 1996).

[f] One of the juvenile males died.

[g] This adult male started out as a solitary male because his son tried to transfer with his mother. When this juvenile male was expeled from the new group, he returned to his father and they disappeared. Two years later this juvenile male joined AMB-LRK.

[h] This young male directly started a bisexual group from his natal group.

[i] The adult male left the all-male band after two weeks with one small juvenile male, who later returned to the all-male band alone.

[j] One of the young males started a bisexual group directly from his natal group (aggressive takeover).

[k] AMB-R came into existence after an aggressive takeover.

[l] When the males of group R were first found, they had already joined AMB-L. AMB-LR later joined AMB-K1. Then we called it AMB-LRK.

[m] AMB-LRK often split up into subgroups. AMB-X also did this when group size was large.

Group names of Steenbeek (1996) and Sterck (1997), in comparison with this table: M=M1, J=J1, K=K1, H=B1, B=B2.

males, coming from four bisexual groups and all-male bands (see * in Table 12.1): three adult breeding/post-prime males, three young adult males and four juvenile males (over three years of age). Two solitary males were included in some of the analyses. One was the post-tenure male from group N which was solitary for a few months until his presumed son joined him. The other solitary male was the oldest juvenile male of all-male band M1; after reaching adulthood, he left the all-male band, and was observed as a solitary male a few months before he aggressively took over group B3.

Details of data collection have been described elsewhere (Steenbeek *et al.*, 1999; Steenbeek, 1999). One focal period of one group consisted of at least 600 minutes for the adult breeding male of a bisexual group, and for post-tenure and young adult males in all-male bands, and at least 400 minutes for young adult males in bisexual groups, and juveniles in both group types. This chapter compares the last focal period of a bisexual group with the first focal period of the same males during the subsequent all-male band phase. All interactions between groups and observed wounds were recorded in detail. A wound was only counted once, and only if it was observed when a group was being followed. A wound was not counted, for example, when it was observed on an individual from a neighboring group. Extra-group males were defined as all males that were not resident in the group. Any attack by an extra-group male was recorded, whether it was by a male from another bisexual group (the majority) or by a male from an all-male band.

Data analyses

Because seasonal changes in the Thomas's langur diet do not influence time spent feeding and allogrooming (Steenbeek, 1999), data from different months could be compared directly. A comparison between bisexual groups and subsequent all-male bands was carried out at two levels. For ten individual males, we collected detailed focal data between one and five months before this male's tenure in a bisexual group ended (this is part of the late tenure phase), and we collected focal samples in the first month after these individuals had ended up in an all-male band (AMB phase). When data for these ten males in bisexual groups were compared with data for the same ten males in all-male bands, statistical tests were first carried out at the individual level (e.g., total interaction frequency with the individual). For all measurements concerning proximity and social interactions, tests were repeated at the dyadic level (e.g., interaction frequency between both males of a dyad). In total there were 15 male–male dyads, and each dyad was first measured in a bisexual group, and again in an all-male band. Results on the dyadic level should be interpreted with caution because some males occurred in more than one dyad (one male in one dyad, three males in two dyads, six males in three dyads). Both at the individual level and at the dyadic level, Wilcoxon Signed Ranks tests were used (Siegel & Castellan, 1988).

Before this test was carried out, differences between the late tenure phase in bisexual groups and the all-male band phase were calculated and these differences were tested with a Kruskal-Wallis one-way ANOVA (Siegel & Castellan, 1988), to determine whether males of different age-classes differed in their reaction. A significant effect of age was found in only one of the behavioral measures: the frequency of 'aggression, given,' in the context of 'feeding.'

Two data points of solitary males were included in the analyses of food competition in all-male bands. Because contact with these males was lost again after 1.5 and 3.5 days respectively, contact time was too short to calculate frequencies of being wounded or being attacked. Spearman rank correlations between group size and time spent feeding were calculated (separately for post-tenure males and young adult males).

In the Thomas's langur diet, the percentage of young leaves shows a strong negative correlation with the percentage of fruit (Sterck, 1995). Because both fruit and flowers mainly consist of energy, whereas leaves are an important source of protein, we chose the percentage of leaves in the diet as the best indicator to test differences between bisexual groups and all-male bands. To test these differences, the average diet of the adult males of each of seven all-male bands was compared with the diet of the adult male of a bisexual group, followed in the same month and the same area (similar or overlapping home range), in order to control for differences between seasons and/or area.

The measurement of time spent alone (or time spent with neighbors at certain distances) addresses the question of safety. To answer this question, it is appropriate to compare the percentage of time spent alone by males in different social settings, because it is the relative amount of time spent alone that indicates the risk-sensitivity of the animal. Because active day length (time between leaving the sleeping tree in the morning and entering in the evening) varied between 661 and 748 minutes, comparisons between individuals on how much time individuals spent on a certain activity, calculated as percentages of time, were not correct. Therefore, feeding and allogrooming were expressed in minutes per day, because the absolute time spent involved in these activities is indicative of the respective cost or benefit.

Because the characteristics of a food patch (i.e., physical space, amount of food) influence behavior when feeding (Sterck & Steenbeek, 1997), all data (except for activities) were analyzed for two different contexts – 'feeding' and 'not feeding' – in order to lower the influence of food patch characteristics on patterns of behavior. Classes of activities, other than resting, in the context of 'not feeding' were all below 5%. The following variables were analyzed: time spent feeding (minutes per day), diet (percentage of leaves), aggression (frequency per hour), displacements (frequency per hour), time spent allogrooming (sum of active grooming and being groomed, minutes per day), affiliative behavior (frequency per hour), tolerant approaches (frequency per hour), time spent alone (percentage of time), time spent with at least one neighbor in contact, between 0 and 1 m, between 1 and 2 m, and between 2 and 5 m (percentage of time), average number of neighbors (number per minute), wounds (wounds per 100 days), and average height (m). Multiple tests on one data set were corrected with a Bonferroni procedure (Hochberg, 1988).

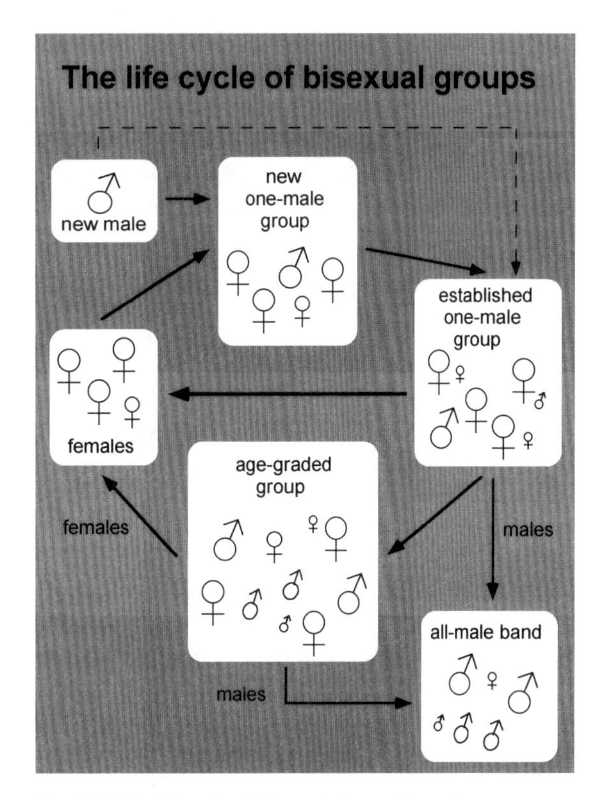

Fig. 12.1. The life cycle of Thomas's langur bisexual groups.

RESULTS

The life cycle of groups

In wild Thomas's langurs, social groups were observed to have a limited life span, lasting as long as the tenure of the respective breeding male. The phases in the life cycle of a group are illustrated in Fig. 12.1. New groups were formed in two ways: gradually and suddenly.

Groups formed gradually when females transferred to a new male, but not all at the same time ($n=6$). Groups formed suddenly when all females were present when the new male started his tenure: in two cases there was an aggressive takeover ($n=2$), and in one case a male joined a group of females after the former male had died (Steenbeek, 1996). In addition to these nine cases, there are four cases for which we do not know the formation history. In the new groups, females usually reproduced and the group grew. We considered such groups to be established groups. A group developed into an age-graded group when the male sired sons early in his tenure, and when his tenure lasted long enough to enable a (presumed) son to reach sexual maturity. Some medium-size male juveniles were expelled by their father (three to five out of 18 juvenile males).

Table 12.1 shows the composition of bisexual groups during the late tenure phase. Male tenure was said to end when all sexually mature females (adult and young adult) had left the male. Male tenure could also end gradually or suddenly. It ended gradually when the females left the male, in favor of a new male, but not all at the same time ($n=8$, including one single-female group where this female left; described in detail in Steenbeek, 1999). Male tenure terminated suddenly when the adult breeding male died ($n=1$), or in the case of an aggressive takeover ($n=2$). After one of these two takeovers, the adult male disappeared (solitary male, group B3); after the other takeover (in group R), the male and his presumed son joined AMB-L, which we then called AMB-LR. After the adult breeding male had lost all sexually mature females, an all-male band remained ($n=10$) or a solitary male ($n=1$). In all cases, this all-male band consisted of all the males of the former group. It always included the last cohort of offspring, which could contain small juvenile females (four out of 11 cases). Known tenure varied from four to 92 months, with a median of 72 months ($n=6$; see Table 12.1).

In ten out of 13 new groups, the age of the new males was known: nine males were young adults, starting their first tenure, and at least seven years of age. One male already had a tenure before (group B1), and was observed with one female during a second tenure, the end of which was not witnessed. Of the three males with unknown ages, one was estimated to be old when he joined a maleless group (group M1, Steenbeek, 1996), and the other two looked quite young.

The age-graded phase in bisexual groups

We will first describe in more detail how groups develop from one-male into age-graded groups in Thomas's langurs, and why some groups do not. We will then discuss factors that were reported to reflect advantages of developing into an age-graded group, that we could measure, namely: shared vigilance for extra-group males, and extra help in keeping extra-group males away from the group.

THE DEVELOPMENT FROM A ONE-MALE INTO AN AGE-GRADED GROUP

Figure 12.2 shows what happened to natal males in bisexual groups. Mortality for male infants (under two years of age and not yet weaned) was 36%. After weaning, infant mortality strongly decreased. Two small juvenile males transferred with their mother. One of these transfers took place into the age-graded group K1, 22 months before male tenure ended. The small juvenile male was accepted in this group. The other small juvenile male transferred with his mother into the newly formed group K2. This juvenile male was immediately evicted from group K2 by the adult breeding male (with the support of some of the females), and rejoined his father. Ten small juvenile males ended up in an all-male band, after the tenure of their father had ended. Of 15 medium-sized juvenile males, three were expelled by their presumed father, two others were found in an all-male band (possibly expelled, the moment of departure from the bisexual group not witnessed), and five males ended up in an all-male band because male tenure had ended. In the two groups in which breeding males expelled juvenile males, there were still younger juvenile males in the group. Figure 12.2 shows that there was no mortality after the age of three years, but this does not account for males that disappeared from all-male bands with unknown fate. We should keep in mind that the counts in Fig. 12.2 were influenced by both the length of male tenure, and when a male was born. Males born in the first cohort of offspring had a higher chance of reaching maturity in the natal group than those born in the second cohort, and males from the third cohort never reached maturity in a bisexual group.

Table 12.1 shows that, in bisexual groups in which the initial breeding male reached the late phase of his tenure, only five out of 11 developed into age-graded groups (45%). Four groups developed into age-graded groups because presumed sons reached sexual maturity (groups K1, J1, N, and L). In the fifth group, an adult male immigrated and joined the resident male when there was still one adult female that had not yet left (group U). The history of this group and the possible relatedness of the group members with this male were unknown. Six groups did not develop into age-graded groups. The adult male of one of these groups had expelled one juvenile male, and he died before another juvenile male could reach adulthood (group M1). In the second group, male offspring were born only in the last cohort of offspring, so it was impossible for this group to develop into an age-graded group before tenure ended (group B2). In the remaining four groups, tenure was too short for male offspring to reach adulthood (B1 and M2: the females left; R and B3: aggressive takeovers). So age-graded groups did not develop if a breeding male's tenure was of insufficient length to allow male offspring to mature.

It seems that, in age-graded groups, there is little or no reproductive competition between the adult breeding male and young adult males. There is only one observation of a copulation attempt between the young adult male of group L and a young adult female of this group. They were assumed to be half-sibs. Their father intervened, and successfully broke off their copulation attempt. Very rarely, young adult males were observed to attempt to copulate with young adult females of a neighboring group.

SHARED VIGILANCE FOR EXTRA-GROUP MALES

Steenbeek *et al.* (1999) deduced that vigilance at low heights reflects predation risk, whereas vigilance above 20 m reflects the chance of being attacked by extra-group males. Young adult males did not show this vigilance pattern: they tended to be *less* vigilant above 20 m than between 10 m and 20 m in overlap areas (Wilcoxon Signed Ranks Test, $n=5$, $T^{+}=14$, one-tailed, $p=0.063$), although this difference was not significant.

HELP TO DEFEND THE GROUP DURING INTERACTIONS WITH EXTRA-GROUP MALES

In age-graded, but not in single-male groups, young adult males participated in aggressive encounters with other groups or with an extra-group male, in this way supporting their father. The participation of young adult males in such intergroup interactions developed gradually. First, young

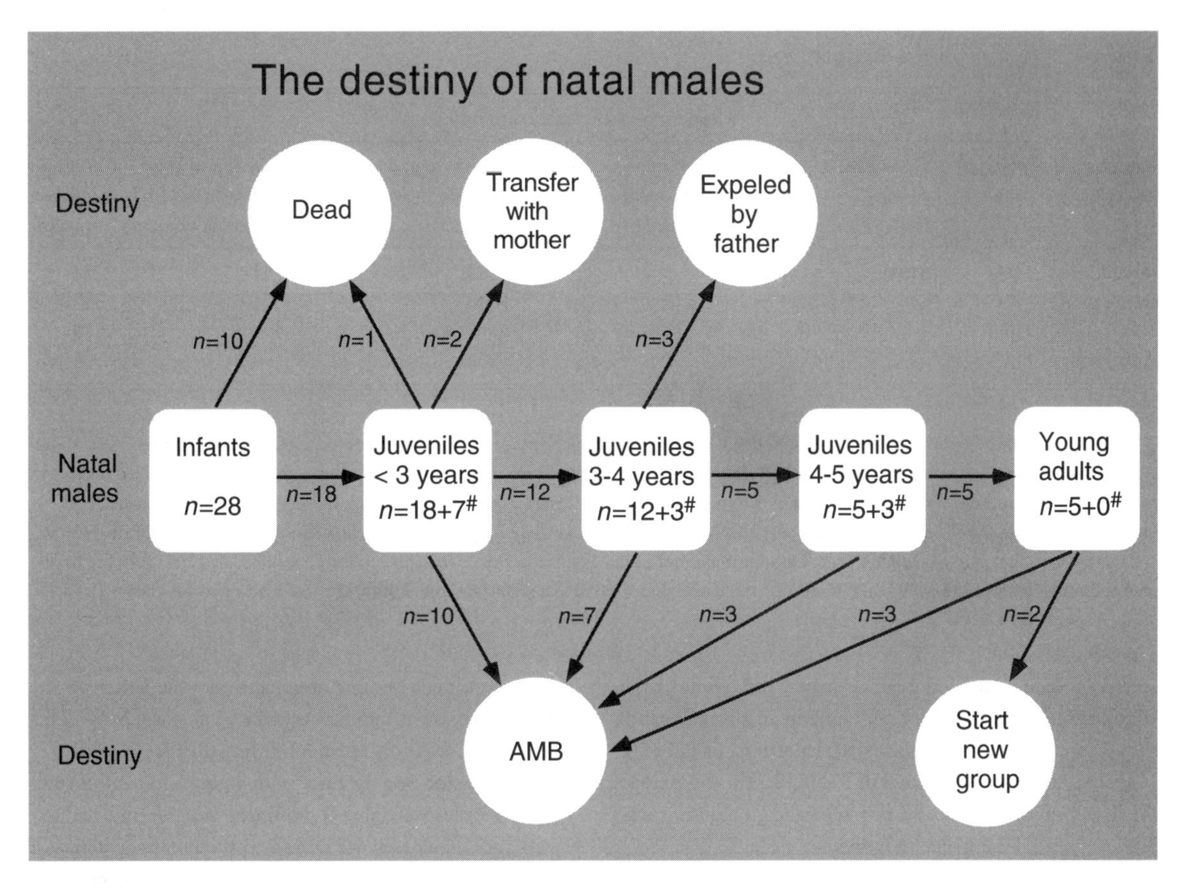

Fig. 12.2. The destiny of natal males. # represents males that were first observed when already in that age-class. (AMB = all-male band.)

adult males followed their father when he attacked other bisexual groups (this could already occur from around 4.5 years of age). However, when it came to fighting, the young adult males were the first to leave. When young adult males were around 5.0 years of age, they started to behave aggressively and they supported their father during aggressive between-group encounters and when being attacked by an extra-group male (33–100% of the interactions). From around 6.5 years of age, young adult males regularly took the initiative in attacking bisexual groups and all-male bands. In age-graded groups, during the late phase of the adult breeding male's tenure, he no longer attacked extra-group males alone. However, the four age-graded groups did not differ from the two single-male groups in the total frequency of group encounters and attacks.

Loud calls showed a similar development. First, young adult males initiated bouts when there was no other group in the vicinity, and these calls could easily be recognized because of the high-pitched frequency. Later, they joined the adult breeding male when he answered loud calls from a group in the vicinity. Only in group K1 did young adult males also initiate loud call bouts when there was another group in the vicinity. The only age-graded group with two young adult males in addition to the adult breeding male, group K1, had the longest tenure we ever observed at Ketambe: 92 months.

A COMPARISON BETWEEN BISEXUAL GROUPS AND ALL-MALE BANDS

Group membership of all-male bands was not as stable as that of bisexual groups. All-male bands showed some kind of a fission–fusion pattern. Neighboring all-male bands joined, large all-male bands split into subgroups (always with one adult or young adult male), and males temporarily left or joined an all-male band. All-male bands typically had a flexible home range, and small groups of males or solitary males left this home range temporarily and traveled large distances in one day. Table 12.2 shows the destiny of all-male band males of different age-classes. Most juvenile males

Table 12.2. *The destiny of all-male band males*

Age–class (males)	The destiny of all-male band males							
	Stay in same age class (%)	Grow into a young adult (%)	Start a new bisexual group (%)	Disappear + later start a new bisexual group (%)	Disappear and not found again (%)	Total (males)	Disappear as all-male band (number of males)	Grand total (males)
Juvenile	2 (14)	9 (64)	—	—	3 (22)	14	11	25
Young adult	1 (7)	—	2 (14)	5 (36)	6 (43)	14	1	15
Adult	0 (0)	—	—	1 (33)	2 (67)	3	1	9
Post-tenure	2 (33)	—	—	1 (17)	3 (50)	6	3	4

Note:
The numbers in the columns represent the numbers of males, with the percentage of the total in parentheses. Stay in same age-class: these males stayed in the all-male band and in their age-class (at least 1.5 years). Grow into a young adult: these juvenile males reached maturity in the all-male band, and were again included in the young adult sample. Start a new bisexual group: these males started a new bisexual group directly from the all-male band. Disappear and later start a new bisexual group: these males first disappeared from an all-male band and started a bisexual group at a later time. Disappear and not found again: these males disappeared from an all-male band and were never seen since. Disappear as all-male band: these all-male bands disappeared as a whole and the destiny of these males was not known.

stayed to reach adulthood in an all-male band. None of the young adult males stayed for very long in an all-male band; they disappeared, started a new bisexual group, or did both in succession. The percentages of Table 12.2 do not include the disappearances of whole all-male bands, because it was not known what happened to individual males.

From the 11 cases for which male age was known when a new group was formed, two males came from a bisexual group, three came directly from an all-male band, and six first disappeared from an all-male band (we could not locate them in the study area), and each was later found as the adult breeding male in a new group (in the study area).

TIME SPENT FEEDING AND DIET

Table 12.1 shows that for new all-male bands, group sizes were always smaller than those of the former bisexual groups. Males in all-male bands spent more time feeding than the same males in bisexual groups (see Table 12.3). We found no difference in the percentage of young leaves in the diet (Wilcoxon Signed Ranks Test, one-tailed, $n=7$ comparisons at the group level; $T^{+}=15$, $p=0.47$).

SOCIAL RELATIONSHIPS: INTRAGROUP AGGRESSION AND DISPLACEMENTS

The adult breeding male was always the most dominant individual in a bisexual group. Of the four age-graded groups, three ended as an all-male band, with both the post-tenure and a young adult male in it. In these three cases, the young adult male became dominant over the father on the very first day the group had become an all-male band. This rank reversal was only apparent during the first week; thereafter, aggression became rare. Both young adult and former breeding males were always dominant over juvenile males.

As for all analyses, we examined the potential effects of age. Only in the analysis of frequency of 'aggression given' within the context 'feeding' did we find a significant age effect: most adult and juvenile males showed a decrease after having ended up in an all-male band, as was expected, while young adult males showed an increase (Kruskal–Wallis ANOVA, $n1=3$, $n2=3$, $n3=4$, $X^2=6.56$, $p<0.05$). Because two young adult males were observed during the first week after the start of the AMB phase, this increase of 'aggression given' within the context 'feeding' might have been caused by the rank reversal between young adult males and former adult breeding males.

All-male bands did not differ significantly from bisexual groups in the frequencies of 'aggression given' and 'aggression received', when tested at the individual level (irrespective of direction and context, one-tailed; Table 12.4). The aggression frequency within male dyads also did not differ significantly between bisexual groups and all-male bands (analyzed at the dyadic level, and irrespective of context, one-tailed, see Table 12.5).

Bisexual groups and all-male bands did not differ significantly in the frequencies of 'displace other' and 'being displaced', when tested at the individual level (irrespective of

Table 12.3. *Median, minimum, and maximum values of activities and proximity, which were analyzed at the individual level (*n *=10 males)*

Variable	Context	WSR test	n	Late tenure phase in bisexual group		AMB phase	
				Median	(minimum–maximum)	Median	(minimum–maximum)
Time spent feeding (minutes/day)	Total	$T^+=55^\dagger$	10	235.2	(141.3–276.1)	271.0	(157.1–311.5)
Time spent grooming (minutes/day)	Total	$T^+=28^\dagger$	7	0.55	(0–5.8)	0	(0–0.7)
Time spent alone (% of time)	Not feeding	$T^+=50^*$	10	62.5	(48.5–83.3)	53.3	(34.1–65.1)
	Feeding	$T^+=46$	10	61.7	(40.0–76.8)	49.6	(41.3–63.8)
Time spent with at least 1 neighbor							
at contact (% of time)	Not feeding	$T^+=16$	6	0	(0–4.4)	0.4	(0–12.8)
at 0–1 m (% of time)	Not feeding	$T^+=33$	10	7.1	(0.3–16.0)	7.7	(0–19.6)
at 1–2 m (% of time)	Not feeding	$T^+=49^*$	10	5.0	(1.9–11.8)	16.9	(2.8–22.5)
at 2–5 m (% of time)	Not feeding	$T^+=31$	10	22.1	(13.7–37.9)	20.7	(8.2–40.4)
Average number of neighbors <5 m	Not feeding	$T^+=28^\dagger$	7	0.6	(0.2–0.9)	0.8	(0.4–1.5)
(per minute)	Feeding	$T^+=43.5$	10	0.7	(0.4–1.2)	0.9	(0.4–1.2)
Average height (m)	Not feeding	$T^+=44^*$	9	14.2	(7.6–23.4)	12.4	(5.8–17.8)
	Feeding	$T^+=49^*$	10	11.9	(8.7–21.6)	9.8	(6.2–15.9)
Time spent alone							
at 0–10 m (% of time)	Not feeding	$T^+=45$	10	70.8	(49.1–88.3)	64.2	(51.8–82.2)
at 10–20 m (% of time)	Not feeding	$T^+=34$	10	55.9	(31.7–95.3)	41.3	(29.9–91.3)
at > 20 m (% of time)	Not feeding	$T^+=45^\dagger$	9	66.3	(46.7–100.0)	30.3	(5.0–61.3)

Note:
$*=p<0.05$; $\dagger=p<0.01$; WSR Test = Wilcoxon Signed Ranks Test; n = the number of pairs without ties that were included in the WSR Test. Tie values were 0 in most cases.

context, two-tailed; see Table 12.4). The displacement frequency within male dyads was significantly higher in all-male bands for the ten (out of 15) dyads that expressed this behavior, but only when not feeding (two-tailed; see Table 12.5).

SOCIAL RELATIONSHIPS: AFFILIATIVE BEHAVIOR

Bisexual groups and all-male bands did not differ in the frequencies of 'tolerant approach' (an approach to within 0.5 m without a reaction by the approached individual within 5 s) and 'tolerant of being approached,' when tested at the individual level (irrespective of context, two-tailed; see Table 12.4). The total frequency of 'tolerant approach,' as a combined measure of dyadic tolerance, tested at the dyadic level, was significantly higher in all-male bands when not feeding (corrected for two tests, two-tailed; see Table 12.5).

Males in all-male bands spent significantly less time allogrooming than males in bisexual groups (two-tailed; see Table 12.3). However, no differences were found at the dyadic level because males hardly ever groomed one another. The decrease in time spent grooming was mainly caused by the fact that females groomed males in bisexual groups.

Bisexual groups and all-male bands did not differ significantly in the frequencies of 'affiliative behavior given' and 'affiliative behavior received,' when tested at the individual level (not feeding, two-tailed; see Table 12.4). The frequency of 'affiliative behavior,' tested at the dyadic level, also showed no significant differences (see Table 12.5).

Proximity patterns

When males from a bisexual group had ended up in an all-male band, time spent alone decreased significantly, but only when males were not feeding (Fig. 12.3; corrected for two tests, two-tailed; see Table 12.3). This means that time spent with neighbors within 5 m when not feeding must have increased, and it is interesting to see at what distance this increase took place. The results for time spent with at least

Table 12.4. *Median, minimum, and maximum values of social frequencies, which were analyzed at the individual level (n=10 males)*

Variable (frequency/hour)	Context	WSR test	n	Late tenure phase in bisexual group		AMB phase	
				Median	(minimum–maximum)	Median	(minimum–maximum)
Aggression given	Not feeding	$T^{+}=15$	6	0.17	(0–0.74)	0.0	(0–0.65)
	Feeding	$T^{+}=33$	9	4.16	(0–7.54)	1.20	(0–9.89)
Aggression received	Not feeding	$T^{+}=6$	4	0	(0–0.54)	0.0	(0–0.65)
	Feeding	$T^{+}=11$	6	0	(0–25.44)	0.0	(0–7.02)
Displace other	Not feeding	$T^{+}=28$	8	0.33	(0–1.68)	0.51	(0–2.18)
	Feeding	$T^{+}=27$	4	1.85	(0–9.59)	0.54	(0–2.97)
Being displaced	Not feeding	$T^{+}=15$*	5	0	(0–0.33)	0.34	(0–2.33)
	Feeding	$T^{+}=21$	8	2.09	(0–11.45)	1.06	(0–6.94)
Affiliative behavior given	Not feeding	$T^{+}=15$	6	0.49	(0–11.23)	0.23	(0–24.22)
Affiliative behavior received	Not feeding	$T^{+}=24$	9	1.45	(0–4.46)	0.30	(0–19.38)
Tolerant approach	Not feeding	$T^{+}=33$	10	0.85	(0.37–2.50)	1.08	(0–2.04)
	Feeding	$T^{+}=29$	8	1.85	(0–9.57)	0.54	(0–2.97)
Tolerant of being approached	Not feeding	$T^{+}=38$	10	0.95	(0–2.35)	1.21	(0.35–2.70)
	Feeding	$T^{+}=33$	9	2.09	(0–11.45)	1.06	(0–6.94)

Note:
* $p<0.05$.
See Table 12.3. for explanation.

Table 12.5. *Median, minimum, and maximum values of the variables, which were analyzed at the dyadic level*

Variable	Context	WSR test	n	Late tenure phase in bisexual group		AMB phase	
				Median	(minimum–maximum)	Median	(minimum–maximum)
Aggression (frequency/hour)	Not feeding	$T^{+}=11$	5	0	(0–0.57)	0	(0–0.47)
	Feeding	$T^{+}=31$	9	0	(0–8.92)	0	(0–5.05)
Displacement (frequency/hour)	Not feeding	$T^{+}=47$*	10	0	(0–0.43)	0.22	(0–0.78)
	Feeding	$T^{+}=28$	8	0	(0–3.55)	0	(0–3.30)
Affiliative behavior (frequency/hour)	Not feeding	$T^{+}=21$	8	0	(0–6.97)	0	(0–2.05)
Tolerant approach (frequency/hour)	Not feeding	$T^{+}=98$*	15	0	(0–1.27)	0.53	(0–2.05)
	Feeding	$T^{+}=21$	8	0	(0–11.58)	0	(0–3.47)
Time spent with dyadic partner between 1 and 2 m (% of time)	Not feeding	$T^{+}=105^{\dagger}$	15	0.8	(0–4.3)	3.4	(0.2–20.3)
Time spent with dyadic partner between 2 and 5 m (% of time)	Not feeding	$T^{+}=97$*	15	8	(1.7–19.4)	11.6	(3.1–29.4)

Note:
Male dyads were compared instead of individual males; $n=15$ male dyads.
* $p<0.05$; † $p<0.01$.
See Table 12.3. for explanation.

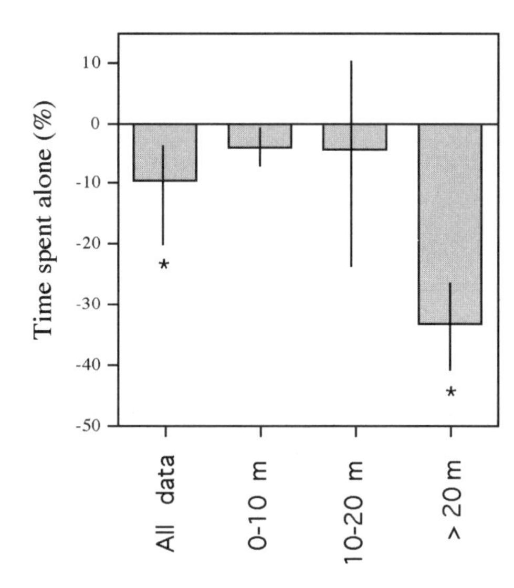

Fig. 12.3. Time spent alone for different height classes, when not feeding (*$p < 0.05$). Median and quartile values of the difference between bisexual groups and all-male bands (AMBs) are shown.

one neighbor differ with distance. When tested at the individual level, and when males were not feeding, there were no differences between bisexual groups and all-male bands for the distances contact, 0–1 m, and 2–5 m (corrected for four tests; see Table 12.3). However, males in all-male bands spent significantly more time with at least one neighbor at the distance 1–2 m (corrected for four tests, one-tailed; see Table 12.3). The average number of neighbors also increased significantly for all-male band males, but only within 5 m (corrected for two tests, two-tailed; see Table 12.3), and not within 2 m.

Because of the results at the individual level, tests at the dyadic level were carried out for both 1–2 m and 2–5 m, for the context 'not feeding.' As expected, time spent with a dyadic partner was significantly higher for all-male bands than for bisexual groups for both the distances 1–2 m and 2–5 m (corrected for two tests, one-tailed; see Table 12.5).

VARIABLES RELATED TO RISK FACTORS: EXTRA-GROUP MALES AND PREDATION

When bisexual groups became all-male bands, the frequency of being attacked by males from bisexual groups significantly increased (Wilcoxon Signed Ranks Test, $n=6$ groups; $T^{+}=21$, $p<0.05$, two-tailed; from Steenbeek, 1999). In addition, males in all-male bands were significantly more often observed to have wounds than males in bisexual groups during the late phase of the tenure (test of observation days with and without a wounded male, $X^2=7.80$, df$=1$, $p<0.01$; Table 12.6). Because of the low expected frequencies of wounding, age-classes could not be tested separately. Most wounds looked like canine wounds and were expected to have resulted from male–male fights. Wounds ranged from superficial flesh wounds (torn-off skin), an immobile leg, to severe head wounds (damaged eye, torn nose and mouth). These results show that the chances of harassment by males from bisexual groups, and of becoming wounded, are higher in all-male bands than in bisexual groups.

Because most predators are ground animals, and large raptors able to catch adult individuals are absent, predation risk is expected to be highest near the ground. Therefore, if predation risk influences the behavior of males in all-male bands, we expect that average height will be higher because group sizes are smaller than in bisexual groups. However, all males showed lower average heights after they had ended up in all-male bands, independent of the context (corrected for two tests, two-tailed; see Table 12.3).

RELATIVE IMPORTANCE OF THE TWO RISK FACTORS

The result that males in all-male bands spent less time alone when not feeding could be caused by both predation and harassment risk. It is very difficult to measure these risks, especially predation risk, but we can try to assess the relative importance of both.

Steenbeek *et al.* (1999) deduced that vigilance at low heights reflects predation risk, while vigilance above 20 m reflects the chance of being attacked by extra-group males. They also showed that vigilance of adult all-male band males reflects both predation risk and the risk of being attacked by extra-group males. The latter was supported by the fact that they found a significant positive correlation for all-male bands between average resting height and the frequency of being attacked by extra-group males.

Because the relative importance of both risks varies with height, we analyzed the proximity data again, but for the three height-classes separately (0–10 m, 10–20 m, and >20 m) and for the not-feeding context. Only above 20 m did males in all-male bands spend significantly less time alone than males in bisexual groups (Fig. 12.3; all three height-classes corrected for three tests, two-tailed; Table 12.3). Furthermore, all-male band members spent significantly less time alone above 20 m than below 10 m (Wilcoxon Signed Ranks Test, $n=9$; $T^{+}=45$, $p<0.01$, two-tailed),

Table 12.6. *Number of wounds per 100 days for males in bisexual groups and all-male bands*

Group type	Age–class (number of males)	Male observation days	Number of wounds (wounded males)	Wounds per 100 days
Bisexual group	Juvenile/young ($n=12$)	486	4 ($n=4$)	0.82
	Adult ($n=6$)	223	2 ($n=2$)	0.90
All-male band	Juvenile/young ($n=16$)	349	10 ($n=6$)	2.86
	Adult ($n=9$)	235	7 ($n=4$)	2.98

whereas no significant difference was found for males in bisexual groups. This strongly suggests that the decrease in time spent alone was caused by the risk of being harassed by extra-group males.

Why do males aggregate in all-male bands, instead of living solitarily?

Both predation risk and the risk of harassment by extra-group males could promote the association among males. An additional observation supports the idea that there is a high risk for solitary individuals: solitary individuals were almost always adult individuals. In the three instances when solitary juvenile males were encountered in the forest, they were traveling with a long-tailed macaque (*Macaca fascicularis*) group and/or emitting soft sounds, that sounded similar to infant distress calls.

When scramble competition for food increases with group size, a positive correlation between group size and time spent feeding is expected. We investigated this for seven all-male bands and two solitary males, and did so for post-tenure males and young adult males separately. There was a significant positive correlation between group size and time spent feeding for young adult males (Spearman rank correlation; $n=6$ all-male bands and 1 solitary male, $r=0.837$, $p=0.019$), whereas there was no such relation for post-tenure males ($n=5$ all-male bands and 1 solitary male, $r=0.232$, $p=0.658$).

One all-male band, AMB-LRK, consisted of 11 males from three different bisexual groups. We analyzed proximity at the dyadic level. Individuals could spend most time with an individual from the former bisexual group (presumed father, son or brother) or with an individual from a strange group. This could be determined for nine out of 11 males. We found that for both the distance class within 2 m, as well as for the distance class within 5 m, three (one of each age class) out of nine males spent most time with an individual from their former bisexual group. These results do not indicate that males prefer to interact with close relatives.

Male alliances, to start or take over a group, seem to be weak or non-existent in Thomas's langurs. Although all-male band males sometimes attacked bisexual groups, alone or in teams, they clearly avoided those most of the time (Steenbeek, 1999). Most young adult males first left an all-male band (alone) and then started a bisexual group (see Table 12.2). Only two young adult males started a bisexual group directly from an all-male band. In the first case, the whole all-male band interacted with group M2. Two of the five females left with the M2 male, and the remaining three females stayed with one of the all-male band males. In the other case, three out of four females of group J1 joined AMB-X. AMB-X was then joined by two new young adult males, and one of these males stayed with the females and started group J2.

DISCUSSION

The age-graded phase: beneficial

In the studied population of Thomas's langurs, one-male groups developed into age-graded groups when male infants were born early in the tenure, when these were not expelled as juveniles by the adult breeding male, and when male tenure lasted long enough for these (presumed) sons to reach maturity (45% of 11 groups). The development into an age-graded group seemed to have no major costs for the adult breeding male, because young adult males never became a reproductive threat to him. Accordingly, the adult breeding male seldom expelled maturing (presumed) sons. This was only observed when the male had other, younger sons in the group, which could still reach maturity at a later time. In contrast with other primate species, young adult Thomas's langur males were never observed to takeover (cf. red howler monkeys: Pope, 1990), or inherit (cf. mountain gorillas: Yamagiwa, 1987a, 1987b; Robbins, 1995) females from the father.

In Thomas's langurs, the development into an age-graded group seemed to have no major costs, but instead was beneficial for the adult breeding male. Young adult males participated in between-group aggression which could have extended the adult breeding male's tenure, and, thereby, possibly increased his reproductive success. Young adult males were not vigilant for attacks by extra-group males. However, a breeding male could still benefit from the vigilance of an additional group member. The development into an age-graded group may also increase inclusive fitness, both for the adult breeding male and for young adult males. It will do so for the adult breeding male because his sons seem to be better off by staying in the natal group. It will do so for the young adult males because a longer tenure of their father may help related infants to survive, and related young juveniles to grow stronger before they end up in an all-male band. In addition, young adult males may get experience in dealing with extra-group males and they may come into contact with females from other groups, with possible support from family members. Thus, two young adult males started a bisexual group with females from a neighboring group, directly from their (presumed) natal group, without spending time in an all-male band. One of these was an aggressive takeover of the neighboring group, and the father and brother were observed to participate in fights with the former breeding male of this group before the takeover.

Robbins (1995) suggested for mountain gorillas that reproductive success is higher in multi-male than in one-male groups because males collaborate in keeping extra-group males away from females and infants, thereby also lowering infanticide risk. In Thomas's langurs, the age-graded phase is only a minor part of male tenure and in two of the four age-graded groups, infants died. It could well be that the presence of young adult males has extended male tenure, and that more infants would have died without the help of young adult males, but more data on life-time reproductive success are needed to answer this question.

Dynamics of the age-graded phase

An important feature of the development of age-graded groups is that additional males are probably closely related to the adult breeding male. In other species with one-male bisexual groups, age-graded groups also occur. In hamadryas baboons (*Papio hamadryas*: e.g., Kummer & Abegglen, 1978; Sigg *et al.*, 1982), gelada baboons (*Theropithecus gelada*: Dunbar & Dunbar, 1975; Dunbar, 1984), and mountain gorillas (Harcourt, 1979b; Harcourt & Stewart, 1981), the adult breeding male may tolerate a 'follower male'. Often, this is a younger, submissive male who buys his right to stay by helping the owner in defense of his females. The 'follower male' may also be an older male, the previous harem holder (Dunbar, 1984; Stammbach, 1987). There is evidence for hamadryas baboons that a second adult male in a harem unit may be related to the present breeding male (Stolba, 1979; Sigg *et al.*, 1982; van Hooff & van Schaik, 1994). In mountain gorillas, males in multi-male groups are thought to be mostly close relatives because male immigration into bisexual groups is rare (Itani, 1977, 1980; Yamagiwa, 1987a; Robbins, 1995).

Other species for which age-graded groups have been described are purple-faced langurs (*Presbytis senex*: Rudran, 1973a), black and white colobus (Dunbar & Dunbar, 1976), red colobus (*Colobus badius*: Struhsaker, 1975; Struhsaker & Leland, 1979, 1985, 1987), silver langurs (*Presbytis cristata*: Wolf, 1984), Nilgiri langurs (*Presbytis johnii*: Hohmann, 1989), and banded langurs (*Presbytis melalophos*: Curtin, 1980). In Hanuman langurs, Hrdy (1977) and Newton (1987) describe a multi-male phase for their study population, but this phase is short lived and seems to occur only during the process of a takeover (Newton, 1987; Sommer, 1988). In other areas, true multi-male groups exist (Srivastava & Dunbar, 1996). In howler monkeys, multi-male groups also arise in different ways, of which the age-graded origin is one (Crockett & Eisenberg, 1987). However, male coalitions of relatives are more successful (higher reproductive success because of a longer tenure) than male coalitions of non-relatives (red howler monkeys: Pope, 1990). An additional variation is that the age-graded phase can be followed by fissioning, with the result that both the breeding male and a young adult male then have reproductive groups (black and white colobus: Dunbar & Dunbar, 1976; Nilgiri langur: Hohmann, 1989). In Nilgiri langurs, all fissions coincided roughly with the onset of loud call vocalization by the second mature male.

The above shows that there is great variation in the formation of age-graded groups. In Thomas's langurs, age-graded groups are promoted by long tenures, which are necessary to allow (presumed) sons to mature.

The AMB phase: making the best of a bad job

For all Thomas's langur males, life in an all-male band involved higher costs than life in a bisexual group. When bisexual groups turned into all-male bands, males were more often attacked by extra-group males and, consequently, more often wounded. Aggression came mainly from extra-group males, which is in contrast with mountain gorillas, for which males in bisexual groups became wounded more often because of intragroup aggression (Robbins, 1996). Males rested lower in the canopy in all-male bands than in bisexual groups, which suggests that they did so in order to lower their chances of being detected by an extra-group male at the cost of increased predation risk. This higher risk for all-male bands may also have caused the decrease in time spent alone above 20 m. Males spent more time with neighbors in all-male bands than in bisexual groups, but only when individuals were more than 1 m apart. When bisexual groups became all-male bands, the displacement and tolerant approach frequencies when not feeding increased, while the aggression and affiliative frequencies showed no difference. These combined results show that all-male band males interacted more with each other without increasing aggression levels: they either showed more tolerance or retreated without overt aggression. This is in contrast with mountain gorillas, for which both aggression and affiliative interaction rates were higher in all-male bands (Robbins, 1996). Robbins (1996) also reported that relatedness did not make a difference for male–male relationships, which is similar to what we found for Thomas's langurs.

Steenbeek (1999) shows that food is not limiting for Thomas's langurs because time spent feeding depends neither on group size nor diet. Therefore, and because group size was smaller in all-male bands than in bisexual groups, it was expected that time spent feeding would be similar or lower for males in all-male bands than in bisexual groups. Instead, time spent feeding was higher, which can only be interpreted as an additional cost of life in an all-male band. Higher risks for males in all-male bands may require higher vigilance levels, which in turn can lower the food intake and increase time spent feeding. However, data on food intake are required to confirm this. Furthermore, a positive correlation between group size and time spent feeding indicated that young adult males in all-male bands experience scramble competition. A higher risk for all-male bands of being attacked and wounded may have caused the decrease in time spent alone, but only when not feeding. This is in contrast with mountain gorillas, where all-male band males spent more time alone when feeding (Robbins, 1996). This difference may be due to a relatively low predation risk for gorillas which allows them to reduce food competition by increasing interindividual distances when feeding. All-male bands in Hanuman langurs (Rajpurohit, 1995), as well as the largest all-male band of our Thomas's langurs, did split into subgroups while foraging, perhaps because this is a less risky way to reduce feeding competition.

In spite of the costs, life in an all-male band still seemed to be better than living solitarily. Solitary adult males were seldom observed, and solitary juvenile males were only seen in the presence of long-tailed macaques or when uttering distress calls. Only young adult males often left the all-male band to travel solitarily, presumably when it was time for them to start a group and become breeding males. In Hanuman langurs, adult males profit from staying in an all-male band, because there they obtain coalition partners in attempts to take over a group (Rajpurohit *et al.*, 1995). Although all-male band males in Thomas's langurs may harass bisexual groups in teams, they do not really collaborate.

For the two other species for which all-male bands have been well documented (i.e., gorillas and Hanuman langurs), the data show that male behavior reflects predation and/or harassment risk and that there is variation in vulnerability to these risks for different age-classes. Yamagiwa (1987a) suggests that an all-male band of mountain gorillas may protect maturing males from hazards within bisexual groups and from the risks of traveling alone; immature and subadult males never travel solitarily, whereas adults do. Similar results were found for Hanuman langurs, although predation risk seems to be far more important for this species (Rajpurohit, 1995; Rajpurohit *et al.*, 1995). For instance, all-male bands sometimes mob predators (Ross, 1993), and the males may profit from the dilution effect, against both predators and hostile breeding males (Rajpurohit *et al.*, 1995; cf. also Bennett, 1983, for banded langurs).

For wild Thomas's langurs, we can conclude that the development of one-male into age-graded groups is beneficial for all males: the adult breeding male obtained help from sexually mature (presumed) sons in keeping extra-group males out of the group. For sexually mature sons, the benefits of staying in the natal group seemed to be increased inclusive fitness, getting experience in fighting extra-group males, coming into contact with females from other groups, and delaying the AMB phase. A long tenure is the most important factor promoting the development of age-graded groups. Males in an all-male band made the best of a bad job. After the tenure in a bisexual group had ended, an all-male band provided a relatively safe place for males to mature, providing safety against both predation and harassment risk. Males only left the all-male band when they reached adulthood, presumably to search for females and start their own bisexual group.

ACKNOWLEDGMENTS

We gratefully acknowledge the Indonesian Institute of Science (LIPI, Jakarta), the Indonesian Nature Conservation Service (PHPA) in Jakarta and Kutacane (Gunung Leuser National Park office), and UNAS (Universitas Nasional, Jakarta) for granting permission to use the Ketambe Research Station facilities and to conduct this field study in the Gunung Leuser National Park, Sumatra, Indonesia. We acknowledge Mike Griffiths and Kathrijn Monk and others from the Leuser Development Project for their help and friendship.

We thank Sri Suci Utami Atmoko for her help in running the project, her never-ending support and personal friendship. We acknowledge the support of Professor Carel van Schaik. We would also like to thank the students and assistants who helped to collect the data: Corine Eising, Jacobine Schouten, Marleen van Buul, Mandy Korstjens, Finley Koolhoven, Ritva Meijdam, Bas duMaine, Hanneke van Ormondt, Judy Bartlett, Ruben Piek, Bahlias, Usman, Rahimin, Samsu and Marlan. We thank Maria van Noordwijk, Tom Struhsaker, and an anonymous referee for useful comments on this chapter. Financial support was obtained from WOTRO (the Netherlands Foundation for the Advancement of Tropical Research), and the Lucie Burgers Foundation for Comparative Behavior Research, Arnhem.

13 • Male dispersal and mating season influxes in Hanuman langurs living in multi-male groups

CAROLA BORRIES

INTRODUCTION

As in most mammals, dispersal in non-human primates is usually male biased (Packer, 1979a; Greenwood, 1980; Pusey & Packer, 1987). Inbreeding avoidance was initially seen as the major cause of dispersal (e.g., Itani, 1972; Packer, 1975, 1979a; Harcourt, 1978). This interpretation was later extended to mate competition (e.g., Moore & Ali, 1984; Pusey & Packer, 1987; Alberts & Altmann, 1995a; for further explanations see Moore, 1993). Because male mammals usually invest less in their offspring compared to females (Trivers, 1972), male reproductive success is mainly influenced by the number of mates. As a consequence, males compete for access to females, and their reproductive success will depend on the number of fertile females and on the number of male competitors. Males may improve their breeding opportunities through dispersal (Shields, 1987; Pusey, 1992), either natal (emigration from the natal group) or secondary (all subsequent dispersal events; cf. Moore & Ali, 1984; Pusey & Packer, 1987) dispersal.

Several studies of primate multi-male groups disclosed benefits of male secondary dispersal. Males dispersed into groups with more (estrous) females (ring-tailed lemur, *Lemur catta*: Sussman, 1992; olive baboon, *Papio anubis*: Packer, 1979a) or into groups with a more favorable adult sex ratio (i.e., fewer males compared to females: rhesus monkey, *Macaca mulatta*: Drickamer & Vessey, 1973; yellow baboon, *Papio cynocephalus*: Alberts & Altmann, 1995a). However, other studies found no improvement (number of females: vervet monkey, *Cercopithecus aethiops*: Cheney, 1983; rhesus monkey: Boelkins & Wilson, 1972; adult sex ratio: vervet monkey: Cheney, 1983; long-tailed macaque, *Macaca fascicularis*: van Noordwijk & van Schaik, 1985; rhesus monkey: Lindburg, 1969). If male mating success as well as reproductive success are related to dominance rank, dispersing males could alternatively or additionally attempt to improve their access to females via rank improval. But again, the available results are equivocal. Some studies reported a positive effect of male dispersal on dominance (vervet monkey: Cheney & Seyfarth, 1983; yellow baboon: Smith, 1992), whereas others found no effect (rhesus monkey: Drickamer & Vessey, 1973). Although all examples refer to multi-male groups (and thus control for social organization), the published results do not present a uniform picture.

Male dispersal may additionally be influenced by seasonal breeding. At least theoretically, males could restrict their attempts at gaining access to females to the mating season and thus disperse accordingly. In some studies male dispersal activities were indeed often higher during the mating season (vervet monkey: Henzi & Lucas, 1980; ring-tailed lemur: Sussman, 1992; but see Jones, 1983; Japanese macaque, *Macaca fuscata*: Norikoshi & Koyama; 1974; Fukuda, 1982; Sprague, 1992; rhesus monkey: Boelkins & Wilson, 1972; Drickamer & Vessey, 1973) and others mention the appearance of extra-troop males in the vicinity of bisexual groups during the mating season (Japanese macaque: Sugiyama, 1976; Sprague, 1992; rhesus monkey: Neville, 1968; Barbary macaque, *Macaca sylvanus*: Mehlman, 1986).

In the following, I shall examine the dispersal of wild male Hanuman langurs (*Presbytis entellus*) from seasonally breeding multi-male, multi-female groups. Although Hanuman langurs are known for their social flexibility and ecological adaptability, data on male dispersal are mainly available for populations in which bisexual groups are one-male, multi-female, and males are assumed to gain access to a group of females only once in their lives (Sommer & Rajpurohit, 1989). Only three studies mentioned male dispersal in seasonally breeding multi-male groups, but they reported conflicting results (Bishop, 1979; Boggess, 1980; Laws & Vonder Haar Laws, 1984). Possible benefits of male dispersal were not investigated in these studies.

This chapter addresses the following questions related to secondary dispersal of individually known males: Do males benefit from secondary dispersal? Do they gain access to estrous females? Do they improve their access (more estrous females, improved sex ratio, improved dominance rank in the new group)? Does seasonal breeding influence male dispersal patterns?

METHODS

Habitat

The study was carried out in a semi-evergreen, primary forest (Wesche, 1997), dominated by Sal trees (*Shorea robusta*) near the village of Ramnagar (Chitwan District, South Nepal; 300 m above sea level; 27°44′N, 84°27′E; Podzuweit, 1994). The forest undergoes seasonal changes with 92% of the annual precipitation (yearly mean: 2279 mm) confined to the monsoon season from May to September (Koenig *et al.*, 1997). With the exception of a few weeks during spring (when most trees shed their leaves), visibility is poor.

Langur population

The langur population at Ramnagar was studied by 22 researchers from August 1990 through June 1997. The langurs lived at a density of about 26 individuals per km^2 and about 1.3 groups per km^2 (Borries & Koenig, unpublished data; calculation based on known home ranges of three groups, considering home range overlap; cf. National Research Council, USA, 1981). Most of the groups (72%) were multi-male, multi-female; 28% were one-male, multi-female. Mean group size was 18.3 and differed in relation to group structure: multi-male groups were larger (group size: mean 20.5, SD 7.16, range 7–41; number of adult females: mean 7.2, SD 3.39, range 2–14; number of adult males: mean 3.1, SD 1.10, range 2–6) than one-male groups (group size: mean 12.8, SD 3.08, range 6–19; number of adult females: mean 4.8, SD 1.95, range 1–9; census data for 18 groups, 176 group counts, see below).

Reproduction

Breeding was seasonal (conceptions: July–November, peak: August, with 49%; births: January–June, peak: March, with 47%; 22 group years, 51 conceptions/births; Koenig *et al.*, 1997). The period when conceptions occurred is referred to as the mating season (Butynski, 1988). The mean interbirth interval was 2.4 years (range 1–4+), but intervals of one year were achieved only if the infant did not survive the mating season following its birth (Borries, 1997). Most females with surviving infants resumed regular cycling in the second mating season post-partum (71%, Borries & Winkler, unpublished data). Thus, the term cycling female refers to all group females older than five years (first conception possible at 5.5 years; Borries & Winkler, unpublished data) who – at the beginning of the mating season – had no surviving infant born in the same year.

Age classification

Because Ramnagar langurs grow very slowly, the established age categories for other Hanuman langur populations (e.g., Rajpurohit & Sommer, 1991) were not applied. Instead, the following categories were used (Borries & Winkler, unpublished data). Infant: from birth to final weaning from nipple contact (mean age: 24.9 months, range: 19–33 months, both sexes). Juvenile male: after weaning until the scrotum appears (as a small piece of bare skin) at about 3.7 years of age. Juvenile female: after weaning until first conception (not before the age of 5.5 years, commonly 6.5 years). Subadult male: from appearance of the scrotum to attainment of adult male head–body length (around 7–8 years). Adult males have full head–body length, adult females are $\geq$5.5 years old. Young adult males (full head–body length but not width, canines of maximum size) were mostly treated as adult males because the intra-observer and the inter-observer reliability of these age estimates was extremely low and adult langurs could appear quite old or quite young within a few months, depending on their actual physical condition. From testes size development, I assumed that males do not reach sexual maturity before the age of 5.5 years. In total, about 100 adult males were known individually.

Natal males

It was impossible to distinguish natal males from other resident males at the beginning of the study. However, at the end of the study we had acquired the following criteria for a retrospective classification: the youngest natal male of known age was 4.2 years when he finally left; immigrating males (i.e., known secondary dispersal) were at least as large as adult females, though most were markedly larger. Exceptions occurred in a few cases when a known natal male who was not fully grown joined a neighboring group for a few days to return to his natal group. Thus, if at the beginning of the study males of less than adult female size resided in a bisexual group and remained with it for at least two months, they were considered natal. The behavior of natal males in comparison to non-natal males (e.g., pattern of interactions with adult females or immatures, play behavior) confirmed our assumptions but has not been quantified.

Secondary dispersal

Adult males residing in bisexual groups were regarded as non-natal males. Dispersal was analyzed for 25 individual males of two multi-male, multi-female groups (see below; note that males from one-male groups were not considered). The duration of male tenure in multi-male groups was calculated in days and later converted to months for all males staying for at least three days. Males who remained for some hours up to two days were excluded from the analysis because it was not always possible to determine the exact number of males present due to poor visibility. For the same reason, an absence of up to seven days was not counted as dispersal. Only complete tenures and those still ongoing at the end of the study were included. Male dominance rank was assessed by the outcome of dyadic displacement interactions (Borries *et al.*, 1991) collected in *ad libitum* sampling during all hours of contact with the langurs. Within about the first three days, the rank position of an immigrant male appeared to be settled, i.e., it could be assessed by the observers.

Extra-troop males

Focal and *ad libitum* data for extra-troop males suggest that most adult males were residents in bisexual groups although some stayed in the same area as – but apart from – the groups. Based on the frequency of males sighted near the two main study groups during 1994 and 1995, extra-troop male density was estimated to be two adult or subadult males per km^2. These males were encountered as solitaries, in duos or in associations of up to five males. These associations never had more than two adult males, were unstable and, according to three periods of dawn-to-dusk follows (18, 9, 4 consecutive days respectively; Borries, unpublished data), membership changed on a day-to-day basis. Because extra-troop males were often inconspicuous, their exact number could not be determined.

Data sets

Two neighboring multi-male groups were observed almost continuously (P-troop, averaging about 20 members from October 1990 to June 1997; O-troop, averaging about 30 members from October 1992 to June 1997), providing data for the age at and the annual distribution of natal dispersal. Furthermore, the general pattern of secondary dispersal, the monthly variation in the number of resident males, tenure length, and dominance ranks were assessed. Additional data concerning natal dispersal were derived from a third focal group (A-troop, averaging about ten members; a one-male group during most of the time, known since 1991, observed from October 1993 to April 1995). The seasonal pattern of secondary dispersal was analyzed for well-documented complete years (P-troop: 1991–1995; O-troop: 1993–1996). In addition to focal observations, as many groups as possible were censused four times a year (March, June, September and December; June 1992–March 1996; 18 groups including A-troop, O-troop, and P-troop as well as their neighboring groups B, J, S, U, X, of which all adult members were known individually), providing mean values on the group composition and data on the actual composition of eight neighboring groups simultaneously during four consecutive mating seasons.

RESULTS

Natal dispersal

In the langur population of Ramnagar, males were the dispersing sex, whereas females were generally philopatric (for exceptions see Koenig *et al.*, 1998). Most natal males left the group before they became adult – at a mean age of 6.0 years (range: 4.2–7.3+ years, $n = 13$, A-troop, O-troop, P-troop; includes three males older than four years still residing in their natal group at the end of the study, two were 6.0 and one 7.3 years old). Most natal males left prior to or at the beginning of the mating season (Fig. 13.1). In addition, two natal males from A-troop dispersed in the second half of the year. They are not included in Fig. 13.1 because the exact month of dispersal is unknown. However, the dispersal of these two males indicates that natal dispersal also occurs *during* the mating season. Of the ten dispersing natal males, two simply disappeared from the natal group, three were later seen as extra-troop males before they finally disappeared, three became residents in different adjacent groups, and two re-immigrated into their natal group (after an absence of 16 months each; in one case the identity of the male was not entirely certain).

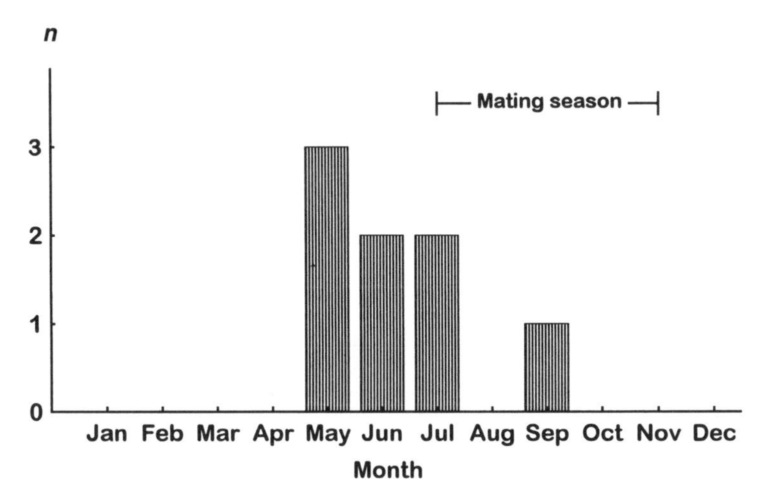

Fig. 13.1. Annual distribution of male natal dispersal events (O-troop, P-troop).

Secondary dispersal

From a bisexual group's perspective, there were four possibilities for male dispersal: males did not come from a neighboring group ($n = 14$), males went beyond neighboring groups (or died; $n = 15$), males left the group and re-immigrated into the same group ($n = 13$), or males came from ($n = 9$) or went to ($n = 9$) a neighboring group. Males dispersing between neighboring groups ($n = 18$) did not necessarily improve their breeding opportunities (Table 13.1; $n = 17$, in one case the condition in the new group was unknown). Neither the number of adult females nor the adult sex ratio was generally higher in the new group, and the number of adult males was not lower (Sign test, Table 13.1). This pattern remained when only dispersal during the mating season was considered ($n = 12$) or when all males were counted only once ($n = 9$, first account for each male). Even the number of cycling females and the ratio of cycling females to adult males were not generally higher in the new group. It seems that males simply changed between groups of different (female) size in a population in which the number of adult males was positively correlated with the number of adult females per group (Spearman rank order correlation, $r_s = 0.757$, $p < 0.002$ for the largest census, June 1993, $n = 14$ groups). Only one male changed between the two main study groups and provided information on the effect of group change on dominance rank.

Adult resident males in multi-male groups dispersed quite frequently, and tenure length averaged 11.7 months ($n = 39$; range: 3–2146 + days; median = 3.1 months). This includes five tenures which still continued at the end of the study (of 10, 11, 21, 21, and 71 months, respectively). About 31% ($n = 12$) of the tenures did not exceed one month (Fig. 13.2). Of these, five were shorter than one week, one shorter than two weeks, but three longer than two weeks and three longer than three weeks. Most of the tenures (62%) were shorter than six months and only 13% exceeded 24 months.

Most males entered a new group during the mating season (Fig. 13.3a). The frequency of immigrations was significantly higher during the mating season compared to the rest of the year ($\chi^2 = 16.03$, df = 1, $p < 0.001$). Emigrations followed a slightly different, though still seasonal, pattern (Fig. 13.3b). The frequency of emigrations was significantly higher during the mating season, compared to the rest of the year ($\chi^2 = 4.18$, df = 1, $p < 0.05$). In contrast to immigrations, however, most emigrations took place in December (25%), the first month after the mating season. When subtracting emigration events from immigrations, the residuals indicate more immigrations during the mating season, especially in August and September, and a clear tendency for emigrations toward the end of the mating season (Fig. 13.3c).

The number of adult males per group varied significantly in the course of the year (four census years with 7, 9, 9, 10 groups per year; Friedman ANOVA $F = 8.85$, df = 3, $p < 0.03$), with the highest number during the mating season in September (medians: March = 2.3; June = 2.4; September = 2.9, December = 2.4 males per group; *post hoc* comparison September *vs.* March: $|R_{Mar} - R_{Sep}| = 10.5 \geq 9.63$ for $\alpha = 0.05$, #c = 6; Siegel & Castellan, 1988). This increase indicates that during the mating season extra-troop males gained residency in bisexual groups.

Table 13.1. *Group composition at the time when males dispersed between neighboring groups*

Male	Old group	New group	Dispersal month	Ad. females[b]	Ad. females[c]	Ad. males[b]	Ad. males[c]	Ad. sex ratio[b]	Ad. sex ratio[c]	Cyc. females[b]	Cyc. females[c]	Cyc. sex ratio[b]	Cyc. sex ratio[c]
M9	A	P	**Jul 91[a]**	**3**	**10**	**2**	**2**	**1.5**	**5.0**	**2**	**8**	**1.0**	**4.0**
	P	A	**Sep 91[a]**	**9**	**3**	**6**	**2**	**1.5**	**1.5**	**8**	**2**	**1.3**	**1.0**
	A	P	**Oct 91[a]**	**3**	**9**	**2**	**7**	**1.5**	**1.3**	**2**	**8**	**1.0**	**1.1**
	P	A	**Nov 91[a]**	**9**	**3**	**7**	**2**	**1.3**	**1.5**	**8**	**2**	**1.1**	**1.0**
	A	P	Dec 91	3	9	2	6	1.5	1.5	2	8	1.0	1.3
	P	A	Jan 92	9	3	5	2	1.8	1.5	8	3	1.6	1.5
	A	O	**Aug 93**	**3**	**14**	**2**	**5**	**1.5**	**2.8**	**2**	**12**	**1.0**	**2.4**
	O	A	Dec 93	14	3	5	2	2.8	1.5	6	2	1.2	1.0
M10	A	P	**Sep 91[a]**	**3**	**9**	**2**	**6**	**1.5**	**1.5**	**2**	**8**	**1.0**	**1.3**
M11	A	P	**Sep 91[a]**	**3**	**9**	**2**	**6**	**1.5**	**1.5**	**2**	**8**	**1.0**	**1.3**
	P	O	Dec 91	9	12	5	6	1.8	2.0	8	10	1.6	1.7
M12	B	P	Jan 92	6	9	4	3	1.5	3.0	6	8	1.5	2.7
M13	P	J	**Aug 95**	**5**	**8**	**3**	**6**	**1.7**	**1.3**	**3**	**6**	**1.0**	**1.0**
M21	O	U	**Sep 93**	**14**	**3**	**5**	**3**	**2.8**	**1.0**	**12**	**3**	**2.4**	**1.0**
M23	O	U	**Sep 93**	**14**	**3**	**5**	**3**	**2.8**	**1.0**	**12**	**3**	**2.4**	**1.0**
M35	U	O	**Sep 95**	**4**	**14**	**3**	**6**	**1.3**	**2.3**	**4**	**13**	**1.3**	**2.2**
M80	O	A	**Jul 96**	**14**	**4**	**4**	**2**	**3.5**	**2.0**	**11**	**2**	**2.8**	**1.0**
Mean				7.4	7.4	3.8	4.1	1.9	1.9	5.8	6.2	1.4	1.6
Sign test[d] n, p				17, 0.315		16, 0.598		13, 0.709		17, 0.315		16, 0.402	

Notes:
Bold = mating season; a = male influx; b = condition in the (old) group left; c = condition in the (new) group entered; Ad. females = all females > 5 years; Ad. sex ratio = adult females/adult males; Cyc. females = all females > 5 years who at the beginning of the mating season had no surviving infant born in the same year, refers to the following mating season if male dispersed in the non-mating season; Cyc. sex ratio = cycling females/adult males; dispersing male included in sex ratio calculations; d = one-tailed, all tied cases excluded from analysis.

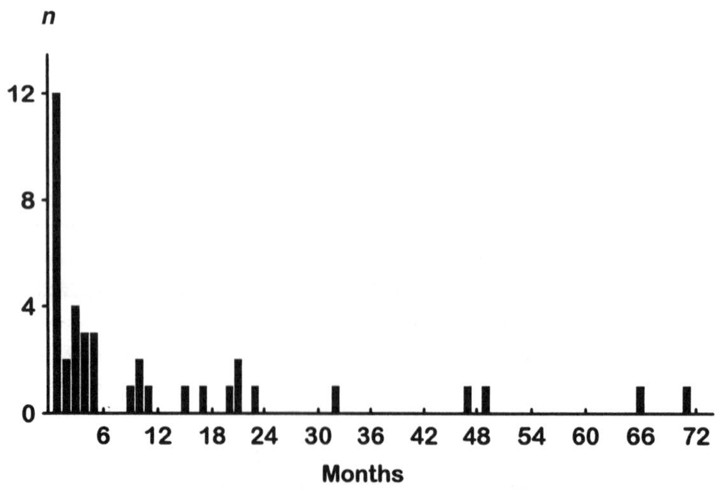

Fig. 13.2. Distribution of male tenure lengths (O-troop, P-troop; 11 group years).

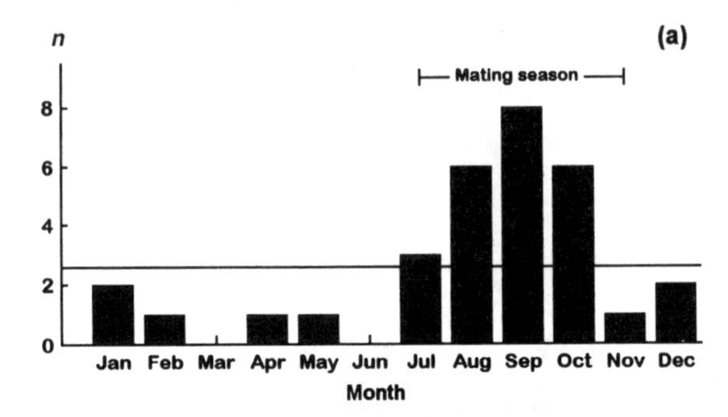

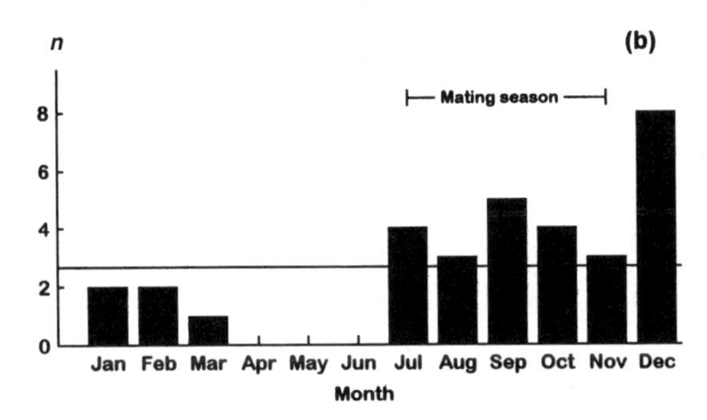

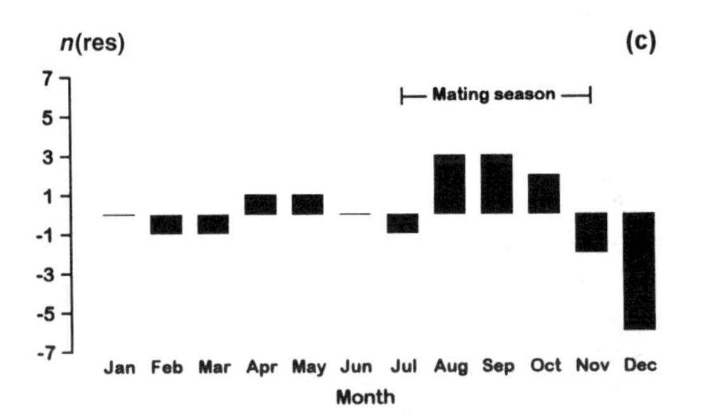

Fig. 13.3. Annual distribution of male secondary dispersal: (a) immigrations, (b) emigrations, (c) residuals = immigrations minus emigrations. Lines indicate an even distribution (O-troop, P-troop, 9 complete group years).

Influxes

Over the years, the maximum number of adult males per month remained more or less constant in the two focal groups (Fig. 13.4). Two out of ten documented mating seasons (P-troop 1991; O-troop 1995) were characterized by a remarkable increase in the number of males, which declined again after the respective mating season. I call this phenomenon *male influx* (i.e., the number of males increased by at least two, cf. also Struhsaker, 1977, 1988, p. 344). The median number of males was significantly higher during influxes compared to mating seasons without influx (Mann-Whitney U test, $U = 1$, $p = 0.024$, one-tailed). Influxes only occurred during mating seasons. Based on all available data (focal plus census, 1991–1996), it was estimated that a group experienced a male influx every 6.4 years (seven influxes in 45 group mating seasons). In other words, in each mating season, one out of less than seven groups was flooded by males.

To examine the factors that might cause a male influx, two measures for a group's attractiveness and two measures for reproductive output were analyzed for non-influx mating seasons versus mating seasons with male influx (Table 13.2). The number of females during male influxes was not significantly different from non-influx periods, but the number of cycling females tended to be higher. However, this trend was not obtained in the larger sample including census data (Table 13.2, bottom lines). In this larger sample, none of the four measures reached statistical significance, leaving unexplained why groups experienced male influxes in some years but not in others.

Particular groups were probably subject to male influxes in a particular year because they were the most attractive groups in the area. Table 13.3 provides information for eight neighboring groups simultaneously, thus possibly simulating the picture presented to (extra-troop) males in the area. Male influxes were not confined to groups with the highest number of cycling females, or to the largest female group size in the area, thus corroborating the above finding that neither female group size nor the number of cycling females was decisive for a male influx. Furthermore, the reproductive output following male influxes did not differ from non-influx periods: neither did the number of conceptions nor the percentage of all cycling females who conceived differ significantly in the two samples (Table 13.2).

In search for a possible benefit for males participating in influxes, the duration of male tenures was examined in relation to the period when they began. Tenure measured in days was longest for males entering during the non-mating season (mean: 603, SD: 639.2, median: 506, range: 7–1994+, $n = 11$), intermediate when it began during mating seasons with influx (mean: 302, SD: 574.8, median: 72, range: 3–2146+, $n = 19$), and shortest if males entered during mating seasons without influx (mean: 166, SD: 150.9, median: 127, range: 18–444, $n = 9$). However, these differences were not statistically significant (KW ANOVA $H = 4.37$, $p = 0.11$, $df = 2$, $n = 39$). Thus, the chances for a long male tenure were independent of the time of immigration.

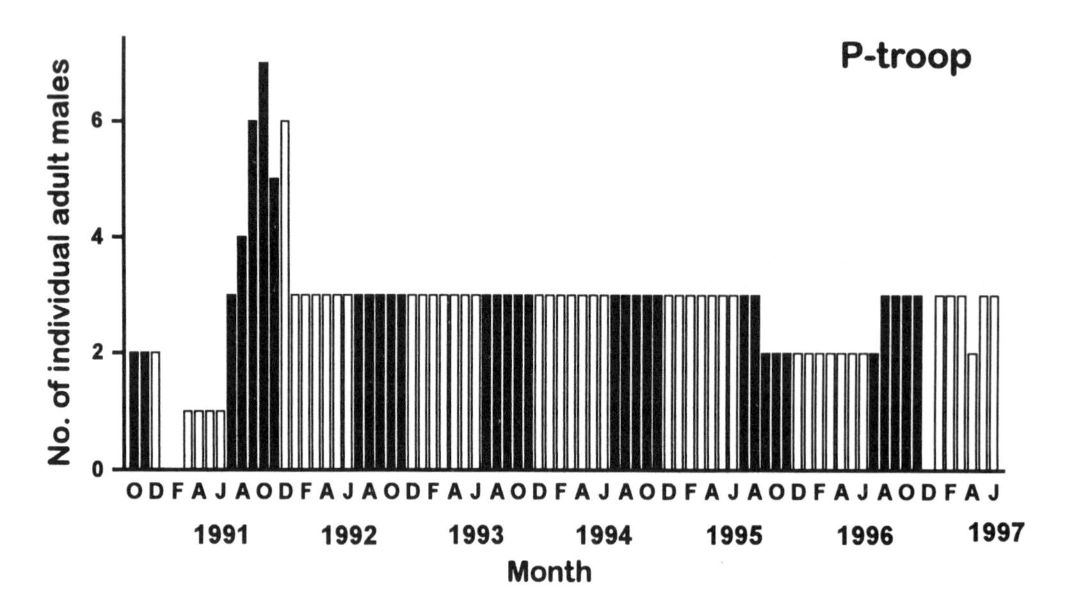

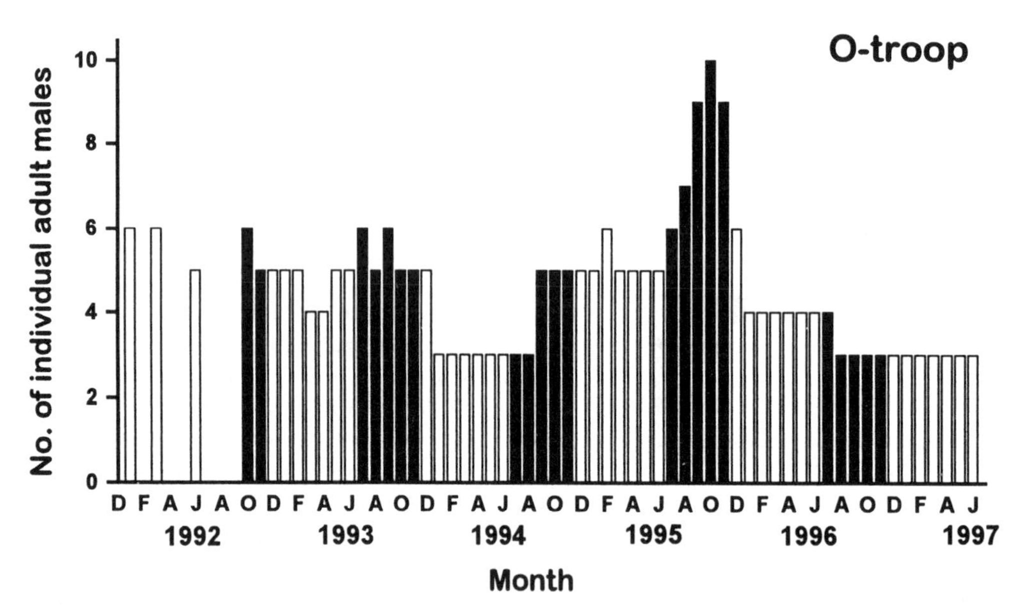

Fig. 13.4. Maximum number of individual adult resident males per month in O-troop and P-troop. Black columns = mating season, i.e., July-November; white columns = non-mating season.

Dominance and dispersal

During the whole observation period, 72 rank changes were witnessed, 93% of which were due to dispersal (44.4% due to emigration; 44.4% due to immigration; 4.2% due to replacement within a rank position, i.e., one male immigrated, another emigrated). If males occupying the omega position after immigration or before emigration were not considered because the rank positions of the other males were not affected, male dispersal still accounted for 90% of all rank changes (total $n = 50$). In other words, only 10% ($n = 5$) of all rank changes occurred between two resident males. Interestingly, in four of these five changes, at least one young adult male was involved, suggesting that older adult males could mainly acquire a change in dominance position by dispersing.

Dispersal seems similarly important in obtaining the alpha position. The rank positions obtained immediately

Table 13.2. *Group composition and reproductive output for mating seasons without male influx and for mating seasons with male influx (bold) for O-troop and P-troop (ten group years) and for eight additional groups (34 group years)*

	Mating season	n^a females	n^b cycling females	n^c females conceiving	Females[d] conceiving (%)
P-troop	**1991**	**10**	**9**	**1**	**11**
P-troop	1992	9	8	7	88
P-troop	1993	9	2	0	0
P-troop	1994	5	5	3	60
P-troop	1995	7	5	3	60
P-troop	1996	8	7	3	43
O-troop	1993	15	13	9	69
O-troop	1994	15	6	3	50
O-troop	**1995**	**15**	**13**	**5**	**38**
O-troop	1996	15	12	7	58
Mean, O-troop + P-troop	$n=8$	10.4	7.3	4.4	54
Mean, O-troop + P-troop	$n=2$	**12.5**	**11.0**	**3.0**	**25**
U, influx *vs.* non-influx[e]		4.5	2.5	6.0	2.0
p, influx *vs.* non-influx[e]		0.172	0.074	0.589[f]	0.116[f]
Mean, all groups	$n=37$	7.3	5.5	2.7	51
Mean, all groups	$n=7$	**8.0**	**6.9**	**2.7**	**37**
U, influx *vs.* non-influx[e]		113.5	96.0	126.0	87.5
p, influx *vs.* non-influx[e]		0.303	0.139	0.909[f]	0.203[f]

Notes:

[a] All females > 5 years.

[b] All females > 5 years without surviving infant born in the same year.

[c] All females who gave birth in the following birth season.

[d] Percentage of females conceiving of all females that were *cycling*.

[e] Mann-Whitney U Test, one-tailed.

[f] Mann-Whitney U Test, two-tailed.

after immigration (Fig. 13.5, black columns) were distributed almost according to expectation, except the alpha position, which tended to be overrepresented (χ^2 (1 vs. 2+) = 2.84, df = 1, $p<0.10$). The impression that immigration might serve to gain the alpha position, is further supported by the fact that 77% of the alpha males gained the position immediately after immigration. Furthermore, the tenure as alpha male was significantly longer for these recent immigrants than for already resident males (136 days, range: 6–1573+, $n=10$ vs. 16 days, range: 10–21, $n=3$; Mann-Whitney U test, U = 3.00, $p=0.04$, two-tailed). The duration of alpha male tenures was analyzed to determine when the tenures began for those males who became alpha immediately after immigration. Alpha tenure was longest for males entering during the non-mating season (mean: 534.8, SD: 610.3, median: 413, range: 40–1573+, $n=5$), intermediate when it began during mating seasons with influx (mean: 194.8, SD: 281.8, median: 79.5, range: 6–614+, $n=4$), and shortest if the male entered during a mating season without influx (126 days, $n=1$). However, these differences were not statistically significant (KW ANOVA H = 1.37, $p=0.50$, df = 2, $n=10$). Thus, the chances for a long alpha tenure were independent of the time of immigration. Even if the analysis is restricted to mating seasons with influx versus non-mating seasons, there was no significant difference (MW U test U = 6, $p=0.33$).

Former alpha males either left (50%) or remained in the group (50%, only males who became alpha immediately after immigration). Former alpha males who stayed had been alpha for at least one *complete* mating season. There were no indications that a male could become an alpha male in different groups in succession, although one male gained the alpha position in the same group twice after an absence of 61 days.

Table 13.3. *Group composition and reproductive output for eight neighboring groups*

Year	Troop	n^a females	n^b cycling females	n^c females conceiving	Females[d] conceiving (%)
1992	O	15	13	2	15
	S	11	10	5	50
	P	9	8	7	88
	B	**7**	**7**	**5**	**71**
	A	4	4	2	50
	X	4	3	2	67
	J	4	2	0	0
	U	2	0	0	
1993	O	15	13	9	69
	S	12	7	1	14
	X	6	6	5	83
	J	**4**	**4**	**1**	**25**
	U	3	3	2	67
	P	9	2	0	0
	B	7	2	1	50
	A	3	2	2	100
1994	B	10	10	2	20
	S	8	7	2	29
	O	15	6	3	50
	P	5	5	3	60
	J	5	4	2	50
	U	4	4	2	50
	A	3	2	2	100
	X	6	1	1	100
1995	**O**	**15**	**13**	**5**	**38**
	S	8	7	5	71
	B	8	7	5	71
	J	**7**	**7**	**5**	**71**
	X	6	5	4	80
	P	7	5	3	60
	U	4	4	3	75
	A	4	2	2	100

Notes:
Census data for September; groups arranged per year in descending number of cycling females; male influx bold.
[a] All females > 5 years.
[b] All females > 5 years without surviving infant born in the same year.
[c] All females who gave birth in the following birth season.
[d] Percentage of females conceiving of all females that were *cycling*.

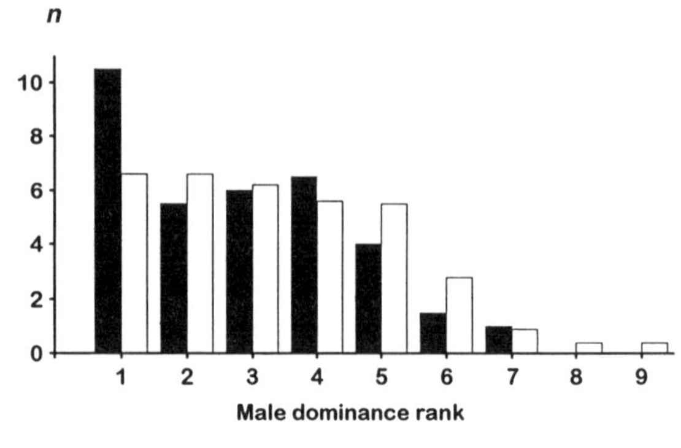

Fig. 13.5. Distribution of male dominance rank immediately after immigration. Black columns = observed; white columns = expected, based on the number of males available after immigration. (O-troop, P-troop, 11 group years.)

DISCUSSION

Sex-biased dispersal

In the langur population investigated, males were the dispersing sex. This result is in accordance with reports for other Hanuman langur populations regardless of social organization (e.g., Hrdy, 1977; Bishop, 1979; Rajpurohit & Sommer, 1993). Only at one study site (Khana), was female dispersal observed quite frequently, although it did not exceed male dispersal activities (Newton, 1987). It seems that most – if not all – natal males leave before they are fully grown, at a mean age of 6.0 years, although after seven years our investigation was still too short to state that *all* natal males dispersed. At the end of the study, three out of 13 natal males older than four years still resided in their natal groups.

The fact that natal males dispersed mostly at the beginning of the mating season (see Fig. 13.1) may indicate increased male–male competition for access to females during this period. Natal males indeed received most aggressive behavior – and six times as often as their female peers – from adult resident males, although the overall frequency of aggressive behavior was low (Nikolei & Borries, 1997). Especially newly immigrant males chased and even wounded natal males, resulting in some of them leaving the group. However, some stayed, even under pressure from resident males, and some left the group without obvious pressure.

Natal males might likewise leave because of reduced mating opportunities. Only one out of 1120 observed copulations involved a male prior to natal dispersal (six mating seasons, O-troop, P-troop, Launhardt, 1998). The chances for inbreeding seem to be very low, even if some copulations at the groups' periphery went unnoticed. However, at least one and perhaps two out of ten natal males later re-immigrated and bred in their natal group. In the one assumed case, the male gained the alpha position and sired infants (Launhardt, 1998). Both cases took place in O-troop, which had an extraordinarily large number of adult females (15 compared to the mean of 7.2 in multi-male groups). Perhaps these natal males still encountered comparatively good mating opportunities even if maternal kin were excluded as mates. The chances that resident males mated with their daughters were generally very low because females were 5.5 years old (in most cases even 6.5 years or older; Borries & Winkler, unpublished data) when they conceived for the first time, whereas of all the 39 male tenures documented, only one approached six years (see Fig. 13.2). Further tendencies for outbreeding can be deduced from secondary dispersal. Thirty-nine percent of all immigrants traveled further than the next group. If re-immigrations are not considered, the proportion rises to 61%.

Seasonal breeding and dispersal

Male secondary dispersal at Ramnagar followed a clear seasonal pattern. Immigrations, as well as emigrations, occurred significantly more frequently during the mating season, compared to the rest of the year (see Fig. 13.3). Males went where cycling females were and they left as soon as the last female had conceived. During the mating season, the number of resident males increased, indicating not only that resident males dispersed between groups, but that extra-troop males also gained residency. At least these latter males acquired access to females via dispersal, and the duration of their tenures was not different from that of males who entered during other periods.

For three other seasonally breeding populations of Hanuman langurs organized in multi-male groups, data on seasonal changes in male group membership are available. The results point in different directions. At Rajaji, the number of males in the main study group increased during the mating season ($n_{max} = 13$, Laws & Vonder Haar Laws, 1984), whereas at Junbesi, only a single male remained in each group during the mating season (Boggess, 1980). At Melemchi, no seasonal effect was apparent ($n = 3$–2; Bishop, 1979). Some of these differences could be connected to ecological differences between the study sites. The langurs at Rajaji inhabited a mixed, deciduous, subtropical forest, of which approximately 50% was 'heavily wooded' (Laws & Vonder Haar Laws, 1984, p.34), and at Melemchi they lived in a steep, nearly continuous broadleaf forest (Bishop, 1979). These two habitats resemble the Ramnagar forest to a larger extent than the cultivated, open valleys predominant at Junbesi (Boggess, 1980). Thus, other factors acting on a male's ability to monopolize a group of females, such as visibility, group cohesiveness, degree of arboreality etc. (cf. Winkler *et al.*, 1993; van Hooff & van Schaik, 1994; Borries, 1996), might have overruled a seasonal influence at Junbesi. Alternatively, all three studies may document just another aspect of a dispersal pattern shaped by seasonal breeding (see Male influxes below).

Several other primate studies on seasonally breeding multi-male groups also documented increased male dispersal activity during the mating season (vervet monkey: Henzi & Lucas, 1980; ring-tailed lemur: Sussman, 1992; Japanese macaque: Norikoshi & Koyama, 1974, Sugiyama & Ohsawa, 1975; rhesus monkey: Boelkins & Wilson, 1972). However, there are a few exceptions. For instance, rhesus monkeys (Pakistan, Melnick *et al.*, 1984, p.238) mainly disperse during the birth season when during 'an 8- to 10-week dry period before the monsoon ... many social groups ... congregate and interact around the few streams where water continues to run.' In a two-year study of Barbary macaques in Morocco, no secondary male dispersal occurred (Mehlman, 1986, p.77), although it was suspected that the continual presence of 'a human observer may have inhibited any visitor males from attempting permanent immigration'. The dispersal pattern of a semi-free-ranging population of Barbary macaques indicates that in this species the rate of secondary male dispersal might indeed be very low (Paul & Kuester, 1985; Kuester & Paul, 1999).

Male influxes

During some mating seasons, the number of resident males increased significantly in some groups and decreased again towards the end of the mating season (see Fig. 13.4). Such male influxes were first described by Struhsaker (1977; for red-tail monkeys, *Cercopithecus ascanius*) and were later observed in other seasonally breeding primates. In contrast to the present study, however, these groups basically were one-male, multi-female (red-tail monkey: Cords, 1984; Struhsaker, 1988; blue monkey, *Cercopithecus mitis*: Henzi & Lawes, 1987; patas monkey, *Erythrocebus patas*: Chism & Rowell, 1986; Cords, 1987b; see also Cords, Chapter 8). In these populations, too, male influxes only occurred during mating seasons, indicating that they might be linked to male access to cycling females. To my knowledge, male influxes have not been reported for primate groups with a year-round multi-male group structure. However, several studies describe the occurrence of extra males who might even associate with bisexual groups for a variable number of days or even months during the mating season (Japanese macaque: Sprague, 1992; rhesus monkey: Neville, 1968; Barbary macaque: Mehlman, 1986). Probably, male influxes occur more often in seasonally breeding multi-male groups than is currently recognized.

Given the variable dispersal pattern in the population investigated, the results from other Hanuman langur multi-male groups mentioned above may not be conflicting after all. The increasing number of male Hanuman langurs at Rajaji during the mating season (Laws & Vonder Haar Laws, 1984) could have been a male influx. Because the study period did not exceed ten months and observations were concentrated on one out of four groups, it is not clear whether male influxes occurred regularly. The constant conditions found at Melemchi in one group during 11 months of study (Bishop, 1979) possibly represent a non-influx mating season. Only the reduced number of adult males during the mating season at Junbesi does not fit here. As indicated above, ecological factors at Junbesi possibly allow for monopolization of a whole group of females by one male, at least during the mating season when it is most important.

The circumstances leading to male influxes at Ramnagar could not exactly be determined. Neither female group size nor the number of cycling females (see Tables 13.2 and 13.3), factors suspected of triggering male influxes in one-male groups (Cords, 1984, 1987b; Tsingalia & Rowell, 1984; Henzi & Lawes, 1987b, 1988; for a detailed evaluation see Struhsaker, 1988: 358 ff.), seemed to be decisive.

Perhaps the behavior, the physical abilities or even the personality of resident males (especially the alpha male) are important in this connection (see also Cords, 1984; Struhsaker, 1988). Resident males were often – though not always – aggressive toward new males. At Ramnagar, the alpha males seemed to play the major part in preventing additional males from immigrating, but no quantitative data are available.

Similarly, it cannot be ruled out that an increasing number of influx males caused females with young dependent infants to resume cycling earlier, thereby increasing the number of cycling females and possibly attracting additional males to the group (see also Cords, Chapter 8).

Dispersal benefits

Adult males dispersed quite frequently and preferentially during the mating season. Did they change group membership opportunistically in search of immediate benefits? At least the males dispersing between neighboring groups could have timed their transfer according to information gathered during earlier encounters. However, none of the parameters tested was generally more favorable in the group entered compared to the group left (see Table 13.1). It is currently not clear why some males left a group to enter another in which breeding conditions seemed less favorable. They may have been forced to leave or may have failed to improve their breeding opportunities.

The rank development of males dispersing between neighboring groups could not be documented unequivocally, so that it cannot be decided whether or not these males benefited via improved dominance rank. Additional information derived from another subsample, however, indicates that acquisition of high dominance rank could be one of the major benefits of dispersal. Rank improvement was very rare for resident males, and newly immigrant males tended to obtain the alpha position more frequently than expected and held it significantly longer than already resident males. Dispersal seems to be crucial for male success in terms of alpha tenure length. Comparable effects were described for Japanese macaques, for which almost all rank changes were due to dispersal (Norikoshi & Koyama, 1974) and the highest rank positions were occupied by prime adult males (Sprague, 1992), although newly immigrant males did not immediately gain the alpha position (Sugiyama, 1976). A positive effect of dispersal on dominance was also found for vervet monkeys (Cheney & Seyfarth, 1983) and long-tailed macaques (van Noordwijk & van Schaik, 1987).

At Ramnagar, rank and mating success were correlated (one mating season: Borries, 1997; eight mating seasons: Launhardt, 1998; Launhardt *et al.*, unpublished data). Moreover, DNA analyses revealed that alpha males in multi-male groups, on average, sired almost 60% of the infants, which is significantly more than the expected 25% based on the number of resident males during conceptions. About 20% of the infants were fathered by other resident males. The remaining 20% were sired by non-resident males who were either extra-troop males or residents from other bisexual groups (Launhardt, 1998; Launhardt *et al.*, unpublished data). Unfortunately, it could not be assessed how many different non-resident males sired these infants. For one mating season, however, it was clear that more than one male was involved (Launhardt, 1998). It would theoretically be possible for a super-dominant male to roam between bisexual groups and enter different groups just to mate, but such a male was never encountered during this study. Ramnagar males should mainly compete for the alpha position because this ultimately provides the highest reward. All other males in the population (with the exception of males in one-male groups) will possibly have better reproductive chances as residents than as extra-troop males. Also, male dispersal itself indicates that males try to become alpha or at least residents rather than extra-troop males.

During male influxes, the number of adult males is significantly higher, compared to non-influx periods, thus increasing male–male competition. In addition, the reproductive output (number of females conceiving or percent of all females conceiving) is quite low (see Table 13.2; see also Struhsaker, 1988). Furthermore, the proportion of infants sired by the alpha male is about as high as during non-influx periods (Launhardt, 1998). So the questions arise: why do males participate in influxes at all? In what way do they benefit? A number of influx males must have been extra-troop males because the absolute number of adult males in bisexual groups increased during the mating season. These males definitely gained by being residents during the mating season, with access to females and a higher probability for reproduction. More generally, all influx males gained residency and some even became alpha males. Their chances to stay in the group and to remain alpha male seem to be as high as for males immigrating during non-influx periods.

Male options

The results presented here indicate a set of male dispersal options related to age and dominance: a male's reproductive career at Ramnagar seems to begin only after natal dispersal. If judged by its effect on reproductive success, the alpha position was most profitable, and all male langurs at Ramnagar should attempt to become alpha. Their chances will be best if they immigrate rather than attempt to improve their rank as residents. Males competing for the alpha position should be physically strong because competition will be intense. Thus, they should be comparatively young because physical abilities decrease beyond prime age. Males will have just a few years, perhaps two to four, during which they can attempt to become alpha. Indeed, 50% of the recent immigrant alpha males were estimated to be young adult, although overall only 9.9% of the resident males belonged to this age-class, and age estimates must be treated with caution (cf. Methods). Once they have become alpha, males should try to hold the position for as long as possible because the chances to become alpha again in another group seem to be very small (if not zero).

Upon losing the alpha position, Ramnagar males seem to follow rather strict rules. Those who completed at least one mating season as alpha male remained in the group, those who did not, left. Probably, infant defense against attacks from newly immigrant males is the main motivation for former alpha males to stay on, although all residents, regardless of rank, defended infants, provided they had been in the group when the infants were sired (Borries, 1997, p.142; cf. also Podzuweit *et al.*, 1992).

All males who do not hold the alpha position should attempt to have access to cycling females. They can attain this by means of residency in a bisexual group, by frequent dispersal between bisexual groups, or by participating in a male influx (factors determining male tenure length in bisexual groups were not examined and remain rather obscure). It seems that, at least during the mating season, males should avoid being extra-troop males.

Toward the end of their reproductive career, males seem to have the freedom to do as they please. Two out of 25 male residents in the two focal groups appeared to be rather old, with ragged fur, worn teeth and a typical cat-like posture. One of these males (M9) is mentioned in Table 13.1. He was seen to disperse frequently between three neighboring groups. Aging males probably seek short-term advantages toward the end of their reproductive lifespans. A comparable 'restlessness' was described for three old male yellow baboons (Alberts & Altmann, 1995a) and two long-tailed macaques (van Noordwijk & van Schaik, 1985).

ACKNOWLEDGMENTS

I would like to thank all 21 colleagues working in the Ramnagar Monkey Research Project for their contribution to the database (Hari Acharya, Ulrike Apelt, Ralf Armbrecht, Jan Beise, Hari Cchetri, Mukesh Chalise, Dud Bikram Ghale, Andreas Koenig, Shiva Lama, Kristin Launhardt, Erik Mittra, Julia Nikolei, Julia Ostner, Doris Podzuweit, Yam Bahadur Rana, Oliver Schülke, Keshab

Thapa, Sylvie Thurnheer, Karsten Wesche, Paul Winkler, and Thomas Ziegler). Without their cooperative efforts this analysis would not have been possible. My co-researchers Dr M.K. Giri, Dr P. Shresta, and Professor S.C. Singh (Natural History Museum, Kathmandu), the Research Division of the Tribhuvan University (Kathmandu), and the Ministry of Education, Culture, and Social Welfare (HMG, Kathmandu) provided constant support, cooperation, and the permission to conduct this study. For valuable comments on the manuscript, I am very grateful to Peter Kappeler, Andreas Koenig, Jutta Küster, Kristin Launhardt, Jim Moore, Julia Nikolei, Andreas Paul, Thelma Rowell, Tom Struhsaker, Paul Winkler, and an anonymous reviewer. Julia Nikolei helped improve the grammar. During data collection, I was a member of the primatological group of Professor C. Vogel and Dr P. Winkler (Institute of Anthropology, University of Göttingen, FRG), who supported the project in countless ways. Financial support by the German Research Council (DFG, Bonn; Vo124/19-1 + 2; Wi966/4–3) is gratefully acknowledged.

14 • Rethinking monogamy: the gibbon case

VOLKER SOMMER & ULRICH REICHARD

INTRODUCTION

Traditional views of gibbon social structure

'The social group of gibbons is a monogamous family structure: an adult male, an adult female, and their offspring. The male and female form a mating pair for their entire lives' (Relethford, 1996, p.191). Such is the prevailing view of gibbons (*Hylobatidae*) – at least in textbooks (see also Sommer, 1989, p.132; Boyd & Silk, 1997, p.171). The highly arboreal hylobatids are confined to south-east Asia, north-east India and Bangladesh (for overviews on their ecology and behavior, see Preuschoft *et al.*, 1984; Tuttle, 1986; Leighton, 1987). These apes live in groups of usually two to six individuals that mostly contain a single adult female, a single adult male and immatures, and gibbon specialists thought that they form permanent pairs (e.g., McCann, 1933; Carpenter, 1940; Chivers, 1971; Tenaza & Hamilton, 1971; Ellefson, 1974; Tilson, 1979; Brockelman & Srikosamatara, 1984).

Traditional descriptions emphasized the following traits of 'monogamous territorial families' (MacKinnon & MacKinnon, 1984, p.291):

(a) *Strict territoriality*: groups live in well-defined ranges with little overlap, which they defend against neighbors.
(b) *Pair formation through natal dispersal*: mature males and females leave their natal groups to settle in a new, uninhabited area with a partner who has similarly dispersed from its natal group.
(c) *Lifelong monogamy*: a pair stays together until one partner dies.
(d) *Mating exclusivity*: all matings take place between the pair partners.
(e) *Nuclear families*: all immatures in a group are full-siblings.
(f) *Pair bond advertized through duetting*: males and females vocalize together in order to broadcast their mutual commitment.

In captivity, male and female gibbons are housed in pairs in an obvious effort to mimic the alleged situation in the wild. Such husbandry does often lead to reproduction, thus reinforcing and perpetuating the idea that 'monogamy' is the natural breeding system. Until recently it was certainly true that the 'monogamous grouping pattern has not been seriously questioned . . . for Hylobatinae' (Anzenberger, 1992, p.204). However, the view of mandatory nuclear families centering around adult pairs has been challenged since the mid-1990s by observations of extra-pair copulations and partner changes (Palombit, 1994a, 1994b; Reichard, 1995; Brockelman *et al.*, 1998). Admittedly, even the earliest observers of wild gibbons found evidence for a more dynamic social structure – but these were and still are often understood as temporary 'breakdowns' (Chivers & Raemaekers, 1980, p.249) or aberrations ('aberrante Einzelfälle', Hoffmann, 1993, p.60).

Nevertheless, such interpretation 'may seriously impede our understanding of hylobatid behavior' (Palombit, 1994b, p.97), not least because it reflects deeply rooted, even if often unconscious, group-selectionist thinking. Particularly the German–Austrian tradition of ethology relied heavily on the belief that species were the units of selection (e.g., Lorenz, 1963). Selection was thought to produce 'ideal' species-typical solutions to environmental challenges, which are reflected in a 'best' form of social organization. Behavioral flexibility made little sense because conspecifics were believed to 'agree' on a common goal. Thus, emphasis was placed on relatively stable 'norms' of behavior, and deviation from the average was dismissed as 'pathological.' Modern behavioral ecology, on the other hand, relies on the concept of gene-centred individualistic selection. Conspecifics are seen primarily as competitors which develop various and, if necessary, antagonistic strategies to achieve maximum reproductive success. The structures of animal societies can thus be expected to be flexible to a certain degree because they reflect the ever-varying compromises amongst individuals.

This chapter employs such an individualistic approach

to the understanding of gibbon social behavior. Our considerations derive largely from the longest field study of wild hylobatids, which focuses on white-handed gibbons (*Hylobates lar*) in the Khao Yai rainforest of Thailand. We believe that the situation for other *Hylobatidae* might be similar. However, we do not attempt to review the existing literature thoroughly because space is limited and because of a lack of exhaustive long-term records for other wild populations. We hope that other researchers will scrutinize our speculations by testing them against the sparse existing data and against newly collected information from the field.

PROBLEMS WITH CATEGORIES: WHICH TYPE OF MONOGAMY ARE WE TALKING ABOUT?

A particular subcategory of *social systems* are male–female associations. It is useful to categorize them in four principle ways, depending on the number of adult males (M) and females (F) involved (cf., for example, Davies & Lundberg, 1984): (a) monogamy (1M, 1F), (b) polygyny (1M, >1F), (c) polygynandry (>1M, >1F), and (d) polyandry (>1M, 1F). Note that the sex-specific viewpoint has to be made clear, under which the male–female association is classified. For example, a constellation in which a single male associates with several females is 'polygyny' for the male, but 'monogamy' for each female. Thus, such classification is equivalent to two types of monandry (female perspective: a, b), two types of monogyny (male perspective: a, d), and three types of polygamy ('neutral' perspective: b, c, d). Moreover, such associations have at least three different dimensions which do not necessarily correspond with each other (cf. also Wickler & Seibt, 1983; Gowaty, 1996): grouping ('who lives with whom?'), mating ('who copulates with whom?') and breeding ('who reproduces with whom?').

An example with hypothetical constellations viewed from the female perspective – which could well refer to a gibbon female – illustrates the complexity such an approach yields (g = grouping, m = mating, b = breeding; mo = monogamy, pa = polyandry):

- nulliparous female lives with one male (g–mo)
  - female copulates
    - only IPC (in-pair copulations) (m–mo)
    - only EPC (extra-pair copulations) (m–mo)
    - both IPC and EPC (m–pa)
  - female conceives first offspring
    - from IPC (b–mo)
    - from EPC (b–mo)
  - female conceives second offspring
    - again from IPC (b–mo)
    - again from previous EPC partner (b–mo)
    - from different father from that of first offspring (b–pa)
- female changes her grouping partner.

The last situation – a change of partner – adds further dimensions to the categories: *cross-sectional*: current constellation (g–mo); *longitudinal*: sequence of constellations (serial g–mo); *cumulative*: total number of partners (pa). Note the various possibilities under which the female – despite being classified as monogamous – could mate or breed monogamously *or* polyandrously. Note further that polyandrous breeding requires a time-depth of at least one interbirth interval for species with single births. (Dyzygotic twins sired by different fathers would be another, albeit rare, possibility.)

Similar considerations apply to males (who could, for example, be classified as monogamous but mate polygynously, or who could be classified as polygynous and mate monogamously). Moreover, the category of polygynandry could only be applied if we shift from the situation of the individual and refer to descriptions of groups or populations. The examples highlight the necessity to be specific about the particular dimension of a given male–female association when employing the term monogamy or any other category, should they have explanatory value. Still, the use of categories is unavoidable in our (however limited) efforts to reconstruct evolutionary processes. Nevertheless, it seems useful to remind us that our categories are only heuristic and sometimes need revision to encompass new data.

KHAO YAI: A LONG-TERM STUDY SITE OF WHITE-HANDED GIBBONS

Knowledge about the various dimensions of male–female associations in gibbons – particularly their mating and breeding system – is relatively scarce, for several reasons.

(a) Groups often flee silently from observers, which is facilitated by their small group size and arboreality.
(b) Recording complex events taking place high up in the canopy is difficult.
(c) Most earlier observers habituated at best a single group.
(d) Life histories and thus kin relations of individuals in adjacent groups are unknown.

We solved these problems partly because we used data from a study of wild gibbons that spans almost two decades and involves several well-habituated neighboring groups (cf. Brockelman *et al.*, 1998). The study site is located at 730–870 m elevation inside the primary rainforest of the Khao Yai National Park, Thailand (2,168 km^2; 101°22′ E, 14°26′ N; 130 aerial km north east of Bangkok). A population of white-handed gibbons immediately west of park headquarters in the Mo Singto area has been studied by various scientists since 1978 (e.g., Raemaekers & Raemaekers, 1984a, 1984b; Treesucon, 1984; Brockelman, 1985; Whitington, 1990; Reichard, 1996). About half a dozen individual trees in the study area were illegally exploited for incense production over the last decade, and one gibbon was presumably shot close to the study area in 1997, but otherwise the gibbons were not threatened by poaching or forest destruction.

ABSENCE OF MULTI-FEMALE GROUPS: COST OF FOOD COMPETITION OUTWEIGHS ANTIPREDATION BENEFITS

Primate social systems are widely thought to be structured by the spatiotemporal distribution of fertile females onto which males map themselves (Emlen & Oring, 1977; overviews in Rubenstein & Wrangham, 1986; Dunbar, 1988). The reproduction of female mammals is largely food limited. Competition for food means that females will not form multi-female groups unless (a) female coalitions secure on average more food for their members than a non-gregarious lifestyle would do (Wrangham, 1979), and/or (b) predation pressure is so high that females benefit from grouping (Alexander, 1974; van Schaik, 1983).

Gibbon social structure has been related to these constraints (e.g., Chivers, 1977a). Hylobatids are extremely selective feeders (Chivers & Raemaekers, 1986) which rely on ripe fruit (e.g., Vellayan, 1981), particularly because brachiation is energetically expensive (Parsons & Taylor, 1977). White-handed gibbons may spend more than 70% of their feeding time eating fruits (Palombit, 1997). Gibbon habitats probably lack a food patch distribution that would allow females to aggregate (cf. Robbins *et al.*, 1991) and simultaneously exploit fruiting trees. This is the probable reason that females are not gregarious. Such an option is believed to be relatively easy for the highly arboreal and agile apes because predation risk has been characterized as negligible (Carpenter, 1940; Ellefson, 1974; Grether *et al.*, 1992) or virtually absent (Raemaekers & Chivers, 1980; Leighton, 1987). However, the latter assumption has recently been questioned (for the following, compare Uhde, 1997; Reichard, 1998; Uhde & Sommer, 1998). A python (Schneider, 1906) and leopards (Rabinowitz, 1989) are known to have eaten gibbons. Such evidence is absent for Khao Yai, but the study area harbors various potential predators of gibbons such as clouded leopard (*Neofelis nebulosa*), reticulated python (*Python reticulatus*), marbled cat (*Felis marmorata*) and raptors, e.g., hawk-eagles (*Spiraetus cirrhatus*). Various observers documented encounters of gibbon groups with some of the potential predators (cf. Uhde, 1997). Gibbons do clearly discriminate against them, whereas harmless animals (e.g., pig-tailed macaques, *Macaca nemestrina*, binturongs, *Arctictis binturong*, hornbills, *Buceros bicornis*) do not elicit alarm calls or mobbing by adults. Gibbons also selected relatively thinner branches while being closer to the ground, presumably to be safer from cats or pythons (Uhde, 1997).

The sleeping habits seem similarly to reflect antipredation strategies (Reichard, 1998). Khao Yai gibbons enter their sleeping trees approximately two to three hours before dusk. Such a schedule has been suggested to reduce feeding competition with other primates (Gittins, 1980), but it is more likely that gibbons try to hide their sleeping places from predators, such as cats which become active when it gets dark. The comparatively slow pythons could also more easily sneak up to gibbons during the night. This may have caused white-handed gibbons not to sleep in groups in one tree, but at individual places, almost always on different trees – except for mothers and infants (see also Srikosamatara, 1984, for pileated gibbon, *Hylobates*). This perhaps lowers the intensity of gibbon-specific odors on which pythons would have to rely. Gibbons selected the tallest trees available (average height 32 m) and settled in an average height of 27 m, well above the overall canopy height of 22 m and 26 m in two selected plots of the study area (Brockelman, 1998). Higher trees are probably more difficult to climb for cursory predators. Adult females with infants and juveniles appeared to be more safety conscious, because they selected taller trees and slept higher than subadult and adult males.

We conclude that gibbons face a considerable predation risk. Nevertheless, we believe that the costs of female–female competition outweigh the antipredatory benefits that would be associated with the formation of same-sex aggregations. We are not convinced, however, that males associate permanently with females in order to provide a predator-monitoring service (Dunbar, 1988). Females cannot count, for example, on male vigilance during group encounters because males are frequently far away (Reichard & Sommer, 1997). It will be necessary to collect specific data

Table 14.1. *White-handed gibbons at Mo Singto, Khao Yai, Thailand: how monogamous?*

	Age and sex of group members (census 1996)													'Ideal' monogamy fulfilled?					
	Adult		Sub-adult			Juvenile			Infant					Duetting of oldest male and female?	Grouping monogamy via natal dispersal?	Grouping monogamy (current)?	Mating monogamy	Breeding monogamy (nuclear family)?	Grouping monogamy (lifelong)?
Group	F	M	F	?	M	F	?	M	F	?	M	Sum	Territoriality						
A	1	1			1	1		1				5	≥76% overlap	Yes	No	Yes	No	Perhaps	No
B	1	2	1			1						5	≥46% overlap	No	Unlikely	Polyandrous	No	Unlikely	No
C	1	1			1	2						5	≥69% overlap	Yes	?	Yes	No	Perhaps	?
D	1	2						1		1		5	?	No	?	Polyandrous	?	?	?
E	1	1			1	1						4	Overlap	Yes (a)	?	Yes (a)	?	?	No
G	1	1			1	1						4	Overlap	Yes	?	Yes	?	?	?
H	1	1			1			1		1		5	Overlap	Yes	No	Yes	?	?	No
I	1	1					1			1		4	?	Yes	?	Yes	?	?	?
K	1	1		1			1			1		5	?	Yes	No	Yes	?	Unlikely	No
L	1	2					1			1		5	Overlap	No	?	Polyandrous	?	?	?
M	1	1			1		1			1		5	?	Yes	No	Yes	?	Unlikely	No
N	1	1						1			1	4	Overlap	Yes	No	Yes	?	?	No
R	1	1						1				3	?	Yes	Yes	Yes	?	?	?
												59							

Notes:

Database, cs = cross-sectional, lt = longitudinal.

Group counts, duetting, current grouping monogamy (cs, 6 weeks, May–June 1996, Reichard & Sommer, unpublished data); territoriality (cs, 3 months; figures from Neudenberger, 1993); all others (lt, 1978–1997; for details see Brockelman *et al.*, 1998, and Reichard, 1998).

(a) From April 1997, grouping polygyny after resident female disappeared and two females immigrated (1 *H. pileatus*, 1 *H. lar*). Later, a former resident male of group B immigrated into group E, rendering a polygynandrous constellation (Uhde, personal communication; Reichard & Sommer, personal observation).

on predator-monitoring budgets of males and females to clarify this issue. Meanwhile, we suggest that the threat of infanticide was the main selective force behind the formation of permanent male–female associations in gibbons (*sensu* van Schaik & Dunbar, 1990; see below).

GIBBON SOCIAL LIFE: FLEXIBLE GROUPING, MATING AND BREEDING

Common generalizations about the social and sexual behavior of hylobatids 'derive from inferences about monogamy, rather than direct observation' (Palombit, 1994b, p.66). We have already pointed out the traditional belief that gibbons have a 'very strict and rigid social system' (Brockelman, 1984, p.285) and 'are extremely constant in their social organization' (MacKinnon & MacKinnon, 1984, p.291), consisting of 'monogamous family groups' (e.g., Carpenter, 1940, p.113; Chivers, 1977b, p.566; Mitani, 1990, p.412). These could only develop if pairs were permanent and mated exclusively. Moreover, mature unpaired individuals are often assumed to leave their natal group to establish a new pair with another previously unmated individual. This view has already been challenged by a six-year study of sympatric wild siamang (*Hylobates syndactylus*) and white-handed gibbons at Ketambe (Sumatra). Seven out of 11 documented pair relationships terminated during the study, but only two by death. Five were the result of mate desertion (Palombit, 1994b). The data from Khao Yai, which span almost two decades, reveal an even more dramatic turnover.

We illustrate this through a census conducted by the authors in 1996, when 60 gibbons at the Mo Singto research site were identified in 13 groups (Table 14.1; Fig. 14.1). The counts could have easily led naive first-time observers to the conclusion that gibbons breed in stable pairs because all

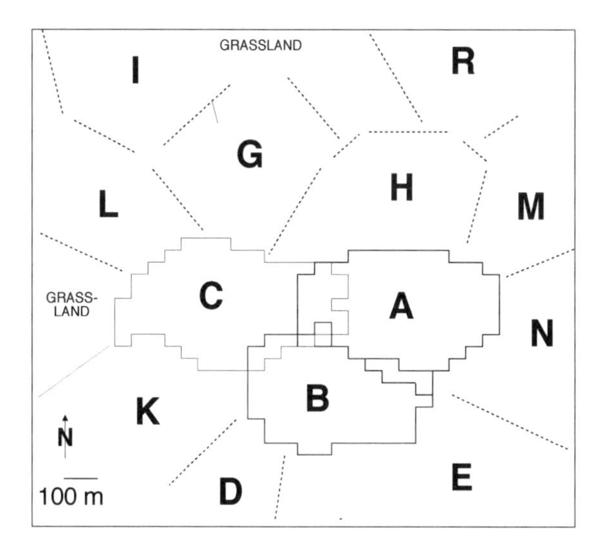

Fig. 14.1. Home ranges of study groups of white-handed gibbons at Khao Yai National Park, Thailand, mid-1996. Borders for main study groups A, B, and C are for 1992–3 and were calculated on the basis of entries into 50 × 50 m quadrats.

groups included an older female and an older male ('the monogamous pair') and one to three immatures. Indeed, three groups (B, D, and L) contained two adult males – but they appeared clearly disparate in age. A tempting common-sense assumption would have been to categorize the younger of the two as a subadult, and thus the older offspring of the pair. However, the longitudinal information for Khao Yai revealed a radically different picture (see Brockelman *et al.*, 1998, for the 1978–95 period).

Grouping monogamy: yes, but only in three-quarters of the cases

The census identified one-quarter of the groups as polyandrous, with an older and a younger adult male. The decision to label this situation as 'grouping polyandry' (and not as 'monogamy' with a mother, father and older son) was partly based on the identity of the duetting individuals (for description, see below). Gibbon duets have been suggested to reflect a paired status between a male and a female (cf. Haimoff, 1984; Raemaekers *et al.*, 1984). We heard duets in all three groups, and *saw* the older male be quiet while the younger male sang with the female. An older male, however, cannot be the son of a younger one. Had we just listened to the calls without seeing the individuals, we could have easily reached the conclusion that the oldest two individuals were duetting while the younger male was quiet. In fact, gibbon researchers will often not attempt to see the singing apes because unhabituated individuals are easily scared away by movements on the ground, but such caution may well have reinforced the perception of monogamous grouping in the past.

We do not have detailed information about groups D and L, but the known history of Group B (Bartlett & Brockelman, 1996; Brockelman, 1997; Reichard & Sommer, personal observation) provides a model for the formation of a polyandrous group.

(a) A prime male immigrates into a group and competes with the older resident for access to the female. (Emigration from group C into group B took place in mid-1994.)
(b) The new male duets. (The immigrant maintained proximity with the female, did most – but not all – grooming with her, and became the sole participant in call duets with the female after some time.)
(c) The young male mate-guards, but the female mates with both males. (Group B's female supported her old mate at times against efforts of the new male to peripheralize him; thus, the older male also copulated.)
(d) The older male emigrates, reverting grouping polyandry to grouping monogamy. (The older male disappeared after group B had been polyandrous for three years. At first, it was thought unlikely that he left to live as a 'floater' in the saturated population at Khao Yai, but he was later found in neighboring group E, which thus became polygynandrous; see Table 14.1.)

There is no agreement between researchers on how to label this situation, but it seems that terms such as 'extra-male' (Brockelman, 1997) or 'takeover' do not describe the constellation well. Both terms are designed to reconcile the observations with the notion of a basically monogamous grouping pattern that just underwent a temporary deviation. Interestingly, other anecdotal reports of polyandrous grouping exist – about so-called 'associated' (Carpenter, 1940, p.109/115) or 'loosely attached' males in white-handed gibbons (Brockelman *et al.*, 1973, p.638) and a group of hoolock gibbons (*Hylobates hoolock*) with two adult males (Choudhury, 1996).

Observations of group B revealed a clear conflict between the males, but also some effort by the female to maintain polyandry. In callitrichids, polyandry is believed to be a relatively stable alternative to monogamy, because multiple males can reduce the energy demands on the female (for review, see Lamprey, 1998). It is difficult to relate the above specified case of polyandry in gibbons to increased offspring survival because a juvenile female of group B was found dead during the polyandrous stage (Reichard, unpublished data). The cost–benefit equation for siamangs – in which males provide direct care by, for example, carrying infants – may be different, and multiple male partners could perhaps shorten the birth interval. On the other hand, additional males will perhaps increase food competition within the group. Nevertheless, white-handed gibbon females at Khao Yai have priority of access to feeding sites (Reichard & Sommer, 1997; Uhde, 1997), which minimizes the cost of an additional male.

During our census, we did not find polygynous groups, but group E developed into one after the former resident female was replaced by two immigrants: one white-handed gibbon female and one pileated female from the nearby population of pileated gibbons at Khao Yai. Both females gave great calls, often at the same time, and the male joined in the calls. The pileated female seemed at times peripheral, although it is known that white-handed and pileated gibbons do regularly interbreed at Khao Yai (Brockelman & Gittins, 1984). At least four 'mixed trios' of a male (white-handed, pileated or hybrid) and two females (white-handed/pileated; white-handed/hybrid) have been previously observed in the contact zone (Brockelman & Srikosamatara, 1984). Rare cases of single-species, polygynous grouping have also been reported (Srikosamatara & Brockelman, 1987, for white-handed gibbons; Ahsan, 1995, for hoolock gibbons; Liu *et al.*, 1989, Haimoff *et al.*, 1986, for white-cheeked gibbons, *Hylobates concolor*). Observations of a polygynous group with two breeding females were first made at Khao Yai in 1998–9. For at least eight months, group J contained one adult male and two females who both carried infants (Reichard, personal observation). Nothing is known about the trade-offs of polygyny for gibbons, but one can assume that females are faced with severely increased female–female competition for food (cf. Brockelman & Srikosamatara, 1984). The formation of mixed-species grouping polygamy may be more common because food competition may be relaxed due to differential exploitation of sources. Males, on the other hand, should want to maintain polygyny as long as the relative breeding success of two females exceeds that of one. This may not always be the case, because at least pileated gibbon males at Khao Yai have been seen to chase away second females (cf. Brockelman & Srikosamatara, 1984).

We hypothesize that polygamous groupings in callitrichids last longer than in hylobatids, relative to the reproductive life span. Post-partum conception, twinning and alloparental care allow mature same-sex callitrichids to enhance birth rate and rearing success within the groups considerably. To take turns in breeding requires a longer-term perspective for gibbons, and the reproductive interests of same-sex group members will therefore not overlap much. Nevertheless, we predict polygamous grouping once a population reaches carrying capacity: gibbons will not stop to compete reproductively at this point, and matured offspring with high resource-holding potential will try to 'squeeze into' neighboring groups, at the expense of established residents. Our figure of about one-quarter of non-monogamous grouping may well represent the expected equilibrium.

Mating monogamy: probably non-existent

Copulations in wild gibbons are notoriously difficult to see, but our longitudinal data for habituated groups (A, B, and C) suggest that mating exclusivity does not exist at Khao Yai. EPCs were noticed on roughly 9% of all days when sexual behavior was observed; for one female, they constituted 12% of observed copulations (Reichard, 1996). Otherwise, EPCs of hylobatids have been documented only for siamangs (Palombit, 1994a), but this is probably an artifact of insufficient data collection.

EPCs may increase a male's chances to leave more offspring, but a male has to reduce mate-guarding in order to philander toward a neighboring female. Thus, he has to balance the pursuit of EPC opportunities with the necessity to ensure paternity of his pair partner's offspring. Khao Yai data clearly show that male gibbons follow females more than *vice versa* (Reichard & Sommer, 1997). During encounters, on the other hand, they position themselves usually between their mate and the neighboring male(s), obviously to guard their mates. The willingness of females to engage in EPCs is probably related to the risk of infanticide (see below).

BREEDING MONOGAMY: CERTAINLY NOT IN ALL GROUPS

The relatively high degree of EPCs adds a cautionary note to attempts to label groups such as A or C 'nuclear families.' Their existence can be virtually ruled out for other groups (B, K, and M) with certain demographic histories that include relatively recent partner changes. Only DNA-analyses can reveal genealogical relationships. Feces have been collected from many of the surveyed individuals to clarify the degree of breeding monogamy (Reichard & Sommer, unpublished data).

LIFELONG GROUPING MONOGAMY: NOT IN A SINGLE GROUP WITH KNOWN HISTORY

Numerous partner changes have been documented at Khao Yai (Brockelman *et al.*, 1998). Of particular interest is a census of 64 groups of white-handed gibbons in Khao Yai (Brockelman, unpublished data, cited in Brockelman *et al.*, 1998) which found that one-third (!) of groups contained young which were estimated to be two years or less apart in age. Thus, their mothers and/or fathers must have previously lived with other partners (if we exclude the extremely rare possibility of twinning). The recent data therefore suggest that the 'most ideal' criterion for monogamy will at best be fulfilled in extraordinary cases only (see Table 14.1).

GROUPING MONOGAMY VIA NATAL DISPERSAL: AN EXCEPTION

At Khao Yai, a male of group A left his natal group during the summer of 1990. He was rediscovered in March 1993 when he had formed a new group (R) with group B's subadult female, who had dispersed in August 1992. The new group occupied a new range at the edge of the study area (cf. Fig. 14.1). The 1996 census found them with a baby. Such 'textbook cases' are necessarily rare, because there are rarely free areas left for unmated individuals to disperse to. Thus, they will typically have to enter established groups, which will lead to the expulsion of residents or to the formation of polyandrous or polygynous groups. The mean distance of natal dispersal of six maturing individuals was 710 m at Khao Yai (range 300–1400 m). These gibbons did therefore disperse into areas only one or two group ranges away (Brockelman *et al.*, 1998). Such a pattern should be considered the rule, not the exception for gibbons, because suitable gibbon habitats tend to be saturated. This was undoubtedly the case until humans started to interfere. Therefore, we expect (and see!) that sexually mature individuals remain in their natal group (Brockelman *et al.*, 1998): offspring will only leave once the costs of further philopatry (such as inbreeding and delayed reproduction) will outweigh the risks associated with dispersal into occupied areas.

TERRITORIALITY: NOT A GOOD TERM GIVEN TREMENDOUS RANGE OVERLAP

A recent textbook definition states that 'primate territories contain all of the sites at which the residents feed, rest, and sleep and the areas in which they travel. Thus, among territorial primates, the boundaries are essentially the same as for their home range, and territories do not overlap' (Boyd & Silk, 1997, p.196). Gibbons are cited as a prime example of territorial primates, but our data from Khao Yai provide evidence to the contrary (Reichard & Sommer, 1997) because – ironically! – the gibbons showed all of the behaviors ascribed to non-territorial species: 'When members of neighboring non-territorial groups meet, they may fight . . . exclude members of lower-ranking groups from resources, avoid one another, or mingle peacefully together. This last option is unusual, but in some species, adult females sexually solicit males from other groups, males attempt to mate with females from other groups, and juveniles from neighboring groups play together when their groups are in proximity' (Boyd & Silk, 1997: 196–7).

Particularly impressive in Khao Yai was the overlap in home-ranges. The average size for groups A, B, and C measured between August and December 1992 comprised 24 ha (Neudenberger, 1993). Non-overlapping areas between the three groups constituted 71% (group A, 64.3% of 24.5 ha; group B, 74.4% of 21.5 ha; group C, 73.6% of 26.5 ha). However, all groups had five to six neighbors and overlap existed with all of them – although it was not measured in the same way as for the fully habituated groups. Instead, the length of borders among groups A, B, and C and the proportion of overlap were used to calculate the expected overlap with other neighbors, yielding a mean of 63.7%. Thus, only about one-third of a group's range was exclusively used (cf. Fig. 14.1). Other studies probably yielded lower overlaps because neighboring groups were not habituated. Non-overlap areas of primate home ranges also become smaller with increasing observation time. Detailed analysis of the range ecology in gibbons is beyond the scope of this chapter. Nevertheless, we believe that the term 'territory' is often applied to gibbons because it is derived from or reinforces the idea of a closely knit family unit. We caution against the unqualified application of the term 'territory' that is frequently used, given such dimensions of overlap and at times non-agonistic or even affiliative interactions between gibbon groups which include co-resting, co-feeding, play, and sexual contact (Reichard & Sommer, 1997). Gibbon territories must, at least, be distinguished from their home ranges.

CALLS: NOT PAIR-BONDING BUT TEST OF MATE QUALITY

Gibbons are famous for their elaborate songs, which can often be heard over several kilometers (review in Haimoff, 1984; Raemaekers *et al.*, 1984; Cowlishaw, 1992). For example, males and females sing so-called 'duets,' mostly from 07:00 to 10:00 h. Duets consist of a warm-up phase by both individuals, followed by a high-pitched female 'great call' and a male 'reply/coda'; the sequence may or may not be repeated many times. Males of most gibbon species also sing 'solos,' typically around dawn.

A first complication is added to the idea of 'duetting' because at Khao Yai these vocalizations are at times transformed into trios. Immatures of groups A and B were heard to sing great calls synchronously with their mothers, for example. Interestingly, such calls were not only sung by juvenile or subadult daughters, but also by a subadult male (Sommer, unpublished data).

Vocal duets are believed to play a major part in 'territorial defense' (see, for example, Boyd & Silk, 1997, p.197). However, it should be noted that such a strategy has an inherent flaw: a group that calls gives away its approximate location – which allows neighboring rivals to move freely into those areas where the callers are *not*. Movements to disputed food trees in overlapping ranging areas which were obviously triggered by distant calls of neighbors were indeed at times observed at Khao Yai (Reichard & Sommer, unpublished data).

Duets in particular are believed to broadcast the existence of a 'pair-bond' throughout the neighborhood (e.g., Haimoff, 1984). This notion has already received some criticism (Cowlishaw, 1992). It is certainly difficult to maintain because (a) unpaired female gibbons utter great calls (e.g., a white-cheeked gibbon female at the London Zoo; Sommer, personal observation, 1996–8), and (b) paired females give solo great calls. The latter were heard during almost one-quarter of group encounters at Khao Yai (Reichard & Sommer, 1997). Most of these solo female calls occur just when the pair mate is *not* close by. If the male replied from far, he would give away that the mate is momentarily unguarded, increasing the risk that another male could copulate with the female. (Gibbons do in all likelihood recognize calls individually because even a trained human observer can tell quite a few of them apart.) Solo female great calls could thus serve a twofold function: they increase the costs for a paired male to seek EPCs, and at the same time attract neighboring males.

Solo female calls bring about a dilemma for a male not within sight. An answer to his pair mate's call will reveal his spatial position and consequently her unguarded status, but *not* to answer will also give away that the male is absent. This dilemma could be solved if the male sometimes did not respond to a female great call even if he *is* spatially close to the mate. In this way, rivals could not conclude from solo great calls that a female is unguarded. Indeed, such a strategy seems to exist: about 10% of great calls remain unanswered by the male – even if he is close to the female (Sommer, unpublished data).

Thus, so-called duets can be interpreted as the outcome of a female test of a mate's quality (cf. Cowlishaw, 1992, for a similar view): females call to advertize their presence; males join in to signal to potential rivals that the calling female is already guarded and in this way male calls provide an indicator of their physical fitness. Future research should disentangle how coordinated and 'cooperative' gibbon calls are, or if individuals in fact call 'alone' even during vocalizations with others.

Clearly, the naive employment of the term 'pair-bond' is rather misleading. Ironically, the benchmark paper that highlighted the dynamics of gibbon social structure used the term without questioning (Palombit, 1994b). In any case, even the existence of a monogamous breeding pair does not necessarily imply 'total psychic commitment' (Brockelman & Srikosamatara, 1984, p.316) between the partners. Research on another allegedly 'monogamous' taxon of primates, the *Callitrichidae*, made this clear more than a decade ago. In fact, one pioneer of marmoset research stated, quite radically, that marmosets were not 'emotionally' monogamous and that a pair-bond did not exist; breeding pairs existed because certain individuals were able to defend their alpha status and thus exclude others from access to potential mates (Rothe, 1975, for common marmoset, *Callithrix jacchus*). Later experiments provided evidence that same-sex aggression was a main factor in maintaining pair structures; if an individual could not prevent interactions between its pair mate and potential rivals, mating exclusivity was unlikely to be maintained (Buchanan-Smith & Jordan, 1992, for white-lipped tamarin *Saguinus labiatus*; see also Price & McGrew, 1991, for cottontop tamarin, *S. oedipus*). That two adult gibbons are found together is therefore not *per se* indicative of an emotional 'bond', but – repeating an argument made for callitrichids – 'following a more parsimonious explanation, could also result from site attachment combined with intrasexual aggression in both sexes' (Anzenberger, 1992, p.206). The term pair-bond should thus be dropped in favor of preferably quantitative measurements of interactive behavioral patterns.

INFANTICIDE AND PERMANENT FEMALE–MALE ASSOCIATIONS IN GIBBONS

We have stated our conviction that female–female agression all but prohibits the formation of multi-female groups. But why do males associate permanently with females? A highly competitive roving male could move from fertile female to fertile female and probably increase his reproductive success considerably. However, indirect evidence exists that there are constraints to such an option. First, increased male–male competition would probably select for heavier males, which in turn would suffer costs by having to eat more and by being less mobile, especially in terminal branches. Second, females at Khao Yai seem to be moderately synchronized in their fertility because 75% of births take place from October to December ($n = 16$; data from Brockelman *et al.*, 1998, Fig. 1, and Reichard, 1998, Fig. 1). Seasonal conditions of the semi-deciduous Khao Yai forest may bring about this pattern. Consequently, it is difficult to monopolize spatially dispersed, but reproductively synchronized, females (Emlen & Oring, 1977). Third, and most importantly, roving males would not be able to protect their offspring from infanticide by rivals. Thus, fewer offspring of roving males are expected to reach maturity than offspring of males who associate permanently with their progeny, providing indirect but vital care to infants through protection.

The risk of infanticide, which exists in a variety of species, and for a variety of reasons (Hausfater & Hrdy, 1984; Parmigiani & vom Saal, 1994), was first postulated as the major selective force behind the evolution of monogamy in gibbons by van Schaik & Dunbar (1990). Is there any factual basis for this assumption? Infanticide should typically occur as a result of (a) temporary invasion by a male neighbor, or (b) a pair-male replacement. Gibbons are equipped with sharp and long canines and are certainly capable of killing conspecifics (Palombit, 1993). Infant killing might be advantageous for several reasons. Killing an unrelated offspring will benefit a male if it increases his likelihood of siring the victimized mother's next infant. The death of an unweaned infant will typically shorten the period of infertility associated with nursing. The death of a weaned offspring could also benefit future progeny of the killer by curtailing maternal investment into another male's offspring. Such investment may continue far beyond nursing, e.g., through the mother bridging gaps in the canopy for her offspring with the own body, or grooming. Killing older, unrelated offspring similarly reduces the competition a future own offspring would face for resources such as food, safe sleeping trees or mates. Finally, the killer of an infant or juvenile might more easily replace the current adult male resident if the female withdraws her support from a mate who failed to protect her progeny (Hood, 1994).

To observe infanticide in the wild is extremely difficult. This might explain why only one case has been reported for gibbons (a putative father killing a newborn male: Alfred & Sati, 1991). Nevertheless, infant disappearances are rare at Khao Yai. However, the absence of observations of infanticide does not necessarily indicate that there is no risk, but may well reflect effective counter-mechanisms (similar to predation risk avoidance versus actual mortality rates from predation; Dunbar, 1988).

At Khao Yai, both aggressive and non-aggressive physical contact between infants and neighboring males has been observed (Reichard & Sommer, 1997). Non-aggressive interactions (play, brief touching and holding) are surprising, given that adult males can certainly harm immatures, but they may well suggest that paternity is confused. In fact, EPCs preceded the observations of friendly male–infant interactions. Female EPCs and the observed long period of preconception copulations could thus reflect counterstrategies to forestall infanticide. We can speculate further that such conditions favor even a slightly lowered degree of mate-guarding: a male can accept a certain proportion of EPCs if his own risk of not siring an offspring with the pair mate is lower than the combined benefit from his own efforts to achieve EPCs and the additional protection from infanticide his (likely) offspring gains as a result of EPCs of his mate. This trade-off might explain the somewhat puzzling observation that males sometimes do not return to their female partner even if they hear or see that a neighbor male is close to her.

CONCLUSIONS

It is our conviction that gibbon social systems are basically driven by female resource competition. Male–male competition for females will lead to agonistic interactions in the overlapping areas of the ranges. Within this framework, gibbons at Khao Yai pursue a variety of social options, which will mostly, but not always, result in monogamous grouping, in sexual contacts with more than one partner, and quite probably also in non-monogamous breeding. These patterns create groups in which young are not always full-siblings but where kinship relations extend well into neighboring groups. Short dispersal distances (Brockelman *et al.*, 1998) reinforce this effect. A positive feedback is thus started, which reduces the level of competition between groups and further promotes the at times affiliative intergroup relationships which can be observed.

Finally, let us attempt to phrase a more appropriate textbook entry for gibbons, largely generalizing the findings from Khao Yai.

The social system of gibbons is flexible, not only with

respect to grouping, but also with respect to mating and breeding patterns. Contrary to previous assumptions, monogamous behavior is not obligate – in neither a cross-sectional nor longitudinal perspective. The majority of groups do, indeed, contain only a single adult male and a single adult female (grouping monogamy), but about one-quarter contain two unrelated adult males (grouping polyandry). Polygynous and polygynandrous grouping is rarer, but also occurs. However, partner changes are so frequent that monogamous grouping is at best serial, and life-long grouping monogamy is perhaps non-existent. This, together with the probably common occurrence of extra-pair copulations, makes it likely that few groups are nuclear families (mother, father, plus their offspring) for very long. Songs, particularly so-called 'duets,' are often thought to be part of territorial defense. But the ranges of gibbon groups overlap so considerably as to call into question the appropriateness of the term 'territory'. Songs may functionally be more related to mate attraction – even (or especially) at the expense of monogamous grouping, mating, and breeding relationships.

ACKNOWLEDGMENTS

We thank Warren Y. Brockelman, Center for Conservation Biology, Mahidol University, Bangkok, for support, criticism, and encouragement as well as observational data from the early years of the long-term gibbon project. The National Park Division of the Royal Forestry Department and the National Research Council of Thailand, Bangkok, kindly granted research permissions. We are grateful to field researchers who contributed observations from the Mo Singto research site, especially J. Neudenberger and N. Uhde. Fieldwork and data analysis were sponsored by the German Academic Exchange Service (UR) and a Heisenberg-Fellowship of the German Research Council (VS; So 218–3 /1).

15 • Causes and consequences of variation in male mountain gorilla life histories and group membership

DAVID P. WATTS

INTRODUCTION

Optimal social dispersion for female mammals depends on food distribution and the resulting potential for feeding competition, often balanced against predation risk. That for males depends on female dispersion, and the outcome of conflicts and confluences of reproductive interest between the sexes then shapes actual social dispersion (Emlen & Oring, 1977; Wrangham, 1979; Vehrencamp & Bradbury, 1984). Females in most primate species live in stable social groups with permanently associated males, and within-species and between-species variation in the number of males per group is well documented (e.g., Andelman, 1986; Altmann, 1990).

Male mating tactics also vary in many mammals, including many primates. The same male may use different tactics at different times (e.g., tending and coursing by bighorn sheep (*Ovis canadensis*): Hogg & Forbes, 1997; consorting, mate guarding, and promiscuous mating by chimpanzees (*Pan troglodytes*): Tutin, 1979). Other variants are alternative strategies, either with equal payoffs (e.g., geladas (*Theropithecus gelada*): Dunbar, 1984) or with poorer competitors constrained to use those with low payoffs. Female choice can reinforce or decrease payoff disparities. For example, female rhesus macaques (*Macaca mulatta*) can decrease the advantages of high male dominance rank by seeking matings with low-ranking males (Manson, 1992). Males not favored by females may aggressively try to prevent females from exercising choice (Smuts & Smuts, 1993); one way for females to counter such coercion is to seek protection from favored males (Smuts & Smuts, 1993; van Schaik, 1996). Variation in female dispersal also influences variation in male tactics. For example, transfer between groups of philopatric females presents challenges to baboon males (Alberts & Altmann, 1995a) unlike those that face male chimpanzees, who cannot migrate between communities, but who can attract female immigrants or gain females by expanding their range at the expense of neighboring males (Wrangham, 1979; Goodall, 1986).

Female mountain gorillas (*Gorilla gorilla beringei*) always reside in groups. Most groups have single fully mature adult males (silverbacks), but some have two or more (Weber & Vedder, 1983; Stewart & Harcourt, 1987). Also, many groups contain sexually mature adolescent males (ages about eight to 14 years: Watts & Pusey, 1993). The mountain gorilla mating system involves female defense polygyny and sequential polygynandry: female natal and secondary transfer is common, males compete intensively to attract and retain females, and many females have offspring in several groups (often including natal groups: Stewart & Harcourt, 1987; Watts, 1990a, 1996; Robbins, 1995). Also, groups with multiple sexually mature males may be polygynandrous.

Male life histories vary considerably in the Virungas population of mountain gorillas in Rwanda, Uganda, and the Democratic Republic of the Congo. Some males stay in their natal groups as adults and breed there, and male philopatry can lead to age-graded, multi-male groups (Harcourt, 1981; Harcourt & Stewart, 1981; Yamagiwa, 1987a, 1987b; Stewart & Harcourt, 1987; Watts, 1989; Robbins, 1995). Others disperse as adolescents and either become solitary or join all-male groups. All-male group members may later emigrate to become solitary, whereas all-male groups can gain females by merging with female groups that have lost their males (Watts, 1989; Robbins, 1995). Here, I review some previous work on demography and male life histories in this population and present some new analyses. Despite small sample sizes (especially for solitary males), the data support the following conclusions: males follow alternative, non-equilibrated reproductive strategies and do better to acquire breeding positions from older males in whose groups they already reside than to become solitary; whether males voluntarily become solitary depends on within-group sex ratios, age of current dominant males, and breeding queue lengths; females are better protected against infanticide in multi-male groups; and females prefer to reside in multi-male groups, within some upper limit on acceptable group size.

The demographic database

Work on the behavior and ecology of the Virungas gorillas started at the Karisoke Research Centre, in Rwanda's Parc National des Volcans, in 1966. Continuous long-term demographic data on part of the population exist from then until the Rwandan civil war interrupted work in the mid-1990s. The sample population includes social units (groups and solitary males) monitored by Karisoke researchers and assistants (two to seven units per year from 1966 through 1993), and four 'tourist' groups monitored by National Parks personnel and expatriate conservation workers between 1981, or (in one case) 1983, and 1993. Relevant demographic data include ages at which males dispersed and at which they gained females or attained dominant breeding status in established groups; the composition of groups from which males emigrated or in which they attained dominant status; how long males maintained tenure as single or dominant breeding males; rates of change in the number of females per group; the composition of groups to and from which females transferred; and birth and infant mortality rates.

Mountain gorilla socioecology

Mountain gorillas are highly folivorous and mostly eat perennial food from sources densely and evenly distributed in most of their habitat (Fossey & Harcourt, 1977; Vedder, 1984; Watts, 1984; McNeilage, 1995). Gregariousness imposes only low costs of scramble and contest feeding competition (Watts, 1985, 1991, 1996, 1998b). Females thus have little to gain either by staying with female relatives to cooperate in within-group or between-group contest competition, or, within limits, by preventing potential immigrants from joining their groups. High home range overlap, limited site fidelity, and relatively low variation in home range quality reduce the importance of familiarity with natal areas for female foraging efficiency (Watts, 1996, 1998b). This makes the ecological and social costs of female transfer low, as shown by the fact that philopatric females and those who disperse have the same median age at first reproduction (ten years: Watts, 1996).

Because females are free to transfer, males compete to attract and retain them (Harcourt, 1978, 1981; Watts, 1990a; Sicotte, 1993). Most transfers are switches from one male to another during encounters between social units. Outside males apparently do not aggressively take over intact groups that contain fully mature males, although female groups whose males have died may merge with intact bisexual or all-male groups or join solitary males. Low female reproductive rates and intense male contest mating competition make infanticide an advantageous male reproductive tactic (Watts, 1989). If low feeding competition permits female gregariousness, infanticide risk probably explains it and explains permanent male–female association: females need male protection against other, potentially infanticidal, males (Wrangham, 1979; Watts, 1990a, 1996).

Male life history variants

Males reach adolescence at about eight years and are fully grown at 15 to 16 years (Watts & Pusey, 1993), and I refer to males 15 years or older as 'adults'. Males are sexually active by eight to nine years, but age at fertility is unknown. Male life histories vary in several ways, but males can conveniently be divided into two classes whose reproductive prospects differ, based on the composition of the social units in which they become adults:

1. 'Followers' reach adulthood in groups that contain older adult males, to whom they are initially subordinate, and adult females. These groups are the males' natal groups in most cases, but can be all-male groups that gained females and that the followers had joined when juveniles or adolescents. Thus 'follower' and 'philopatric' are not synonyms. Nor is this definition of follower identical to its meaning for hamadryas baboons (*Papio hamadryas*) and geladas: follower males in these species also reach adulthood in groups that contain older, dominant adult males, but these are non-natal breeding groups into which the followers have immigrated. Male immigration into breeding groups is rare in mountain gorillas (Harcourt & Stewart, 1981; Robbins, 1995; Watts, 1996).
2. 'Bachelors' are males who reach adulthood in social units that do not have females. Many are males who emigrate from natal groups when sexually mature, but not fully grown, and become solitary; six males with known birth dates who did so, left natal groups at a median age of 13 years. Others are males who reach adulthood as members of all-male groups that they joined as juveniles or adolescents (most of whom subsequently emigrate and become solitary). One bachelor was an adult who became solitary when females deserted him and who had earlier been a follower. A male can thus be a follower and a bachelor at different times, although most fall into only one category.

Followers wait in queues (Wiley & Rabenold, 1984) and have limited access to estrous females while subordinate

(Harcourt & Stewart, 1981; Stewart & Harcourt, 1987; Watts, 1990b; Robbins, 1995). True solitaries have uncontested access to any females who join them, although bachelors in all-male groups contest group membership if females join them, and not all gain access to the females (Watts, 1989; Robbins, 1995). However, followers may achieve dominant status in multi-female groups while young, whereas solitaries may never gain females, and, if they do, start with few. Both may have opportunities to breed as subordinates in their natal groups, but followers probably have more because of their longer residence. Followers should more often gain indirect fitness benefits by protecting the infants of male relatives, especially after those males die; they stay in natal groups long enough to achieve full body size, needed to protect infants effectively (Watts, 1989).

Reproductive payoff from following or becoming a bachelor

Given variation in these and other factors (e.g., tenure length), can a male expect to do as well reproductively if he becomes a bachelor as he would as a follower? One indication that these are not alternative strategies with equal expected payoffs, and that 'bachelor' is poorer, is that many males lose the option to become followers in natal groups when the single mature males in their groups die and the groups dissolve (Fig. 15.1). Perhaps expected reproductive payoffs for a male who voluntarily emigrates from his natal group are low, but he becomes solitary because his prospects for reproducing in his natal group are also poor. This could be the case if the group contains several other males or few females, for example, or if the current dominant male is young.

MODELING EXPECTED PAYOFFS

To explore the issue of relative payoffs, I used long-term demographic data to model expected payoffs for solitaries and followers. The logic follows Dunbar's (1984) model of male gelada reproductive strategies and similar work on golden lion tamarins (*Leontopithecus rosalia*) by Baker *et al.* (1993). All values for followers were empirically derived. We have reasonably complete life histories for six bachelors; others left the study population. For some analyses, I also included data on males whose groups, when first seen, had no immatures older than juveniles and who thus probably started as bachelors. I defined expected payoffs, G (gain in offspring equivalents), as:

$$G=[S(\sum_{i=1}^{x} F_i \times m_F)] - I + O_{sub} + P$$

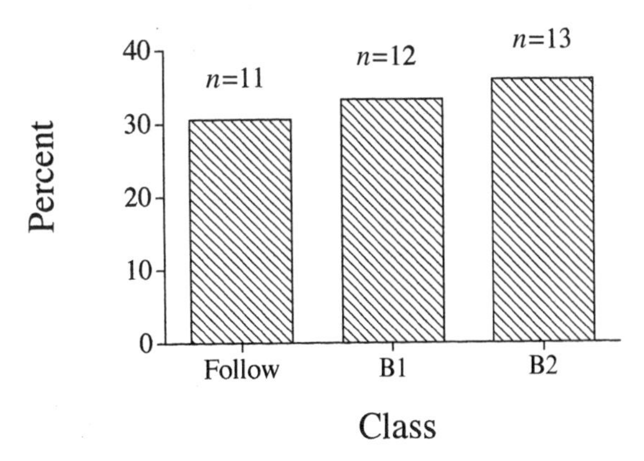

Fig. 15.1. Sample sizes and proportions of males who fell into three life history classes. Follow, followers; B1, sexually mature males who dispersed voluntarily, plus one silverback who became solitary when females deserted him and who did not breed successfully. Data from these two classes were used in analyses of demographic influences on strategy choice and in modeling expected reproductive payoffs. B2, juvenile and adolescent males who joined all-male groups after the breeding groups to which they belonged (usually natal groups) disintegrated.

1. S = The probability that a male becomes the dominant male in a group with females. For followers, this requires reversing rank with an older, dominant male and/or successfully retaining females at his death. For a bachelor, it depends on success at inducing females to transfer and to stay long enough to breed. I estimated S for followers as 0.9 (n = 10) or 0.82 (n = 11), depending on classification of one male who may have stayed in his natal group until adult, but then emigrated without reversing rank with an older male (records were unclear). The sample of successful followers includes one who attained dominant mating status there after four years as a follower in an initially all-male group that had gained females. I set S at 0.5 for bachelors because three of six with reasonably complete histories had surviving offspring. Indirect evidence that S was not higher, and may have been lower, comes from the observation that 12 males in the study population voluntarily dispersed to become solitary bachelors (Fig. 15.1), and eight others had left all-male groups to become solitary by 1992, but only five new breeding groups formed during sample years (of which three involved mergers with all-male groups). At the least, S is considerably lower for bachelors than for followers.
2. F_i = Female group size for each year of a male's tenure. F at the start of tenure was the number of females in a follower's group when he became the dominant/breeding

Table 15.1. *Number of females in a male's group at the start of his breeding tenure (Initial n females) and yearly increase in that number (Change) for four versions of the reproductive payoffs model*

Version	Initial *n* females	Change
I	Census mean:	Constant
	Follower = 4.2	
	Bachelor = 3.6	
II	Census mean:	
	Follower = 4.2	<0.18/year
III	Study population mean:	Constant
	Follower = 6.09	
IV	Study population mean:	
	Follower = 6.09	<0.18/year
	Bachelor = 1.80	<0.41/year

Note:
Census mean = mean number of females per group with two or more silverbacks (for followers) or with only one (for bachelors), from census data. Study population mean = mean initial number of females for followers and bachelors, from long-term data on research and tourist groups. Annual increases are based on long-term data. Versions II and III apply to followers only.

male, or the initial number that joined a bachelor. In alternate versions of the model (Table 15.1), I used two sets of estimates of F: (a) mean values for the study population (6.09 females for followers, 1.80 for bachelors); (b) mean values for groups with two or more silverbacks (4.2 females; taken as F for followers) and for one-male groups (3.6 females; taken as F for bachelors). For some model versions, I assumed that female group size remained constant; for others, I used data on research groups to estimate the yearly rate at which males gained females (Table 15.1).

3. X = Expected tenure length, defined as the median observed length of time that males who attained dominant status maintained it. X was nine years for followers and ten for bachelors (n = 7, including those first seen in one-male groups with no immatures older than juveniles).
4. m_F = The group size-specific yearly rate at which females have offspring that survive at least one year (values from Watts, 1996).
5. I = The risk that a male loses infants to infanticide because he dies and his group has no other male sufficiently mature to protect them. For bachelors, this was the probability that a male in an originally one-male group died, or had a known tenure of at least ten years, with no mature follower in his group. All infants born in the last three years of his tenure are at risk; these 'vulnerable' infants are almost certain infanticide victims (Watts, 1989). I used two estimates of I for followers (0.29, 'high' infanticide risk in Table 15.1, and 0.20, 'low' risk): only one follower lost known infants, but sample sizes differed depending on whether I assumed that followers who were overthrown by younger followers stopped reproducing at that point (so that if they lived another three years, they had zero chance of losing infants).
6. O_{sub} = The chance that followers breed with non-dispersing daughters of dominant males. I assumed that the oldest follower in a group sires all offspring of these females. I estimated the number of offspring by multiplying waiting time to become dominant by 0.25 times the number of the dominant male's offspring aged nine or older. This assumes a 50:50 sex ratio (Watts, 1990b) among surviving offspring, and emigration by 50% of natal females even when mates other than presumed fathers are available (Watts, 1996). Waiting time started when males turned 13 (the median age at which voluntary emigrants became solitary) and was 4.7 years.
7. P = Indirect fitness gains that followers get by defending related non-offspring infants against infanticide if the dominant male dies (P in Table 15.1). This was:

$$P = 2r\,(F_i \times m_F)\ \text{prob(dominant male dies)}$$

where r, the coefficient of relatedness, was 0.19 (because followers could be defending infants of fathers or paternal brothers). The factor 2 accounts for infants born in the last two years of the older male's tenure, because followers are probably not fully effective defenders until age 15 to 16 years (i.e., during the last two years of their waiting period). I set F at the number of females in a group at the end of a follower's waiting period.

RESULTS

Four versions of the model (Table 15.1) that involved different assumptions about or estimates of initial female group size, rate of change in female group size, and infanticide risk gave the same result (Table 15.2; Fig. 15.2): followers can expect a far higher payoff than bachelors, even higher than a bachelor could expect with the longest observed

Table 15.2. *Expected fitness payoffs for followers and bachelors*

Model	Follower, I_{low}		Follower, I_{high}		Bachelor		
Success	0.9	0.82	0.9	0.82	0.5	0.5	0.5
X	9y	9y	9y	9y	10y	15y	Equal
I	6.22	5.56	5.87	5.35	2.7	4.8	16y
II	8.8	7.22	8.59	7.71	—	—	18y
III	7.80	7.59	7.57	6.90	—	—	18y
IV	9.20	9.02	8.85	8.09	2.82	5.92	18y

Note:
Models as in Table 15.1. I_{low} = lower estimated infanticide risk; I_{high} = higher estimated infanticide risk (see text).
Success = probability of success (see text). X = tenure length.
'10y' gives number of offspring for a bachelor after ten years (median tenure length) and '15y' after 15 years (maximum observed tenure); 'Equal' gives tenure length needed to gain same expected payoff as a follower.

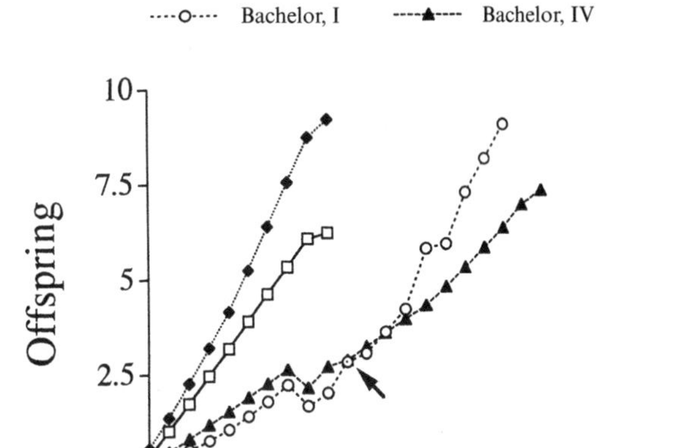

Fig. 15.2. Number of offspring (including indirect fitness gains) that a male could expect from each of two versions of the alternative strategies, 'follower' and 'bachelor.' See text and Table 15.1 for model parameters. I: model version I; IV: model version IV (see Table 15.1). *X* value for last point on curves for followers indicates expected tenure length. Arrow indicates expected tenure for bachelors.

tenure as a dominant male (15 years; although this was an open-ended interval). Bachelors would need 16–18-year tenures to equal followers; these are unrealistically long expected values (cf. Robbins, 1995).

Two factors account for most of the follower–bachelor disparity. Most important is the low chance of success for bachelors. Data on the careers of more bachelors and census counts of solitary males would improve the estimate of S for bachelors. However, it would have to be 1.0 to close the gap unless the sample used here is biased in some way that seriously underestimates the rate at which bachelors typically gain females and their typical tenure length.

The other factor is that followers start with more females and have more, on average, during each year of their reproductive careers. This advantage would diminish if followers, after they have attained dominant breeding status, lose fertilizations to other males who have become their followers. Seven successful followers had their own followers during at least parts of their tenures; these were either younger males ($n = 4$), or the older males with whom they had reversed rank ($n = 3$), but who still tried to mate. Only one successful follower had died with no follower of his own as of 1993. The incidence of multiple paternity within groups (excluding cases in which natal females mate with males other than presumed fathers) is unknown. Subordinate males may steal some fertilizations, but probably not many. Females are often proceptive towards subordinate males, and 13 of 18 females in multi-male research groups who cycled during 1985 to 1987 copulated with more than one male. However, dominant males often separate subordinates from estrous females and interrupt their copulation attempts (Stewart & Harcourt, 1987; Watts, 1990b; Sicotte, 1994; Robbins, 1995). Subordinates consequently seem to have poor fertilization success: all 15 females for whom relevant data exist copulated with their groups' respective dominant males during presumed conception cycles in 1985 to 1987, but only three copulated with more than one male. Even if subordinates gain all fertilizations when they mate during conception cycles, reducing the number of offspring that successful followers could expect accordingly (by a factor of 0.8) would not eliminate their advantage. Also, many such losses would be to other followers; the summed term in the equation given above would decrease for followers, but the O_{sub} term would increase.

Neither gains from mating with daughters of dominant males nor indirect fitness gains from protecting the infants of relatives added much to the advantage of followers. Higher expected losses to infanticide for solitaries contributed modestly, and compromise their potential gains from exclusive breeding access to females.

Using Karisoke data, Robbins (1995) concluded that successful bachelors started groups at the same average age at which followers attained dominant status. Followers still had more reproductive opportunities at younger ages, while they waited in natal groups, and started with more females. However, the fact that successful bachelors do not 'wait' longer than successful followers probably helps to explain their relatively long tenures. Tenure length in turn is positively correlated with male reproductive success (Robbins, 1995). The follower–bachelor disparity in reproductive success would be even greater except that followers also can have followers, and males can jump queues: position in a queue depends on fighting ability, not age, and a successful follower can lose his position to a younger male after only a few years. This happened in one group in 1989, when a follower successfully challenged an older male whose tenure had started only four years earlier (when a female group merged with an all-male group; cf. Sicotte, 1994; Robbins, 1995).

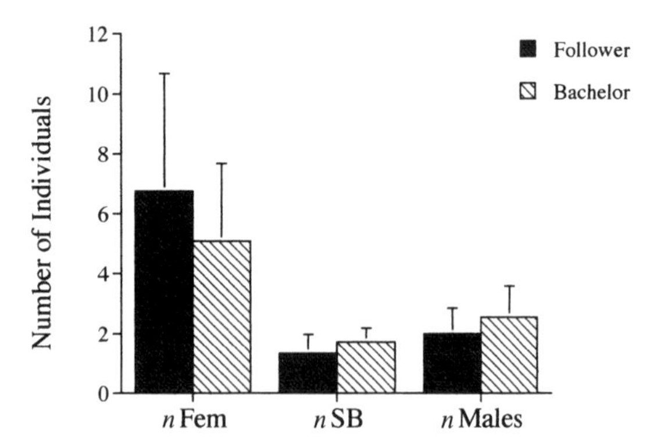

Fig. 15.3. Numbers of females, of silverbacks, and of all sexually mature males in the groups of followers and bachelors. Column height shows mean; bar gives 1 SD. Differences in group composition were non-significant for the two male classes (t-tests, $p > 0.05$ in all cases).

Why do males disperse?

Few individuals encounter average conditions. Demographic variation should influence whether males (besides those whose groups dissolve) become bachelors or followers, just as body size variation can influence how long successful males stay dominant to followers. Male yellow baboon (*Papio cynocephalus*; Alberts & Altmann, 1995a) and chacma baboons (*Papio ursinus*; Henzi *et al.*, 1998) tend to leave groups with few females and with many males per female, and to join those with the opposite characteristics. Variation in within-group sex ratios could have similar influence on dispersal decisions by maturing male mountain gorillas, although they do not have the baboon's option of immigration into breeding groups. Maturing males may be more likely to disperse the fewer females their groups contain, the more other males (all sexually mature males or adult males only) their groups contain, and the fewer females per male in their groups. Dispersal may also become more likely the lower males are in breeding queues (cf. Harcourt, 1979b) and the younger their groups' current dominant males. All else being equal, reversing rank should be easier with relatively old dominants than with young ones still at (or closer to) their peak fighting ability. Also, young dominants should live longer.

Demographic data support most of these predictions. Groups that bachelors left had as many females, silverbacks, and sexually mature males as those in which followers stayed (Fig. 15.3; cf. Robbins, 1995). Operational sex ratio (OSR) can be specified for bachelors at emigration. No comparable point exists for followers, but OSR for a follower can be estimated from the number of females in his group who cycled when he was 13 to 14 years old. OSR did not differ significantly for bachelors and followers (Fig. 15.4). However, the ratio of females to males was slightly but significantly higher for followers (Fig. 15.4). Given slow and probably asynchronous female reproduction and long waiting periods either to become dominant or to acquire females, the number of females per group may be a better predictor of future mating opportunities than immediate OSR. Also, bachelors had longer queues ahead of them, on average (Fig. 15.4), and thus probably longer waiting time to become dominant and greater risk of not becoming dominant. Finally, followers were more likely to reach age 13 to 14 in groups whose dominant males had already been fully grown for at least ten years (Fig. 15.5). This reinforced the discrepancy in queue length: males were more likely to stay when their waiting time to become dominant would probably be short because they had few other males ahead and the current dominant male was old.

Most males at the head of queues became followers: nine of 13 subordinate males who were first in queues stayed in their groups; two of these were still subordinate as adults in 1993, while the other seven became dominant males. Two males left Karisoke Group 5 in 1970, when the group's dominant male was relatively young (a presumed son reversed rank with him in 1984). One male left a group that had only two females, one of whom was old and died within a year of his emigration. Being a bachelor may be as good an option,

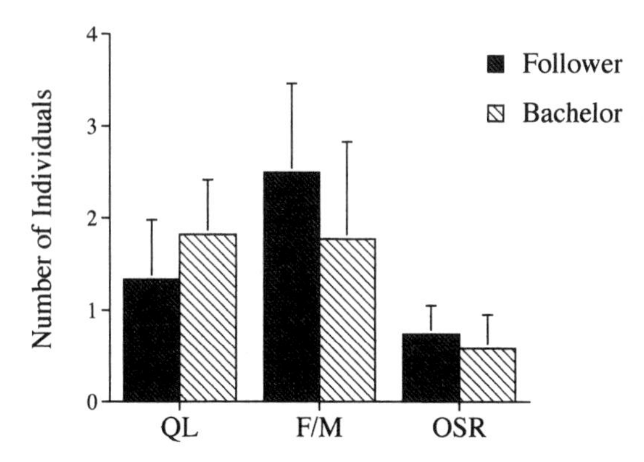

Fig. 15.4. Queue length (QL, the number of males in a group), sex ratio (F/M ratio of adult females to all sexually mature males), and estimated operational sex ratio (OSR) for followers and bachelors. Column height gives mean; bar shows 1 SD. Sex ratios were higher for followers (t-test: $t=2.16$, df$=21$, $p<0.05$), but OSRs were not significantly different (t-test: $t=1.14$ df$=21$, $p>0.05$). Bachelors faced longer queues (t-test: $t=1.85$, df$=21$, $p>0.05$, one-tailed).

or a better one, than staying in such a group. Finally, a male at the head of a queue in a formerly all-male group was evicted during a period of intense competition among males after a group merged with a female group that lacked a male. Another formerly all-male group had two adults who did not have an established dominance relationship, one of whom was evicted (along with two adolescent males lower in queues and three juvenile males).

Old dominant/young (senior) follower pairs would be expected if males in multi-male groups are fathers and sons, and followers are usually oldest sons whose younger brothers disperse (Harcourt & Stewart, 1981; Yamagiwa, 1987a; Dunbar, 1988; Robbins, 1995). Oldest sons may accept younger brothers as followers, however. Of 17 males who had females and reproduced successfully during the sample period, nine had followers (eight died without followers); two of these had two followers, and one had four, including the older male with whom he had reversed ranks. Seven pairs comprised a male who had been adult for at least ten years plus a young male. Of these, six were presumed or possible fathers and sons (the pair in Karisoke Group B were unrelated), and the only two for which relevant data exist were presumed father/oldest son pairs. Four pairs were putative paternal siblings (one of these consisted of maternal brothers).

Social relationships between males

Mature followers and older adolescents cooperate with dominant males against extra-group males (Sicotte, 1993). Relationships between males are otherwise mostly neutral to antagonistic, and dominant males try to prevent subordinates from mating with females other than the dominants' daughters. Dominants sometimes harass followers or potential followers (Watts, 1995), and adolescent males receive more aggression from dominants as they grow older (Watts & Pusey, 1993). However, dominants do not seem to target them for eviction with persistent, high levels of aggression. Coalitions between males are possible in groups with three or more males: a previously dominant old male in Group 5 sometimes supported his adolescent son (a follower) against his older, currently dominant son (a presumed half-brother); that dominant son later occasionally supported his presumed full brother (a third-ranking follower) against the same half-brother (Watts, 1997). Coalitions were infrequent in both of these dyads.

Why tolerate a follower?

An all-male group probably contains multiple sexually mature males, and some may try to evict others if the group gains females (above). In all three documented mergers, however, the dominant males tolerated one or more adolescent males from the original all-male group. Two of these who were first in their queues later reversed ranks with the dominant males (younger subordinates emigrated from both groups); the male who was first in the queue in the third group was still subordinate as of 1993. This again raises the question of why males in heterosexual groups do not evict maturing subordinates by targeting them with high levels of aggression, especially younger brothers who may threaten their tenure. Perhaps moderate aggression from large and vigorous dominant males who allow adolescents little chance to interact with females and few, if any, copulations, can induce voluntary dispersal, especially from groups with relatively few females and many older males. However, tolerance of followers can offer several benefits:

1. Better protection of infants, particularly those of older dominants past their peak fighting ability.
2. Consequently, better ability to retain, and perhaps to attract, mates. Particularly relevant to old males, this could also apply to males with large groups from which females might otherwise emigrate to reduce feeding competition. Karisoke Group 5 from 1984 to 1993 might have

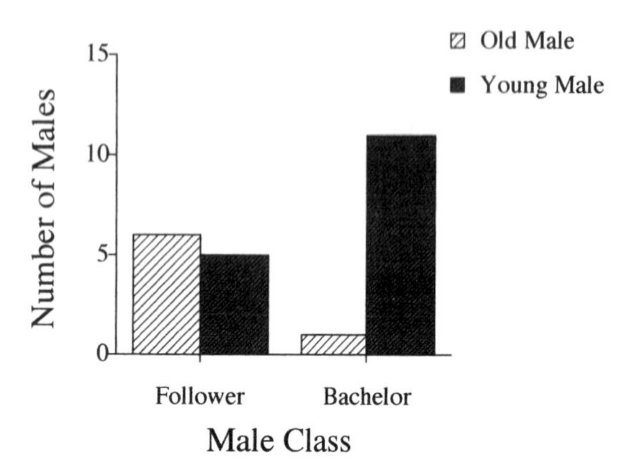

Fig. 15.5. Number of followers and bachelors who reached age 13 in groups in which the current dominant males had been silverbacks for at least ten years (old males) or for less than ten years (young males). Bachelors were more likely than followers to be in groups with young males (Fisher's exact test: $p=0.014$, one-tailed).

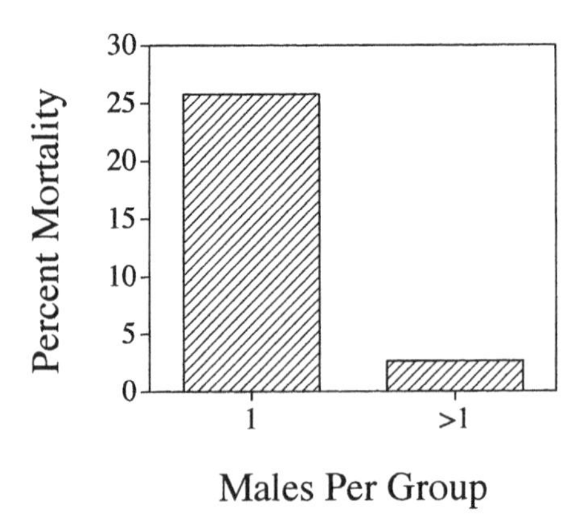

Fig. 15.6. The percentage of infant mortality due to infanticide in groups with a single adult male and those with more than one (based on Karisoke data through 1992).

been an example: the number of females grew from four to 14, and the dominant male (a successful follower) allowed three other followers to stay in his group.

3. Indirect fitness gains achieved by helping sons to retain mates and/or by protecting grand-offspring.

Implications for females

Females in multi-male groups often copulate with more than one male, despite interference by dominants (see above; Stewart & Harcourt, 1987; Watts, 1990b; Sicotte, 1994). This may confuse paternity and induce those males to protect infants against infanticide, or at least forestall within-group infanticide (Hrdy & Whitten, 1987). Physically mature males are better protectors than adolescents (Watts, 1989), but all mating partners, including bachelors before they emigrate, should protect infants, regardless of when they copulate. Protecting infants may later help subordinates to retain females as mates, as Smuts (1985) argued for baboons. Also, followers are usually at least indirectly related to infants in their groups.

Infanticide risk and males per group

Data on infanticide as a cause of infant mortality show that the risk is lower in multi-male groups than in one-male groups (Fig. 15.6; cf. Robbins, 1995). This is mainly because most infanticides occur when males with no mature followers die, and females and any immature males cannot defend infants adequately (most 'one-male groups' in Fig. 15.6 had no adult males when attacked; Watts, 1989). Also, extra-group males sometimes attempt infanticide when group males are present (Fossey, 1984; Watts, 1989). Subordinate males commonly participate in aggressive encounters between social units (Sicotte, 1993; personal observation), and two males should be better able than one to thwart such attempts. Females thus benefit from membership in multi-male groups and can benefit from the presence of followers.

Data on the number of males in source and destination groups for a sample of female transfers support the hypothesis that females prefer to reside in multi-male groups. The null hypothesis is that females went from one-male groups to other one-male social units or to multi-male groups, and from multi-male groups to one-male social units, in proportion to the availability of one-male and multi-male social units. I used two availability estimates.

1. I calculated the mean proportions of multi-male groups and single-male breeding groups in censuses, then estimated the proportion of males who were solitary by dividing the number of male-years that emigrants from Karisoke groups and from two tourist groups were known to have been solitary by the total number of male-years for all males at least 13 years old in these groups. This gave a figure of 30% of all males as bachelors.
2. For females in one tourist and four research groups, I counted the numbers of one-male and multi-male groups and of solitary males whom they sometimes met. This

directly measures opportunities and is more realistic than population-wide averages, but may undercount solitary males because their presence near groups is not always obvious to observers.

In both cases, results support the hypothesis that females join multi-male groups more than expected by chance (Table 15.3). A major reason is that females who emigrated from multi-male groups usually joined other multi-male groups, not one-male social units.

Table 15.3. *The number of cases in which a female transferred from either a one-male or multi-male group (Origin) to either a one-male social unit (group or solitary male) or a multi-male group (Destination), with expected values given in parentheses*

Version:		
(1)	Origin	
Destination	One-male	Multi-male
One-male	22 (29.0)	8 (16.2)
Multi-male	15 (8.0)	12 (3.8)
(2)		
Destination		
One-male	12 (16.8)	1 (10.9)
Multi-male	11 (6.7)	13 (2.5)

Note:
Expected values in version 1 ($n=57$ transfers) from census data; those in version 2 ($n=37$ transfers) from data on research groups and one tourist group only (see text). Females preferentially joined multi-male groups (G-tests; version 1: $G_{adj}=21.66$, $p<0.005$; version 2: $G_{adj}=40.02$, $p<0.001$).

Male–female social relationships

Most females are in close proximity to silverbacks more than to any female partner and have more affiliative interactions with silverbacks, although females who live with close female relatives can be exceptions (Harcourt, 1979a; Stewart & Harcourt, 1987; Watts, 1994a). Male–female relationships in a multi-male group are differentiated. Most or all females may still be close to the dominant male more than to other males and interact more with him (Watts, 1992a; Sicotte, 1994). However, females may compete for access to the dominant male, and subordinates try to stay close to females and to groom with them (Watts, 1992; Robbins, 1995; Sicotte, 1994). Consequently, some females may have a subordinate male as their main social partner. Subordinate males sometimes regularly groom infants (which may or may not be paternal care). Grooming females and infants may help them to retain the females after they reverse ranks with dominant males or after a group fissions, as Henzi *et al.* (1998) suggest for mountain baboons. For example, when the largest research group fissioned in 1993, two followers became the respective dominant males in the two daughter groups, between which females divided evenly (Robbins, 1995).

Dominant males commonly make control interventions in contests between females, which then usually end without clear winners (even when a single female faces a female coalition: Watts, 1997). This may help males to retain mates by evening competitive differentials among females (Watts, 1994b, 1997). Followers also make control interventions, but any benefits or disadvantages that policing offers females are not directly proportional to the number of males per group because dominant males try to prevent followers from engaging in it (Watts, 1997).

Comparisons with other gorilla subspecies

Female transfer and dispersal by maturing males who become solitary have been documented in eastern lowland gorillas (*Gorilla gorilla graueri*: Yamagiwa, 1983) and western lowland gorillas (*G. gorilla gorilla*: Tutin, 1996). In some populations of western lowland gorillas, the frequency of groups with two or more silverbacks is comparable to that in the Virungas (Harcourt *et al.*, 1981a). While these findings suggest that similar general variation in male life histories occurs across populations and subspecies, no other long-term demographic and life history data are available.

Comparisons with other gregarious primates and gregarious equids

Alternative male reproductive tactics are common in primates (Smuts, 1987; Cowlishaw & Dunbar, 1991; Newton & Dunbar, 1994), but are not combined with female transfer in most well-known cases. Conversely, female transfer is not necessarily combined with alternative male strategies like those in mountain gorillas. For example, Thomas's langurs (*Presbytis thomasi*) show egalitarian female dominance relationships, female transfer, and merger of female groups with solitary males, and seem to lack aggressive group takeovers by extra-group males. Males need to incite female transfer or mergers, and their ability to protect females against infanticide probably influences transfer decisions. However, resident males evict maturing males from their groups, and male followers are unknown (Steenbeeck *et al.*, Chapter 12; Sterck, 1997, Sterck & van Hooff, Chapter 11; Sterck & Steenbeeck, 1997). Many other Asian colobines show female transfer and form one-male reproductive units that in some cases coalesce into larger 'bands' (reviewed in Kirkpatrick, 1998; Yeager *et al.*, 1998) that could potentially include followers (although Kirkpatrick *et al.*, 1998, did not identify any in a one-year study of black snub-nosed monkeys, *Rhinopithecus bieti*), but little is known about male life history tactics in these taxa.

Callitrichids have cooperative breeding systems in which males can stay in natal groups as subordinate helpers while waiting for opportunities to breed there or to transfer to neighboring groups, or can 'prospect', i.e., disperse and try either to found new groups or to join established groups as subordinates or as dominant breeders (Goldizen, 1990; Baker *et al.*, 1993). Waiting can delay reproduction (although some subordinates may breed: Goldizen, 1990). However, prospecting was a poorer option in golden lion tamarins: prospectors had poor odds of establishing breeding groups, like gorilla bachelors, and faced high mortality risks (Baker *et al.* 1993).

Geladas (Dunbar, 1984) and hamadryas baboons (Kummer, 1968, 1980; Abegglen, 1984) also have male life history variants. One is for a male to become a follower in a non-natal reproductive (one-male) unit, an option closed to most male gorillas (although hamadryas followers also mostly join relatives). 'Follower' and 'takeover' are equilibrated and mutually exclusive strategy sets for geladas, mostly because followers have higher chances of success (like followers in gorillas), but males who attempt takeovers gain more females when successful (Dunbar, 1984). Gelada females are philopatric, and followers gain females via unit fission, not by reversing rank with dominants (Dunbar, 1984). Hamadryas alternative tactics are not mutually exclusive, although their frequency varies within, and perhaps across, populations (Kummer *et al.*, 1981; Abegglen, 1984): some followers who try to incorporate juvenile females into their 'initial units' without engaging in aggression with dominant male unit-holders later challenge those males and try to take over units aggressively. Followers form age-graded queues and the oldest followers of leader males are most likely to mount challenges and to acquire females in this way (Kummer, 1980; Abegglen, 1984). Males also try to recruit or steal females from other males, and most female transfer between one-male units involves coercion (*ibid.*). Hamadryas male reproductive success depends on male tenure length and on the size and stability of one-male units (Colmenares, unpublished data), but the relative contributions of following and takeover attempts to variation in male reproductive success have not been analyzed in detail.

As in gorillas, female natal and secondary transfer are common in gregarious equids. Most groups have single males, and males typically emigrate from natal groups (e.g., horses: Berger, 1986; Monard & Duncan, 1996; mountain zebra: Lloyd & Rasa, 1989), but male followers as well as bachelors occur in at least some species or populations. Male success at holding females can depend on protecting them against harassment by extra-group males; harassment can decrease foraging efficiency and reproductive performance (Rubenstein, 1986; Rutberg, 1990) and can even induce abortions (Berger, 1986). Followers in horse bands help to defend females against intruder males and may help to prevent female emigration (Berger, 1986; Feh, 1990; Stevens, 1990), although male alliances are likely to be unstable (Berger, 1986). Two-male horse bands on Shackleford Island retained females better than one-male bands (Stevens, 1990). Males also sometimes co-operatively defend bands against intruders in plains zebra (Rubenstein, 1986). Followers probably do little breeding (Stevens, 1990), but may steal copulations (Berger, 1986), and sometimes acquire bands from dominants (Stevens, 1990). However, in the one explicit analysis of the question, Monard & Duncan (1996) found that Camargue horse females preferred to join one-male bands, contrary to expectation if they benefit from membership in multi-male groups. Males in that population do little to protect immigrant females from harassment by resident females (Monard & Duncan, 1996; but see Berger, 1986).

Variation in demography and population density can influence the costs and benefits of primate male life history tactics, regardless of whether reproductive units typically have single or multiple males (Dunbar, 1984, 1988; Alberts & Altmann, 1995a). For example, dispersal costs for male yellow baboons are higher at low population densities than at high densities, and males in low-density populations should be less likely to emigrate from groups in which their mating success is poor and/or sex ratios are unfavorable (Alberts & Altmann, 1995a). Most male hamadryas baboons are philopatric, but unfavorable sex ratios could make transfer between bands, or even transfer into anubis baboon (*Papio anubis*) groups, advantageous for young males whose prospects for obtaining one-male units in natal clans or bands are poor (Phillips-Conroy *et al.*, 1992). In geladas, population growth and resulting increases in the size of one-male units give followers more opportunities to gain females via group fissions and should change the relative frequencies of followers and of males who try takeovers (Dunbar, 1984). Interactions among population growth and demographic variation could influence male gorilla life history tactics in similarly complex ways. For example, the Virungas population increased between 1978 and 1989, while the proportion of groups with multiple silverbacks decreased from 42.3% to 27%. The mean adult sex ratio within groups changed little (2.44 to 2.22 females per male), while the number of groups increased from 25 to 29. Population growth could have improved prospects for bachelors to attract females from large groups, or could have allowed some followers to gain females via group fissions. Male mortality in multimale groups also contributed to the changes (cf. Yamagiwa, 1987b; Robbins, 1995). The recent disruption of long-term data collection unfortunately hinders investigation of such complexity.

ACKNOWLEDGMENTS

I thank l'Office Rwandaise du Tourisme et des Parcs Nationaux for permission to work in the Parc National des Virungas, and l'Institut Zairois pour la Conservation de la Nature for permission to work in the Parc National des Virungas. Many people have contributed to long-term demographic records; particular thanks to D. Fossey, R. Campbell, A.H. Harcourt, K.J. Stewart, R. Barnes, P. Sicotte, D. Doran, M. Robbins, D. Steklis, J.P. von der Becke, M. Condiotti, C. Shollay, and J. Kalpers, and apologies to everyone not named. The information base only exists because of the dedicated and heroic efforts of A. Nemeye, E. Rwelekana, A. Banyangandora, F. Barabgiriza, L. Kananira, E. Rukera, L. Munyanshoza, F. Nshogoza, K. Munyanganga, C. Nkeramugaba, S. Kwiha, J.B. Bizumuremye, A. Vatiri, J. Sekaryongo, other Karisoke staff, and ORTPN guides and guards. My research at Karisoke has been supported by NIMH grant 5T32 MH 15181–03 and by The L.S.B. Leakey Foundation, The Dian Fossey Gorilla Fund, The Eppley Foundation for Research, The World Wildlife Fund-U.S., The Wildlife Preservation Trust International, and the Chicago Zoological Society. Many thanks to Peter Kappeler for organizing the Göttinger Freilandtage Conference and for inviting me to participate, and to The Wenner Gren Foundation for Anthropological Research, the Ministerium fur Wissenschaft und Kultur, Hannover, and the Deutsche Forschungsgemeinschaft, Bonn, for sponsoring the conference. Peter Kappeler, Carel van Schaik, and an anonymous reviewer provided useful criticisms of an earlier version of this chapter.

Part IV
Behavioral aspects of male coexistence

Whenever groups of multiple males exist, co-resident males must cope with the permanent presence of rivals. As males are typically equipped with large bodies and weapons, this constellation harbors potential risks for escalated fights. Behavioral mechanisms to mitigate this risk, and more intense scramble competition are therefore expected in most multi-male groups. If subordinate males can provide dominants with benefits, more complicated relationships involving staying incentives for subordinate males should be observed. Similarly, if co-resident males cooperate, interesting questions about the distribution of costs and benefits emerge. The chapters in this section explore some of these topics in detail.

Jan van Hooff opens this section with a review of male social relationships. He shows that competitive regimes among females set the stage for male options. In a few species, the option for male philopatry exists and provides these males with additional benefits for cooperative behavior. In the majority of primates, however, males are the dispersing sex and their relations are primarily competitive. In the interesting cases in which dominants benefit from the presence of other males, the latter have more leverage power and are expected to receive staying incentives in the form of shared reproduction. Van Hooff uses the concept of market principles to explore how these males negotiate their relationships.

The chapter by Charles Nunn takes up this idea and investigates the resolution of one such problem, using male extra-group conflict as an example. He argues convincingly that whenever the benefits of a particular action are shared among males, individuals have the option of reducing or withholding their contribution at the expense of others. Using comparative methods, he goes on to show that several measures of male group defense against neighboring groups vary in the predicted manner among species. The promising results of this refreshing approach should encourage students of primate behavior to apply concepts developed in other disciplines, such as economics, more widely to analyze behavioral variation within groups and among species.

Some less commonly used concepts are also employed by Signe Preuschoft and Andreas Paul to study conflict resolution among Barbary macaques. Because of similar competitive abilities among adult males, stalemate may arise to avoid escalated aggression. Using an experimental approach, they show that seemingly egalitarian relationships can be the result of fear and respect. This delicate balance can be easily tipped by coalitions among males. By extending their observations to the more common competition for estrous females, Preuschoft and Paul provide one of the most fascinating examples of subtle social behavior of primate males balancing competition and cooperation as a result of their co-residence in the same groups.

16 • Relationships among non-human primate males: a deductive framework

JAN A.R.A.M. VAN HOOFF

INTRODUCTION

Male relationships, the refinement of a picture

Our understanding of male relationships in primates has undergone considerable refinement in the past decades. Initially, the emphasis was on the antagonistic aspects of between-male relationships. Zuckerman's (1932) early study of a captive colony of hamadryas baboons fostered the view that adult males were mutually intolerant in their competition for sexual access to fertile females. This view was reinforced by many of the early observational studies: between-male conflicts were a salient feature, especially of cercopithecine groups (see Melnick & Pearl, 1987), and these were prominently represented in the early studies (Hill & van Hooff, 1994). These studies reinforced the view that the unrelenting inter-male competition resulted either in the exclusion of rival males, giving rise to one-male groups or harem systems, or in the formation of strong hierarchical structures guaranteeing priority of access to the most dominant male(s).

The theoretical explanation is still valid, be it only in part. Male and female primates differ in their strategies to maximize their inclusive fitness. In this respect they are like other species in which female reproductive output is limited (Trivers, 1972; Emlen & Oring, 1977; Wrangham, 1980; Clutton-Brock & Albon, 1982; Clutton-Brock & Vincent, 1991). Especially in primates with their long interbirth intervals, female life time reproductive success is determined by a long life in good condition. So female fitness-maximizing strategies should be directed at two matters in particular: achieving safety for themselves and their offspring, and securing critical resources, in particular food. Consequently, between-female competition will also be directed especially at the latter (van Hooff & van Schaik, 1992).

Since the fitness of males depends largely on the number of females they have successfully inseminated, males should give high priority to competition for fertilizations. Unlike food and many other resources, fertilizations cannot be shared. Access to a fertile female yields an all-or-none success. This must reduce the possibilities to form coalitions and collaborative bonds (van Hooff & van Schaik, 1992).

Skew in male reproductive success

There is, indeed, ample evidence that competition between males for access to females is strong and that there is a correspondingly strong skew in reproductive output. The evidence has come especially from studies of multi-male groups of cercopithecines. These showed that females are a contested resource (e.g., Altmann, 1962; Wrangham, 1980; Smuts, 1987), that aggression between males increases in the presence of estrous females (Vandenbergh & Vessey, 1968; Wilson & Boelkins, 1970; Bercovitch, 1986; Janson, 1984), and that in many cases males of high rank had a larger proportion of the copulations (Kaufmann, 1965; Packer, 1979b; Smith, 1981; Berenstain & Wade, 1983; Chapais, 1983; Hill, 1987; Cowlishaw & Dunbar, 1991; Bulger, 1993; Reed *et al.*, 1997), although this has not been found in all studies (e.g., Loy, 1971; Smuts, 1985; Bercovitch, 1986, 1987; McMillan, 1989). It has been a matter of debate whether this also means that dominant males achieve a higher reproductive output (e.g., Fedigan, 1983). Although some DNA studies on captive colonies have failed to demonstrate this (for a review see de Ruiter & van Hooff, 1993), recent studies (Paul *et al.*, 1993; Smith, 1993), especially for wild primates (e.g., Pope, 1990; de Ruiter *et al.*, 1992; Berard *et al.*, 1993; Keane *et al.*, 1997), confirm an effect of male dominance. De Ruiter's studies have shown that in wild groups of long-tailed macaques (*Macaca fascicularis*) the highest ranking sired many more offspring than could be expected given his share of copulations. This implies that males of high rank succeed in monopolizing sexual access at the moments when it matters (cf. Packer, 1979b; Chapais, 1983). This can both be due to selective monopolization when the male notices the estrous peak of the females, and to a female preference for the highest ranking male at this moment.

At any rate, a preference of adult females to associate with dominant males has been found in a number of studies (e.g,. Smuts, 1985; Hill, 1987; Reed *et al.*, 1997). The proximate reasons for which females prefer the company of dominant males may be diverse (e.g., protection against hostile conspecifics; reduced feeding competition and access to preferred resources – see below), but if this implies a mating preference, then the female benefits also in terms of obtaining 'best genes'.

Types of competition and reproductive skew

Whether competition for fertilizations results in strong reproductive skew depends on the nature of inter-male competition. It is useful to distinguish two components of competition: contest or exclusion competition, and scramble competition (van Schaik & van Noordwijk, 1988). Which component prevails depends on the availability and distribution of the resource items for which competition takes place, be they food, social/sexual partners, safe places or other desired matters. These variables determine the monopolizability of the resource. Two aspects of the distribution of the items matter most, namely:

1. Concentration *in space*: the more compact and clumped the resource items, the easier it is for a powerful animal to claim access to them all and to exclude others, i.e., the *higher* the monopolizability.
2. Concentrated availability *in time*: the more simultaneous the occurrence of the items and the shorter the time they remain available, the more difficult it is for a powerful animal to claim them all at once, i.e., the *lower* the monopolizability.

Van Schaik (1989) has pointed out how important the degree of monopolizability of resource items is for the nature of social relationships between members of a social group. Dominance and reproductive skew are associated with the component of contest competition. Such skew is not necessarily expected when male competition is mainly by scramble.

A major cost of social grouping is competition for the same finite and vital resources (van Schaik & van Hooff, 1983). When their monopolizability is high, there will be a strong skew in the degree to which group members can exploit the resource. This skew is proportional to the differences in contest potential or power. It then pays an individual who is able to do so to invest in behavioral dispositions and structural characteristics which increase her or his superiority in contest. Consequently, there will be a selective pressure toward individual dispositions which are dominance oriented ('despotic') and nepotistic (because of its inclusive fitness benefits).

When monopolizability is low or absent, there will be little or no skew in what group members can appropriate. Competition is mainly or solely by scramble. Also, selection for 'expensive' features which afford contest superiority and power will be low or absent. Relationships between group members will tend to be egalitarian. If a resource is limited, it makes a difference for everyone how many there are who scramble for a morsel. Individuals might, therefore, be excluded from group membership. Who will be excluded will certainly depend on the 'value' of the individual, that is her or his contribution to the advantages that everyone derives from being socially together (this value concept also plays a role in the skew of contested exploitation – see below).

Female competition, contest or scramble for resources

Since males and females have different priorities in competing, this will lead to differences in relationships (Trivers, 1972; van Schaik, 1989; van Hooff & van Schaik, 1992). There is abundant evidence that females profit from associating mainly in that it increases their safety, but also that there is an increased competition for the resources they exploit together (van Schaik, 1983; Dunbar, 1988; Janson, 1988). Females will be subjected to competition for food, in the first place.

1. If competition is mainly of the contest type, we expect females to be dominance oriented and to form social hierarchies. As a result, they will also be nepotistic; they promote their inclusive fitness when they favor relatives selectively in tolerance and support. Consequently, they will tend to stick to their natal groups, in which they find relatives who support them (Wrangham, 1980; Pusey & Packer, 1987; van Schaik, 1989). However, this also means that if females refuse to leave their natal group, inbreeding avoidance will force males to do so (e.g., Packer, 1985).
2. If scramble prevails, female relationships will be more egalitarian and individualistic; bonds between female relatives will be comparatively weak and there will be no great necessity to stick with relatives. Both males and females may now disperse. This means that males are no longer forced to be the exogamic sex. It opens the option

that males can stay in their natal group (van Schaik & van Noordwijk, 1989).

These two syndromes of traits have long been recognized as 'female-bonded' and 'non-female-bonded' societies. Comparing a great number of taxa, Sterck *et al.* (1997) could show that a number of the traits do occur together, and that there is always male dispersal when there is female bondedness. When there is no female bondedness, either sex may disperse. Although the array of taxa carries conviction, there is the possibility that the distribution of the traits is determined by phylogenetic dependencies (DiFiore & Rendall, 1994). The most convincing evidence, therefore, comes from cases in which closely related species show an associated divergence in ecology (presence or absence of resource monopolization) and in the social traits. Examples are the studies by Mitchell and Boinski enabling a comparison of Bolivian and Costa Rican squirrel monkeys (*Saimiri oerstedi*: Boinski, 1994; *S. sciureus*: Mitchell *et al.* 1991; Mitchell, 1994). The Bolivian species has clumped, monopolizable fruit resources and it shows all aspects of female bondedness, including female philopatry. The Costa Rican species shows the opposite: dispersed fruit resources, non-female bondedness, and male philopatry. Another illustrative example is the analysis of the variations in ecological factors, competition regimes, and social structure in the different species of the taxon *Papio* (Barton *et al.*, 1996; Barton, Chapter 9).

Type of competition between males, sexual dimorphism, and sperm competition

When males compete, this will be primarily for fertilizations, so, in fact, for access to fertile females. Male strategies will, therefore, be determined by the distribution of fertile females (Wittenberger, 1980; Rubenstein & Wrangham, 1986; Clutton-Brock, 1989). Again, the distinction between contest and scramble competition is helpful (van Hooff & van Schaik, 1994; cf. precopulatory and postcopulatory competition: Parker, 1984).

1. If access to (groups of) females can be monopolized, then male contest may be expected. As Darwin (1871) and others since then have noted (e.g., Clutton-Brock *et al.*, 1977), reproductive skew can be so great that high investment in fighting armory and motivation is selected for, leading to a corresponding sexual dimorphism. A comparison of taxa by Alexander *et al.* (1979) seemed to support this, justifying the assumption that there was a positive correlation between sexual dimorphism and the number of females a male could monopolize (cf. Clutton-Brock, 1985). Yet there have been doubts whether the relation would remain if corrections for phylogenetic dependencies had been made (for references see Mitani *et al.*, 1996b). However, a recent study by the latter authors has confirmed the correlation after taking the operational sex ratio into account.
2. The distribution of fertile females does not always allow monopolization, in which case males will scramble for their share in copulations, and possibly in fertilizations. This scramble may lead to frequent matings in order to outcompete rivals in sperm competition. If so, this might be reflected in a comparatively high fertilization potential, such as a high relative testis size (Short, 1979; Harcourt *et al.*, 1981b; Møller, 1988; Kappeler, 1997b). These authors did, indeed, find that relative testis size tends to be higher in taxa for which more than one adult, sexually active male resides in a group. At the same time, the need to invest in the dimorphic aspects of contest power must have lessened.

A two-dimensional rendition of these relationships obviously is too simple. For example, an expected negative correlation, over primate taxa, between sexual dimorphism and relative testis size (e.g., Dunbar & Cowlishaw, 1992) was not found (van Hooff & van Schaik, 1994). Within some taxa, differences covaried as predicted, especially in the great apes. The species of other taxa differed strongly in one characteristic but not in the other. For example, both the langurs and the macaques showed considerable within-taxon variation in sexual dimorphism. There was, however, no great within-taxon variation in relative testis size, whereas the taxa differed markedly from one another in relative testis size. Clearly, there are strong phylogenetic dependencies. Apart from these, however, more specific, as yet unknown factors must be responsible. For example, the role of females in the allocation of partners has to be considered.

One-male versus multi-male groups

To what extent can the variation in male group membership be related to monopolizability of fertile females, and what determines this monopolizability? Primate species in general differ in the presence of adult, potentially reproductive males in a group. Although in some species there also appears to be great intraspecific variability in this respect (e.g., Hanuman langurs (*Presbytis entellus*): Newton, 1988; callitrichids: Goldizen, 1987a, 1987b, 1988), most polygynous species can be characterized as predominantly having either one-male or multi-male groups (Smuts *et al.*, 1987). Especially when species vary within a taxon, and, therefore, phylogenetic constraints play a minor role, this offers a possibility to test the competition model and the influence of environmental, ecological, and social variables.

The presence of more than one adult, sexually active male in a group can be due to two major reasons: first, the inability of dominant contestants to keep rivals out, and second, (selective) tolerance of other males who buy their stay by providing some benefits (Sterck & van Hooff, Chapter 11)

Monopolizability of access to fertile females

Monopolizability can either not be effectively possible or it makes no sense in view of the costs that have to be invested (Mitani *et al.*, 1996a). Different factors have been proposed that may influence the degree to which males can monopolize access to fertile females.

1. Females become fertile more or less simultaneously during a short breeding period, because there is an influence of ecological seasonality (Berenstain & Wade, 1983; Ridley, 1986). Under these circumstances, a single male may not be able to mate with all females. There is even the hypothetical possibility that females decrease their monopolizability by actively synchronizing their periods of estrus, but evidence for this is still lacking.
2. Females maintain a dispersed and/or flexible distribution in space, making it impossible for a dominant male to exclude others from contact.
3. The number of females and consequently the size of a group are too great for one male to keep under control (van Schaik & van Hooff, 1983; Terborgh, 1983; Andelman, 1986; Crockett & Eisenberg, 1987; Dunbar, 1988; Newton, 1988; Altmann 1990; Janson, 1992).

In a recent study, which controlled for phylogenetic dependencies, Mitani *et al.* (1996b) investigated the correlated occurrence over many primate taxa of, on the one hand, number of males per group, and, on the other hand, duration of breeding seasons and number of females per group. The number of males was not related to breeding season duration. It was related, however, to the number of females.

Srivastava & Dunbar (1996) investigated the variation in the number of adult males in groups in one taxon, the Hanuman langur. They analyzed 23 populations and concluded that the percentage of one-male groups in a population does, indeed, depend on the monopolizability of females. The number of males was related to the number of females (group size), but also to the duration of the breeding season. In addition, a greater distance between adjacent groups decreased the willingness of males to leave a group, indicating a trade-off with migration costs. Also the intactness of the predator community influenced the number of males (Treves & Chapman, 1996). This latter factor is a component of the disturbance index of Bishop *et al.* (1981), which has also been linked to the rate of male takeovers (Rudran, 1973a) and the dynamics in the social processes (Sterck, 1998).

When males cannot exclude one another, they have to tolerate one another and more egalitarian relationships may form. A particularly interesting variant of 'egalitarian tolerance' has been proposed by Preuschoft & Paul (Chapter 18), who conclude that tolerant and even egalitarian relationships can be the result of extreme fighting power. This would cause both parties to run great risks when allowing a confrontation to take place and thus there is a stalemate.

Benefits from supernumerary males

Even where group size would permit a dominant male to exclude others, this does not always happen. This suggests that some or all group members may derive some benefits from the presence of 'supernumerary males'. These males could render services in various ways (Noë *et al.*, 1991). For example, males tend to invest more in vigilance than females (e.g., Rose, 1994; Koenig, 1998). There may be two reasons for them to do so. First, their vigilance may make for earlier detection of predators, and extra males may be tolerated for this benefit (van Schaik & van Noordwijk, 1989). Pertinent evidence comes from a study by van Schaik & Hörstermann (1994) in which they compared Asian, African and New World folivorous monkeys. They found that the number of adult males per adult female is greater where there are real predators which hunt monkeys. Equally suggestive is the finding for some wild callitrichids that males invest more in vigilance and sentinel behavior than do females (e.g,. Goldizen, 1989; Savage *et al.*, 1996b), and also that in groups containing more adult males, relatively more infants are found (Sussman & Garber, 1987; Koenig, 1995).

The second reason is that male primates may also be more watchful for rivals from outside. In other words, the vigilance of males may be primarily directed toward extra-group males. In that case, one would expect such vigilance to be greatest in areas where the chance of meeting rivals is also greatest, namely where ranges of different groups overlap (in Thomas's langurs; *Presbytis thomasi*: Steenbeek, 1999; white-faced capuchins, *Cebus capucinus*: Rose & Fedigan, 1995; Gould *et al.*, 1997). Whatever the motive for male vigilance, females may profit by increased protection against both predators and infanticidal attacks.

One of the most pertinent data sets is the one on langurs. A comparison of the Hanuman langur data compiled by Srivastava & Dunbar (1996) with those of other langur taxa by Sterck & van Hooff (Chapter 11) revealed that also in the other langurs multi-male groups do occur, although groups are much smaller and birth seasonality is reduced. In these taxa, neither the number of females nor the risk of predation influenced the number of males. The extra males were often natal males or a superseded dominant who stayed. These then were mostly age-graded groups which seemed to continue because the direct and inclusive fitness costs of expulsion and leaving might be too high, and because there may be direct and inclusive fitness benefits of staying. The work on Thomas's langurs (Steenbeek *et al.*, Chapter 12) suggests that the benefits are: extra males offer increased protection against infanticidal attacks by outside males; support by natals can extend the longevity of dominant tenure and thus the longevity of a bisexual group; matured natal males can linger until there is a chance to establish themselves independently (for an analogous situation in gorillas see Robbins, 1995; Watts, Chapter 15).

The importance of infanticide and its possible role in structuring social relationships have received increased interest recently (e.g., Newton, 1988; Hrdy *et al.*, 1995; van Schaik, 1996; Sterck *et al.*, 1997). The fact that primates are comparatively exceptional among mammals in showing year-round association of females with males has been interpreted as a protective response against infanticide (van Schaik & Kappeler, 1997). Females and offspring would profit by remaining associated with a strong male who, in turn, has an interest in staying as a protector because he is the most likely father.

Whatever the motive for male vigilance, the increased protection it brings to females and, as a result, also to a resident male who has already reproduced successfully with these females, may tip the balance toward tolerance of a supernumerary male as a group member. By such service they might 'buy' their 'right to stay'.

The iterative interaction of cost–benefit trade-offs

This example illustrates the possible complexity of the interactions involved. If, indeed, a resident, dominant male profits from the presence of one or more other males, this presents a trade-off problem to him. He risks a loss of fertilizations to the other males. His tolerance should remain below the level at which this risk offsets the benefits accruing from the presence of the subordinate males. This trade-off can in turn be influenced by female strategies. The opportunities for males to monopolize access to fertile females are the result of the interaction of both male *and* female socio-sexual strategies. Females may have an interest to decrease monopolizability. They can break the monopoly of access to many fertile females in two ways, which each has its specific consequences. (1) Females may either blur their estruses, thus decreasing the predictability of their fertile period and inducing a situation of scramble competition between males. (2) Females can also make a polygynous male strategy become pointless by sharply synchronizing their estruses. Either method may have evolved to increase the option of female choice and seduce males to investment in services. In this way, females may tip the balance toward special relationships or even monogamous bonds (e.g., Smuts, 1985, 1987; Noë *et al.*, 1991). It is in the interplay of such strategies that males form their relationships with one another.

The 'monopolizability model' and the 'benefits from extra males model' do not explain all cases. In a number of Madagascan lemur species, polygynandrous groups occur. The asynchrony in female fertility and the distribution of females seem to allow investment in exclusion competition, and indeed contest between males during the mating season appears to be common. Yet females do engage in multi-male matings. However, there are no indications that the females derive any service benefits from the extra males. There is no evidence of antipredator services, or of support in intergroup conflicts, and the males do not ward off strange males who might form an infanticidal risk (Kappeler, Chapter 5). Therefore, Kappeler concludes that neither processes of intrasexual selection nor of benefit attribution to the females can give satisfactory explanations for the existence of multi-male groups in this taxon.

Variation in male relationships

In many primate species, neutrality or even hostility is the most striking aspect of the relationships between adult males (e.g., Cords, 1987b). Although there has been a traditional emphasis on the competitive nature and antagonistic character of adult male relationships, recently there have been an increasing number of references to tolerant and affiliative aspects of their relationships (Hill, 1994; Hill &

van Hooff, 1994). We have become aware of the enormous variation in male relationships, both their relationships with females and those between the males themselves. There are quite a few species in which adult males live together peacefully in groups, both in bisexual and in all-male groups. In fact it is remarkable that in species in which males are typically severely antagonistic to each other, such males can live comparatively peacefully together when outside a bisexual group (e.g., Pusey & Packer, 1987). This indicates that adult males of the respective species do not have an intrinsic hostile motivation towards their same-sex conspecifics. Their mutual intolerance is contextually determined. The presence of adult, receptive females and the resulting competition are the specific motivational determinants (cf. Robbins, 1996).

For more and more species, however, not only between-male tolerance in bisexual groups has been described, but also collaboration and affiliative bonding. Examples are red colobus (*Colobus badius*: Struhsaker, 1975; Struhsaker & Leland, 1987; Starin, 1994), spider monkeys (*Ateles* spp.: van Roosmalen, 1980; Fedigan & Baxter, 1984; McFarland, 1986; Chapman *et al.*, 1989), woolly spider monkeys (*Brachyteles*: Milton, 1985; Strier, 1994b), red howler monkeys (*Alouatta seniculus*: Pope, 1990), squirrel monkeys (*Saimiri oerstedi*: Boinski, 1994; and *S. sciureus*: Mitchell, 1994), hamadryas baboons (*Papio hamadryas*: Kummer, 1968), and chimpanzees (*Pan troglodytes*: e.g., Goodall, 1965, 1986; de Waal, 1982). Note, though, that cooperation between same-group males, e.g., in group defense, need not automatically be associated with high levels of affiliative interaction, but instead can be tension loaded (capuchins: Perry, 1998; mountain gorillas: Robbins, 1996). This reveals the dilemma between common interests in group defense and inter-male competition and coercion within the group.

Bonding and kinship

Such male bonding can only be explained if it yields a mutual advantage. There are theoretical arguments to expect such cooperative bonding only under specific conditions. We have noted already that these conditions must be met more easily in female–female relationships than in male–male relationships. The resource for which females compete most – food – can much more easily be shared after collaborative efforts to secure it than the resource for which males compete most – namely, fertilizations (van Hooff & van Schaik, 1992).

Individuals can share an inclusive fitness interest. This will facilitate 'altruistic' investments, if such investments can be selectively directed to close relatives. Then, conspecifics must be able to ascertain their relatedness with one another. Mother and offspring have an assured relatedness of 0.50. Offspring of the same mother are on average at least 0.25 related if fathered by different males. In a female-nepotistic society, where females are philopatric and stay with their sisters, the highest degrees of relatedness may be expected if also there is strong male contest, so that different females conceive from the same male. The longer a male can maintain an exclusive tenure, the closer the average relatedness of maternal kin can creep toward 0.50. If infants of different females are fathered by the same males, average relatedness might approach 0.25, and, if mothers are maternal sisters, up to 0.375. This is most likely for age cohort members sired by the same tenure-holding male. It is true for both male and female offspring. Recently, de Ruiter & Geffen (1998) analyzed blood samples from about 230 wild long-tailed macaques living in adjacent groups in a large, undisturbed and continuous population. There was a surprisingly good fit of these expected patterns to the relatedness patterns actually found. Surprisingly, there was even a higher level of paternal relatedness among high-ranking group members. This is due to the observed fact that the alpha males monopolized mating preferably with the dominant females.

Evolutionary logic thus tells us that male kinship must be a contributing factor to male bonding. There is evidence also that in some tolerant-male and bonded-male species, males tend to be more than averagely related (e.g., see review in Gouzoules & Gouzoules, 1987; hamadryas baboon: Kummer, 1968; Stolba, 1979; Sigg *et al.*, 1982; Stammbach, 1987; woolly spider monkey: Strier, 1994b; squirrel monkey: Mitchell, 1994; chimpanzee: Morin *et al.*,1994).

Relatedness between male sibs can be maintained by two processes: male philopatry and parallel dispersal. We have noted that the first, male philopatry, can be expected when the females are not strongly bonded and nepotistic, and, therefore, are not philopatric.

Male philopatry is indeed a characteristic of a number of species for which male bonding has been documented, namely red colobus (Struhsaker, 1975; Struhsaker & Leland, 1987), spider monkeys (van Roosmalen, 1980; Fedigan & Baxter, 1984; McFarland, 1986; Chapman *et al.*, 1989), woolly spider monkeys (Milton, 1985; Strier, 1994b), hamadryas baboons (Kummer, 1968), and the chimpanzee (Pusey, 1979; Goodall, 1983; Nishida & Hiraiwa-Hasegawa, 1987).

The second process, parallel dispersal, has also been documented. It can occur when paternal sibs, members of the same age cohort, migrate together. Especially in species with strong male exclusion competition, the relatedness between paternal sibs will be high. The proximate factor would then be familiarity and predictability in their relationships due to membership of common peer-play groups in their youth. A factor leading to distinct age cohorts which may facilitate peer familiarity is birth seasonality (in as far as this does not inhibit male exclusion competition – see above). Parallel dispersal can also occur when young males move preferably to groups where they meet older males with whom they were familiar before these left their common natal group. This requires that young males recognize maternal brothers in the group they go to. There is evidence of kin-linked migration of males for some species, for instance macaques and vervets (e.g., Cheney & Seyfarth, 1983; van Noordwijk & van Schaik, 1985; see Melnick & Pearl, 1987, for a review).

There is still no unambiguous answer to the question of whether genetic relatedness is an important factor explaining the occurrence of male bonding. There is evidence for fraternal nepotism also in some non-philopatric species (Meikle & Vessey, 1981). Male red howler monkeys (Pope, 1990) and male bonnet macaques (*Macaca cynica*: Silk, 1992, 1994) supported each other more often when they were related. By contrast, for another well-studied primate taxon, the savannah baboons, for which male alliances are common, no evidence for a role of kinship has so far been reported (Noë, 1986, 1992).

Much of our present knowledge about the possible relatedness of males has been based on demographic and life history data. We are much in need of molecular–genetic data on relatedness for populations of species with differing social systems. For some species, we are beginning to obtain these data. Thus, male chimpanzees in a community appeared to be more related than the females (Morin *et al.*, 1994). Yet, in contrast to expectations (e.g., Goodall, 1986), within a community, measures of social bonding were not correlated with genetic relatedness (Goldberg & Wrangham, 1997). Male chimpanzees clearly select their alliance partners on the basis of political opportunities (cf. de Waal, 1982).

The chimpanzee data resolve a question raised in particular by te Boekhorst (1991) and te Boekhorst & Hogeweg (1994), namely whether kin-selected adaptations are necessary to explain the association patterns of chimpanzees. They modeled the behavior of 'artificial CHIMPS' in a 'MIRROR WORLD'. They equipped their 'INDIVIDUALS' with simple, general rules concerning foraging and mate searching, and they let the 'CHIMPS' live in a 'CHIMP-like habitat'. From the interactions of these 'CHIMPS' emerged a pattern of spatial relationships, associations and movements resembling in detail those of real chimpanzees. They concluded that parsimonious carefulness is required to prevent the unjustified acceptance of unnecessarily complex traits and processes, i.e., selection of specific adaptive traits. Thus, the necessity of kinship as a factor explaining so-called male bonding in chimpanzees became doubtful. This doubt was fed by empirical data indicating that dispersal of males (e.g., young males with their mother: Wrangham, 1986) and promiscuity might dilute paternal relatedness. The example of maternal brothers helping one another in coalitions (Goodall, 1986) may be a special case.

Reciprocity and negotiation

In one other taxon at least, the savannah baboons, relatedness and inclusive fitness do not play a role in the formation of male coalitions (Packer, 1977, 1985; Noë, 1986, 1992; Bercovitch, 1988).

Baboon males have at least three mating strategies. More dominant males rely on their monopolization potential in defending exclusive consorts. Subordinate males may induce a female's preference in their favor by rendering services and thus forming special friendly relationships (Smuts, 1985). In addition, subdominant males may form coalitions to steal an estrous female from a consorting dominant male (Noë *et al.*, 1991). These coalitions have been interpreted as an example of reciprocal altruism (Packer, 1977). But then there is a clear distribution problem: who is going to get the new consort? There appear to be strong deviations from reciprocity, in that a particular male almost always obtains the female, and the other coalition partners almost never do. This suggests that a more complex system of 'trading' operates (Noë, 1990; Noë *et al.*, 1991). The asymmetries in who gets the profits appear to depend on the contest potential which each of the partners brings to the coalition. The male whose contribution is essential for making the coalition a winning one can play off the other animals one against another as a 'veto player' in a 'veto game' (Noë, 1990, 1992).

It is remarkable that the most dominant male is the one who has entered the group most recently. The subdominant animals have already been together for a longer time in the same group. This may have allowed them to acquire a familiarity which is necessary for developing trustworthy expectancies about the input of others in collaborative exchanges.

Here, familiarity is not a motivational factor *per se*, but it is a condition under which the expectations of opportunistic bargaining processes are most reliable. The same is indicated by the studies suggesting negotiation in chimpanzees (de Waal, 1989b, 1992, 1996a).

In general, relatedness seems less important for male association and cooperation than it is for female relationships. Whereas devotion to kin is fundamental for female-bonded societies (Gouzoules & Gouzoules, 1987), males can form opportunistic bonds. Clearly, familiarity facilitates successful bargaining with other males, because it forges a certain predictability in terms of (mis)trusting expectancies. But it is striking that even in a reputedly male-bonded species such as the chimpanzee, this sex difference is visible in coalition formation (de Waal, 1978, 1984; but see bonobos, *Pan paniscus*: Hashimoto *et al.*, 1996). The difference, therefore, seems to reflect a more fundamental heritage, which may be characteristic for all non-human primates. Female devotion to kin may in part be deducible from an emancipation of maternal affiliative tendencies.

FEMALE INFLUENCE ON MALE RELATIONSHIPS

An aspect which has received little attention so far is the possible influence which females may exert on the nature of male relationships. If females try to establish particular relationships *with* males, this must have an effect also on relationships *between* males.

Recently, we have come to appreciate the enormous influence that male infanticide can have on the way females steer their own relationships with males (van Schaik, 1996; van Schaik & Kappeler, 1997). Males can eliminate the offspring of females, but they can also help them and support their mothers. For this to be beneficial to themselves, males have to 'know the truth' about their paternity. It is obviously in the interest of the female to create a paternity illusion in all males that might be a threat to her infant but could also be a supporter of it. The active promiscuity of females and the blurring of estrus in some species have been interpreted as a strategy to create such illusions (van Noordwijk, 1985; Hrdy & Whitten, 1987; Martin, 1992; de Ruiter *et al.*, 1994).

An unsolved question is why females succeed in this promiscuous strategy in some species, and not in others in which infanticide occurs. At first sight one is inclined to think that males who are able to maintain exclusive access to a group of females would ruin their own paternity prospect if they permitted their females to create paternity illusions in other males. This seems to point to failure of control of the dominant male. However, it is not that simple. A dominant male should not object to sexual contacts of other males with 'his' females as long as he can be sure that these will be to little avail. If this promiscuity reduces the risk of infanticide of his own offspring, he would benefit enormously as well. Thus, he should be permissive if he can be sure that a female is exclusively accessible to him when it matters, either because, at the time of ovulation, he can effectively constrain her, or because at this time she prefers her, proven, male as the supplier of good genes.

Indeed, the mating pattern of the long-tailed macaques, studied by de Ruiter *et al.* (1994), has this effect. The fact that the females of this species blur their estrous signals (van Noordwijk, 1985) indicates that they are out to deceive their mating partners. Also, if the rivals are fooled, why is the dominant male not? This makes it likely that the female controls the selective access. Only a precise behavioral analysis of the temporal pattern of constraining efforts and sexual solicitations in relation to the endocrinological cycle of the females can answer this question.

Logically, for a dominant male, the decision to allow a female a certain promiscuity must be a trade-off between the risks of not having an offspring by spoiling a chance or of losing an offspring by having it killed. This will be the case, especially, in species which bear singletons, as in almost all primates. In this respect, the twin-bearing callitrichids with their inclination toward polyandry are intriguing.

Male maturational stages: bimaturism and female choice

In species in which contest is decisive in the competition between males for mates, there is often also a pronounced sexual dimorphism (see above). Males of such species usually reach their full size and dimorphic characteristics some time after they have reached sexual maturity (Walters, 1987). As a rule, the development to the fully mature form proceeds rather gradually.

In at least one species, the orangutan (*Pongo pygmaeus*), this development takes place quite abruptly. Moreover, the transition from the submature to the mature form with its conspicuous secondary sexual characteristics occurs rather unpredictably. It can be arrested for a considerable number of years (Kingsley, 1982; Galdikas, 1985b; Schürmann & van Hooff, 1986). This leads to a remarkable bimaturism. There is some evidence that the transition is partly socially determined, the presence of a fully mature male inhibiting the maturation in a subordinate young adult (Kingsley, 1982; Maggioncalda, 1995).

Subadult and 'arrested' adult males roam in areas which overlap with those of the mature males. The latter are highly intolerant toward other mature males, who make themselves detectable by their far-reaching long calls (Mitani, 1985a). The matures are much less so toward submatures, although the submatures are also sexually very active, be it more often in the form of forced matings (e.g., MacKinnon, 1974; Rijksen, 1978; Schürmann, 1982; Mitani, 1985b; Schürmann & van Hooff, 1986). However, not all copulation attempts are resisted by the females (Galdikas, 1979, 1985b; Schürmann & van Hooff, 1986), so the sub-matures are real sexual competitors for the mature males. This suggests the existence of alternative male strategies coupled to bimaturism (cf. Rodman & Mitani, 1987), bearing resemblance to the sneaker and satellite strategies described for a number of taxa (Gross, 1996), e.g., insects (Alcock *et al.*, 1977), fish (Gross, 1985), amphibians (Arak, 1988), and birds (van Rhijn, 1985).

Rodman & Mitani (1987) have argued that the orangutan social system with its strong sexual dimorphism must be the result of strong male–male competition, and they deny that female mate choice plays a role. Schürmann & van Hooff (1986), on the other hand, have emphasized the pivotal role of female choice. Females are active in establishing consorts with fully mature males. It is hard to understand how the pronounced bimaturism among adult males could exist as a stable evolutionary situation if female preference for the mature males were not so strong that these could relax their antagonism against younger males (van Hooff, 1996). If females directed their proceptive engagement to the mature male when they were fertile, and if they were most reluctant toward the arrested males in this respect, then both females and mature males would profit in two ways. First, the females would assure themselves of the genes of males of proven quality. Also, in spite of their sexual efforts, the arrested males would have so little reproductive success that the dominant resident would not need to worry. Second, both the female and the father of infants born to her might profit from an infanticide inhibition generated in future matures. So far, the orangutan is the only great ape for which infanticide has never been reported, even though mothers with infants often move so independently as to be easy victims for a surprise attack (van Schaik & van Hooff, 1996).

In order to settle whether it is female preference which is pushing the evolution of bimaturism, data on the correlation of the sexual patterns with the females' endocrinological state and on the actual paternity distributions are needed. Then it would be a striking example of the fact, implicit in much that has been treated above, that female strategies not only determine their relations *with* adult males, but also the patterns of relationships *between* adult males.

ACKNOWLEDGMENTS

I am greatly indebted to Peter Kappeler for his efforts in organizing the Göttinger Freilandtage on this fascinating topic of male relationships.

17 • Collective benefits, free-riders, and male extra-group conflict

CHARLES L. NUNN

INTRODUCTION

Many behavioral acts in animal societies confer *collective benefits*: when an individual acts to obtain some benefit, usually at a cost to the actor, other individuals also benefit. For example, when an individual in a primate social group defends the group's territory, other individuals in the group will benefit, regardless of whether these individuals contribute to defense. Collective benefits therefore commonly lead to *free-riders*, who are individuals that benefit from a collective act without contributing to its costs. Collective benefits and free-riding are the essence of economic models of collective action, which examine individual cost–benefit decisions in human societies in the context of shared gains (Olson, 1965; Sandler, 1992). When free-riding leads to a suboptimal level of the collective benefit, a *collective action problem* (CAP) is said to exist.

Principles derived from these models have greatly improved our understanding of human cooperation in a variety of institutional structures (e.g., Ostrom, 1990). However, the extensive theoretical and empirical research on collective action in humans has only recently been applied to animal behavior (van Schaik, 1996; Nunn & Lewis, 1999).

This chapter examines how the general framework of collective action applies to animal behavior by examining the case of male extra-group conflict in primates. Here, male extra-group conflict refers to male–male competition in which a resident male prevents outside males from entering his group. Male behavior in extra-group conflict is known to vary, and it appears to be at least partly a function of social context (e.g., the number of males: Hamilton & Bulger, 1992; van Schaik, 1996). Economic models of collective action may account for some of this variation in male behavior; thus, such models may explain cases in which male cooperation is expected, but not observed.

First, the economic concept of collective action is reviewed. Then, evidence is provided for CAPs in male extra-group conflict based on reports of intraspecific variation in male behavior. After this basic introduction to collective action and how it can be studied in non-human societies, I derive four predictions from the collective action literature, and test these predictions using phylogenetic comparative methods. The comparative evidence presented here is an indirect test of the CAP; more direct evidence is therefore needed, especially evidence in the form of experiments and observations. One goal of this chapter is to stimulate further research on collective action in biological systems, as this approach to studying animal cooperation differs in subtle but important ways from previous approaches (Nunn & Lewis, 1999).

Social rules, like the CAP, structure behavioral options at the individual level, and these individual interactions then result in group-level patterns. This chapter therefore concludes with a discussion of the socioecological model, and how 'interaction effects' and 'social rules' fit into our general understanding of primate socioecology (van Schaik, 1996).

ECONOMIC MODELS OF COLLECTIVE ACTION

Economic models of collective action have at least three general benefits for studies of animal cooperation (although some of these benefits are shared by existing models of animal cooperation, including game theory approaches; see Dugatkin, 1997; Nunn & Lewis, 1999). First, by explicitly labeling some benefits as collective, the framework of collective action focuses on individual motivations in the context of shared benefits. This focus on the individual differs from many previous approaches to studying the acquisition of collective benefits in animal societies. For example, in comparative studies of territoriality, individuals in the social group are implicitly assumed to share a common set of costs and benefits; hence, the focus of inquiry is typically at the group level (e.g., Mitani & Rodman, 1979). In contrast, models of collective action investigate how the distribution of benefits within groups motivates individual action.

Collective action models may therefore explain why some individuals, but not others, are involved in territorial encounters.

Second, theories of collective action are commonly used to design policies that facilitate the acquisition of collective benefits in human societies (e.g., Ostrom, 1990). Therefore, the economic literature often focuses on mechanisms that reduce free-riding. Economists have generally documented more ways of overcoming free-riders than have biologists, and confirmation of similar mechanisms in animal societies would constitute strong evidence for the selective force of free-riding.

Finally, the models needed to study collective action in animal societies already exist in economics. Hence, biologists can borrow these economic models, and the principles derived from them, to examine cooperation in non-human societies. In this cross-disciplinary approach, then, an existing theoretical framework can be applied to a different system with a similar set of problems.

Background

I begin with a formal definition of the CAP: when costs are incurred by a subset of the group, but the benefits accrue to all, individuals may avoid contributing, while benefiting from the production of others. From this definition, it is obvious that CAPs only apply to certain types of benefits, or *goods*; in economics, these goods are known as *collective goods* (also called *public goods*). A collective good is a good whose benefits are not excludable (i.e., they are not monopolizable by excluding others), so that an individual cannot be denied benefits when the good is obtained by others (Samuelson, 1954). Collective goods are contrasted with *private goods*, whose benefits are excludable (and therefore monopolizable).

Many examples of collective goods exist in human societies. For example, public radio in the USA is supported largely through charitable contributions. However, anyone with a radio can listen to the broadcasts, and this 'non-excludability' gives rise to extensive free-riding. In animal societies, collective goods also exist, including benefits related to territorial defense, predator mobbing, and defense against infanticide; in these cases, individuals can withhold participation, yet still benefit from the efforts of others.

As mentioned above, collective goods commonly lead to free-riding. However, even individuals that participate at a low level may qualify as free-riders. Thus, *weak free-riders* contribute to a collective good, but they contribute less than the true marginal value they derive from the good; in contrast, *strong free-riders* contribute nothing toward production of a collective good from which they benefit (Marwell & Ames, 1981; Isaac *et al.*, 1984). Returning to the example of public radio, the individual who listens without contributing money is a strong free-rider, while the individual who contributes money, but provides less than his or her true value, is a weak free-rider.

The benefits of a collective good may be *rival*, meaning that use by one individual reduces the amount available for others (also called congestion or crowding), or *non-rival*, meaning that use by one individual does not reduce the amount available to others (Sandler, 1992). In human societies, roads are an example of a rival collective good: when more individuals drive, this reduces the benefit acquired by others (through traffic jams, a form of congestion). In group-living species, a rival collective good might involve defense of food resources from extra-group animals; in such a case, use by one individual typically reduces the amount available to others within the group. In contrast, public radio and predator alarm calls are non-rival, in that all individuals gain approximately equally, regardless of how many others obtain the benefit. CAPs may be most serious when a good is non-excludable and non-rival (i.e., a *pure collective good*). However, even partially excludable rival goods can lead to CAPs (*impure collective goods*: Hardin, 1982; Ostrom, 1990; Sandler, 1992).

Overcoming collective action problems

Early models of collective action focused on the difficulty of achieving collective action (e.g., Olson, 1965). However, collective goods are actually quite common in human societies, indicating that CAPs are frequently overcome (e.g., Ostrom, 1990). Economists have identified a number of factors relevant to overcoming CAPs in human societies, and similar principles should apply to animal societies. Here, I identify three general classes of factors that are likely to be important for overcoming CAPs in animal societies. In addition to these factors, however, 'design principles' may be important, including the ability of individuals to monitor the actions of others (Ostrom, 1990).

ASYMMETRICAL BENEFITS AND PRIVILEGED GROUPS

Asymmetrical benefits are common in animal societies. When asymmetries occur in the context of collective benefits, this skew in the rewards may result in some individuals being more willing to provide the good. Asymmetrical

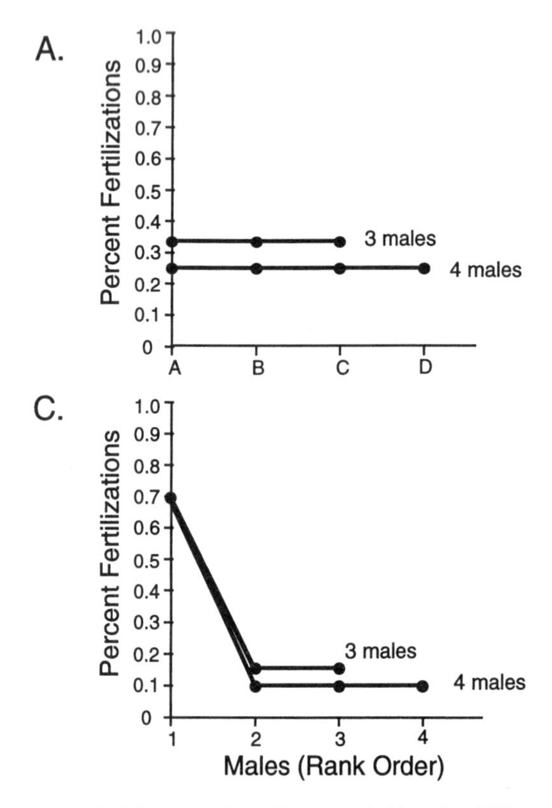

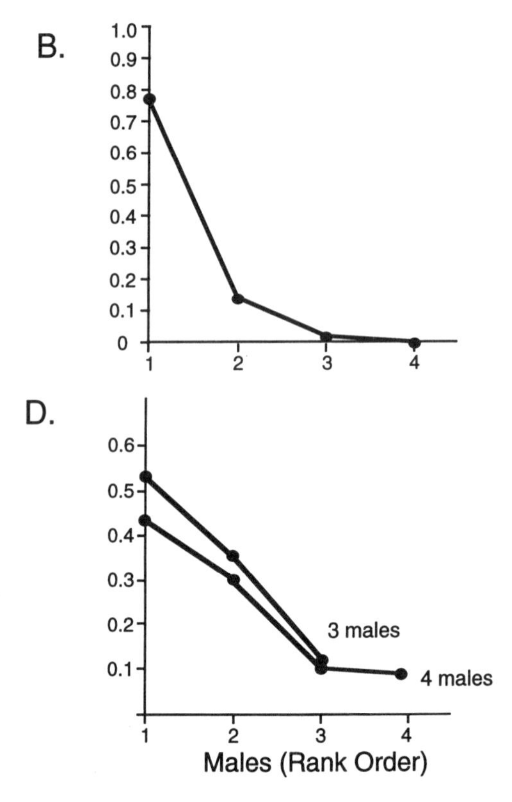

Fig. 17.1. Collective action and asymmetrical benefits. CAPs may be overcome when one individual benefits more than others. The hypothetical example shown here examines the effect of immigration of a fourth male to a group of three males under various forms of within-group competition (different lines represent scramble effects, while the slope of each line represents the contest effect; see van Schaik, 1989). Numbers refer to dominance rank, with 1 indicating the highest-ranking male, while letters mean that a rank order does not exist. (A) In pure scramble competition (sperm competition in the case of males), the benefits are symmetrical. Thus, action by one male to prevent the entry of a fourth male benefits all three resident males equally, and individual males may attempt to free-ride on these collective benefits. (B) In pure contest competition, adding another male does not shift the lines. However, the new male could alter the rank of the resident male, and males expected to be displaced in the hierarchy should act to prevent immigration. (C) Dominants receive a set number of fertilizations regardless of the number of other males in the group, while subordinates divide the remaining fertilizations among themselves (pure scramble). Dominant males should act if the immigrant would become alpha, while CAPs may exist among the subordinate males. (D) Adding a male increases the scramble component disproportionately for the alpha male. Hence, the alpha male is predicted to act in preventing entry of extra-group males (i.e., a privileged group).

benefits may therefore result in *privileged groups*, where an individual (or coalition of individuals) acts to provide the good (e.g., Olson, 1965; Olson & Zeckhauser, 1966). In such a case, privileged individuals benefit to the extent that they are better off providing the good for all, even in the context of strong free-riding by less privileged individuals.

The concept of asymmetrical benefits and privileged groups may be especially relevant to animal behavior, where asymmetries are common. For example, in male extra-group conflict, the male who benefits most from preventing another male's entry is most likely to act. The distribution of benefits within the group (measured using reproductive skew or dominance ranks) thus predicts who will be most active in achieving a collective benefit.

As it applies to animal behavior, the general principle of asymmetrical benefits is illustrated by the examples in Fig. 17.1. This hypothetical example involves a group of three males in which the individuals must decide whether to exclude a fourth male. The different panels show the payoffs under different distributions of benefits (obviously, these four panels represent only a fraction of the possible patterns). Some males clearly benefit more than others; however, in most cases, each male benefits when the outsider is excluded. Because most of the males benefit when any male acts to prevent an immigrant's entry, extra-group conflict probably qualifies as a collective good.

The questions to be addressed are: will the collective good be provided? If so, who will provide the good (i.e., who will act to keep the potential immigrant out)? When all males benefit equally, CAPs might be more likely (pure scramble competition in Fig. 17.1A). Assuming the good is provided, the individual(s) with the largest magnitude difference with changing group size is expected to invest most heavily in extra-group defense (e.g., the alpha male in Fig. 17.1D). In pure contest competition (Fig. 17.1B), males whose dominance rank would change following immigration of an outsider should be most involved; in other words, some assessment of the immigrant's resource-holding potential is probably going to occur in this (and the other) examples. These different scenarios are considered again below.

PRIVATE INCENTIVES AND COERCION

CAPs may be overcome more directly in human societies through *private incentives* and *coercion* (Olson, 1965). Private incentives involve private goods that are distributed to those who contribute to the collective good. Because the rewards are private, only contributors benefit from them, even though others might benefit from the collective good (i.e., free-riders). Returning to the example of public radio, T-shirts are commonly given to contributors; as another example from human societies, magazines or journals frequently accompany membership in special interest groups (e.g., environmental or academic organizations).

The use of private incentives may also facilitate collective action in non-human societies (e.g., de Waal, 1996b). For example, in the context of male extra-group conflict, one individual could exchange grooming or food resources (essentially private goods) for contributions toward extra-group defense (a collective good). This situation involves interchange (the exchange of one type of act for a different act) rather than reciprocity (the exchange of the same act: Hemelrijk & Ek, 1991). Some evidence exists for interchange between two private goods (i.e., food or grooming interchanged for coalitionary support: Seyfarth & Cheney, 1984; de Waal, 1989a; Hemelrijk & Ek, 1991). While there is no evidence for interchange between private and collective goods, this scenario is certainly plausible and is worthy of further study.

In contrast to private incentives that *encourage* individual contributions, coercion can be used to *force* individual participation. For example, in human societies, 'closed shops' force workers to join labor unions (Olson, 1965). In a similar manner, threats of financial penalties or imprisonment coerce citizens into paying taxes or contributing to war efforts (i.e., conscription into the armed forces).

The use of coercion, or punishment, is also observed in animal societies, and such force (or threat of force) could be used to motivate collective action (Clutton-Brock & Parker, 1995a; Ruxton & van der Meer, 1997). The limited evidence in mammals is mixed on this issue. Heinsohn & Packer (1995) found no evidence of punishment in their study of lions (in the form of retaliatory non-cooperation); however, Hauser & Marler (1993) found that rhesus macaques (*Macaca mulatta*) who fail to give food-related calls are more likely to receive aggression from group mates.

Private incentives and coercion are themselves costly, and these acts provide benefits to the group as a whole. In other words, enforcement is also a collective good, leading to *second-order CAPs* (Boone, 1992; Sandler, 1992). However, second-order CAPs may be less of a problem in animal societies, where asymmetrical benefits are common: individuals that benefit most will not only be motivated to provide the good, but will also be more likely to spur free-riders to action (Bates, 1988; Ostrom, 1990; Ruxton & van der Meer, 1997).

MUTUALISM AND KINSHIP

Finally, mutualism and kinship are probably relevant to overcoming CAPs in animal societies. It is important to note, however, that both factors have been considered in previous models of animal cooperation (Dugatkin, 1997).

In mutualistic situations, cooperation may arise because the benefits of cooperating exceed the benefits of free-riding on another's action. Dugatkin (1997) calls such cases 'no-cost cooperation', which reflects the fact that cooperation requires no further explanation (i.e., it would be more costly to defect than to cooperate). Economic terms for the same effect might be *synergies* or *economies of scale*; in these cases, individuals acting together are worth more than the sum of their parts.

When economies of scale arise in human societies, cooperation is expected (Sandler, 1992). However, *diminishing returns* with increasing group size may place limits on mutualistic benefits as groups get larger. Thus, diminishing returns may lead to some level of free-riding, as the collective good may be provided relatively efficiently with fewer than the number of individuals who benefit. Departure from optimal provision levels may therefore be small.

Kinship is also thought to help overcome CAPs: by providing collective benefits to kin, an individual increases his or her inclusive fitness (e.g., Hawkes, 1992; van Schaik, 1996). However, this issue is not as simple as it first appears. For example, in social groups in which relatedness varies among individuals, how would kin prevent non-kin from sharing in the collective benefits obtained? How do diminishing returns, discussed in relation to economies of scale, affect opportunities for free-riding in the context of relatedness? Finally, how are coercion and private incentives used to motivate kin versus non-kin, since effects on fitness may differ (with coercion being potentially more costly: Nunn & Lewis, 1999)? Clearly, more theoretical work is needed on how CAPs are overcome in animal societies (see also Ruxton & van der Meer, 1997).

EVIDENCE FOR COLLECTIVE ACTION PROBLEMS IN ANIMAL BEHAVIOR

Several possible cases of free-riding in the context of collective goods have been documented in non-human societies. Perhaps the best example comes from Heinsohn & Packer's (1995) experimental research on lion territoriality. These authors reported on a series of playback experiments in which they played back neighboring roars to a pride of lions, and then recorded individual involvement in the simulated territorial intrusions. In their experiments, Heinsohn & Packer (1995) showed that some females consistently failed to participate. They called these females 'laggards'. Lion territoriality probably qualifies as a collective good: when one female contributes to territorial defense, all the females in that pride benefit from the exclusive access that is obtained. Furthermore, territorial defense is costly for female lions, as injuries are common.

Other possible examples of collective goods in animal societies include predator mobbing and burrow construction. However, the focus here is on male extra-group conflict: males that defend the group from extra-group males potentially share the benefits of defense (i.e., fertilizations) with other males in the group. Male extra-group conflict provides abundant data for comparative tests, and, as shown below, it is possible to use proxies for the 'collectiveness' of the benefits. Moreover, variation in male extra-group conflict exists at several hierarchical levels: at the lowest level, individual variation occurs, with certain males taking a more active role in extra-group conflict; at an intermediate level, populations differ in their expression of male extra-group conflict; finally, at a higher level, species differ in how males respond to intruders.

CAPs will only arise when providing the good is costly. Male extra-group conflict is probably costly for two reasons: first, extra-group conflict typically involves energetic expenditures; second, injury is possible (Cheney, 1987). In some cases, I use proxies, such as male loud calls, as a measure of extra-group conflict. However, only the underlying behavior needs to be costly, not the proxies themselves.

I start with evidence at the individual and group level; then, in the next section, I develop and test a series of predictions at the species level. The conclusion that CAPs exist is potentially weakened by alternative hypotheses, and, in some cases, obvious alternatives are discussed. However, whereas distinguishing between alternatives is crucial for strong inference (Platt, 1964), it is not always possible to identify and test all alternatives, especially when multiple interacting factors coexist (Quinn & Dunham, 1983). Instead, I attempt indirectly to eliminate alternative hypotheses by using evidence from multiple hierarchical levels (within-species versus between-species comparisons). I also use multiple proxies for the hypothesized causal factors. If patterns are consistent at multiple levels and with multiple proxies, alternative hypotheses become much less likely.

Evidence for collective action problems in male extra-group conflict

Using intraspecific variation in sifaka and langur social groups, van Schaik (1996) showed that male aggressive extra-group encounters occur at a higher rate in groups that contain only a single male. Van Schaik (1996) argued that this pattern provides evidence for the CAP: when males share the benefits of extra-group conflict with other males, they are less willing to endure the costs of defense; thus, in single-male groups, where one male obtains all the benefits, this male takes a more active role in extra-group defense.

Additional evidence for van Schaik's (1996) hypothesis comes from detailed field observations in baboons. Hamilton & Bulger (1992) compared a chacma baboon (*Papio ursinus*) group with a single male to the normal situation of groups with multiple males (see also Cowlishaw, 1997a). The authors found several patterns consistent with a CAP. For example, the male in the one-male group was more vigilant toward outside males, he took a more active role in group movements, and he was more active in intergroup encounters.

Because CAPs are thought to increase with the number

Table 17.1. *Intraspecific variation in encounter frequency in vervets*

Location (year)	Rate of encounters[1]	Number of males	References
Amboseli (1964)	0.2	4.0	Struhsaker (1967)
Amboseli (1977–78)	0.5556	4.2	Cheney & Seyfarth (1981)
Bakossi	0.5581	2.5	Kavanagh (1981)
Senegal	0.25	4.5	Harrison (1983)

Note:
[1] Number of encounters per day. Data taken from Cheney (1987).

of individuals involved (Olson, 1965; cf. Sandler, 1992), variation in group size can be used to test for CAPs in animal societies. The logic here is straightforward: in larger groups, there are more opportunities to free-ride because there are more 'others' to provide the collective good. In addition, as mentioned above, diminishing returns may result in a low level of free-riding as group size increases, even under synergistic benefits. One way to test this prediction uses interpopulational variation in the expression of male loud calls in long-tailed macaques (*Macaca fascicularis*). On mainland Sumatra, groups tend to contain an average of 6.4 males per group (de Ruiter, 1992), and males do not give loud calls. However, on the island of Simelue, groups typically contain fewer males (2.2 adults per group: Sugardjito *et al.*, 1989), and males have been reported to give loud calls (van Schaik & van Noordwijk, 1985). Hence, the evidence is consistent with more extra-group conflict in smaller groups, although the absence of predators may also account for this pattern.

Another case of intraspecific variation involves vervets (*Cercopithecus aethiops*): rates of extra-group conflict decline in social groups that contain more males (Kavanagh, 1981). If the data and references from Cheney (1987) are used, and the number of males taken from these and other references (Table 17.1), a negative (but non-significant) relationship is found between the number of males and rates of extra-group conflict (least squares regression on log-transformed data: $b = -1.09$, $p = 0.45$).

The data from which these conclusions regarding group size are drawn come from different populations, making it difficult to rule out ecological effects as alternative hypotheses. However, some comparisons are available using variation among groups within the same population, for which ecological factors should be more similar. For example, in red howler monkeys (*Alouatta seniculus*), Sekulic (1982b) found that males howl more frequently when there are fewer other males in the social group. Similarly, in white-faced capuchins (*Cebus capucinus*), male vigilance is directed toward males outside the group, and overall vigilance declines when there are more males in the social group ($n =$ 4 groups; Rose & Fedigan, 1995). Thus, while alternatives remain possible, the consistency of the patterns makes these alternatives less likely. I return to the issue of male number in the context of cross-species patterns below.

Overcoming collective action problems in male extra-group conflict

Testing for privileged groups requires analysis at the individual level, as information is needed to assess differential involvement relative to how the benefits vary among individuals. In primates, male dominance rank is one means of estimating how the benefits are distributed. As expected, then, dominant males are frequently more involved in extra-group conflict (e.g., gray-cheeked mangabeys (*Cercocebus albigena*): Waser, 1976; pig-tailed macaques (*Macaca nemestrina*): Oi, 1990; Japanese macaques (*Macaca fuscata*): Yamagiwa, 1985; chacma baboons: Cheney & Seyfarth, 1977; gorillas (*Gorilla gorilla*): Schaller, 1963).

Reproductive skew is another means of testing for privileged groups. For example, Launhardt & Borries (1997) suggest that gray langurs (*Presbytis entellus*) resemble the pattern shown in Fig. 17.1c. Capuchins might also fit this pattern, as the dominant male probably obtains a high percentage of fertilizations (Janson, 1984). In contrast, the results of Cowlishaw & Dunbar (1991) suggest the pattern in Fig. 17.1d for macaques and baboons: adding additional males tends to lessen the dominant male's monopolization of females. In the context of privileged groups, it is therefore important that dominant males are often more involved in extra-group conflict in these species.

As mentioned above, asymmetrical benefits can overcome second-order CAPs (i.e., who provides coercion and private incentives). It should be easy to document such patterns, if they exist, by examining male within-group benefits relative to involvement in extra-group conflict.

Both mutualism and kinship are probably important in some cases of male extra-group conflict. For example, common chimpanzee (*Pan troglodytes*) males cooperatively defend a common boundary (Wrangham, 1979; Nishida & Hiraiwa-Hasegawa, 1987). While mutualistic synergies probably exist, kinship may also play a role. The synergies may in fact drive the high kinship, making it difficult to disentangle these two factors.

CROSS-SPECIES COMPARATIVE TESTS

In this section, the general framework of collective action is used to develop cross-species predictions. I first consider how cross-species studies of male extra-group conflict can be conducted, and briefly outline the methods used here. Then, predictions are developed, and for each prediction, the results of phylogenetic comparative tests are provided.

Methods

In the comparative tests that follow, the dependent variable is the presence of extra-group conflict. No straightforward method exists for comparing the expression of extra-group conflict across species (Waser & Wiley, 1980; Cheney, 1987). Here, my main proxy is the presence of male loud calls. Loud calls are distinctive and thus easy to identify across a broad array of species. Furthermore, these vocalizations are present in all primate radiations (Cheney, 1987). To qualify as a loud call, I required that such calls alter the behavior of males in other groups, usually by generating countercalling in a 'contagious' fashion (see also Marler, 1968). Male loud calls are also reported as reactions to falling trees or branches (Marler, 1972; MacKinnon, 1974; Baldwin & Baldwin, 1976), or as alarm calls to potential predators (Chalmers, 1968; Zuberbühler *et al.*, 1997). While this might seem to contradict their function in between-group encounters, their use at these times explains neither countercalling by neighboring males, nor the sex specificity usually reported for these calls (S.M. Green, 1981).

Here, I assume that the absence of loud calls reflects CAPs. Therefore, the focus is on strong free-riding, and mechanisms aimed at overcoming free-riding (about which little information presently exists) are ignored. In addition, I assume that males compete for fertilizations rather than other reproductive resources, such as food (Emlen & Oring, 1977). Evidence suggests that males are more interested in mate defense than resource defense (Ims, 1988; van Schaik *et al.*, 1992), even in monogamous species (French & Snowdon, 1981; Cowlishaw, 1992). Thus, measures relevant to defending a geographical boundary, which probably concern females more than males (e.g., defensibility indices and home range overlap: Mitani & Rodman, 1979; Cheney, 1987), were not used.

The independent variables in these tests proxy the risk of free-riding. For this, I used several variables, including social system and the number of males (as in the above intra-specific comparative tests), and two proxies for scramble competition (testes mass and reproductive seasonality). Information on social system was taken from Appendix A-1 and relevant chapters in Smuts *et al.* (1987). References were also obtained from Rowe (1996). The number of males was taken from several other comparative projects currently in progress (unpublished comparative database).

Scramble competition in males was proxied using two correlates. Because male scramble competition basically involves sperm competition, I first used relative testes mass (Harcourt *et al.*, 1981b, 1995). Data were taken from Harcourt *et al.* (1995) for anthropoids, and from Kappeler (1997b) for prosimians (converted to grams using the formula in Harcourt *et al.*, 1995). Data on body masses were taken from Smith & Jungers (1997). Relative testes mass was calculated as residuals from the regression of testes mass on body mass (with the regression coefficient calculated using phylogenetically independent data).

Seasonality can also be used as a proxy for scramble competition (Ridley, 1986). However, this analysis must be restricted to those cases in which breeding seasonality clearly affects male monopolization potential, as this issue is currently debated (Mitani *et al.*, 1996a; Nunn, unpublished data). Therefore, I restricted my analysis of seasonality to the macaques (Oi, 1996; Paul, 1997). Data on seasonality was taken mainly from van Schaik *et al.* (1999).

Phylogenetic comparative methods

A number of studies have demonstrated the importance of including phylogenetic relationships in cross-species comparative studies (Harvey & Pagel, 1991). Here, I used Purvis' (1995) composite estimate of primate phylogeny. Phylogenetically independent contrasts were used to examine associations between continuous characters

(Felsenstein, 1985). Contrasts were calculated using the computer package CAIC (Purvis & Rambaut, 1994). For analyses involving one discrete and one continuous character, the methods outlined in the CAIC manual were used (i.e., the 'brunch' algorithm). For two discrete characters, I used either Maddison's Concentrated Changes Test (Maddison, 1990), as implemented by the computer program MacClade (Maddison & Maddison, 1992), or Pagel's (1994a) method for discrete data, as implemented by the computer program discrete.appl (distributed by M. Pagel). The computer program MacClade (Maddison & Maddison, 1992) was used to reconstruct ancestral states using parsimony.

Comparative predictions and results

PREDICTION 1. SINGLE-MALE VERSUS MULTI-MALE SOCIAL SYSTEMS

In a true single-male situation, only one male is present in the group at the time of fertile matings, so the resident male reaps all the potential benefits of extra-group conflict. The benefits are thus private, and CAPs are not possible. In contrast, in multi-male groups, extra-group conflict should be more collective. Using loud calls as a proxy for extra-group conflict, the following prediction is therefore possible: loud calls should be more common in single-male than in multi-male groups.

The comparative evidence supports this prediction: males of single-male species usually give loud calls (87% of species), while these calls are less common (but not absent) in multi-male species (47% of species; e.g., ring-tailed lemurs, *Lemur catta*: Jolly, 1966; gray-cheeked mangabeys: Chalmers, 1968; Waser, 1976). Using Maddison's Concentrated Changes Test (Maddison, 1990), losses of loud calls are significantly concentrated on branches of the phylogeny reconstructed as multi-male ($p = 0.05$). Intraspecific variation from above provides additional support for this cross-species pattern (i.e., Hamilton & Bulger, 1992; van Schaik, 1996; Cowlishaw, 1997a). Exceptions to this pattern may relate to how CAPs are overcome and, partly to understand these exceptions, Predictions 2–4 examine patterns in multi-male species only.

Alternative hypotheses may account for patterns consistent with Prediction 1. For example, species that contain multiple males usually also contain multiple females (Andelman, 1986); thus, multi-male groups will tend to be larger. Larger groups require larger home ranges (Clutton-Brock & Harvey, 1977b), and it is possible that the costs of producing calls loud enough to cover this larger home range simply outweigh the benefits. In support of the hypothesis that home range size matters, males of the single-male patas monkey (*Erythrocebus patas*) do not possess loud calls, and this species lives in a large home range (2770 ha: Chism & Rowell, 1988). However, Guinea baboons (*Papio papio*) also live in large home ranges, and alarm calls of this species carry up to 3 km (Byrne, 1981). Assuming a roughly circular home range of 1295 ha for Guinea baboons (Rowe, 1996), alarm calls therefore carry nearly one home range diameter (using the formulas in Mitani & Rodman, 1979). In addition, given home range overlap and patterns of male dispersal (i.e., following a new group at relatively close range: Pusey & Packer, 1987), baboon male loud calls need not carry across the entire home range to be effective.

Another alternative hypothesis exists: when comparing single-male and multi-male groups, the level of competition might switch from between males in different single-male groups to competition among males within a multi-male group. Thus, extra-group conflict need not arise in multi-male groups, simply because competition occurs at a different level. By restricting analyses to multi-male taxa, the remaining predictions help rule out this alternative.

PREDICTION 2. SCRAMBLE COMPETITION, TESTES MASS, AND COLLECTIVE ACTION PROBLEMS

In animal behavior, competition describes how a benefit is distributed among individuals in a group; thus, competition can be used to measure the collectiveness of some benefit. To review briefly, competition can occur within groups and between groups; here, unless otherwise stated, I am referring to within-group competition. Competition at either level can be broken down into two components: contest and scramble. Contest competition describes a good that can be monopolized by excluding others, whereas scramble describes a good that cannot be monopolized through exclusion. Thus, when an individual acts in extra-group defense of a good characterized by within-group scramble, other individuals in the group will also benefit from this extra-group defense. In other words, a combination of between-group contest and some degree of within-group scramble will tend to result in a collective good, and should thus be correlated with CAPs in animal societies. In contrast, within-group contest competition describes asymmetrical benefits, so this information may predict which male (or males) is most likely to contribute toward the collective good.

To show how scramble competition relates to CAPs, consider the extreme case of pure scramble competition (see

Fig. 17.1a). Under pure scramble, adding additional males reduces the amount of fertilizations available for all males. Thus, if a group contains two males, they are expected to split the fertilizations equally; if it contains three males, each obtains, on average, one-third of the fertilizations, and so on. In the case of three males, if one of these males prevents the entry of a fourth male, then all three benefit: they each obtain one-third of the available fertilizations rather than one-fourth. Hence, they should all prefer to avoid this costly conflict, while benefiting from the actions of their group mates. In some cases, two males will be required to cooperate in order to keep out an extra-group male. However, CAPs can still arise in this situation: one of the three males can hold back, and free-ride on the benefits obtained by the other two.

Male extra-group conflict is an excellent situation to examine CAPs because it is possible to measure quantitatively male intrasexual scramble competition using several proxies. For example, testes mass is thought to reflect sperm competition (Harcourt *et al.*, 1981b, 1995), which is a form of scramble competition. This leads to the following prediction: among multi-male species, those with relatively large testes (relative to body mass) should not give loud calls.

This prediction is supported. Multi-male species were grouped into those whose testes residuals lie below the median for multi-male species (median = 0.293), and those whose testes residuals lie above this median. Males of species with relatively large testes tend to lack loud calls, and this difference is significant in phylogenetic comparative tests: in four of the five contrasts in the presence of loud calls, testes residuals decline, and the mean of all five contrasts (−0.0124) differs significantly from zero in a one-tailed t-test ($t = 2.19$, $p = 0.047$). In only one of these contrasts is the social system reconstructed as changing simultaneously (in the contrast between cercopithecines and colobines), so this is not likely to be a major confounding factor.

PREDICTION 3. SCRAMBLE COMPETITION, BREEDING SEASONALITY, AND COLLECTIVE ACTION PROBLEMS IN MACAQUES

Reproductive seasonality can also be used as a proxy for scramble competition: in seasonal species, females are more likely to overlap in their receptivity, so that scramble competition should be high (Ridley, 1986). However, breeding seasonality may not play a role in all species of primates (Mitani *et al.*, 1996a). Here, I focus on macaques because research has shown that male macaques are less able to monopolize matings in species in which females mate seasonally rather than non-seasonally (Oi, 1996; Paul, 1997). This leads to the following prediction: in the more seasonal macaques, scramble competition is high, so male loud calls should be absent.

Neither seasonality nor loud calls can be reconstructed unambiguously on the phylogeny in Fig. 17.2. The mapping of seasonality depends on which value is assigned to long-tailed macaques: when they are assumed to be seasonal breeders, two gains are reconstructed, but when they are non-seasonal, three gains are present, and the trait cannot be reconstructed unambiguously. The ancestral value is equivocal, although outgroup information would suggest a non-seasonal ancestor. For loud calls, it is equally parsimonious to assume two gains of loud calls, shown with gray bars, or one gain and one loss, shown with black bars.

These ambiguities in reconstruction can perhaps be eliminated, especially through a detailed analysis of the homology of loud vocalizations in the different macaque species. But this is not helpful here, because under either reconstruction in Fig. 17.2, loud calls are not significantly concentrated on the non-seasonal branches (concentrated changes test; $p = 0.44$ for two gains of loud calls, and $p = 0.63$ for one gain and one loss).

One possible solution is to use Pagel's (1994a) comparative method for discrete data. Pagel's (1994a) method uses maximum likelihood to determine the correlation between two traits, and therefore escapes from the assumptions of equally weighted parsimony. Pagel's (1994a) method also makes use of branch lengths, while Maddison's (1990) test does not. With this method, a significant relationship is found between loud calls and non-seasonal breeding (likelihood ratio = 4.59; $p = 0.04$, based on 100 simulations of the null hypothesis). Of course, issues of phylogenetic uncertainty remain, both in terms of the topology of evolutionary relationships and in branch length estimates.

PREDICTION 4. THE NUMBER OF MALES

As noted above, economic models suggest that CAPs are more serious in large groups (Olson, 1965; cf. Sandler, 1992). This observation leads to a testable prediction for animal societies: species with more males should lack loud calls. This prediction is supported by the comparative data (Fig. 17.3), and is statistically significant in tests that control for phylogeny (seven of eight contrasts go in the predicted direction; $t = 2.49$, $p = 0.04$, two-tailed). This prediction is not entirely independent of the pattern in Prediction 2, as species with more males may have more sperm competition. Nevertheless, both predictions address the same issue, namely, the shareability of benefits, and opportunities for free-riding on the benefits provided by others.

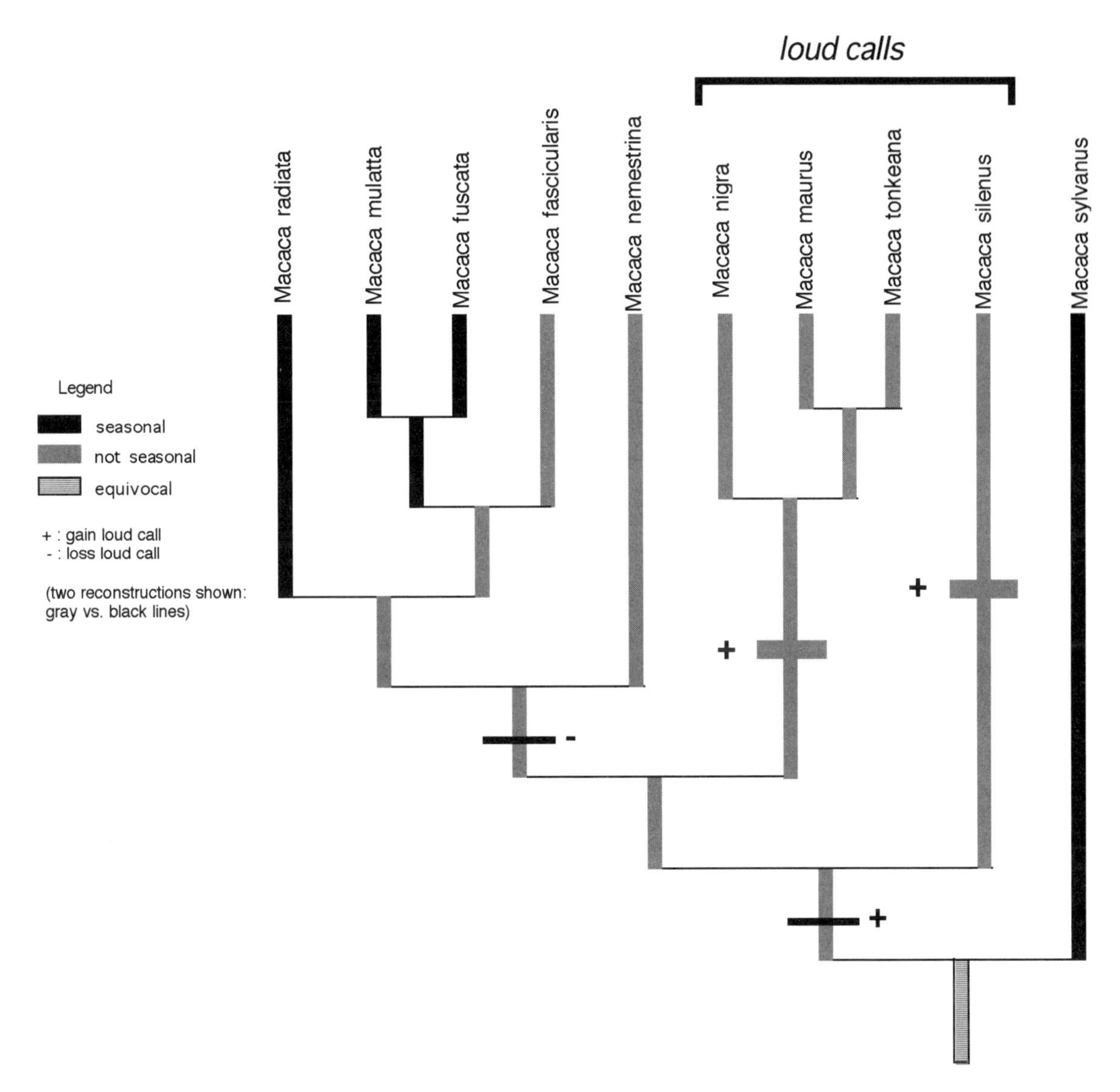

Fig. 17.2. Seasonality in macaques. Breeding seaonality and the presence of loud calls are mapped onto Purvis' (1995) phylogeny. Using equally weighted parsimony, three changes of breeding seasonality are reconstructed. However, the ancestral condition is equivocal, and reconstruction depends on whether long-tailed macaques are treated as seasonal or non-seasonal. Also, loud calls do not map unambiguously; it is equally parsimonious to assume two gains of loud calls, shown in gray, or one gain and one loss, shown in black. Because of this uncertainty and small sample size, Pagel's (1994) method for the analysis of discrete data was used.

DISCUSSION

This chapter reviews the economic framework of collective action and applies this framework to the example of male extra-group conflict in primates. Using proxies for male extra-group conflict, intraspecific and interspecific patterns are generally consistent with the existence of CAPs among males: CAPs appear to be more common in multi-male taxa, and, among multi-male taxa, CAPs appear to be more serious in cases of high scramble competition and when there are more males in the group.

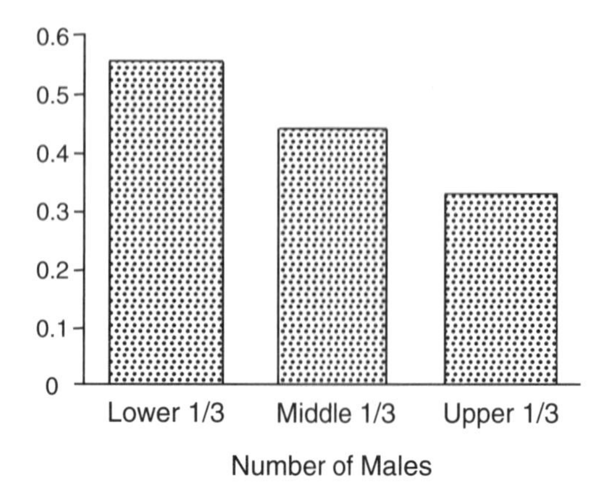

Fig. 17.3. Loud calls in relation to the number of males in primate groups. Multi-male species were broken down into three classes with an equal number of species in each class: those with a small number of males (<3.3), those with an intermediate number of males (3.3 to 4.8), and those with a large number of males (>4.8). Consistent with models of collective action, loud calls are less common in groups with more males.

Alternative hypotheses may account for some of these patterns, and more direct evidence for CAPs in animal societies is therefore needed. In addition, little is known about the dynamic interactions between participants and free-riders, which will require observations and experiments rather than comparative tests (e.g., Heinsohn & Packer, 1995). Experimental research is currently in progress (in collaboration with R. Deaner; see also Nunn, 1996). What follows considers how factors like the CAP fit into our general socioecological model.

The socioecological model

A major goal of socioecology is to understand the selective factors that explain variation in social systems. These factors can then be represented as a general model, or a 'network of causation' (Wrangham, 1979), that summarizes the major selective forces. The standard version of the socioecological model (Fig. 17.4) has directionality: environmental risks and resources set female strategies, and the resulting spatiotemporal distribution of females then structures options for males to monopolize matings (Emlen & Oring, 1977; Wrangham, 1979). There is good evidence for this directionality in mammals, including experimental (Ims, 1988) and comparative tests (Andelman, 1986; Mitani *et al.*, 1996a; Nunn, unpublished data).

Social systems arise from the behavior of interdependent actors (Wilson, 1975; Hinde, 1983; Lee, 1994). Thus, in attempts to explain variation in social systems, researchers should also investigate the effects of interactions among the individuals involved. In addition to the usual factors relating the environment to females, and then females to males, the socioecological model should therefore also examine *social rules* and *interaction effects* (see Wrangham, 1979; van Schaik, 1996).

The model in Fig. 17.4 includes one type of interaction effect involving inter-sexual conflict. Males generally benefit from increasing their number of fertile matings, but females may prefer to mate with only certain males. Males therefore manipulate females directly, perhaps by herding them away from other males, or by committing infanticide, forced copulations, or harassment (i.e., sexual coercion; Smuts & Smuts, 1993). Females are not passive reproductive resources, and they respond to this selection with counter-strategies (van Schaik, 1996, Chapter 4).

Identifying the selective factors in this network of causation will be limited by several obstacles. First, many behaviors in this model are not amenable to experimental manipulation; instead, a comparative approach is needed. In order to obtain sufficient variation in this comparative approach, comparisons will usually focus on the species level. However, at the species level there might not be enough degrees of freedom (i.e., species) to disentangle all the selective factors important in social evolution. The degrees of freedom will be further reduced because phylogeny must be considered, where only independent evolutionary origins of different suites of traits are analyzed (Harvey & Pagel, 1991). Finally, systems with many interactions often behave in complex, non-linear ways (Kauffman, 1993).

The collective action problem in primate socioecology

The CAP results from interindividual interactions, and thus qualifies as a social rule in the socioecological model. The selective effects of the CAP and free-riding should be important at several levels in this model.

This chapter focuses on CAPs among males in defending access to fertile females. However CAPs probably also exist among females (e.g., Heinsohn & Packer, 1995). CAPs may differ between the sexes, especially because food resources may be more 'shareable' for females than fertile mating opportunities are for males (van Hooff & van Schaik, 1994). In either case, however, free-riding should be a strong selective force on individual behavior within the social group.

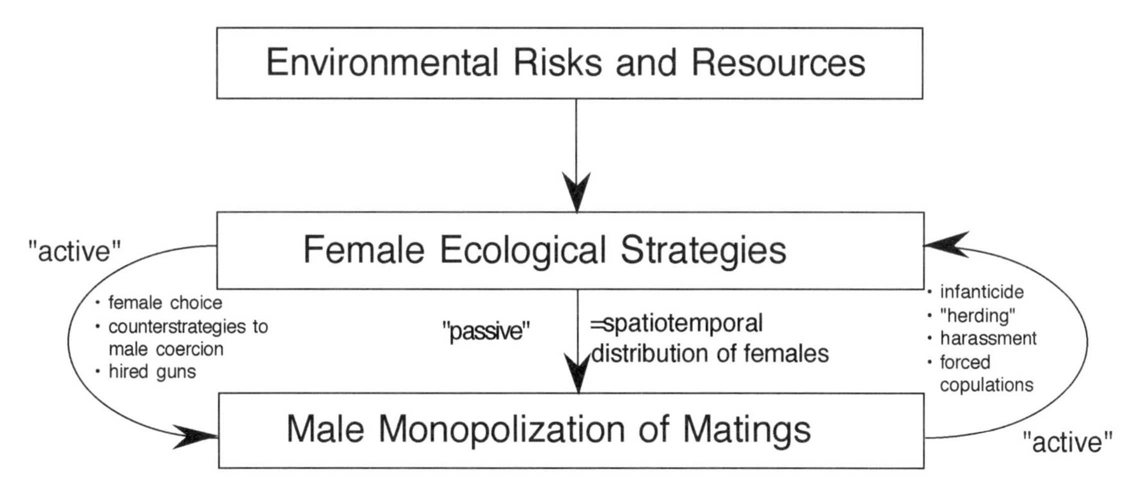

Fig. 17.4. The socioecological model. In the standard socioecological model, environmental risks and resources are the first round of inputs, and these factors set female strategies. The spatiotemporal distribution of females then structures options for males to monopolize matings. The 'expanded' portion arises from interaction effects (such as the inter-sexual interactions shown here).

CAPs may also arise in the intersexual interactions shown in Fig. 17.4. One important case involves 'hired gun' models of social evolution, which refers to cases in which females are thought to manipulate male membership in social groups to obtain certain benefits. Some benefits might be ecological, for example reduced predation (van Schaik & Hörstermann, 1994). Other benefits involve interaction effects, for example when females pair up with one or more powerful males to gain protection from other, potentially coercive males (Wrangham, 1979; Smuts & Smuts, 1993).

An example of hired gun models relevant here involves resource defense. Wrangham (1980, 1987) proposed that females increase male membership in social groups to obtain assistance in resource defense. In principle, adding more males should increase the ability of a group to defend resources. However, as shown above, male behavior changes when there are more males in the group. Unless females have some means to enforce participation in territorial defense, males are not likely to participate. In other words, adding more males does not necessarily lead to more vigilance or more territoriality, and some sort of private incentive may be needed to obtain male assistance in territorial defense.

Finally, females might benefit from collective *inaction* in males. For example, with less extra-group conflict, it should be easier for females to mate with males outside the group. Extra-group mating might be an adaptation to forestall infanticide should one of the extra-group males immigrate after the female's fertile period (van Schaik *et al.*, 1999). Because of the directionality in the socioecological model, females can probably set the conditions for collective action (and inaction) in males. For example, the number of males in a primate social group is a function of female grouping patterns and reproductive synchrony (Andelman, 1986; Ridley, 1986; Mitani *et al.*, 1996a; Nunn, unpublished data). Because male behavior toward extra-group individuals is probably a function of how many males are present, females potentially can control the expression of CAPs in males.

CONCLUSIONS

The framework of collective action has several benefits for studying cooperation in animal behavior. First, these models focus attention on individual cost–benefit decisions in a group context. Second, methods of overcoming CAPs should also apply to animal societies and may therefore provide new ways of understanding how cooperation is achieved and maintained. Finally, the models needed to study collective action are already widely available in the economics literature.

The results presented here indicate that CAPs exist in animal societies. However, additional studies are needed, including: (1) theoretical models that apply more detailed concepts from economics, and that link these concepts to game theory approaches in animal behavior (e.g., Nunn & Lewis, 1999); (2) additional intra-specific and inter-specific comparative tests in primates and other taxa; and (3) empirical field research aimed at documenting CAPs in animal societies (e.g., Heinsohn & Packer, 1995; Nunn, 1996). Experimental approaches are perhaps most important, as they may be the most direct tests of the causality hypothesized here.

ACKNOWLEDGMENTS

I especially thank C. van Schaik, R. Deaner, and B. Lewis for their discussions of issues related to collective action. D. Brockman, L. Digby, B. Fox, K. Hawkes, S. Nowicki, S. Patek, and R. Wrangham provided many useful suggestions. This research also benefited from comments at informal talks given in the Department of Biological Anthropology and Anatomy and at the Triangle Behavior Group. This research was supported by an NSF Graduate Student Fellowship and an NSF Dissertation Improvement Grant (#SBR-9711806).

18 • Dominance, egalitarianism, and stalemate: an experimental approach to male–male competition in Barbary macaques

SIGNE PREUSCHOFT & ANDREAS PAUL

INTRODUCTION

Ecology, social mechanisms, and the number of males

Socioecological models endeavor to predict male spacing patterns on the basis of the spatiotemporal distribution of fertilizable females and their fitness interests, which are, evidently, not independent of the males' strategies. The ecological predictions are relatively straightforward. Female reproductive success is limited by food and safety. The spatial distribution of these resources is therefore used to predict female sociability (e.g., Emlen & Oring, 1977; Wrangham, 1980; van Schaik, 1983). The temporal pattern of food availability determines the degree of reproductive seasonality in females. In seasonally reproducing species, females come into estrus within a relatively brief period of the year. Births are timed such that lactation coincides with a period of relative food abundance. It is predicted that the monopolizability of estrous females determines the number of males per female group. Large numbers of females, lack of female cohesiveness, and seasonal reproduction are factors that lead to relatively large numbers of males per female (e.g., Wrangham, 1980; Berenstain & Wade, 1983; Ridley, 1986; Altmann, 1990; Mitani *et al.*, 1996a; Paul, 1997). The causality in this model is straightforward. As female parental investment by far exceeds that of males, female reproduction is more tightly constrained. Consequently, females are the limiting resource for male reproductive success, and males have to adjust themselves to the distribution of females (Darwin, 1871; Trivers, 1972; Andelman, 1986; Altmann, 1990).

A number of features characteristic of primates are not satisfactorily explained by this model, however. For instance, why do males associate year-round with females even in species with strictly seasonal reproduction? Similarly, why do females seek to copulate with more than one male, and why do females in many species copulate even when they are not fertile, and even after they have already conceived? These, and similar observations, suggest that intrasexual competition and intersexual conflict interact in complex ways, thus producing feedback loops that set constraints for future reproductive options (e.g., Wrangham, 1987; van Schaik, 1996; Sterck *et al.*, 1997). A number of social mechanisms have been detailed that influence intersexual conflict, e.g., coercion of females by males (Mitani, 1985b; Smuts & Smuts, 1993, Clutton-Brock & Parker, 1995b), including infanticide (Hrdy, 1979; van Schaik, 1996), or affect modes of intrasexual competition, e.g., market effects (Noë, 1990; Noë & Hammerstein; 1994, 1995), and problems associated with collective action (e.g., Hawkes, 1992; Nunn, Chapter 17).

In this chapter, our intention is to call attention to yet another factor, 'stalemate,' that is of potentially far-reaching consequence for inter-male relationships and spacing patterns, for male–female interactions, and for female sexual options. 'Stalemate' describes a situation in which competitors of symmetric power mutually neutralize each other, because for both rivals the risks of escalated contest outweigh its potential benefits. This phenomenon has thus far largely been ignored in socioecological theorizing, although related phenomena, such as ownership conventions, have been recognized by ethologists for a long time (e.g., Kummer *et al.*, 1974; Packer & Pusey, 1985; Krebs & Davies, 1984).

Macaque and baboon societies

Like many other primates – but unlike most other mammals – most macaque and baboon species live in multi-male, multi-female societies with a promiscuous mating system (Melnick & Pearl, 1987). In these societies, females usually live in philopatric, matrilineally organized groups (Wrangham, 1980; Gouzoules & Gouzoules, 1987; van Schaik, 1989). This self-clumping tendency of females provides opportunities for intense competition among males

Table 18.1. *Conditions from which relatively symmetrical power relations may arise*

Source of contest	Rank order?	Mechanism	Indicators
No contest	No	None	No status signaling, no escalation
Between groups + within groups *type A*	Yes	Leverge power + formal subordination	Recruitment, bonding and coalitions, status signals
Within groups *type B*	No	Stalemate	De-escalation and escalated outbursts

Note:
Within-group contest *type A*: resource value is high relative to risks of escalation (this reflects van Schaik's (1989, 1996) within-group contest condition). Within-group contest *type B*: risks of escalation are high relative to resource value. No contest: corresponds to scramble competition. See text for further explanation.

(Emlen & Oring, 1977). Indeed, for several of these polygynandrous species, clear-cut and stable dominance relationships among males have been reported (e.g., Manson, 1996; Reed *et al.*, 1997), which, sometimes, coincide with a reproductive advantage of high-ranking over low-ranking males (e.g., de Ruiter *et al.*, 1992). In a few other species, this takes to the extreme of female defense polygyny with harems nested in larger, probably male-bonded, fission–fusion units (Stammbach, 1987; Barton, Chapter 9). The fact that papionines are characterized by a relatively large sexual dimorphism in body size and canines (Clutton-Brock *et al.*, 1977; Leutenegger, 1978; Plavcan & van Schaik, 1992) is in line with this behavioral evidence. However, not only does intense inter-male competition mark this taxon, but also present is a conspicuous capacity for inter-male bonding, as evident from coalitions (Kummer, 1968; Packer, 1977; Bercovitch, 1988; Kuester & Paul, 1992; Hill, 1994), and elaborate bonding rituals (Hanby, 1974; Busse, 1984; Reinhardt *et al.*, 1986; Colmenares, 1991; Ogawa, 1995).

In Barbary macaques (*Macaca sylvanus*), males join females in such numbers that the adult sex ratio is almost even (10.5 females and 9 males, cf. Ménard & Vallet, 1993; see also Mehlmann, 1989; Kuester & Paul, 1997). Male Barbary macaques have a reputation for being egalitarian and male-bonded (Taub, 1980; Small, 1990; Ménard *et al.*, 1992). Affiliative interactions among males, e.g., 'agonistic buffering,' occur regularly (Deag & Crook, 1971; Taub, 1984; Kuester & Paul, 1986; Paul *et al.*, 1992, 1996). Females are extraordinarily promiscuous, changing their sexual partners seemingly unrestrained by males (Taub, 1980; Small, 1990; Kuester & Paul, 1992; Ménard *et al.*, 1992). Yet, among males dominance reversals, escalated fighting, and severe wounds peak in the mating season (Paul, 1989; Kuester & Paul, 1992). Barbary macaque males thus present us with a puzzling combination of egalitarianism and violence.

Three causes of egalitarianism

If males do not exclude all other males from the females with whom they associate, but instead co-reside with other males, this calls for an explanation. Is it because they cannot help it? Or do they want to associate – perhaps because they need or even like one another? Or is it perhaps simply that they do not mind? Socioecological reasoning predicts a lack of overt competition, and a relatively even distribution of wins and losses among males as a consequence of the spatiotemporal distribution of fertilizable females if (1) estrous females are spatially scattered or temporally clumped and competition for them is by scramble; or (2) there is a possibility for contest but only alliances can monopolize a group of females, hence the leverage power of inferior males places limits on the power exertion by dominants.

Ovulating females are not economically defensible when they are scattered in space, when their cycles are highly synchronized, or when sperm competition is intense. Under these conditions, escalation does not pay and competition is exerted by scramble (line 1, Table 18.1). Here, mating with one female is as good as mating with another. Hence, no attempts at monopolization are made; males just do not mind having other males around.

Female Barbary macaques are, indeed, strictly seasonal breeders. The mating season averages 76 days (Ménard *et al.*, 1982) and 70% of conceptions take place within four weeks (Paul, 1989; Kuester & Paul, 1989). With Ménard & Vallet's data (1993) this results in about seven ovulating females for nine males over a period of four weeks. It therefore appears that seasonality alone would not present males with a time budget problem precluding the monopolization of individual estrous females.

Alternatively, if fertilizable females occur in groups, there is a possibility for contest competition, yet only groups of males can monopolize them. As a consequence of this

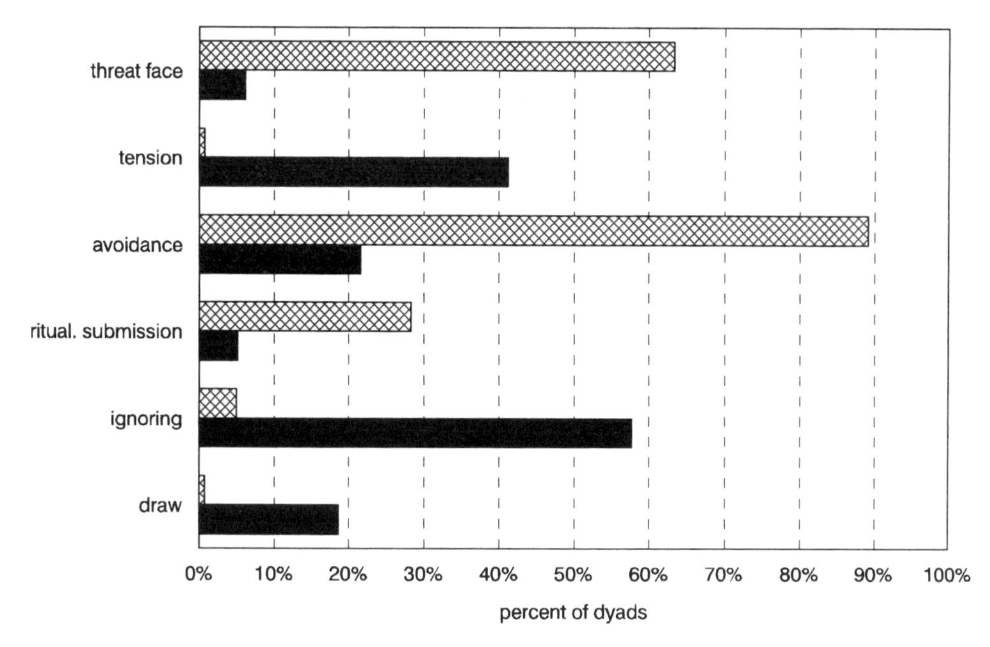

Fig. 18.1. Competition over peanuts among adult females ($n = 117$, cross-hatched) and among adult males ($n = 98$, black). Only first tests of different dyads are shown.

need to cooperate in female defense polygyny, leverage power of coalition partners can result (Hand, 1986; but see Noë, 1990). Cooperative defense of female groups may complement within-group contest for individual females when the probability of fertilization makes it worth risking a fight with a rival. In this case (line 2, Table 18.1), the leverage power of inferior males should put a limit to monopolization of females by dominants and thus act to dilute the existing power asymmetries (Vehrencamp, 1983a; van Schaik, 1989). It is under this condition that we can expect formal status signals and conditional reassurance (*sensu* de Waal, 1986), mechanisms by which dominants make 'concessions' to inferior coalition partners – provided these show formal subordination (de Waal, 1986). In short, if there is contest over females both within and between groups of males, competition is balanced by bonding: opting for the lesser evil, males prefer staying with 'comrades.'

There is yet another possibility to derive egalitarian relationships among males: even if estrous females are defensible, the trade-off between probability of fertilization and risks of escalation may yield a 'stalemate,' when males have dangerous weapons or when variance of resource-holding power between males is small (line 3, Table 18.1). In this situation of effectively symmetrical power distribution, we expect no uni-directional status signals, but sporadic outbursts of severe aggression complemented by powerful de-escalation mechanisms, like respect of ownership (Kummer *et al.*, 1974; Maynard Smith, 1976; Bachmann & Kummer, 1980; Packer & Pusey, 1985; Sigg & Falett, 1985), war of attrition (Maynard Smith, 1974), or war of nerves (Caryl, 1981). Thus, under the condition of within-group contest competition, the absence of overt aggression may have two rather different causes: it may occur as a consequence of formalization of a relationship among rivals with undisputed power asymmetry (van Rhijn & Vodegel, 1980; de Waal, 1986; Preuschoft, 1999), or it may be the result of a stalemate between well-armed rivals of symmetrical power (Packer & Pusey, 1985; Sigg & Falett, 1985). In the latter case, the high risks of escalation create a balance of power between rivals: males actually do mind the presence of rivals but they simply cannot help it.

Dominance styles of male and female Barbary macaques

In a previous study (Preuschoft *et al.*, 1998), we found significant differences in the way in which adult males and adult females compete over peanuts (Fig. 18.1). When two adult females competed (AF–AF), the loser usually retreated, while the winner showed a threat face. In virtually all cases, one rival, the 'winner,' ate the nut. By contrast, among adult males (AM–AM), threat faces and avoidance were rare. Instead, ignoring was observed in every second dyad and, in contrast to females, undecided encounters did occur. Ritualized submission was unusual among adult males. This is noteworthy, because if adult males followed the conditional reassurance strategy described by de Waal (1986), one rival should signal submission to be 'rewarded' for subordination with tolerance by the dominant.

When adult males competed with other group members such as adult females, immatures, or subadult males, they behaved just like females competing amongst each other. They did show aggression, i.e., threat faces, and the loser avoided the winner (usually the adult male). At the same time, ignoring and ritualized submission were rare. This pattern emerged no matter if it were adult or subadult males that competed, and no matter if they competed over peanuts or over females in estrus (Fig. 18.2a). As a result, the behavior of an adult male competing with a subadult male over a female in estrus resembled that of two females competing over a peanut.

Also, when competing against one another, other adult males (AM–AM) did not profoundly change their competitive tactic when an estrous female, rather than a peanut, was the incentive (Fig. 18.2b). Scarcity of aggression and avoidance characterized competition over both kinds of incentives. The ignoring in the peanut tests, however, tended to be substituted for appeal aggression (displays used to recruit agonistic support, cf. de Waal, 1977) in contests over estrous females. With both kinds of incentives, contests ended undecided in about 20% of the dyads. In nut tests, draws, i.e., encounters without a winner, resulted when nuts were 'stolen' by third parties. These were juveniles rushing in and snatching the nut away. This was possible when rivals showed 'ignoring;' they suddenly seemed overcome by intense sleepiness, shoulders and eyelids drooping, observing each other only from the corner of their eye, their impassiveness only interrupted to yawn or scratch. In this situation the youngsters struck – and suddenly the males were wide awake, threatening the juvenile, often in concert.

This response latency of adult males may be interpreted as reluctance to engage in overt competition. If so, the stealing by juveniles appears possible because of a mutual neutralization of power when the two adult males are uncertain in their assessments, and de-escalate by ignoring one another and even the nut. This interpretation implies also that power is distributed relatively evenly, and not asymmetrically, in adult male dyads. If this is the case, the same pattern should emerge in the consistency of winning per dyad. In other words, both partners should eat more or less the same numbers of peanuts in serial tests. Furthermore, reluctance to take priority of access might be related to other mechanisms preventing escalated competition, such as respect of ownership. Finally, if, indeed, caution is at the root of Barbary macaque male egalitarianism, there should be evidence of impending danger in male–male competitive encounters.

In the subsequent, empirical part of this chapter, data are presented on consistency of winning, respect of ownership, and the associated behavioral actions of males competing over peanuts. These are complemented with qualitative results on 'ownership' and female behavior in inter-male competition over estrous females. These results are discussed in the light of the above-described alternative causes of egalitarian relationships among males. Emphasizing the stalemate aspect, we intend to explain why Barbary macaque male relationships are egalitarian and violent at the same time. Finally, the focus is again expanded to show that stalemate is only one of several aspects characterizing inter-male relationships in this seasonally breeding polygynandrous primate.

METHODS

Subjects

Results are based on observations of semi-free-ranging Barbary macaques at Affenberg Salem, Germany. The colony was established in 1976. The monkeys live outdoors in a 14.5–ha outdoor enclosure throughout the year. Once a day, the monkeys are provided with widely dispersed wheat grains and apples. Water and commercial monkey pellets are available *ad libitum*. Naturally occurring food sources of the forested terrain are also routinely eaten. As in the wild, the monkeys are organized in several multi-male, multi-female groups including all age–sex classes (see Preuschoft *et al.*, 1998). At Salem, the groups are larger, but the almost even adult sex ratio is similar to that in the wild (Paul & Kuester, 1988; Ménard & Vallet, 1993). Reproduction is seasonal, with a mating season in late autumn and a birth season in spring. Male migration occurs regularly and at a rate similar to that in the Moyen Atlas (Paul & Kuester, 1988, Ménard & Vallet, 1996; Kuester & Paul, 1997). All subjects were known individually and completely habituated to the observers.

Peanut tests

For a detailed investigation of dyadic competition, 598 peanut tests were conducted between May 1987 and April

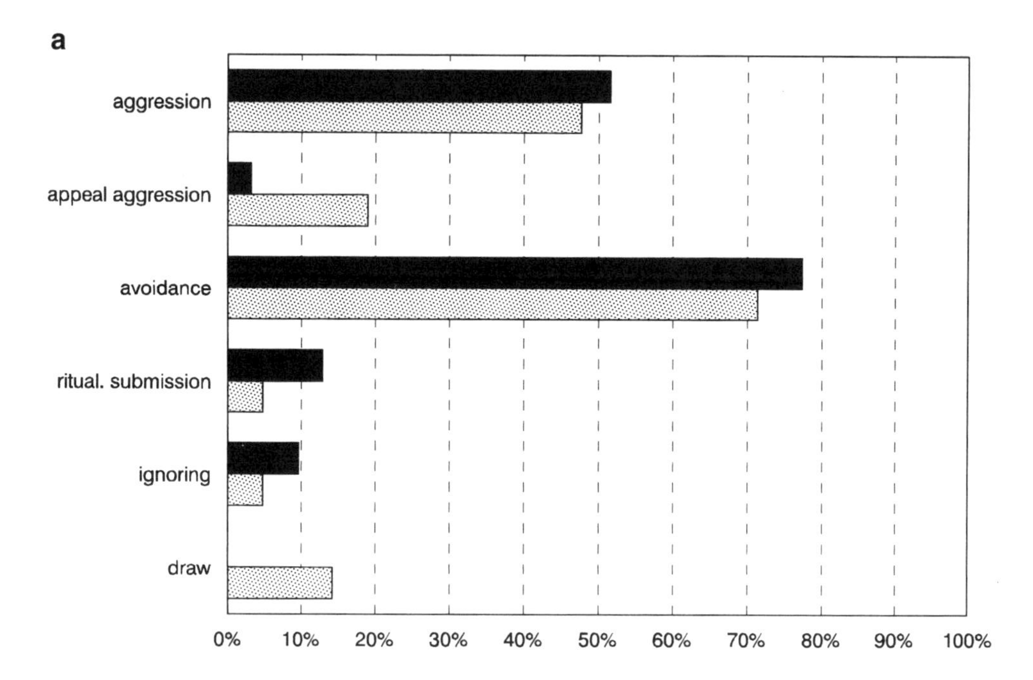

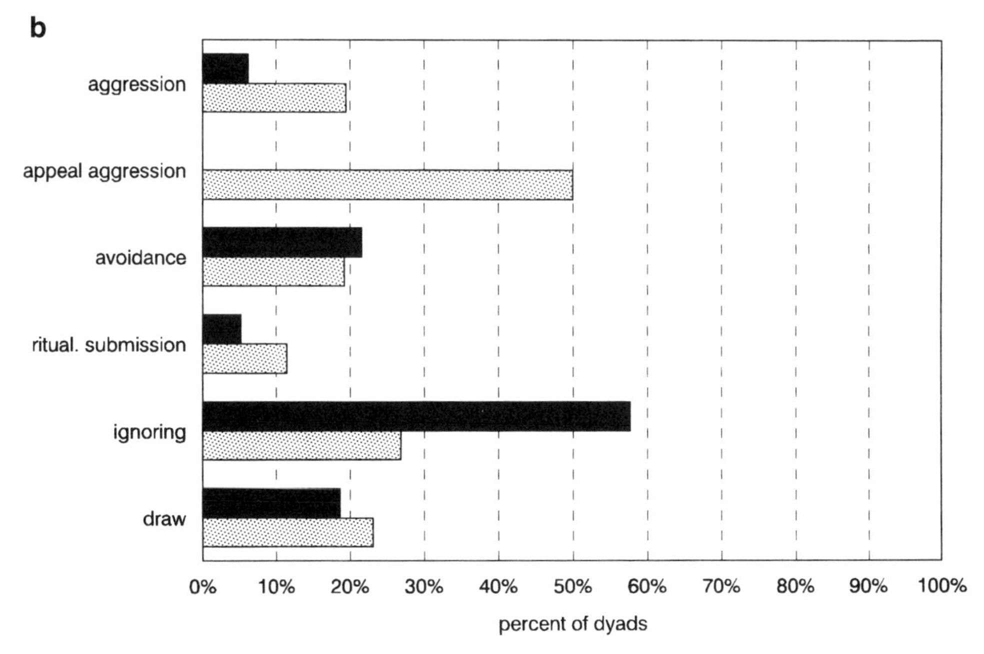

Fig. 18.2. Competition over peanuts (black) and females in estrus (stippled): (a) in AM–SM dyads (peanut: $n=31$; female: $n=21$); (b) in AM–AM dyads (peanut: $n=98$; female: $n=26$). Only first tests of different dyads are shown.

Table 18.2: *Database of peanut tests*

	All tests	All dyads	Repeated-tests dyads	Mean number of tests/ repeated-tests dyad	Distance/all tests: Mean	Distance/all tests: Range
AM–AM	372	98	74	4.7	2.4	0–10
'Other'	*226*	*164*	*34*	*2.8*	*2.6*	*0–6*
AF–AF	158	117	22	2.9	2.5	0–6
AM–SM	46	31	7	2.8	2.9	0.5–4
SM–SM	22	14	5	2.6	2.6	1–5

Note:
AM: adult male, AF: adult female, SM: subadult male; 'other' = AF–AF + AM-SM + SM–SM.

1988, and in May 1995. Tests were carried out whenever two individuals were within a distance of 6 m or less from each other with no potentially superior group members attending or close by. The purpose of these tests was to understand the dyadic power relations among adult males (AM–AM), but we also tested dyads of adult females (AF–AF), of subadult males (SM–SM), and of adult versus subadult males (AM–SM; Table 18.2). These 'other dyads' served as reference points to discern the peculiarities of competition in AM–AM dyads.

Each test consisted of throwing one peanut between two rivals. The individual that consumed the nut was termed the winner. A test ended in a draw if neither one of the competitors ate the nut. Nuts that landed in the middle between two rivals were labeled equidistant; those that landed closer to one of the two rivals were labeled asymmetrical. In those cases in which we had a prediction about the dominance relationship in a given dyad, the nut was thrown closer to the presumably inferior rival. The average distances between rivals was 2.4 m for AM–AM dyads (range: 0–10 m) and 2.6 m for 'other dyads' (range 0–6 m).

When the outcome of a nut test was ambiguous (draw, ignoring or tension; see below), or when the presumably inferior rival ate the nut while the rivals stayed close, additional tests were conducted until at least one rival left, or a clear outcome was achieved. This procedure yielded 74 adult male dyads and 34 'other dyads' that were tested repeatedly (Table 18.2).

The observations made in the context of the peanut tests were validated with observations on competition over estrous females. Competition over estrous females was not induced experimentally, but was recorded in focal animal observations conducted during the mating season 1988/9 (see Preuschoft *et al.*, 1998).

Behavioral categories

Aggression: physical assault, charge, rounded-mouth threat face.

Avoidance: includes flight and retreat after non-aggressive approach.

Ritualized submission: teeth-chattering, lip-smacking, silent bared-teeth display, or ano-genital presentation – if it occurred in combination with withdrawal.

Ignore: conspicuous lack of interest.

Steal: a third individual snatches the nut away.

Tension: scratch, piloerection, yawn, tree shake, and redirected aggression.

Appeal aggression: directed screaming, and unvocalized scream face accompanied by show-looking (cf. de Waal *et al.*, 1976).

Counter aggression: rival retaliates.

'Danger:' tension, appeal aggression, and contact aggression.

Affiliation: teeth-chattering, lip-smacking, or ano-genital presentation – if reciprocal or in combination with approach or non-hostile body contact with the rival or an infant in the rival's proximity; agonistic buffering, mounting, allo-grooming.

Winner index: for repeatedly tested dyads, the winner index is the proportion of nuts eaten by the rival who ate more nuts (n_s) out of the total number of nuts offered to the rivals (T), expressed as a percentage ($n_s/T*100$), modified after Hand (1986). In cases in which nuts were 'stolen,' this index can fall below 50%; if all nuts were eaten by the same rival, the winner index is 100%.

Superiority index: the proportion of nuts eaten by the rival who ate more nuts (n_s) out of the number of nuts eaten by any of the rivals (R), i.e., excluding those nuts that were eaten by third individuals ($n_s/R*100$). This purified index ranges from 50% to 100%.

Respect of 'ownership:' when a nut lying closer to rival A is not eaten by rival B.

Disrespect: when a nut lying closer to rival A is eaten by rival B.

Taker: an individual eating a nut lying equidistant to both rivals.

Tarrier: an individual hesitating (longer) to take an equidistant nut.

RESULTS

Consistency

The sample consists of 74 AM–AM dyads that were tested repeatedly. The mean number of tests per dyad was 4.7, with a range of 2–16. Only in 24 of the repeatedly tested dyads (32%) did one rival eat all available nuts (winner index = 100%). In 28 dyads (38%), at least one nut was lost to a third individual, this includes one dyad which lost all four nuts offered to them. Comparing the superiority indices of dyads with and without nuts lost to third individuals revealed no systematic differences between these two samples ($t = -0.88$, $p = 0.38$, df = 71), i.e., winning was not more balanced in dyads that lost nuts to third individuals. There was a significant negative correlation between number of tests per dyad and the superiority index ($R = -0.52$, $p < 0.001$), indicating that dyads tested less often were characterized by higher superiority indices. Note that we tested dyads more often when they produced ambiguous results but remained in proximity (see Methods). Hence, 'cohesive' dyads behaving ambiguously in the first test were tested more often, and eating of peanuts was distributed more evenly over both rivals.

We therefore standardized the data such that only the first two tests of all 74 multiply tested dyads were considered. In 41 dyads (55%), the same rival ate the first and the second nut (clear superiority). In the remaining 45% of dyads, winning was inconsistent. In 18 dyads, either rival ate one nut. In the remaining dyads, one or both nuts were lost to a third individual (12 and three dyads, respectively).

Respect of 'ownership'

Tests involved an asymmetrically placed nut 136 times in 'other dyads' and 213 times in AM–AM dyads. Among adult

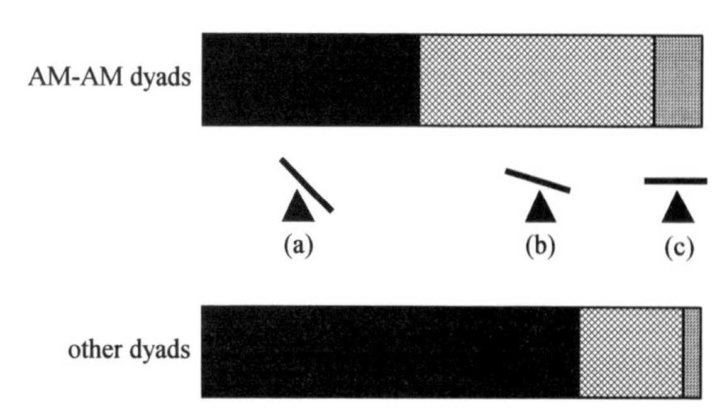

Fig. 18.3. Respect of 'ownership' of asymmetrically placed nuts in 74 AM–AM dyads and 29 'other dyads' (22 AF–AF + 7 AM–SM). (a) power asymmetry: 'precedence taker' disrespects asymmetrically placed nuts; (b) power bias: 'precedence taker' respects asymmetrical nuts; (c) power balance: rivals share or ignore equidistant nuts and respect asymmetrical ones. See text for further explanation.

males, 93 nuts (44%) were obtained because the rival farther away from it respected the 'owner's' closer proximity to the nut. By contrast, in the 'other dyads', 28 nuts (21%) were obtained by respecting the 'ownership' of the rival closer to the nut. This difference is significant ($X^2 = 19.88$, $p = 0.00001$), but these data are partially dependent. We then tested per dyad whether disrespect of 'ownership' of asymmetrically placed nuts was associated with precedence on equidistant nuts (Fig. 18.3). If power is distributed strictly asymmetrically in a dyad, the rival taking precedence with equidistant nuts ('taker') should disrespect when nuts lie closer to the non-preceding rival ('tarrier'). On the other hand, when power is distributed more evenly in a dyad, the 'taker' should show respect if the 'tarrier' eats nuts which are lying closer to him, i.e., in the case of asymmetrically placed nuts, 'taker' and 'tarrier' should mutually respect nuts lying closer to the other.

Strictly asymmetrical relationships (Fig. 18.3) in which the 'precedence taker' disrespected 'ownership' of all nuts lying closer to his rival characterized 76% of 'other dyads' but only 43% of the AM–AM dyads. In 47% of the AM–AM dyads, 'ownership' was respected at least once – this contrasts with only 21% 'other dyads' in which respect occurred (Fig. 18.3). Among adult males, there were seven dyads in which respect of asymmetrical nuts was mutual and precedence with equidistant nuts was equivocal (Fig. 18.3). This happened only once in the 'other dyads'. In sum, respect of 'ownership' was significantly more frequent among adult males than among adult females or among adult and sub-adult males ($X^2 = 8.88$, $p = 0.0028$).

Table 18.3. *Respect of ownership in dyads tested only once*

	Single-test dyads	Asymmetric nuts	Disrespected nuts*	Stolen nuts
AM–AM	24	20	18 (90)	1
'Other'	*128*	*73*	*66 (90)*	*1*
AF–AF	95	47	43 (91)	0
AM–SM	24	18	16 (89)	1
SM–SM	9	8	7 (88)	0

Note:
* Numbers in parentheses are percentages of asymmetrically placed nuts.

In addition, there were five dyads of sub-adult males that were tested repeatedly. In all cases, clear precedence with equidistant nuts was combined with at least one instance of respect, where the 'taker' respected that the 'tarrier' ate a nut lying closer to him. By contrast, in the SM–SM dyads tested only once, seven of eight asymmetrical nuts were obtained by disrespect (Table 18.3). This resembled the proportions for other single-test dyads (Table 18.3): AF–AF disrespected 44 of 48, AM–SM disrespected 16 of 18, and also single-test AM–AM dyads disrespected 18 of 20 asymmetrical nuts. Thus, single-test and repeated-test dyads were equally characterized by disrespect when they consisted of two adult females or of an adult male and a subadult male. In adult male dyads and in subadult male dyads, however, single-test dyads frequently disrespected, whereas repeated-tests dyads usually respected 'ownership' of asymmetrically placed nuts.

The large proportion of AM–AM dyads which exhibited respect, and for which winning was inconsistent, suggests that the power relations among adult males tend to be more symmetrical than those in 'other dyads.' But what is the reason for that? Do adult males volunteer to share nuts – and estrous females? Are there any behaviors that differentiate between nuts acquired by disrespect and by respect on the part of the rival?

Respect and 'danger'

Males gained their rival's respect through threat faces five times (5.4% of 93 respected nuts). In two tests, an adult male ate a nut lying closer to him and withdrew subsequently, though never in combination with ritualized submission. So, clearly, neither threatening nor appeasement was predictive of respect in adult males. Furthermore, affiliation coincided with respect in only six tests.

Among the behaviors characterizing AM–AM confrontations were tension and appeal aggression – both were hardly observed in 'other dyads.' *Tension* was frequently evident in both partners simultaneously (34% of 59 incidents of tension) and also in losers alone (27%) but was not typical of winners alone (8%). Another 30% of tension behaviors occurred in undecided tests. *Appeal aggression* was typically reciprocal (43% of 28 appeals), or performed by the loser alone (39%). It was not typical of winners alone (7%, two cases), or of undecided tests (10%, three cases). *Physical aggression* was very rare in nut tests ('other dyads': seven of 266 tests; AM–AM: six of 372 tests). However, when aggression occurred it was clearly more escalated in AM–AM dyads. In 'other dyads', aggression consisted basically of charges by which individuals backed up their threat faces, either when they took nuts lying closer to their rival or when the threatened individual had taken the nut nevertheless. Among adult males, aggression consisted always of attack and was almost always answered by counteraggression. Only once did physical assault help an adult male to take a nut; in the other cases, fighting resulted in the loss of the nut to a third party. When males competed over estrous females, aggression was expressed as a rounded-mouth threat face in only six cases (10% of encounters), while physical assault was observed seven times (11%).

Lumping tension, appeal aggression and contact aggression into one category, 'danger,' revealed this as characteristic of adult male dyads (49 of 98 AM–AM dyads versus six of 67 'other dyads': $X^2 = 30.1$, $p < 0.0001$). Furthermore, among adult males, 'danger' was significantly associated with respect of 'ownership' (41 of 128 respect versus 18 of 120 disrespect cases: $X^2 = 71.0$, $p < 0.0001$) in individual tests, and evidence of 'danger' allows prediction of dyads in which both partners ate similar numbers of nuts, i.e., dyads

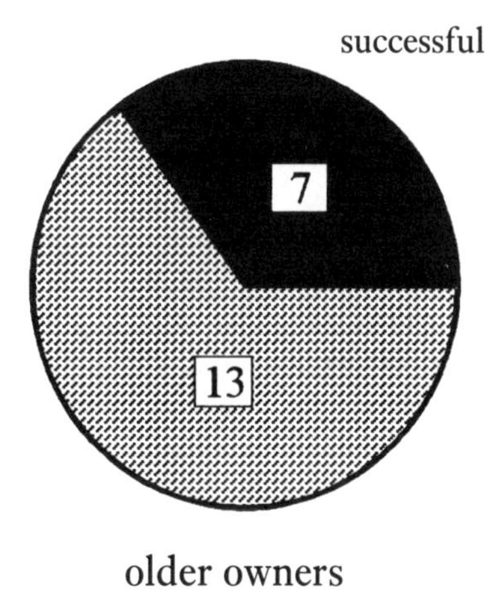

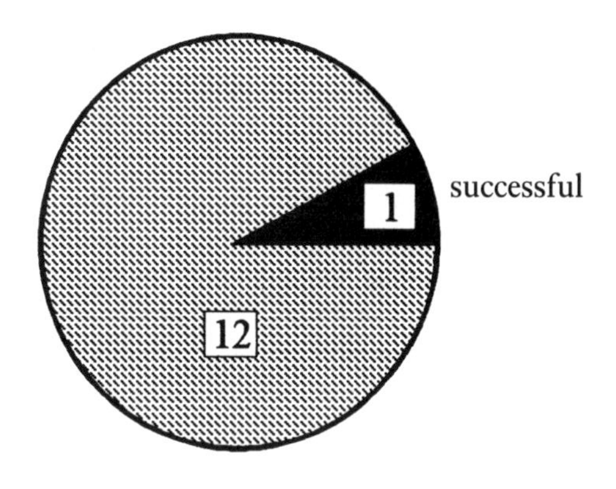

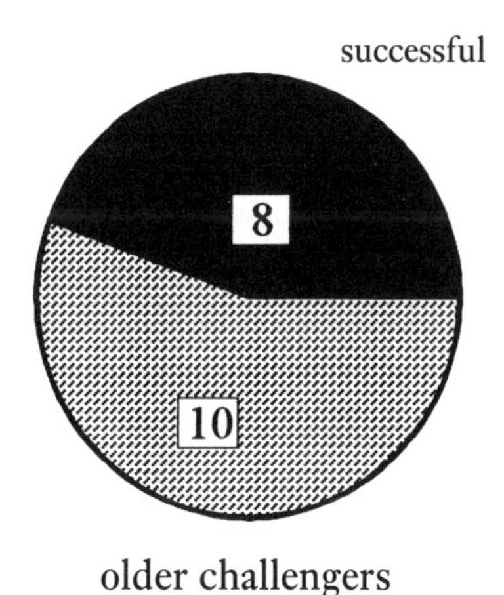

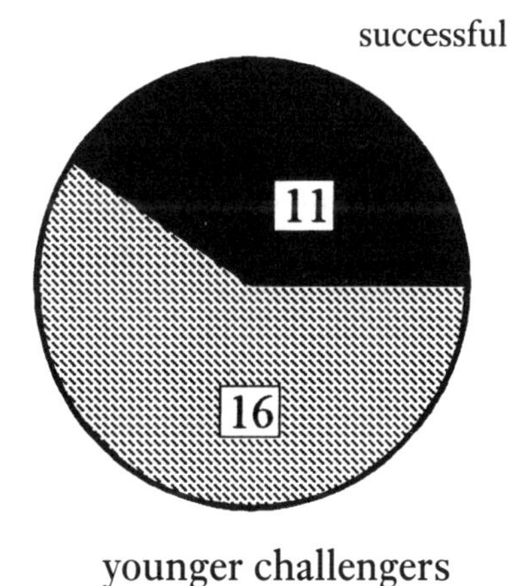

Fig. 18.4. Success (black portions) of 'owners' and 'challengers' as a function of relative age in AM–AM dyads competing over estrous females. Hatched portions represent owners and challengers who lost the competition.

characterized by low winner indices ($t = -4.81$, $p < 0.001$, df = 72).

Timing of actions

The parsimony of acting that characterized adult male contests was also reflected in the timing of their actions. In 'other dyads', threat faces, physical aggression, and appeal aggression occurred typically in the first test (83% of 118 threats, six of seven physical, and four of five appeal aggressions). By contrast, adult males showed these behaviors only in later tests (12 of 18 threats, all six physical, and 23 of 28 appeal aggressions). In the first tests, only tension was frequently evident (44% of 59 instances of tension behavior).

Sexual competition among males

OWNERSHIP

Males engaged in contests over estrous females in either the role of the 'owner', i.e., the current consort of the female, or the role of a 'challenger.' In contrast to nut tests, 'owners' of an estrous female usually lost the competition. Among adult males, owners had only a 24% chance of winning, but also challengers had only a 40% chance of winning a confrontation. Not surprisingly, older rivals usually won against younger rivals in SM–SM and AM–SM dyads. In AM–AM dyads, basically only older owners were successful (Fig. 18.4), whereas relative age did not strongly affect winning by challengers (Fig. 18.4).

Challengers usually simply came into sight and sat down anywhere from about 15 m to close proximity of the female.

Owners usually responded by staying on. Many times, several rivals sat dispersed around a female, deliberately not looking at one another. Gaze contact frequently resulted in appeal aggression (compare with Fig. 18.2b). 'Ignore duels' could last several minutes, and in 56% of all confrontations one male left after a while. In contrast to 'avoiding,' this leaving seemed casual; it was not a clear-cut response to the rival's behavior. Leaving was regularly combined with approach of, or starting or joining into aggressive interactions with other group members, including females. In 81% of the spontaneous-leave cases, the male staying on became the female's new consort. Thus, it seemed that, after having entered into the competition, future losers suddenly gave up, or lost their interest.

FEMALE INITIATIVE

In 44% of all competitive encounters a preference of the female for one of the males could be inferred from her behavior (approach, present, follow, look into face). Although in 70% of these encounters, the female managed to stay with the preferred male, the exceptions prove illuminating. When the preferred male was higher ranking than his rival, she was successful in 15 of 16 cases. However, when a female chose the lower-ranking rival (ten times), she was successful four times but ended up with the unpreferred higher-ranking male five times. In 11 confrontations the rank relations among the males were unclear to the observer. Here, the female obtained the preferred male seven times, and the competition ended undecided four times. When males stayed put, females often contacted additional males, frequently returning to males who failed to follow. When newly contacted males followed females to previous partners, the females' going to and fro resulted in reduced distances between males. Females also sometimes waited in the vicinity of males engaged in 'ignore duels,' eventually uttering estrous calls. In these situations, rivals often threatened or chased the female away, occasionally in concert.

DISCUSSION

Dominance and egalitarianism as characteristics of dyads

The results show that dyads of competing adult males differed from all other age–sex classes. When two adult females competed, one performed a threat face and the other retreated, leaving the incentive to her rival. Females thus established clear-cut results in virtually all encounters. By contrast, among adult males, typically, neither partner threatened or avoided the other. Instead, rivals showed evidence of tension while ignoring one another and, frequently, even the incentive. When adult males competed with sub-adult males, their competitive tactic was indistinguishable from that of adult females, suggesting that tactics depended on the relative fighting abilities of the rivals. When clear differences in body size and weaponry existed – hence the risk of encountering resistance or becoming injured was low, as in the case of female and sub-adult male rivals – priority of access and behavioral actions were unambiguous. Thus, the seemingly indifferent behavior among adult males might be understood as a reluctance to engage in potentially dangerous overt competition. When two males are uncertain in their assessment of relative resource-holding power, or when power is distributed symmetrically among both rivals, they de-escalate by ignoring. Such restraint is compatible with the idea of a 'stalemate' among the competitors. This interpretation is further supported by the finding that third parties were able to 'steal' a nut only when two adult males competed. This 'stealing' strongly resembles the manner in which sub-adult males 'sneak' copulations when adult males are locked in a competition over an estrous female (Kuester & Paul, 1989).

However, a balance of power does not affect each and every competitive encounter, nor does it characterize all adult male dyads. In the present sample, in almost 50% of all dyads competing over peanuts, one male was clearly dominant over the other; that is, winning was biased in favor of one rival. Note that our sample represents only one-fourth of all possible adult male dyads of the study group at that time. This is a consequence of our opportunistic testing procedure. We tested only those males who came close enough to one another. A subgrouping among the males is evident from these data, and this subgrouping is in accordance with group membership of males four years later, after the group had fissioned into three daughter groups (Kuester & Paul, 1997).

Nevertheless, inconsistent winning was exhibited by almost 70% of the dyads that had a chance to do so because they were tested more than once. There was a significant difference between repeated-test dyads of adult males and 'other dyads,' indicating that predictability of winning was higher in AF–AF, SM–SM and AM–SM dyads. These were also the dyads in which winners employed a threat face. In dyads of adult males, threat faces were used ten times less frequently than in 'other dyads'. Clearly, cost–benefit ratios for the performance of threat faces differed between adult male and 'other dyads'. Whereas females regularly complied with threats, the reactions of adult male competitors exhibited a readiness to 'debate,' by affiliation or appeasement, or to rebel, by escalation or appeal aggression (Preuschoft *et al.*, 1998). Moreover, whereas females performed threat faces when taking precedence on equidistant nuts or when disrespecting 'ownership,' adult males usually threatened to claim nuts that lay closer to them already. As adult males hardly ever showed ritualized agonistic signals, it is unlikely that their egalitarian relationships could be the result of formalized power relationships and tolerance on part of the dominants (cf. de Waal, 1986).

Stalemate and the war of nerves

The behavior of males can be understood as reflecting hesitation to engage in overt competition.

1. They did not show decisive behaviors such as assertion or yielding to one another.
2. They persistently ignored not only one another but also the entire competitive situation, thus losing nuts and estrous females to third parties
3. If they employed assertive tactics in the nuts tests, be it threat face, physical, or appeal aggression, they did so only in later tests, not in the first encounter.
4. They did give evidence of tension by redirecting aggression against bystanders, by tree shaking, yawning, scratching and piloerection.
5. In many dyads, an adult male hesitated to take a nut that lay closer to his rival, and tolerated his rival eating the nut even when he did not hesitate to eat those lying in the middle.

Respect of 'ownership' was not found in contests over estrous females. But even though males tended to treat females and peanuts alike, females, evidently, are no peanuts. They actively sought the proximity of other males, frequently preferring a newly contacted male to the current consort. As a matter of fact, females might even defy the males' attempts at avoiding overt competition by leading them into proximity and instigating competition through proceptive behavior, including spontaneously emitted estrous calls. Thus, even if Barbary macaque females mate promiscuously with as many males as they can possibly get, they may implicitly select for certain qualities in males, such as persistence, self-control, and strong nerves.

In our view it is justified to interpret ignoring, response latencies, respect of ownership and inconsistent winning as de-escalation mechanisms preventing potentially dangerous escalation. Even with a moderate incentive such as a peanut we were able to induce (i) escalated competition, (ii) side-directed behavior aimed at recruiting agonistic support, and (iii) social tension even in rivals that refrained from overt competition to the point of missing a consummatory act altogether.

That a higher resource value may tip the balance toward incurring greater risks is suggested by the facts that: (i) adult males already showed appeal aggression during first runs, and were also more ready to escalate, using physical aggression as frequently as threat faces, when an estrous female was the incentive; (ii) the mating season is also the period in which canine slashes and puncture wounds on males are most frequent; and (iii) polyadic fights among adult males occur most often. Usually, these polyadic fights break out as a consequence of the recruitment behavior, described above for the peanut tests.

The danger of escalation – as evident in tension behavior, recruitment aimed at enlisting agonistic support, and overt aggression – shows that adult males do not voluntarily relinquish nuts, or estrous females, to their rivals. Instead, power asymmetry is so small that one rival can force the other to share.

The power relationships among male Barbary macaques thus deviate considerably from the clear power asymmetries found among male rhesus (*Macaca mulatta*), long-tailed (*M. fascicularis*), and Japanese macaques (*M. fuscata*) (Chapais, 1983; van Noordwijk & van Schaik,1988; de Ruiter *et al.*, 1992; McMillan, 1989; Bercovitch, 1991; Inoue *et al.*, 1993; Paul *et al.*, 1993; Paul,1997). The contrast to the equally seasonally reproducing rhesus and Japanese macaques is perhaps attributable to differences in sexual dimorphism. Of the three species, Barbary macaques exhibit the greatest sexual dimorphism in body mass (male/female: Barbary macaques, 1.55: Fa, 1984; Japanese macaques, 1.28; rhesus monkeys, 1.31: Plavcan & van Schaik, 1992). This can be interpreted as evidence for severe contest competition among male Barbary macaques, which drove an arms race leading to the rather pronounced sexual dimorphism (Crook, 1972; Clutton-Brock, 1985; but see Martin *et al.*, 1994). This dimorphism in turn makes overt aggression more risky in this species, thus producing a stalemate.

Another possibility is that the stakes are higher in the more 'despotic' rhesus and Japanese macaques, although the mating seasonality is similar to that of Barbary macaques (Mitani *et al.*, 1996a). Whereas the former species follow a multi-mount-to-ejaculation pattern (MME: Enomoto,

1974; Manson, 1996), Barbary macaques complete copulations in a single mount (single-mount-to-ejaculation, SME: Taub, 1982). This copulation pattern favors sneaking and sperm competition. Proximately, males compete for matings, not for fertilizations. In a species like Barbary macaques, in which 300–600 ejaculatory copulations precede a conception (J. Kuester, unpublished data), males may be designed to compete less vigorously for single copulations. But note that it remains ambiguous which of the two – competitive style or mating pattern – is cause and which is effect.

Beyond the dyad: breaking stalemates and compensating power asymmetries

The regular occurrence of appeal aggression and polyadic fighting indicates that the stalemate explanation is, again, not the entire story. Coalition formation also plays an important role among adult males of this species (Kuester & Paul, 1992). Post-prime males compensate their disadvantageous position *vis à vis* a prime male very efficiently by 'unhinging' the intrinsic power of their opponents with superior social power, that is, by allying against them (Kuester & Preuschoft, unpublished data). As a result, post-prime males have the highest mating success and a reproductive success similar to prime males (Kuester *et al.*, 1995; see also Fig. 18.4). Within the circle of post-prime coalition partners, a male has leverage power (see Table 18.1) over his age mates and allies. He could, in principle, deny support to another post-prime male in case this male attempted to outcompete him relentlessly on other occasions (Noë & Hammerstein, 1994). If this interpretation is correct, it may also explain why the competitive strategies are generalized across different incentives. It is the limiting resource (fertilizable females) which shapes the competitive style (dyadic stalemate, polyadic coalitions) and, as a side-effect, leverage among post-prime males results, which in turn can be exerted in conflicts over anything else. In other words, adult males must maintain cooperative relationships, and these necessitate conciliatory attitudes in other contexts as well. Thus, in the temporary cease-fire among coalition partners, lack of overt competition may be understood as calculated generosity. In non-cooperating dyads, distrustful inhibition seems to result from a war of nerves.

Altogether, Barbary macaque males employ three interdependent competitive strategies: (1) dyadic stalemates lead to war of nerves/war of attrition strategies; (2) on the polyadic level, coalitions function to break the stalemate and overwhelm dyadically superior males, thus leading to an 'oligopolization' of estrous females by post-prime males; (3) this seemingly hermetic oligopoly of the older males is punctuated by opportunistic sneaking of copulations while other competitors are engaged in 'ignore duels' or 'scream fights'.

ACKNOWLEDGMENTS

We gratefully acknowledge the insights and inspiration gained in discussions with Jutta Kuester and Carel van Schaik. We are also greatly indebted to Walter Angst, Ellen Merz, and Gilbert de Turckheim for permission to study the Barbary macaques at Affenberg Salem.

Part V
Evolutionary determinants and consequences

This part of the volume features four chapters that examine various determinants and consequences of variation in the number of males.

Whether males remain in their natal group or whether they disperse to groups with unrelated females is one important determinant of male reproductive strategies and, thus, male numbers. Theresa Pope examines this question in great detail in New World primates. Using long-term records from well-studied taxa, she shows that population density is a main determinant of male and female dispersal options. By also focusing on the genetic consequences of these decisions, she explains a great part of the observed sex differences in behavior. Because ecological conditions in the New World apparently often limit female breeding opportunities, males have more opportunities for group membership, resulting in the large proportion of species with male philopatry or polyandry. It will be exciting to see how well these variables explain the social systems of primates in the Old World.

Jeanne Altmann returns to the basic socioecological prediction that males should distribute themselves in response to the distribution of females. She makes the important distinction between the mechanisms and the outcome of this process. Unlike many previous studies which focused on interspecific variation, Altmann analyzes male distribution among groups and populations. Using the unique data set from Amboseli baboons, she shows that there is more variation within species than predicted by existing univariate models. This study therefore focuses attention at the level of individuals, both males and females, for which different group sizes and compositions may be optimal at different times of their reproductive careers. This focus on individuals naturally leads to a closer examination of behavioral aspects of dispersal, which constitutes the main proximate determinant of variable numbers of males in most species. The detailed insights Altmann and her colleagues obtained from their male baboons will be difficult to match for others not studying savannah-dwelling species, but they will certainly provide an inspiring goal.

The chapter by Richard Wrangham examines sex differences in gregariousness in multi-male, multi-female species, using chimpanzees as a particularly intruiging example. Male chimpanzees are much more gregarious than females, a rare constellation among mammals. Instead of examining male relationships, Wrangham presents a simple and elegant ecological explanantion based on the behavior of females. His review of the chimpanzee literature shows that females are subject to strong scramble competition for food, and that they are severely handicaped by costs of infant carrying. As a result, females reduce their travel costs by moving alone or in small groups. Male gregariousness is therefore only an indirect consequence of female behavior. By examining other species with similar social structures, Wrangham argues convincingly that this phenomenon may be more widespread.

In the final chapter of this section, Robin Dunbar revisits the original central question concerning male reproductive strategies. The main issue of contention in many studies of variation in the number of males has been whether female numbers or the synchrony of their cycles limit male monopolization potential. Dunbar re-examines this question from scratch by developing an explicitly simplistic model of alternative male mating strategies, which explains variation among great apes surprisingly well. In another set of modeling calculations, he shows that the number of females themselves is not a sufficient predictor of male numbers, but that information about the temporal distribution of their receptive periods improves the predictive power of these models. This reconfirmation of the verbal Emlen–Oring model indicates that we should now begin to illuminate biologically meaningful dimensions of estrus synchrony.

19 • The evolution of male philopatry in neotropical monkeys

THERESA R. POPE

INTRODUCTION

Most New World monkey mating systems exhibit either female-biased dispersal from social groups or dispersal by both sexes (Moore, 1992; Strier, 1994a). Differences between the sexes in degree of philopatry shape the direction of genetic lineage formation within social groups, such that the potential for nepotism and kin-based cooperation is usually greater in the non-dispersing sex. Male kin-bonded mating systems are thus prevalent among polygamous platyrrhines. This forms a distinct contrast to the female kin-bonded, matrifocal mating systems found in most Old World monkeys, in which dispersal is predominantly male biased (Pusey & Packer, 1987). Behaviors that comprise mating systems have been widely viewed as highly evolutionarily labile, and based primarily upon factors such as resource distribution and life history characteristics. New World primates, however, exhibit a highly divergent array of feeding ecologies and life history attributes, ranging from small-bodied insectivores to large folivores, and occupy feeding niches that are broadly convergent with those of Old World monkeys.

Differences between these two phylogenetic lineages in direction of sex-bias in kin bonding imply that phylogenetic constraints have played a major role in shaping this pattern. Phylogenetic inertia predicts that pleiotropy, epistasis, and developmental canalization limit the possible array of changes to genomic rearrangements that have resulted from past adaptation. The manner in which these processes mediate the response of behavior to environmental selection factors remains largely unknown. However, recent research on the role of such factors as neuropeptide receptor site distribution in the modulation of species-specific social behaviors like intraspecific aggression (Ferris & Potegal, 1988; Young, 1997), partner preference formation (Winslow *et al.*, 1993), and paternal behavior (Wang *et al.*, 1994), suggest mechanisms whereby complex behaviors involved in mating systems are neurologically and genetically modulated. Once a complex behavior has evolved, regulatory mechanisms such as these may impose strong limits on subsequent adaptive pathways.

Clues to the origin of male philopatry are sought here in those platyrrhines in which both sexes disperse, a mating system characteristic that is primarily confined to monogamous and small-group polygamous species. Monogamous breeding has been proposed to be the primitive condition for New World monkeys (Eisenberg, 1981; Wright, 1984; Kinzey, 1987), a view that is supported by the distribution of monogamy among the major clades identified by recent molecular phylogenetic data (Barroso *et al.*, 1997; Porter *et al.*, 1997). Small-group polygamy occurs mainly in species that exhibit flexible mating systems in which the proportion of monogamous to polygamous groups can vary among populations (e.g., many callitrichines: Ferrari & Ferrari, 1989; Goldizen & Terborgh, 1989; Baker & Dietz, 1996; Rylands, 1996), or between population growth phases (red howler monkey, *Alouatta seniculus*:Pope, 1998; Soini, 1995a).

Although both sexes disperse in these species, one sex is usually much more philopatric than the other. For any given population, the sex that remains nearest the place of birth will, on average, share a higher coefficient of relatedness than the long-distance dispersing sex . Thus, any competitive advantages that might accrue to cooperative alliances between individuals will have the added benefit for the resident sex of increasing inclusive fitness. This will lower the benefit-to-cost threshold above which cooperative behavior becomes advantageous. The formation of cooperative alliances in order to acquire or defend resources should thus arise more frequently in the resident sex.

I focus here on the manner in which intrasexual competition and differences in the distribution of resources limiting to reproduction for each sex affect differences between the sexes in relative philopatry. Trends in this relationship are examined by comparing three ecologically divergent species with flexible mating systems and dispersal by both sexes: red howler monkeys, golden lion tamarins (*Leontopithecus rosalia*), and saddleback tamarins (*Saguinus fuscicollis*). The ecological correlates of population variation in group breeding structure may illustrate the manner in which evolutionary shifts from monogamy to polygyny took place, thereby providing a model for the incipient radiation of the highly diverse mating systems represented in New World monkeys.

RED HOWLER MONKEYS

Red howler monkeys are large-bodied cebids that are characterized as folivorous, although they include a wide variety of plant parts in their diet when available, including unripe fruits, flowers, and seed pods. They live in territorial social groups containing two to five adult females, one to three adult males, and their offspring. Both males and females emigrate from their natal social groups, but who leaves, how far they go, and their chances of successful reproduction after emigration differ between the sexes. These outcomes are primarily the consequence of differences in patterns of intrasexual competition over resources that limit reproduction, which are distributed differently for males and females. The overview of male and female intrasexual competition and dispersal presented below is based on my own long-term studies of 78 social groups distributed among three populations in different stages of growth (Mata, El Frio, and Pinero), each of which was monitored for a minimum of five years, plus long-term demographic monitoring of a fourth population (Gallery) by C. Crockett, and additional years of demographic and other research on the Mata population by C. Crockett, R. Rudran, and R. Sekulic.

Female intrasexual competition and dispersal

Female dispersal from the natal social group depends on how many adult females are already breeding within the group when a natal female approaches sexual maturity (Pope, 1989; Crockett & Pope, 1993). In groups with only one adult female, a natal female reaching sexual maturity always remains in the group and reproduces there. Approximately 50% of natal females emigrate from groups containing two breeding females, 90% from groups with three breeding females, and 100% from groups with four breeding females. This pattern was consistent across three populations in different growth stages (Crockett & Pope, 1993; Pope, 1998). All exhibited different ecological densities as stable population size was approached, but mean group size (approximately nine to ten) and composition were similar once this stage was reached (Crockett, 1996; Pope, 1998). Once a female reproduces within a group, she remains there for the rest of her reproductive lifespan unless the group fails and disintegrates. Thus, emigration of prereproductive females is dependent on population density relative to carrying capacity, whereas emigration of parous females from extant groups has not been observed.

Females that emigrate from their natal groups are not a random sample. Adult females try to evict one another's daughters, such that usually only the daughters of a single presumably dominant adult female are successful at remaining to breed. Emigration is frequently preceded by aggression between the maturing natal female and a female who is not her mother (Crockett, 1984; Crockett & Pope, 1993), resulting in physical injury to both parties as well as to the natal female's mother, who may also engage in aggressive interactions with the adult female challenging her daughter. In those cases in which the pedigrees of successive natal recruitments were known, they were the daughters/granddaughters of only one of the two adult females resident in each group (Pope 1989). These observations are supported by extensive genetic data at both the nuclear and mitochondrial genomes (Pope, 1992, 1996a, 1998; see below). Differential success in recruiting daughters may be a manifestation of an otherwise seldom-expressed dominance relationship, or the outcome of whoever 'wins' the first recruitment battle: a mother–daughter coalition would be difficult for the third female to overcome in future recruitment conflicts.

Group females are also extremely aggressive towards extra-group females, which are howled at, chased, and severely wounded if caught (Sekulic, 1982a; Crockett & Pope, 1988, 1993). Attacks on extra-group females are highly coordinated and result in the prevention of immigration into groups by migrant females virtually 100% of the time (Sekulic, 1982a; Pope, 1989, 1998; Crockett & Pope, 1993). So, although females frequently emigrate from their natal groups, they are prevented by resident group females from transferring into established groups through rigorous exclusion from the group territory.

Female territoriality is accompanied by dietary advantages for group females. Results of paired activity budget samples between solitary and group females and males indicated that diets consumed by group females were significantly higher in nutrients limiting to reproduction (Pope, 1989). Group and solitary animals were matched for foraging area, age class, sex, and reproductive condition (all females were nulliparous and non-pregnant). Samples of all foods observed to be eaten were collected, dried, and analyzed for content of crude protein, neutral detergent (non-digestible)

Table 19.1. *Comparison of nutrient component analysis of diets eaten by solitary and group-living males and females*

	Crude protein %	IVOMD %	NDFaf %	Phosphorus %
Group females	20.4***	50.4	43.4*	0.288***
	(9.2)	(18.0)	(9.3)	(0.140)
Solitary females	13.5	44.5	47.0	0.187
	(6.9)	(14.1)	(10.1)	(0.082)
Group males	13.6	45.1	50.7	0.213
	(7.1)	(14.5)	(10.0)	(0.105)
Solitary males	14.1	46.9	50.5	0.191
	(7.3)	(17.3)	(10.3)	(0.110)

Note:
IVOMD = *in vitro* digestibility; NDFaf = ash-free neutral detergent fiber. Numbers in parentheses beneath values are standard deviations, and values followed by asterisks are statistically different from other means for that nutrient. * $p<0.05$; *** $p<0.001$

fiber, *in vitro* digestibility, and phosphorus (Table 19.1). Group female diets had 50% more crude protein and phosphorus and significantly lower non-digestible fiber than the diets of solitary females. Phosphorus, protein, calcium, and kilocalories are the four nutrient requirements that increase most for females during reproduction (Lloyd *et al.*, 1978). Phosphorus is the most common nutrient deficiency in herbivore diets comprised of leafy plant parts, which are generally low in usable phosphorus (Lloyd *et al.*, 1978). Tropical soils and leaf mass in particular are typically poor in phosphorus (Walker & Syers, 1976; Irion, 1978), and the soils that prevail in the llanos of Venezuela are among the poorest (Vitouseck & Sanford, 1986). Differences between the solitary and group female diets in phosphorus and protein were apparently due to fewer flowers and leaf buds in the solitary female diets, both of which exhibited highly ephemeral and clumped distributions. Consequently, it seems reasonable to conclude that both nutrients are extremely important, if not limiting, variables in terms of female red howler reproduction. Although extra-group, migrant females regularly copulate with both group and extra-group males, they do not produce offspring (Sekulic, 1982a; Pope, 1989). Whether this is a consequence of failure to conceive or to bring a pregnancy to term is unknown.

The only means whereby a migrant female can establish membership in a social group is to form a new group with other solitary migrants. New groups form when solitary females meet, form social bonds, join solitary males, successfully defend a territory from surrounding groups, and produce offspring (Pope, 1992, 1998). Extra-group associations of females generally do not howl or attempt to establish a territory until joined by male(s), which are apparently required for successful territory formation. Rate of new group formation and new group failure rate (defined as dissolution and/or 100% offspring mortality) vary with habitat saturation (Pope, 1992, 1998; Crockett, 1996). New group failure rate was lowest (7%) in a growing population with high rates of new group formation in unoccupied areas, and highest (50–100%) in saturated habitats in which new groups were limited to establishing territories at the very edge of the forest or in areas of overlap between established group ranges. The latter was a lengthy process, involving almost daily confrontation with resident groups, including extended chases, prolonged howling bouts (>one hour), and physical aggression between males that resulted in severe physical injury of the new group male by resident group males in at least one case. In addition to high group failure rate, new group females at Mata and El Frio had 44% longer interbirth intervals and 200% higher infant mortality than females in established groups. Mean age of first birth for those females at Mata that succeeded in breeding in new groups (seven years) was significantly higher than age of first birth for females that bred in their natal groups (5.1 years; Crockett & Pope, 1993).

Pedigree data, coefficients of relatedness estimated from allozyme data, partitioning of genetic variance among new groups, and mitochondrial DNA (mtDNA) haplotypes all indicate that females who join together to form new groups are nearly always unrelated (Pope, 1992, 1996a, 1998). Natal females have not been observed to emigrate together, possibly as a consequence of small group size and staggered interbirth intervals. In addition to lower-quality diets and inability to reproduce, solitary females are exposed to higher mortality risks and range over larger distances than group females (Pope, 1989; Crockett & Pope, 1993). The benefits of cooperation are therefore high, probably more so in a high-density population in which the majority of quality resources are located within cooperatively defended territories. Once groups are established, however, female competition and nepotism within groups result in the development of closely related matrilines. Mean coefficient of relatedness among breeding females within groups in four red howler populations in different stages of growth increased from 0 in the most rapidly growing population, in which rate of new

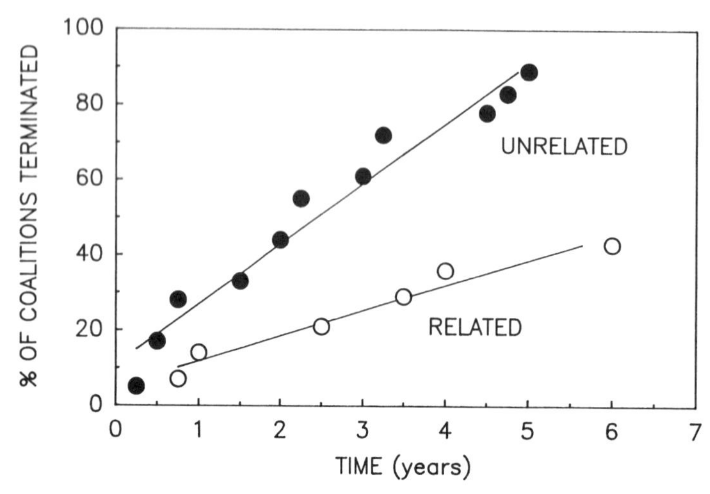

Fig. 19.1. Comparative termination rates of coalitions of related versus unrelated red howler males ($r > 0.97$ for both).

group formation was high, to 0.43 in the population with the oldest groups (Pope, 1998). Mean coefficient of relatedness among female coalitions is therefore correlated with population growth stage and rate of social group formation.

Male intrasexual competition and dispersal

Males compete for reproductive access to females by aggressively attacking resident group males and attempting to evict them. Like females, males can also gain group membership by forming a new group with extra-group migrants. Once successful at becoming resident within a group, males must then successfully defend their positions from incursion attempts by other males. Infanticide is a frequent outcome of both successful and unsuccessful incursion attempts, accounting for at least 44% of all infant mortality at Mata (Crockett & Rudran, 1987b). Infants under six months of age rarely survive male invasions (Rudran, 1979; Sekulic, 1983a; Crockett & Sekulic, 1984; Izawa & Lozano, 1991, 1994; Agoramoorthy, 1994b). As natal males mature, they become active and effective participants in aiding their father defend his position from incursive males, thereby also preventing the infanticide of group siblings (Pope, 1990). The majority of males eventually emigrate from their natal groups as subadults or young adults between four and six years of age, considerably older than the emigration age of females (two to three years; Crockett, 1984; Crockett & Pope, 1993).

As groups age in populations approaching stable population size, most groups eventually contain one or more subadult males. Number of subadult/adult males per group is correlated with mean group size (Crockett & Eisenberg, 1987), such that opportunities for successful invasion as a solitary competitor are reduced. As the opportunities for invasion become fewer, males tend to accumulate in their natal groups, often remaining well into adulthood (Pope, 1990). The percentage of multi-male and age-graded male groups in the four study populations was positively correlated with population density relative to stable population size. In the low-density, growing populations at Pinero and Gallery the majority of groups were still single male, whereas in the high-density, territory-saturated populations at Mata and El Frio only 0–14% of all groups were single male.

Males frequently form coalitions of two to four males that cooperatively invade a group, evict resident males, and collectively maintain group tenure. Coalitions form between both related and unrelated males, but those between related males last significantly longer ($\bar{x} = 8.2$ years) than those between unrelated males ($\bar{x} = 2.3$ years) and are more stable with regard to dominance status changes (Fig. 19.1; Pope, 1990). Observations of copulations in multi-male groups (Sekulic, 1983a; Pope, 1990), and paternity exclusion using genetic markers (Pope, 1990), indicate that all or most offspring in a multi-male group are fathered by only one of the males (i.e., in no case could the male genetically identified as fathering the majority of group offspring be excluded from fathering all). Asynchronous estrus (Crockett & Rudran, 1987a) and interbirth intervals of 16.6 months (Crockett & Sekulic, 1984) among the two to four adult females that occupy a typical group facilitate monopolization by the alpha male of each female as she comes into estrus.

A direct advantage of forming a coalition for both dominant and subordinate red howler males results from the superior competitive ability of coalitions over single males in establishing and maintaining: (1) tenure in the limited number of female groups in the population, and (2) territorial boundaries in saturated habitats (see above). No case of a single male being able to evict males from a multi-male or age-graded male group has ever been observed. Thus the number of groups that a single male would be able to invade without help becomes increasingly small as populations approach equilibrium density. Unrelated coalitions frequently form through single-male groups being joined by single outside males who then co-inhabit the group with the resident male (eight of the 18 unrelated coalitions described above: Pope, 1990). This was only observed in the Mata population, in which the majority of groups were already multi-male or age-graded male.

Although the mating success of a subordinate male in a coalition appears to be extremely low, his chances of establishing and maintaining membership in a group as a single male may be even lower. The relatively short duration of coalitions among non-relatives suggests that after some critical period of time in which no mating success has been achieved (e.g., successful contest for the dominant position), the subordinate male gives up and tries elsewhere. In coalitions in which males are related, the subordinate male will benefit from inclusive fitness, thereby mitigating some of the costs associated with cooperation. A dominant male that forms a coalition with a relative can expect to have longer reproductive tenure than in a coalition with a non-relative, thus producing more offspring and directly enhancing his fitness. Anecdotal evidence suggests that males choose relatives over non-relatives whenever possible (i.e., males and their sons evicting unrelated coalition partners when the sons reach subadult age: Pope, 1990). The small size of red howler groups, in conjunction with relatively long and staggered interbirth intervals, may preclude the possibility of males emigrating with sibs or half-sibs much of the time. However, the opportunity for a young male to leave his natal group with another male of comparable age should increase as group size goes up, thereby coinciding with those population conditions that favor male coalitions.

When the likelihood of obtaining reproductive status in a group after dispersal is low, a young male may increase his fitness more by helping his father to produce full and half-siblings than by trying to reproduce on his own, particularly when a father would be left as the only group male – thereby reducing his chances of being able to maintain tenure. In the highly saturated Mata habitat, groups with more than two breeding females that were reduced to a single adult male via emigration or death of adult/subadult males were usually invaded within days (sometimes hours) by males from neighboring groups. Subadult males are active and effective participants in group defense, in several cases helping their father to drive out males that invaded prior to their reaching maturity, or while they were gone on an exploratory foray (Pope, 1990). Prior to emigration young males typically make brief forays away from the group, sometimes joining an extra-group association for several days, or attempting to invade another group, after which they return to their natal group if they are unsuccessful. Thus young males frequently 'wait' in their natal groups for breeding opportunities to open up in the surrounding area (e.g., death or dispersal of a male). The majority of all male dispersal events involved immigration from an immediately adjacent group with no time spent in the solitary phase (Pope, 1989, 1990; see below). Red howler males that remain in their natal groups as sub-adults and adults are, in this respect, similar to males that remain as helpers at the nest in communally breeding birds.

A relatively old male will eventually lose his ability to win in aggressive encounters with younger, more robust animals. He may be able to increase his fitness more at this point by helping his son to establish and maintain breeding tenure than by trying to gain reproductive status in another group on his own. I observed two cases of a male inheriting a group from his father at Mata. Three suspected cases were observed at El Frio in which a subadult/large juvenile male upon reaching adulthood became dominant over the breeding male, with which he shared a genetically determined mean coefficient of relatedness of 0.4. All former breeding males exhibited heavily worn teeth, consistent with older age. In all five cases, the son became dominant and the father remained in the group as a coalition partner.

Differences in dispersal distance for males and females

Resources that limit reproduction are distributed differently in space for each sex. Because they are unable to transfer into established groups, females that emigrate must find suitable unoccupied territory that contains adequate food resources for reproduction, or extra-group male(s) that enable them to carve-out a territory in areas of home range overlap. Males must successfully take over and defend a group of females from other males. Those females may either be members of an established group, or an association of extra-group females. Differences in dispersal distance and ranging patterns of extra-group males and females correspond to differences between the distribution of unoccupied forest suitable for colonization, and the distribution of groups of females.

At both Mata and El Frio study areas, wherein population density was approaching equilibrium, very little unoccupied forest was available. Group territories were either completely bordered by those of other groups, or bordered by a combination of other groups and forest edge. Mean home range size was approximately 7 ha at Masaguaral and 10 ha at El Frio. Thus for males, potential breeding opportunities were located every 300–500 m (i.e., one home range diameter), whereas emigrant females would either have to travel outside the study area or attempt to carve a territory out of areas of overlap between existing home ranges. Radio-collared, solitary females at Mata consistently ranged over a three to five times greater area than solitary males during each month of telemetry monitoring (Fig. 19.2a; Pope, 1989). Mean cumulative area covered per month increased steadily from 35 ha to 160.5 ha over four months for three solitary females, and from 6.9 ha to 45 ha for four solitary males during the same period. Mean displacement distance per day was significantly larger for solitary females (316 m) than for solitary males (208 m). All three of the radio-collared males whose natal groups were known ranged over small areas that included all or most of their natal home range and the area immediately surrounding it. This behavior was observed in eight additional ear-tagged males, six of which formed new groups with other solitary migrants in areas immediately adjacent to their natal home ranges. The majority of females that emigrated disappeared from the study area (Crockett & Pope, 1993). None relocated near her place of birth (see below). The only radio-collared female of known natal origin spent most of her time located approximately 1 km away from her natal group, where she died after 4.5 months of observation, apparently due to a pathogen (Pope, 1989).

The majority of males move next door when they emigrate. Mean dispersal distance per generation for immigrants of known origin was one home range diameter for both natal transfer and subsequent breeding transfer (Fig. 19.2b). The percentage of males that were individually identifiable (75% at Mata and 71% at El Frio) was similar to the number of immigrants that were identified (67% at Mata and 70% at El Frio). Thirty-two percent of males remained in their groups long enough potentially to breed with their mother or daughter and had a per generation dispersal distance of zero. At Mata, 86% of immigration events for which the exact time of natal emigration was known involved immediate transfer with no time spent in the costly extra-group stage (Pope, 1990).

In contrast to males, only 10% of recognizable females that emigrated from their natal groups transferred into new groups within the Mata. Only five females could be relocated in either extra-group associations or new groups. Mean dispersal distance for these females from place of birth to place of first attempted reproduction (associations did not bear offspring) was six home range diameters (Pope, 1992). As the majority of females that emigrated either died or relocated at much farther distances, this is considered a mean minimum dispersal distance. At El Frio, only two natal females reached reproductive age while the population was at equilibrium during the first two years of the study. Both remained in their natal groups (one of which already contained one adult female and the other two adult females) and bred.

Female dispersal patterns change with population growth state. Both Pinero, during the first two years of study (Pope, 1998), and Gallery (Crockett, 1985, 1996; Pope, 1992) were undergoing population growth through a high rate of new group formation accompanied by an increase in group size. Both populations had undergone recent declines, but Gallery was approximately five years further along in the recovery process and exhibited a different age structure from Pinero. Very few juveniles in the medium to large age classes were present at Pinero, and two females emigrated from the 16 study groups during this relatively brief period of expansion. At Gallery, 10% more natal females bred in their natal groups than at Mata (Crockett & Pope, 1993). New groups at Gallery had a much higher success rate (93%, Crockett, 1996) than in the saturated habitats at Mata and El Frio (50% and 0%, respectively; see above), such that females immigrating into low-density areas had higher reproductive potential than those in saturated habitats.

The social consequences of population decline depend on how mortality is distributed across age–sex classes. Both Pinero and El Frio underwent a population decline during the last four years of study (Pope, 1998), apparently as a consequence of an unidentified pathogen that also affected the Mata and Gallery at Masaguaral (Rudran, personal communication). At Pinero males and females disappeared at essentially the same rate for all ages, and most groups went extinct before any females matured. At El Frio, mortality during the first two years of decline was strongly biased toward adult females. Seventy-two percent of adult females died in this

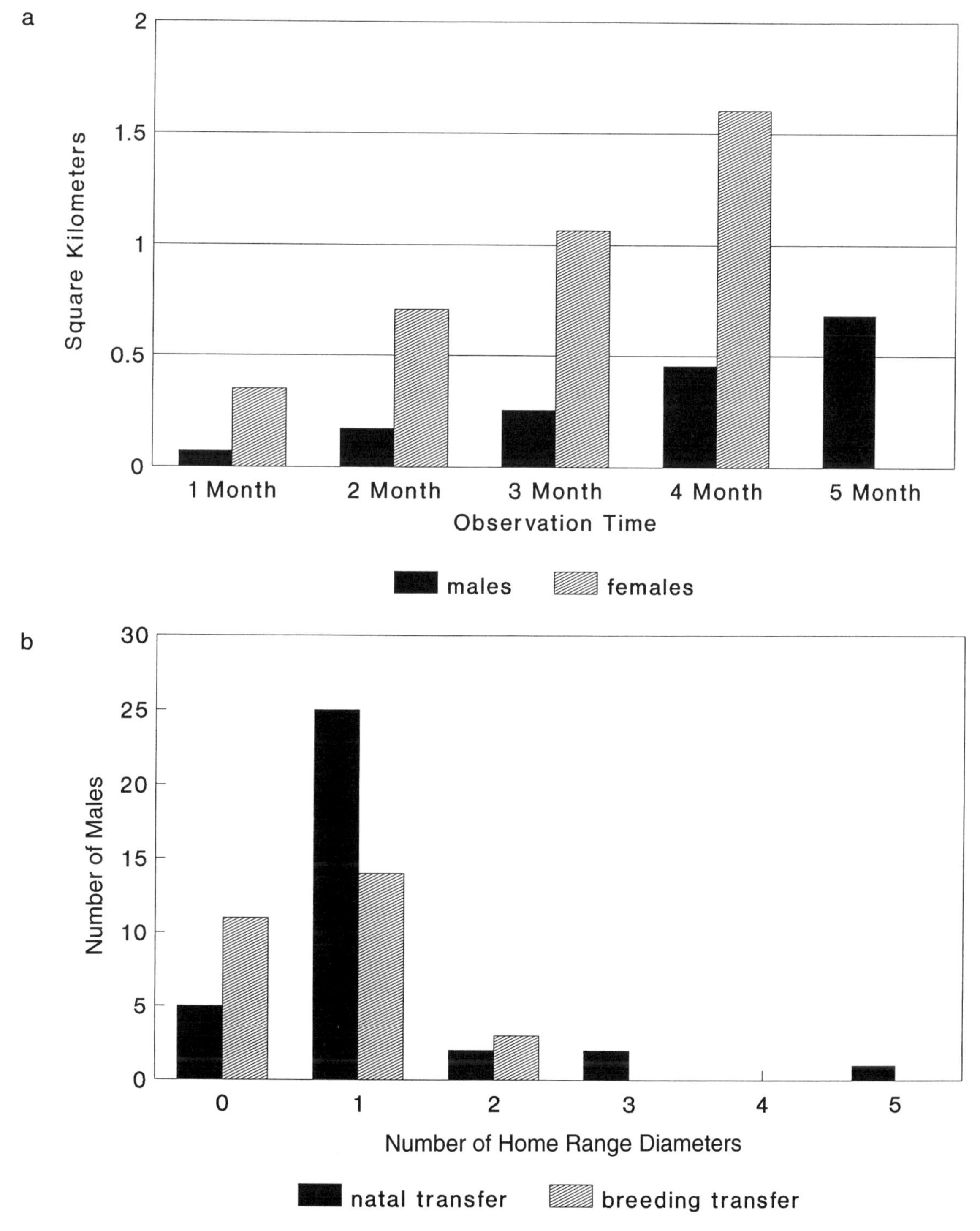

Fig. 19.2. (a) Mean cumulative ranging area increase as a function of observation time for solitary, radio-collared red howler males and females. (b) Distances traveled per generation by known red howler immigrant males from their group of origin. Breeding transfer refers to males transferring from non-natal groups. Zero dispersal distance was assigned to 'non-emigration' events in which males remained in a group long enough potentially to mate with their mother or daughter (i.e., longer than one generation; see text).

time span, leaving most groups with only one or zero adult females. All eight females that matured during this period remained in their natal groups, five in groups with zero adult females, and three in groups with one. All age–sex classes were affected equally during the last two years of the study, in which 40% of the population was eventually lost (Pope, 1998). The coefficient of relatedness among group adult females, however, remained the highest of all four populations (0.43).

In summary, although both sexes emigrate, differences in patterns of intrasexual competition result in large intersexual differences in vagility. For males, familiarity with what is happening in local groups and extra-group associations is important for the successful takeover of groups of females. The majority of males sit and wait either in their natal group or as a floater in and adjacent to their natal territory, some of which subsequently form a new social group with migrant females that exhibits overlap with the natal territory. There is no evidence that they are forced from their natal groups, in which they frequently remain well into adulthood, and to which they return after failed attempts to establish membership in other groups. Female emigration is frequently accompanied by aggression from group females. In order to reproduce successfully, emigrant females require suitable habitat unoccupied by other groups. Consequently, they usually have to travel much farther than males to find reproductive opportunities, particularly in an already saturated habitat. The majority of males that emigrate transfer into adjacent groups, whereas the majority of emigrant females range over large areas and eventually disappear from the study area.

CALLITRICHINES: GOLDEN LION TAMARIN AND SADDLEBACK TAMARIN

Intrasexual competition for group membership and patterns of dispersal displayed by red howlers are most closely resembled among the platyrrhines by those reported for members of the Callitrichinae (marmosets and tamarins). The two taxa are highly divergent ecologically. Whereas the folivorous howlers are among the largest of New World primates, with males weighing up to 8 kg (Pope, unpublished data), the primarily insectivorous–frugivorous–gummivorous callitrichines are among the smallest of primates, weighing between 110 g and 750 g.

Callitrichines are the only anthropoid primates that typically bear twins instead of singleton offspring. A litter weighs from 14% to 25% of the mother's weight at birth, and up to 50% of her weight at weaning (Goldizen 1987b). Most species exhibit a cooperative breeding system in which usually only one female reproduces, regardless of the number of other adult females in the group, and adult males, non-reproductive females, and offspring of previous years share extensively in the carrying of offspring (Goldizen, 1987b). Those species for which extensive field studies are available exhibit highly flexible mating systems that include monogamous pairs, polyandrous groups containing one breeding female and multiple males that copulate with her, and multi-male/multi-female groups that can be facultatively monogamous, polyandrous, or more rarely polygynous/polygynandrous – in which more than one female bears a litter (Goldizen, 1987b; Rylands, 1996). Although two or more of these patterns are frequently exhibited by groups in a single population (Ferrari & Ferrari, 1989; Goldizen & Terborgh, 1989; Garber *et al.*, 1993a; Baker & Dietz, 1996), the modal group composition appears to be primarily affected by demographic factors, specifically habitat saturation and age of groups (Baker *et al.*, 1993; Garber *et al.*, 1993a; Goldizen *et al.*, 1996). In saddleback tamarins there is evidence that monogamous pairs without offspring of previous years to act as helpers have low probability of successfully raising twin offspring, such that new and young groups should contain more than one adult male (Goldizen, 1987a, 1987b). Breeding opportunities within groups are limited for both sexes, and both males and females emigrate from their natal groups (Garber, 1994). As habitat becomes increasingly territory saturated and population density increases, both the possibilities for new group formation and reproductive opportunities in surrounding groups decrease, at which point males and sometimes females begin to accumulate in their natal groups as they wait for breeding vacancies to arise (Goldizen & Terborgh, 1989; Baker *et al.*, 1993; Garber *et al.*, 1993a).

Long-term population data on dispersal patterns of known individuals are available for only two species: the golden lion tamarin (Baker & Dietz, 1996), and the saddleback tamarin (Goldizen & Terborgh, 1989; Goldizen *et al.*, 1996). Both populations were territory saturated, such that breeding opportunities were primarily limited to deaths or disappearances of breeding individuals in established groups. These populations were thus demographically comparable to the Mata and El Frio populations described above for red howlers. Mean group sizes were 5.13 for saddleback tamarins and 5.4 for golden lion tamarins. Patterns of migration and intrasexual competition for group membership are summarized below for each sex.

Males

Both species exhibited single-male and multi-male groups in the study population, with multi-male groups predominating (70% for golden lion tamarins: Baker *et al.*, 1993; approximately 78% for saddleback tamarins: Goldizen *et al.*, 1996, their Appendix I). Both males in multi-male groups copulate with the breeding female and subsequently help to rear her infants, but Baker *et al.* (1993) demonstrated that dominant golden lion tamarin males performed almost all copulations when she was likely to be fertile. As for most callitrichines, it is not known to what extent saddleback tamarin males share paternity in multi-male groups, but mate guarding of up to several days has been observed in wild populations (Soini, 1987b; Goldizen, 1989), and in captive groups subordinate males exhibit very little sexual behavior (Abbott *et al.*, 1993). Dominance relationships between group males have been reported for all four callitrichine genera (Epple, 1972, 1975; Abbott, 1984; Soini, 1987b; Baker *et al.*, 1993), providing a potential mechanism for monopolization of paternity (Baker *et al.*, 1993). Evidence of physiological reproductive suppression of subordinate males by dominants has been found in common marmosets (*Callithrix jacchus*), in which low levels of sexual behavior by subordinate males were accompanied by endocrine deficiencies (Abbott *et al.*, 1993), and in moustached tamarins (*S. mystax*), for which Garber *et al.* (1996) found large differences in testes volume of up to 174% between males in multi-male groups.

The circumstances surrounding male replacement are better documented for golden lion tamarins than for saddleback tamarins, but were rarely observed for either species. Immigration following deaths of breeding males was believed to account for some unknown proportion of immigration events in both populations. Three out of the 19 male immigration events that occurred in the golden lion tamarin population were consistent with aggressive male takeovers: two involved apparent eviction of single resident males by male coalitions, and one an eviction by a single male of a single resident that resulted in the latter's death. Five additional immigration events involved single males immigrating into groups already containing a resident breeding male, after which both males co-inhabited the group for variable lengths of time. Thus 16% of immigration events were consistent with male takeover, 26% involved immigration into groups already containing a breeding male, and 58% involved replacements in which it was unknown whether the disappearance of the resident males occurred before or after immigration. Male floaters and extra-group associations that appeared near territory boundaries in the golden lion tamarin population were actively chased by resident group males (15/21 encounters), suggesting that direct competition among males for group membership operates as a strong deterrent to male immigration (Baker & Dietz, 1996).

In the saddleback tamarin population, 7/27 (26%) of immigrant males listed in Table IX of Goldizen *et al.* (1996) entered groups already containing a resident male, and four males (15%) were found as singletons and 16 males (59%) as duos (eight pairs) in groups in which previous male breeders were not present. Four of the eight duos were actually colonizing new social groups, as opposed to immigrating into established groups. The other four pairs and the singletons apparently replaced previous resident males, but whether the resident males disappeared before or after the immigration events is not reported. Interactions with extra-group males were not described.

Cooperative male duos consist of both related and unrelated individuals in both species. Polyandrous male pairs in the saddleback tamarin population consisted of both suspected relatives (father–son and sibs), and probably unrelated duos resulting from an immigrant male co-inhabiting a group with the resident male (Goldizen *et al.*, 1996; see above). At least two and possibly four new groups were formed by pairs of unrelated males. Of the eight multiple male immigration events observed in the golden lion tamarin population, seven of the immigrating pairs were close relatives (sibs or parent–offspring), whereas the five immigration events followed by co-inhabitancy with the resident male probably resulted in non-closely related male pairs. Seven of the 17 male duos of known relatedness listed in Table I of Baker & Dietz (1996) are related and ten are unrelated. These data excluded males that remained in their natal groups into adulthood, which accounted for about a third of multi-male groups. As in red howlers, related male coalitions lasted nearly three times longer ($\bar{x}$= 13 months) than those between unrelated males ($\bar{x}$=4.8 months), and appeared to be more stable: 60% of unrelated coalitions ended when one male emigrated, an event that did not occur in the related duos. Four (57%) of the related coalitions were intact and ongoing by the end of the study, whereas only one (10%) of the unrelated coalitions remained.

Wandering in search of a breeding opportunity was demonstrated to be the least successful reproductive tactic for male golden lion tamarins (Baker *et al.*, 1993). The predominant tactic for both golden lion and saddleback tamarin males appears to be one of remaining in the natal group or accepting subordinate status in another group while waiting for breeding vacancies to open (Goldizen *et al.*, 1996). The majority of breeding vacancies in both populations were filled by males from immediately adjacent groups. In the saddleback tamarin population, 37.5% of natal males disappeared from their natal group without having been observed copulating, 37.5 % dispersed into another study group, and 12.5% bred or attempted to breed in their natal groups and were never observed to transfer into another group. Goldizen *et al.* (1996) considered this to be an underestimate of the number of males breeding in their natal groups. The majority of natal males (67%) that dispersed into other groups moved one home range diameter, and the remainder moved two to three home range diameters away (Goldizen & Terborgh, 1989; Goldizen *et al.*, 1996). Dispersal distance was not quantified for golden lion tamarins, but of the eight replacement events and five non-replacement events described, ten (77%) involved immigration from adjacent groups (Baker *et al.*, 1993; Baker & Dietz, 1996). Natal males were rarely evicted from their birth group, and emigration was apparently voluntary.

Females

The majority of both golden lion and saddleback tamarin groups contain only one breeding female, with multi-female groups usually comprised of mother–daughter pairs (Dietz & Baker, 1993; Goldizen *et al.*, 1996). In most callitrichines, the number of females breeding in a group at the same time is limited by a highly specialized behavioral–endocrine mechanism in which the ovarian cycle of subordinate females is suppressed by the presence of the alpha female. Dominant golden lion tamarin females do not appear to be able to control reproduction in this manner (Abbot *et al.*, 1993), but rather limit group size by aggression and low intrasexual tolerance (see below). Mother–daughter polygyny in golden lion tamarins occurred in 10% of social groups, and typically occurred when new males immigrated into groups containing adult mother–daughter pairs (Dietz & Baker, 1993). The daughter was usually evicted after one breeding season by her dominant mother (Dietz & Baker, 1993), but four cases were observed of a daughter inheriting the primary breeding position after the disappearance of her mother (Baker & Dietz, 1996). Polygyny was significantly correlated with the amount of swampland in the territory, suggesting territory quality may be involved in females allowing their daughters to breed (Dietz & Baker, 1993). Saddleback tamarin groups contained more than one adult female in only 15 of 47 group-years, and all but one of the duos included a natal female (Goldizen *et al.*, 1996). The oldest female was the primary breeder in all multi-female groups, although six cases of the secondary female breeding were observed. Four of these were mother–daughter pairs and two were of unknown genetic relationship. Of the seven to eight (23.3–26.7%) saddleback tamarin females that bred within their natal groups, five to six inherited the primary breeding position.

Female emigration and immigration patterns are distinctly different from those of males. Whereas males frequently dispersed with same-sex partners and exhibited both natal dispersal and subsequent breeding dispersal, females of both species dispersed alone and were not observed to change groups more than once in their lifetime (Baker & Dietz, 1996; Goldizen *et al.*, 1996). Female emigration in golden lion tamarins was usually preceded by aggressive eviction by the reproductive female, and natal females not closely related to the primary female were expelled at a very early age (Baker, 1991). As in red howlers, resident group females were extremely aggressive toward extra-group floater females, chasing them in 81% of encounters when they appeared at territory boundaries (Baker & Dietz, 1996). No female was ever successful at joining a group already containing a reproductive female (Dietz & Baker, 1993). In 70.4 group-years of observation, only five females immigrated into established groups (21% of all immigration events: Baker & Dietz, 1996). All involved filling a breeding vacancy created by the disappearance of the reproductive female; none resulted in polygyny. In the saddleback tamarin population, 11 of the 23–24 females that emigrated from their natal groups immigrated into other study population groups (Goldizen *et al.*, 1996). As in golden lion tamarins, the majority of these immigration events (80%) were into groups that contained no other female immigrants or previous female breeders. The remaining two immigration events involved replacement of the resident female by two immigrants that were both present at the next census, but the circumstances were not observed.

Approximately 76% of golden lion tamarin females and 52–54% of saddleback tamarin females that emigrated from their natal groups either died or disappeared without relocating into a group in the study population (Dietz & Baker, 1993; Goldizen *et al.*, 1996). Opportunities for group membership for females of both species were far more limited than those for males. This was due to: (1) the inability of females to transfer into groups containing other same-sex individuals; (2) the apparent inability of females to evict resident females; and (3) the fact that the majority of groups in both populations contained two or more adult males and only one adult female. In both species males accounted for the preponderance of immigration events. In golden lion tamarins, far more natal males than females transferred into local groups (Baker & Dietz, 1996). This proportion was nearly equal for saddleback tamarin natal males and females, but more females than males disappeared from the study population at the age at which most individuals seek breeding positions (Goldizen *et al.*, 1996). Most dispersing females from both study populations were unable to immigrate into established groups. A better success rate among saddleback tamarin females was associated with a relatively high rate of new social group formation in the population (see below).

Although females tolerated adult daughters at least some of the time, same-sex tolerance of natal individuals was generally lower for females than for males. This would preclude a sit-and-wait tactic for most natal females, but regular encounters with surrounding groups are probably important for assessing local breeding opportunities, and probably account for the fact that the majority of breeding vacancies are filled by females from adjacent groups (Goldizen *et al.*, 1996). In contrast, lower mortality and multi-female groups prevented this from being a frequent option for red howler females, and it was observed primarily in declining populations in which groups had been reduced to one adult female (see above). The majority of red howler females that relocated in the study population transferred into new social groups. This was also the case in the saddleback tamarin population, in which approximately 70% of female immigration events were into groups that were newly formed since the study began (Table IX, Goldizen *et al.*, 1996). Newly formed groups appeared to have a higher frequency of female vacancies than long-established groups (Tables V and IX, Goldizen *et al.*, 1996), suggesting that, as in red howlers, newly formed groups are generally more socially unstable. In red howler monkeys this was in part due to the fact that new groups in the saturated habitats were limited to settling in marginal territories. This may have been a contributing factor in the saturated saddleback tamarin study area, but no data are available.

New group formation

As in red howlers, new groups form in both tamarin species when extra-group floaters meet, establish social bonds, and attempt to establish a territory (Goldizen & Terborgh, 1989; Baker *et al.*, 1993; Goldizen *et al.*, 1996). Evidence from saddleback tamarins suggests that territories are established by coalitions of males that subsequently attempt to attract a breeding female (Goldizen *et al.*, 1996). Both long-term tamarin studies report a relatively high rate of new social group formation (Baker *et al.*, 1993; Goldizen *et al.*, 1996). Because almost no unoccupied habitat was available in the golden lion tamarin population, newly formed groups usually had relatively small, difficult to defend territories located between established groups (Baker *et al.*, 1993). It was hypothesized that due to their small size, these groups were more vulnerable to predation and would be less successful in intergroup competition, but no data were reported on the relative persistence of new versus established groups. At least four new groups formed in the saddleback tamarin population (approximately 36% of all study groups that were described, Goldizen *et al.*, 1996). Group membership and persistence appeared to be highly changeable for most of the groups, with at least three groups (27%) becoming extinct and one experiencing a complete replacement of all members in less than a year.

COMMON FEATURES OF RED HOWLER, GOLDEN LION TAMARIN, AND SADDLEBACK TAMARIN MATING SYSTEMS

Patterns of intrasexual competition and dispersal for red howlers, golden lion tamarins, and saddleback tamarins exhibit many common features, which are summarized in Table 19.2. All three species are territorial, with strong intrasexual competition for group membership resulting in large variances in reproductive success for both males and females. Breeding sex ratio within groups is highly variable and depends primarily on the degree of territory saturation and group age. New groups are formed when migrant extra-group males and females establish social bonds, successfully defend a territory, and produce offspring. New group males are more likely to be related than females.

Resident breeding females limit reproduction, thereby controlling group size. Emigrant females are aggressively prevented by resident females from immigrating into established groups in red howler monkeys and golden lion tamarins. In saddleback tamarin groups, reproduction is hormonally suppressed in subordinate females, such that there is usually only one breeding position in any given group; no females were observed to immigrate into a group

Table 19.2. *Summary comparison of patterns of intrasexual competition and dispersal in* Alouatta seniculus, Leontopithecus rosalia, *and* Saguinus fuscicollis

	Females	Males
Intrasexual competition	Territoriality Female alliances *S. fuscicollis*: reproductive suppression	Territoriality Male alliances Aggressive replacement Breeding dominance hierarchy
Natal emigration	Common: density dependent	Usually
Breeder replacement	Daughters	Usually immigrants
Immigration	Few opportunities: rare	Common: density dependent
Dispersal distance	Depends on possibilities for new social group formation Most disappear from population	1 home range diameter ($\bar{x}$)
New group formation	Floater males and females establish social bonds and attempt to defend a territory	

already containing a resident breeding female. Female intrasexual competition limits the extent of polygyny possible for males in all three species. Red howler females are highly successful at evicting floater females from their territory even when group males are actively attempting to mount the floaters (Pope, unpublished observations), despite a high degree of sexual dimorphism. Extra-group golden lion tamarin females can actually jeopardize resident male reproductive success if they succeed in replacing a pregnant or lactating female (Baker & Dietz, 1996). All cases of male golden lion tamarins chasing floater females occurred when the resident female was either pregnant or probably pregnant.

Breeding opportunities for emigrant females are therefore rare in territory-saturated populations in which the possibilities of new group formation are low. Adult daughters are often tolerated by resident females when they can be an asset to reproduction. Daughters of both callitrichines participate in cooperative infant care, and may be critical to infant survival in monogamous groups (Goldizen & Terborgh, 1989). In red howler monkeys and golden lion tamarins, they can form competitive alliances with their mothers that operate in the exclusion of extra-group females. The number of breeding daughters that are tolerated, however, is limited (one for the callitrichines, two to three for red howlers), such that the majority of natal females eventually emigrate. Events preceding female emigration in saddleback tamarins have not been reported, but both golden lion tamarin and red howler monkey natal females are often forcibly evicted by non-related resident females. *L. rosalia* is the only one of the three species in which there is evidence that mothers regularly evict their daughters. The threat of reproductive competition from daughters is potentially higher in golden lion tamarin than in saddleback tamarin groups due to the inability of the dominant female to physiologically suppress the reproduction of helper daughters; thus they may be less tolerated in golden lion tamarins. In all three species, breeder replacement is usually by adult daughters.

Emigrant females probably have very low reproductive potential unless they can transfer relatively quickly into another group. Their ability to do so depends on population growth and demography. Higher mortality rates and the preponderance of one-female groups led to the more frequent occurrence of female breeding vacancies in the callitrichine populations than in those of red howlers. In both the saddleback tamarin and the red howler monkey populations, nearly all female immigration was into recently formed social groups, indicating that female breeding opportunities must increase with the rate of new social group formation. Successful group formation was dependent primarily on population growth state in red howler monkeys, but in the saddleback tamarin population a high rate of group formation appeared to be associated with a high rate of group extinction (i.e., group turnover) in a population that was believed to be saturated. In saturated golden lion tamarin and red howler monkey populations in which the rate of successful new group formation was low, the large majority of emigrant females disappeared from the study population. In the saddleback tamarin population, slightly more than half of the emigrant females disappeared, and the majority of those that remained immigrated into social groups formed after the beginning of the study. In all three species, newly formed groups are less stable than established groups, suffering higher extinction rates and member turnover.

Unlike females, males can transfer into established groups via aggressive eviction of resident group males or by accepting subordinate status in groups already containing a resident male. They frequently form alliances with other males that cooperatively defend group territories, and in red howler monkeys and golden lion tamarins these coalitions aggressively evict resident group males and then defend the group from subsequent incursions by challenging males. This may also occur in saddleback tamarins, but the circumstances of male replacement have not been observed in this species. These alliances are necessary for successful competition in saturated populations, aid in the prevention of infanticide in red howler groups, and increase reproductive success in callitrichine groups through cooperative paternal care of offspring. In the two species in which paternity has been assayed (*L. rosalia* and *A. seniculus*), the dominant male fathers most or all of the offspring. No paternity data are available for saddleback tamarins, but field observations of mate-guarding and highly skewed copulation frequencies in captive groups suggest that paternity is not equally shared among group males. In all three species, subordinate males in saturated populations probably have the highest chance of reproductive success by waiting as a non-reproductive group member for breeding positions to become vacant in a surrounding group, rather than dispersing in search of reproductive opportunities.

Coalitions are formed between related and unrelated males, but those between related males are more stable and last longer. Subordinate males in related coalitions increase their inclusive fitness when they contribute to the survival of group offspring, either through direct care or by defending them from infanticidal males. The predominant reproductive tactic for natal males in saturated habitats appears to be one of remaining in the natal group or accepting subordinate status in another group while waiting for breeding vacancies to open. Many males remain within their natal groups well into adulthood, sometimes breeding there, and there is no evidence of eviction. Consequently, a large proportion of natal males breed within the study population, and the majority of them transfer into immediately adjacent groups.

THE EVOLUTION OF PHILOPATRY IN PLATYRRHINES

All three species exhibit variable social group structure that can range from monogamous family groups to polygamous

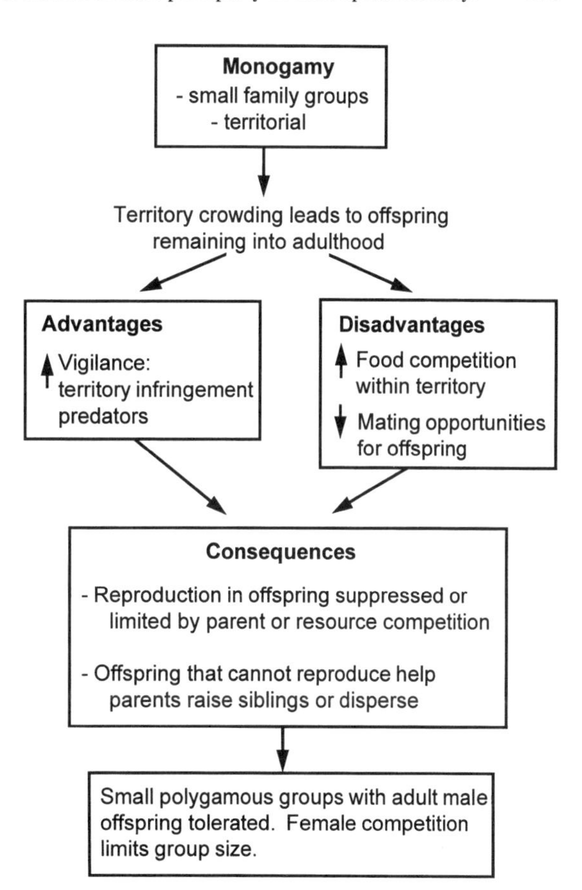

Fig. 19.3. Model depicting transition between monogamy and small group polygamy in hypothetical early platyrrhines.

multi-male/multi-female groups, depending on demographic conditions. The ecological transition observed between these group structures suggests a mechanism whereby these complex mating systems evolved from the monogamous family groups hypothesized to be ancestral in platyrrhines (Fig. 19.3). Both sexes emigrate from their natal group in monogamous species, but in saturated habitats it becomes increasingly advantageous for offspring to delay dispersal as they wait for breeding vacancies to arise within other groups in the population. Advantages also accrue to parents in the form of increased vigilance for predators and superior competitive ability against surrounding groups. Disadvantages include decreased mating opportunities for offspring and increased competition for food within the territory. Reproduction in offspring is suppressed or limited by the parent, by incest avoidance, or by resource competition. Offspring that cannot reproduce either remain and help parents raise siblings, or disperse in search of long-distance breeding opportunities. Ultimately, this can result

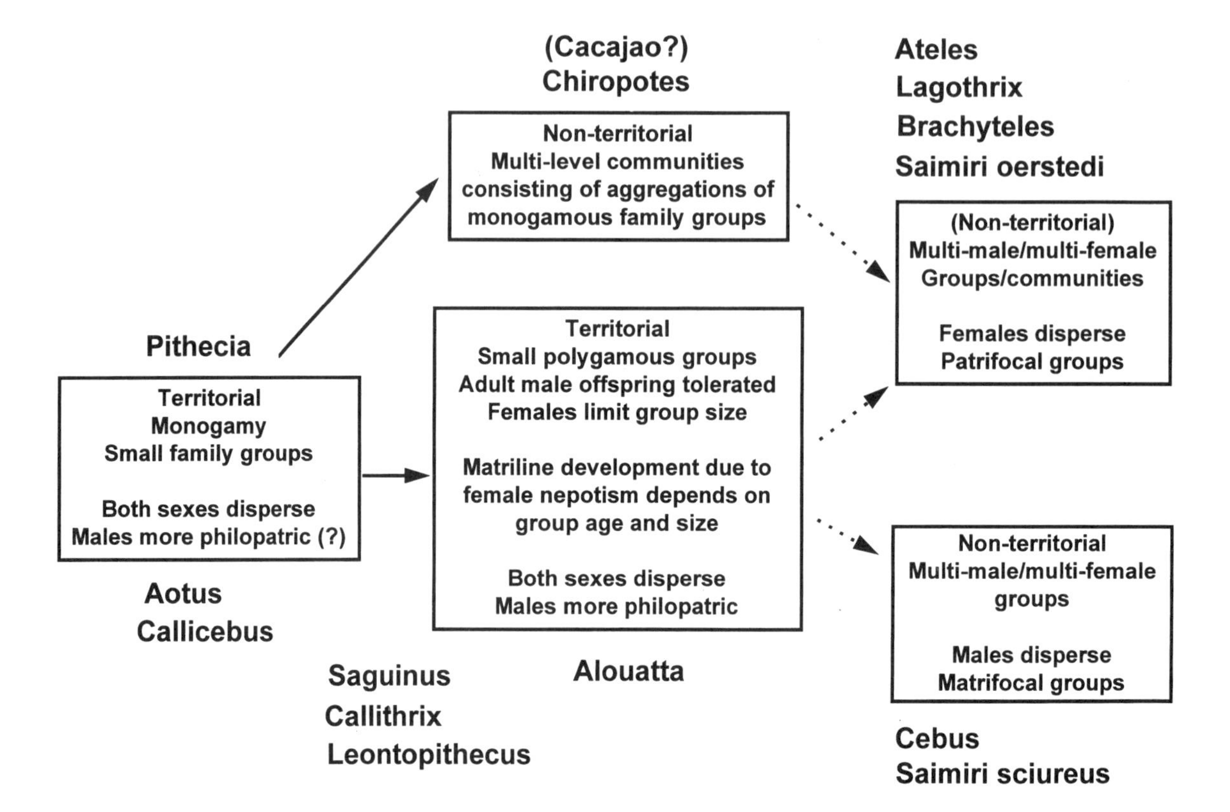

Fig. 19.4. Proposed modes of transition between neotropical monkey mating systems, assuming territorial monogamy/small group polygamy as ancestral condition (see text). Examples of extant species exhibiting each type of modal breeding pattern are mapped onto the flow diagram. Not intended to imply phylogenetic relationships (see text).

in small polygynous groups in which adult male offspring are tolerated and female competition limits group size by controlling the number of daughters allowed to breed. Territories are limited in size by factors such as daily path length, body size, and adjacent territories, thereby imposing concomitant limitations on group size. The reproductive success of females is more likely than that of males to be limited by feeding competition, so females suffer the greatest costs when group size exceeds territory carrying capacity. If aid from offspring increases or is required for reproductive success, sons are potentially less costly than daughters to the breeding female because sons cannot increase group size through the birth of competitors. This is a typical demographic scenario of the type proposed for the evolution of cooperative breeding in many birds (Stacey, 1979; Koenig & Pielka, 1981). Note, however, that cooperative care of offspring is not required for the final outcome. There simply needs to be an advantage to the parents for the offspring to remain in the group (e.g., red howler natal males).

Territoriality is what ultimately mediates the effects of demographic change on group social structure in the social systems described above, and is the most important constant influencing their similarities. Diet *per se* is not as important as whether territoriality is an economically feasible form of competing for dietary resources. Territorial monogamy or small group polygamy such as described here is the most widespread mating system among platyrrhines, and is present in most of the major clades (Fig. 19.4). Territorial monogamy characterizes at least three platyrrhine genera: *Aotus*, *Callicebus*, and *Pithecia*. Very little is known about the dispersal characteristics of any *Aotus* or *Callicebus* species, but a recent long-term study of monk saki (*Pithecia monachus*) suggests that males are more philopatric than females (Soini, 1995b). Males remained in their natal groups longer than females, sometimes well into adulthood. Known and suspected sons often settled at the periphery of their parents' territory, where they formed new groups with solitary females. Adjacent groups sometimes foraged together peacefully in the same tree, and Soini suspected that these were usually parent–son groups.

Although both males and females disperse in these mating systems, there is a clear pattern of male philopatry and female dispersal or disappearance from the study area. This pattern is also typical of territorial monogamous and cooperative breeding birds, and has been proposed to be the expected outcome of male resource defense mating systems (Greenwood, 1980). Very few data are available on effective dispersal distances of males and females in other species of monogamous primates, or on how new pairs are established on territories. A summary of dispersal and family formation patterns in five species of territorial monogamous gibbons indicated that more males remained in or adjacent to their natal territory than females (Leighton, 1987, Table 12–4), but lack of data on the proportion of the total subadult population represented for each sex makes these results difficult to evaluate. Among solitary gibbons, only males have been observed to call and defend a territory . The most successful means of group formation in Kloss's gibbon (*Hylobates klossi*) occurred when parents assisted offspring in establishing adjacent territories through budding-off of the natal territory, similar to the pattern described above for some red howler males. Most gibbon studies, however, have reported very limited breeding opportunities for both sexes due to territory saturation with long-lived pairs (Leighton, 1987; Mitani, 1990; Sommer & Reichard, Chapter 14).

The pattern of differential philopatry between the sexes described above will ultimately result in higher mean relatedness among males within a given population than among females. If this is indeed the primitive condition, then it may be responsible for the fact that a shift to a different pattern of resource utilization was accompanied by the development of patrifocal as opposed to matrifocal mating systems in the Atelidae and at least one species of *Saimiri* (Fig. 19.4). Species in these taxa typically exhibit a pattern of resource use involving large home ranges in which territoriality is not feasible and large multi-male/multi-female groups frequently display varying degrees of fusion–fission behavior (Strier, 1994a, 1996a).

A somewhat different social system seems to characterize the Pitheciinae genera *Chiropotes* and possibly *Cacajao* (Fig. 19.4). Groups of 10–30 are typical, but *Cacajao* can occur in large aggregations of up to 100. Groups typically contain equal numbers of adult males and females (Robinson *et al.*, 1987). White-nosed saki (*Chiropotes albinasus*) groups have been observed to segregate spatially into subgroups of a single male, a single female, and several juveniles that moved and fed together in home ranges of 200–350 ha, suggesting more or less permanent aggregations of monogamous family units. In this case, a shift away from the territoriality exhibited by the monogamous *Pithecia* to a different type of resource utilization may have been accompanied by an increase in the feeding tolerance among adjacent groups such as was observed in monk sakis (Soini, 1995a). This would be facilitated by the patrifilial relationships between males in adjacent groups, and would result in patrifocal social aggregations.

Cebus and at least one species of *Saimiri* (the common squirrel monkey *S. sciureus*) are the only platyrrhine genera to exhibit matrifocal social groups, in which nearly all females remain to breed in their natal group and all natal males emigrate. Members of both taxa typically live in multi-male/multi-female groups that range over large, undefended home ranges. Clues as to how this pattern may have arisen can be found in intrageneric differences in patterns of intrasexual competition in *Alouatta*, *Saimiri*, and *Saguinus*.

In some callitrichine species with female reproductive suppression it appears that female immigration is tolerated at least occasionally by the resident female(s) (e.g., cottontop tamarins, *Saguinus oedipus*: Savage *et al.*, 1996a; moustached tamarins, *S. mystax*: Garber *et al.*, 1993a), after which the immigrant may attempt to compete for the reproductive position (cottontop tamarins), or assume a beta position and presumably wait for the breeding position to become vacant in the resident or a neighboring group (cottontop tamarins and moustached tamarins). The high-density, saturated moustached tamarin population contained up to four adult females per group. Evolution of female reproductive suppression may thus relax constraints on female immigration and the number of adult females tolerated per group if additional females can be reliably suppressed and will help care for offspring. As in red howlers, relatives should be preferred over non-relatives when available. Future research would be well directed toward discovering how this phenomenon is influenced by factors such as population density and the number and relatedness among adult/subadult females already residing in the recipient group.

With the exception of mantled howlers (*Alouatta palliata*), group composition and the little that is known about dispersal and intrasexual aggression in other species of howlers are very similar to those of red howlers. Mantled howler patterns of intrasexual competition differ from those of red howlers, resulting in very different population and group kin structures. Red howler females cooperate with non-related females when forming new social groups, after which nepotistic behavior among group females rapidly selects for closely related matrilines. In this regard they more closely resemble *Cebus* and common squirrel monkeys than the Atelidae to which they are related. In mantled howlers groups are much larger, with eight to ten adult females per group. All females disperse from their natal groups and transfer into other groups as singletons, such that all group females are immigrants (Glander, 1992). Whereas in red howlers the majority of female competition occurs between cooperative coalitions that are generally successful at excluding other females from their territory, in mantled howlers the majority of female aggression occurs between females within groups, in which there is a pronounced dominance hierarchy (Jones, 1980; Glander, 1992). Females defend their territory from surrounding groups and there is considerable aggression directed toward immigrant females. Unlike red howlers, this aggression is not coordinated and not effective at preventing transfer (Glander, 1992). Males more frequently exhibit philopatry than females, with sons sometimes inheriting groups from their fathers (Glander, 1992). Thus mantled howler groups more closely resemble those of their ateline relatives, with two to four adult males often (but not always) born in or near the group, while all females are unrelated immigrants.

A similar divergence occurs in *Saimiri*. While the Central American squirrel monkey (*S. oerstedi*) exhibits patrifocal social groups similar in structure to those of the Atelidae (Boinski & Mitchell, 1994), the common squirrel monkey exhibits matrifocal groups . In fact, males are doing similar things in both species. Common squirrel monkey males form cooperative migration alliances within their natal group that leave and transfer together over multiple migrations between groups, and that support each other in aggressive interactions with resident male coalitions (Mitchell, 1994). In this regard they resemble Central American squirrel monkey coalitions, except that they move among groups of females instead of staying in their natal home range. As in the howlers, intrasexual competition among females is again the pattern that differs most. Central American squirrel monkey females do not cooperate in defending food patches, whereas common squirrel monkey females do (Mitchell *et al.*, 1991). Mitchell *et al.* (1991) hypothesize that in Central America, the smaller size of fruit patches made cooperative patch defense uneconomical.

Cebus and *Saimiri* are the only two members of the Cebidae clade (based on Barroso *et al.*, 1997; Porter *et al.*, 1997) that live in large multi-male/multi-females groups with large undefended home ranges in which the general pattern is defense of the fruit patch being currently exploited rather than a territory (Robinson & Janson, 1987; Mitchell *et al.*, 1991). The matrifocal groups typical in *Cebus* and common squirrel monkeys may have resulted from a shift in diet or resource exploitation that enhanced the benefit to females of cooperatively defending food patches, in which case cooperating with relatives would ameliorate some of the costs . As illustrated by the process of group formation in red howlers, nepotistic behavior within groups of unrelated females cooperatively defending food patches can eventually result in stable female kin groups. In red howler and common squirrel monkeys, both males and females maintain strong kin-affiliation in their movements and patterns of association within a population, blurring the distinction between 'male' and 'female kin-bonded' societies.

CONCLUSIONS

The proposed mechanisms of transition between mating systems described here and in Fig. 19.4 are not intended to imply phylogenetic relationships among the platyrrhine taxa expressing them. However, some of the trends summarized here may have been widespread among their common ancestors and influenced the direction that mating systems took in different lineages. The ecological and life history conditions promoting male philopatry and female dispersal in small, territorial social groups may have set the stage for the evolution of the variable array of patrifocal social systems represented in New World primates. The influence of ecological variables on mating patterns is amply demonstrated in numerous empirical studies, including those described above, but multiple adaptive peaks within a species gene pool are predicted in a shifting balance model of evolutionary change (Wright, 1980). Thus there may be more than one evolutionarily stable 'mating system' for a given set of environmental and life history parameters. Which one comes to predominate in an evolutionary lineage can be the result of both genetic architecture ('phylogenetic inertia') and stochastic population events.

Understanding the relationship between ecological conditions, life history attributes, and mating patterns helps us to understand the selection factors operating on mating systems, but it does not tell us how evolutionary change takes place. The mechanisms involved in moving the large constellation of genetically differentiated populations that comprise a species gene pool from one genetic state to another are far more complex and less well understood than sociobiological theory implies. The proportion of the genome exposed to selection, for example, depends on the distribution of genetic variance within and between social groups (Templeton, 1980; Wright, 1980), which is highly variable among primate social systems (Pope, 1996b). Those attributes that have the largest effect on this phenomenon are group size, and the degree to which the sexes differ in reproductive variance and philopatry (Chesser, 1991a, 1991b; Pope, 1992; Sugg *et al.*, 1996). Mating system thus mediates changes in genetic architecture as well as being constrained by it. One of the consequences of this is that some mating systems should be less constrained by phylogenetic inertia than others. Investigation of the relationship between microevolutionary processes and sociobiological change is a potentially fruitful area of future research.

ACKNOWLEDGMENTS

Many thanks to Dr P. Kappeler for organizing this conference, and for providing funding for me to attend. The red howler monkey research that started me on this path would not have been feasible without the help of many people. R. Brooks, A. Bruni, C. Cadman, C. Crockett, D. Daneke, J. Eisenberg, R. Rudran, C.V. Salas, M. Sanchez, T. Struhsaker, R. Thorington, and D. Wilson all provided assistance in the field and elsewhere. Tomas Blohm, Dr Ivan Maldonado, and Antonio Branger kindly provided their ranches as research sites, and funding was granted by the Harry Frank Guggenheim Foundation, the Smithsonian Institution Environmental Sciences Program, the Scott Neotropic Fund, and Friends of the National Zoo. I thank two anonymous reviewers for useful comments on the manuscript.

20 • Models of outcome and process: predicting the number of males in primate groups

JEANNE ALTMANN

INTRODUCTION

Over 65 years ago, Solly Zuckerman promulgated the idea that sexual attraction and 'uninterrupted sexual life' were the main factors determining the social grouping of non-human primates. In his view, environmental factors serve merely as a limit on the expansion of group size (e.g., Zuckerman, 1932, pp. 31, 55). In contrast, 40 years later, ecological factors began to take center stage as the primary determinants of mammalian grouping patterns (Fretwell, 1972; Bradbury & Vehrencamp, 1977; Emlen & Oring, 1977), although whether food (Wrangham, 1980) or predation (e.g., van Schaik, 1983; Barton *et al.*, 1996; and Barton, Chapter 9) is the more important among ecological variables for primates remains an active debate for which resolving evidence is lacking. Emphasis during the subsequent decades has focused on a two-step process, one applying to each sex (Wrangham, 1980; Rubenstein & Wrangham, 1986; Wrangham & Rubenstein, 1986; Andelman, 1986). In this view, the distribution and grouping of females are determined by ecological factors as above, and the distribution of males is then determined by the distribution of females, or of sexually active females (e.g., for primate males see Ridley, 1986; Dunbar, 1988; Altmann, 1990). This chapter explores several questions that are motivated by that perspective.

The first set of questions involves the distributions of males themselves. Although reproductive models of male distribution are formulated as models of process at the level of individual action, they have been examined almost exclusively with data on outcome, not on process, and at much higher demographic levels than the individual. They have been tested with data on normative differences in the distribution of male and female numbers in groups across species or higher taxonomic levels. To what extent do these models explain distributions at lower taxonomic or demographic levels – among closely related species, populations, or individual groups over time, i.e., within units that more directly reflect the postulated processes? Is the salience of other factors, such as ecological, demographic, familial, or social ones, more evident at these lower levels of demographic and temporal variability than at higher ones?

The second set of questions, then, begins to focus more directly on the behavioral processes whose outcomes are observed as distributions of males and females in space. Analyses of variability at the level of groups or individuals over short time spans seek to identify proximate factors that facilitate and limit those behavioral processes. Ultimately, a successful model of demographic outcome that purports to distinguish among determinants of individual action requires elucidation both of the behavioral processes themselves and of the factors that impinge on them. As the major behavioral process whereby distributions can rapidly change within populations, dispersal and the proximate factors affecting it are particularly critical to understanding male distributions in predominantly female-matrilocal species and perhaps female distributions as well when their dispersal is common. These analyses lead to new formal models of dispersal based simultaneously and continuously on both reproductive and ecological factors, with a shifting balance of importance between the two even for the dispersing sex, rather than ecological factors being the deciding ones for females and reproductive factors for males.

Lastly, to what extent do the grouping interests of males and females coincide, and how reasonable is the assumption that single-male groups are always in a male's best interest, even in the short term but especially if a lifetime or intergenerational perspective is considered?

In the analyses that follow as each of these topics is explored, new data come from longitudinal research on a number of study groups (1–6, depending on the year) in the Amboseli baboon population just north and west of Mt Kilimanjaro; analyses of other data are based on the literature as noted. In Amboseli, data were available for 612 group-months from 1975 to 1996 for study groups (Alto's, Hook's, and Lodge, plus groups that were formed by the fis-

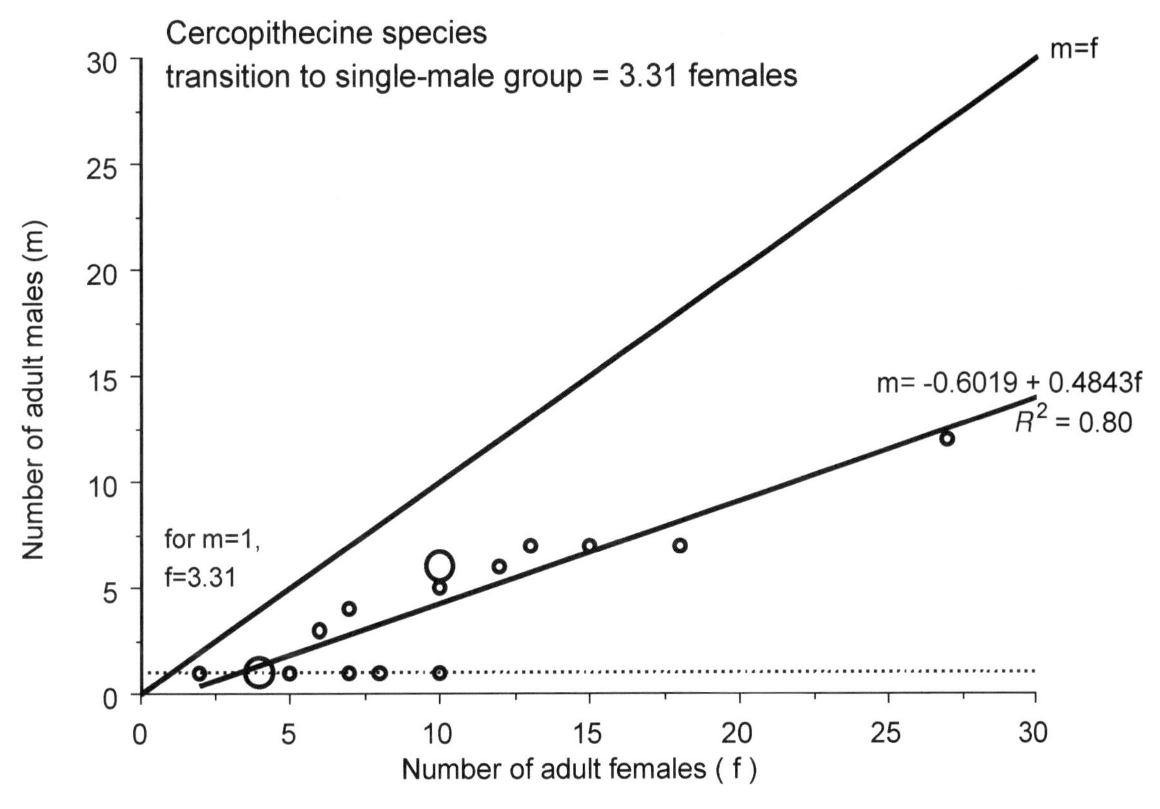

Fig. 20.1. Predicting the number of males in a group (m) as a function of the number of adult females in a group (f) across Cercopithecine species (analyzed and redrawn from Andelman, 1987, Fig. 10.1, to which the reader is referred; see also the figure caption therein for data sources). As is the case for many comparative studies, each species is represented in that data set simply by a single point, and neither intraspecific variability nor sample size was considered. The size of points indicates multiple identical data values.

sions of these). These analyses start with the next four sections that examine the preponderance and distribution of males relative to females in groups and do so at various demographic levels from interspecific to intragroup.

Distribution of Cercopithecine males and females in groups

Andelman (1986) observed that Cercopithecine species with groups greater than ten adult females contained several males, those with fewer than five females characteristically contained only a single male, and those species with intermediate-sized female groups varied in whether they were single-male or multi-male in composition. In Fig. 20.1, the data compiled by Andelman are plotted, and linear regression analysis is applied on the full data set (using each point without attempting to evaluate independence) to examine the ratio of adult males to females in groups and to identify the female group size below which we would expect groups to have only a single male (taken as the female group size at which the fit line intercepts the horizontal line representing a single male). This heuristic use of regression analysis suggests a slope, or male-to-female ratio (because the y intercept is very close to 0), of slightly less than half, an average transition between single-male and multi-male groups between three and four females, and a nominal $r^2 = 0.80$ (Fig. 20.1). In other words, the number of adult females in groups is a good predictor of the number of males found in groups with those females when Cercopithecine species are compared to each other. In addition, the overall adult sex ratio is about two females per male across these species, and a little over three females seems to be the threshold below which single-male groups would be expected. How well does this pattern hold when examined at lower taxonomic levels?

Table 20.1. *Predicting the number of adult males in a group from the number of adult females*

Taxonomic/ demographic unit	Number of years	Number of females	Intercept	Slope	Nominal R^2	Transition[1]	Source
Cercopithecines	Survey	2–27	−0.60	0.48	0.80	3.31	Andelman (1986, Fig. 10.1)
Baboons							
Anubis	Survey	1–25	2.52	0.40	0.46	−3.79	Altmann & Altmann (1970, Table VIII)
Chacma	Survey	8–31	−1.23	0.63	0.55	3.53	Altmann & Altmann (1970, Table IX)
Chacma-D	4	2–16	−0.65	0.61	0.68	2.72	Henzi & Lycett (1995, Fig. 4)
Yellow Amboseli 1963	Survey	6–59	2.06	0.63	0.70	−1.68	Altmann & Altmann (1970, Table V)
Yellow Amboseli 1975–1996	22	4–27	0.49	0.39	0.45	1.30	Current study
Yellow Alto's group	14	13–24	−1.21	0.58	0.15	3.81	Current study
Yellow Hook's group	14	12–24	3.10	0.26	0.30	−7.94	Current study
Yellow Lodge group	11	9–27	2.16	0.21	0.56	−5.50	Current study

Note:
[1] Predicted numbers of females below which groups will contain only a single male

What are the proximate processes by which males go where the females are? What sources of variability might affect the tightness of this relationship? The so-called savannah species or subspecies of baboons offer an ideal opportunity to focus on these questions with increasing 'magnification' because of their broad geographical distribution, relatively extensive history of study, and documented variability in group size.

Baboon males: group living and the preponderance of each sex

In many species of mammals, including Cercopithecine primates, the distribution of males in bisexual groups reflects not just decisions about which group to join but also decisions about whether to live in a bisexual group at all. Some, or even all, males of these species spend considerable portions of adulthood alone or in groups consisting only of other males. However, in common with the majority of primate species (Pusey & Packer, 1987), baboon males spend almost all of their adulthood living in groups that contain females. They do not live in groups consisting only of other males, and they spend relatively little time living alone. Averaged over 92 males in Amboseli, mean time spent alone was 8.5% (SD = 22.8) of the time during which we could confirm group or solitary living. To obtain bounds on this value, apportioning 'unknown time' (time in which location could not be confirmed), first, entirely to group living and, then, entirely to solitary living gives a minimum estimate for mean time living alone of 8.1% (SD = 22.4) and maximum estimate of 16.5% (SD = 28.3). The distribution of time spent alone deviated greatly from normality, however, and variability among males was high, as evidenced by the high standard deviations. The median time spent alone was 0%, with minimum and maximum estimates of the median of 0% and 1.8%, respectively. That is, more than half the males never lived alone, and the distribution of time spent alone was highly skewed among males. Nonetheless, as a consequence of baboon males tending to spend the overwhelming majority of their lives in groups with females, the pooled sex ratio of adults living in groups will be essentially the overall population adult sex ratio, in contrast to those species of primates or other mammals in which all-male groups are formed or in which solitary living is common for males. How, then, are adult male and female baboons distributed among groups?

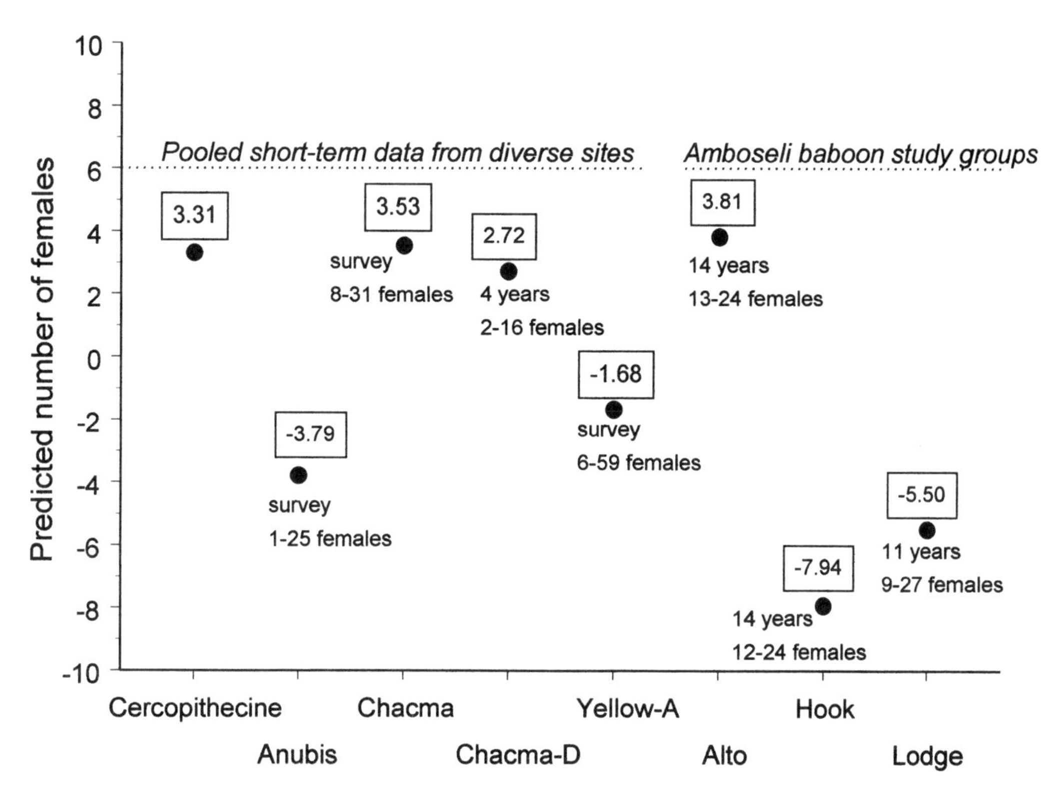

Fig. 20.2. Predicted numbers of females for transition to a single-male group, i.e., the point at which the regression line intersects the horizontal line that represents groups with only a single male, m = 1. Points are presented for data sets from diverse Cercopithecine species (see Fig. 20.1.), several baboon populations, and separate groups over time in Amboseli. Because data from any single group for few years or with small range of females yield inconsistent and sometimes nonsensical results (unpublished analyses of Amboseli data for such groups and for partitions into units of just a few years of longitudinal data for the longest studied groups), no such data sets were included. Data from Table 20.1 Chacma-D = Drakensberg population of chacma baboons; Yellow-A = Amboseli population of yellow baboons 1963–4.

Are baboon males where the females are? Distributions of baboon males and females

Published compilations of data on group compositions are available for three of the four savannah baboon (sub)species – yellow (*Papio cynocephalus*), anubis (*P. anubis*), and chacma (*P. ursinus*), but not Guinea (*P. papio*), baboons – at one or more sites. These are based on surveys or on pooled data for several groups over a few years. In addition, within the Amboseli population, the longitudinal data from over two decades enable examination of adult sex ratios within groups of a single population and even within individual groups over long time periods, and a sufficiently great range of female group sizes. For these baboon data, the use of regressions heuristically as described above yields interesting patterns that can be broadly compared to those from the interspecific Cercopithecine analysis.

First, in each baboon population, the predicted transition to single-male groups is at female group sizes below those that have usually been observed (Table 20.1, Fig. 20.2). That is, single-male groups probably occur so rarely and persist relatively briefly simply because the females of these (sub)species, wherever they are found throughout the continent (with the possible exception of the most marginal habitats), tend to form groups that are larger than those that would promote stable single-male groups. The female groupings of these multi-male (sub)species stand in contrast to those of their relatives the Ethiopian geladas (*Theropithecus gelada*) and hamadryas (*Papio hamadryas*), in which single-male groups predominate. As expected, those single-male species are characterized by small female group sizes and/or all-male groupings. In general, single-male groups will be demographically likely only: (1) if females of a species form very small groups; (2) if they outnumber

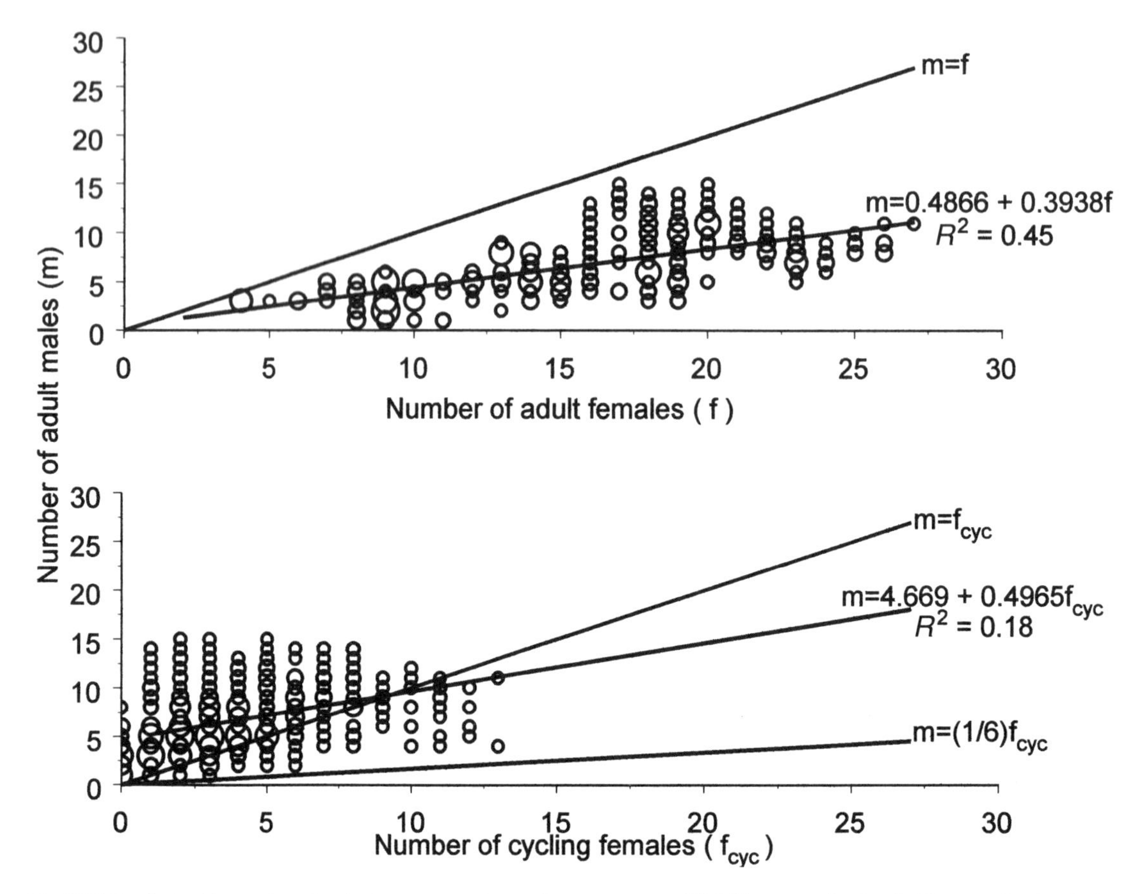

Fig. 20.3. Predicting the number of males in a group each month as a function of the number of adult females (*top*) and the number of cycling females (*bottom*) that month. Amboseli data, 1–6 study groups, 1975–96; see text for details. The size of points indicates multiple identical data values.

adult males by more than two or three to one; or (3) if males commonly live alone or in all-male groups. Females of most Cercopithecine primate species, including chacma, anubis, and yellow baboons (e.g., Smuts *et al.*, 1987), do not live in very small groups. Larger female groups may be driven primarily by predation pressure (e.g., van Schaik, 1983; Barton *et al.*, 1996) or by food distribution and competition (Wrangham, 1980), the two factors on which debate has focused. Alternatively, the advantage of larger female groups may accrue because such groups will in general provide more like-age playmates, alternative and supplementary care-givers for immature animals, demographic buffers against group extinction in the face of significant mortality and other stochastic effects, and greater access to a diversity of potential mates or to a larger set of conspecifics that will increase social acquisition of knowledge. In each case, the advantages of larger female groupings will be further enhanced if these groups consist of female kin, as is the situation when the predominant dispersal pattern is matrilocal with male dispersal.

The second finding relates to the extent to which the number of females predicts the number of males in the group. Using multi-male baboon groups, this model based on female number has less explanatory value at lower taxonomic or demographic levels than at higher ones or in the cross-Cercopithecine analysis (see Fig. 20.1). This effect of taxonomic level is consistent with most comparative studies (e.g., Harvey *et al.*, 1987). In the present analyses, the effect of taxonomic level is seen in the decline of nominal r^2 values in going from Cercopithecines to baboon (sub)species to the pooled Amboseli longitudinal data among groups and finally to the data on variability over time for each of the three Amboseli study groups with long histories (Table 20.1). Emerson & Arnold (1989) explored the relationships

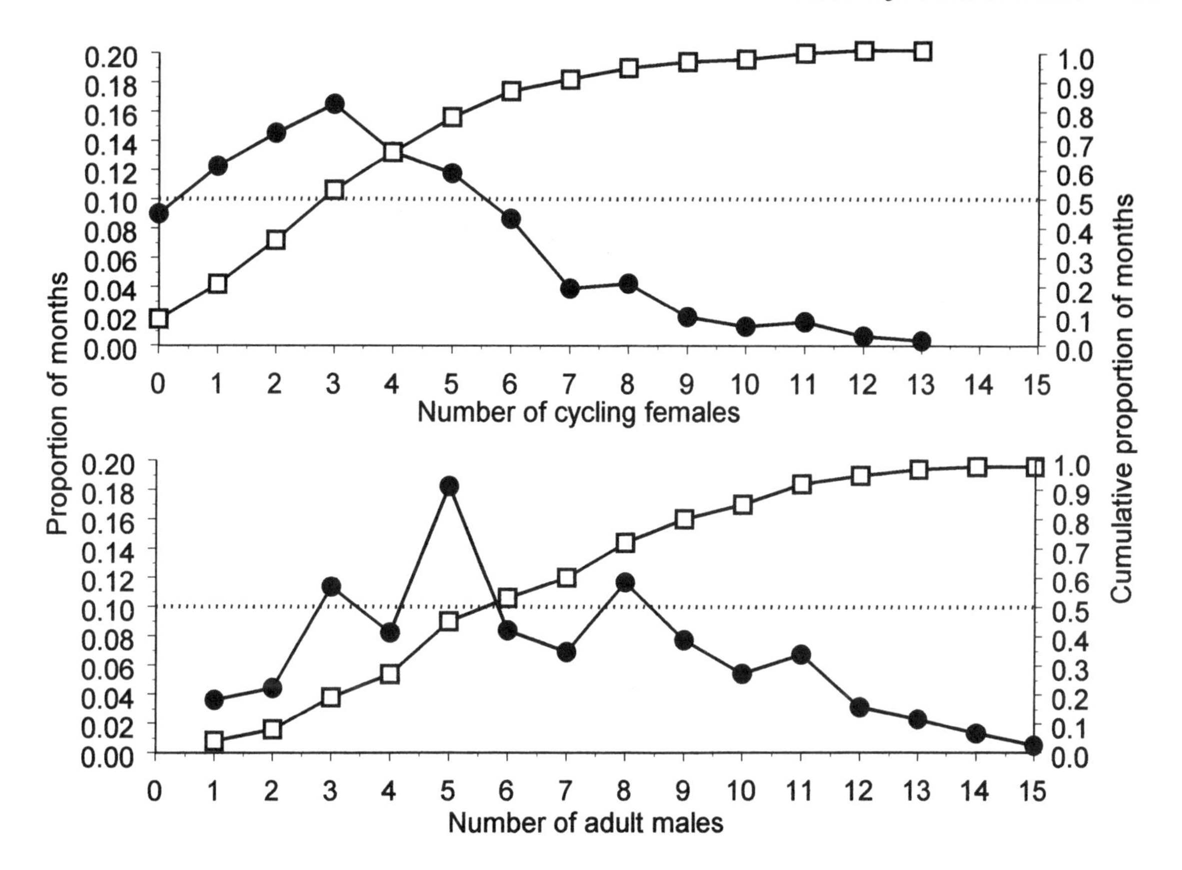

Fig. 20.4. Relative frequency of cycling females (*top*) or adult males (*bottom*) per month. The axis for the cumulative graph is on the right. Amboseli data 1–6 study groups, 1975–96. The size of points indicates multiple identical data values. Redrawn and analyzed from Andelman, 1986, Fig. 10.1.

between comparative patterns and intraspecific forces from a theoretical perspective. With respect to male numbers and grouping patterns, was the number of females in groups important on a broad historical scale but not, as most authors assume, of importance in understanding individual behavior and current processes? If sexual access is the highest priority in determining current male grouping patterns, why does the availability of females not better predict the number of males?

Female reproductive state and male distributions

Perhaps females are not all equivalent to primate males, one important factor being female reproductive state. The presence of mature females is a necessary but not a sufficient condition for sexual access. Sexual access will, in general at least, require fertile females in most primate species. Female primates spend the major portion of their reproductive years either pregnant or in post-partum amenorrhea, and only a relatively small proportion (approximately 25% for baboons: Altmann *et al.*, 1977) is spent experiencing sexual cycles and being sexually active. The implications for sexual access can be seen in the Amboseli data. In all 612 group-months in Amboseli, groups contained more adult females than adult males (Fig. 20.3, top). However, in only 21% of these group-months were cycling females as abundant as adult males (Fig. 20.3, bottom); 53% of the months had three or fewer cycling females in a group (Fig. 20.4, top), whereas the median number of males in groups was between five and six (Fig. 20.4, bottom). Not only were males usually in groups with fewer cycling females than adult males, but because females are fertile (conception is likely) for only approximately one-sixth of each cycle, daily availability of fertile females was even lower (see Fig. 20.3, bottom). Within individual groups, on over half of the days in a month, no fertile females were present, and among the days

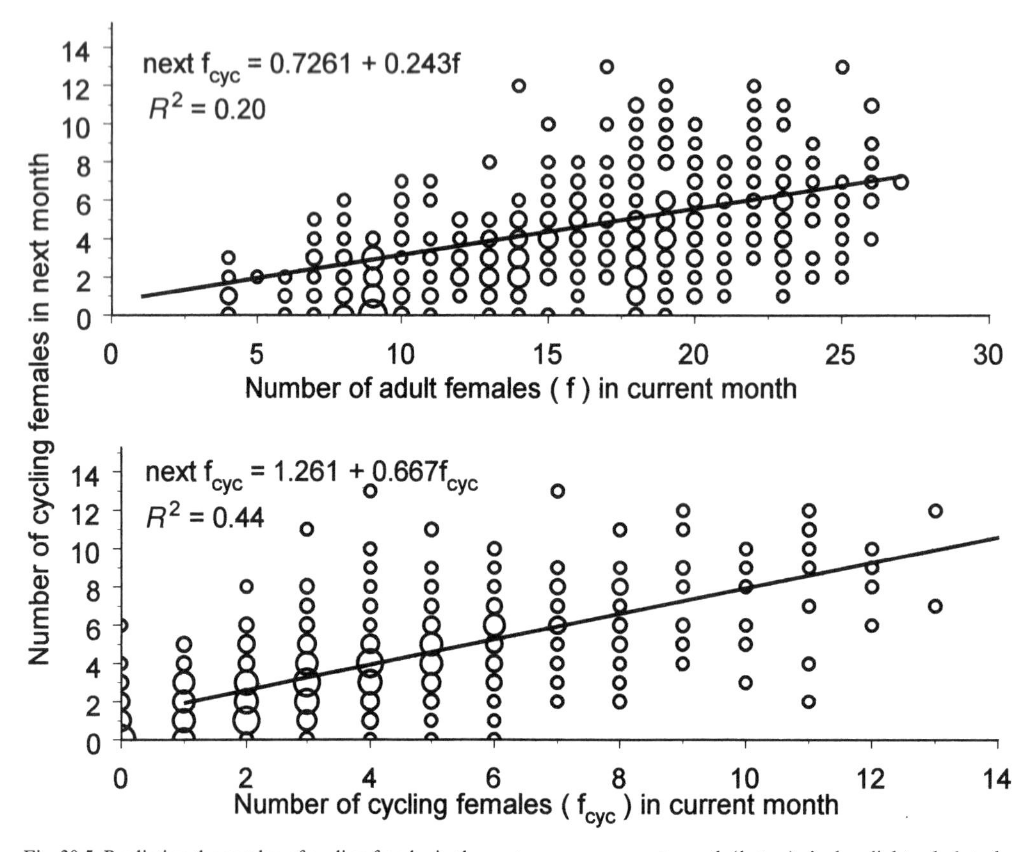

Fig. 20.5. Predicting the number of cycling females in the next month as a function of the number of adult females present in the current month (*top*) or the number of cycling females in the current month (*bottom*). Amboseli data, 1–6 study groups, 1975–96. The size of points indicates multiple identical data values.

with at least one fertile female, approximately half had only a single fertile female, and the remaining had more than one (Altmann *et al.*, 1997; Altmann, 1997, and unpublished data).

How might these distributions of fertile females be related to the decision processes of males? At the level of individual decision-making, the expected number of cycling females in the next month is presumably the important value for a male that is deciding to stay or leave a group, yet in Amboseli the number of females in a group in a month is a relatively weak predictor of the number of cycling females that will be in the group in the next month (Fig. 20.5, top). The number of cycling females the next month is, not surprisingly, much better predicted from the number of cycling females in the group in the current month (Fig. 20.5, bottom) than it is by the current total number of adult females in the group (0.44 vs. 0.20, Fig. 20.5). Therefore, one would expect that the number of cycling females rather than the total number of adult females in a baboon group would be a better predictor of the number of males in the group if reproductive access is the key issue. Such a finding could be seen as a parallel to that in which bisexual groups of some species experience a great influx of males during the mating season, especially in those species that are highly seasonal or in which most males live alone or in all-male groups (e.g., Pusey & Packer, 1987; Pereira, 1991). Yet, in Amboseli the number of cycling females in a group was an even poorer predictor of the number of males than was the total number of adult females (0.18 vs. 0.45; see Fig. 20.3): the adult males were not well distributed among cycling females. In other words, the males do not seem to be using the presence of cycling females as the indicator of sexual access. Why is a relatively poor measure (the number of adult females in a group) apparently used by the monkeys as well as by most researchers to estimate the baboon males' reproductive opportunities? The next sections explore the processes that

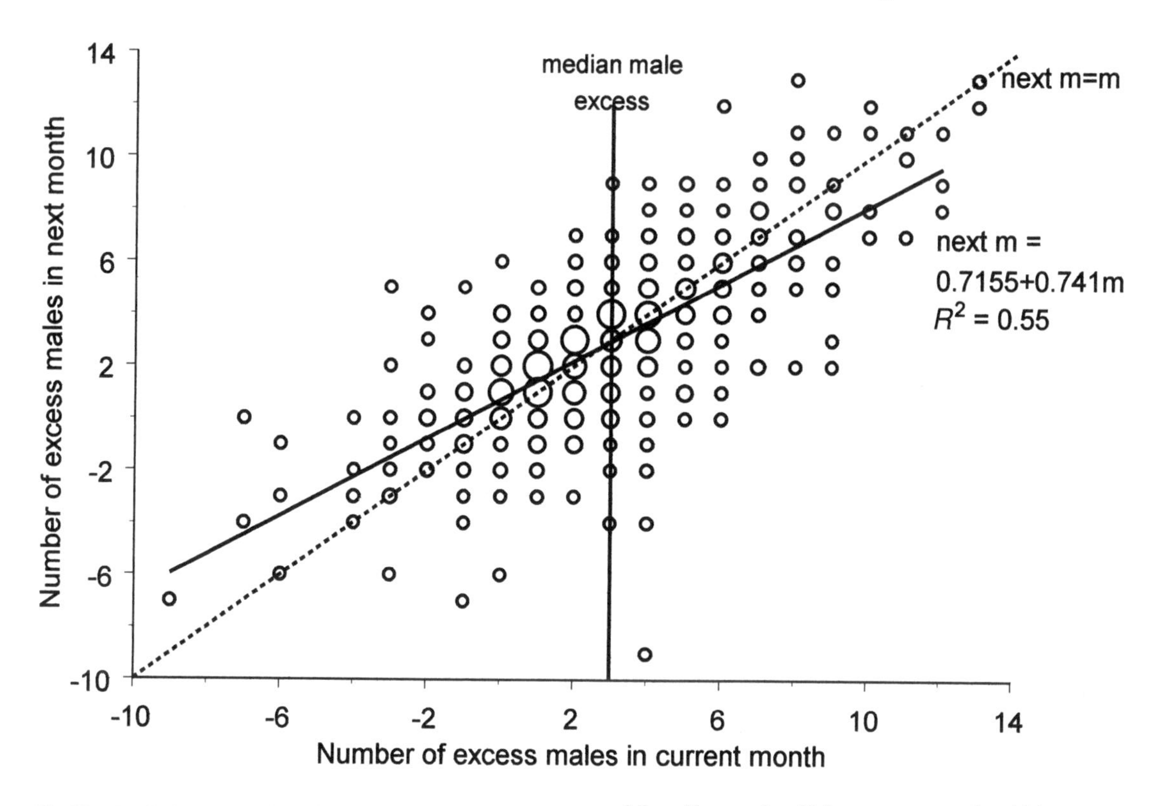

Fig. 20.6. Predicting the number of excess males, i.e. $m - f_{cyc}$, in the next month as a function of the excess in the current month. The median number of excess males per month, 3, is indicated by a vertical line at m = 3. In general, months with high excess tended to be followed by months with lower excess; months with lower excess tended to be followed by months with greater excess. See text for details. The size of points indicates multiple identical data values.

result in the outcomes we observe as distributions of males among females or among fertile females.

Males' tendency to go where the cycling females are: between pattern and process

Amboseli data on the movements of individual males do demonstrate considerable sensitivity of males to the current availability of cycling females. In previous reports from Amboseli, we found that both the availability of cycling females and an individual male's own mating success were powerful predictors of male dispersal patterns (Alberts, 1992; Alberts & Altmann, 1995a; Fig. 20.3, Fig. 20.4). In that study, both males born into a group (natal males) and those whose first membership resulted from immigration into it (non-natal males) were much more likely to emigrate in months when the number of males exceeded the number of cycling females, and the males virtually never emigrated when cycling females were in excess. Similarly, the proportion of new immigrants in a group was approximately four times greater in months when cycling females were in excess than otherwise.

The consequence of these excess-male-based dispersal probabilities can be seen at the level of group compositions in Fig. 20.6, in which the excess number of males in a group (number of males minus number of cycling females) in a given month is examined as a predictor of the excess number in the next month. If the number of excess males was highly stable from one month to the next, the best fit line would go through the origin and would have a slope of one with a tight clustering of points around that line. The observed slope of 0.741 is one indicator, and partially the result, of the tendency of males to emigrate from groups with high excess and into those with low. Further, when a group had greater than the median excess of three males in one month, then it tended to have lower excess in the next month (123/170 ignoring ties), and the opposite was the case when groups had less than median male excess (165/233 ignoring ties), i.e., for over 70% of the months the direction of change was as expected by dispersal from groups with a great excess of males and toward groups with a low excess of males. Moreover, the average deviation from median excess was reduced from one month to the next, from 3.84 to 2.47, i.e., in the short term, movements of males tended to decrease the variability in ratio of adult males to cycling females despite the fairly weak overall match between number of males and number of cycling females (see Fig. 20.3, bottom).

How are these two groups of findings to be reconciled? On the one hand is the evidence of demographic processes that should promote a fit between the number of males and cycling females. On the other are the data from the same site on the product of such processes that indicate the poor ability of male numbers to be predicted by the number of cycling females. The contradiction is more apparent than real. First, although males were much more likely to leave those groups with great excess of males, the proportion of both natal and non-natal males emigrating from such groups was still appreciably less than the excess, and similarly the number of new immigrants did not usually completely eliminate the excess of cycling females. Similarly, although Amboseli males that were reproductively less successful had shorter tenure in groups than did males that were more successful (Altmann *et al.*, 1988; Alberts & Altmann, 1995a), most males did not immediately leave groups in which their current mating success was low. Similarly, at the group level, (see Fig. 20.5), the very groups that improved in fit of males to females one month might move in the opposite direction the next month, leading in the long term to the modest overall fits that are observed in Fig. 20.3. These findings, separately and together, point to important factors that constrain male dispersal despite the powerful impact of the relative preponderance of cycling females in a group.

Why males do not always go where the cycling females are: counterforces to dispersal

Why did the match between males and cycling females not always improve in successive months, and why was the improvement usually imperfect? The various reasons can be framed in terms of: (1) mating costs and mortality costs of dispersal that are primarily entailed by time alone during dispersal; (2) additional direct and indirect costs of dispersal entailed by leaving familiar associates and relatives; (3) individual differences; and (4) the difficulty of decision-making in the face of imperfect information and a changing context of decision-making.

1. First, then, are the mating and mortality costs of dispersal itself, a topic recently treated in some detail by Alberts & Altmann (1995a). Even when the resident group provides a male with few mating opportunities, emigration entails immediate mating costs as well as potential benefits. Emigration usually involves a period of living alone before a male can enter a new group. Such solitary periods entail the cost of lost mating opportunities: low though non-zero mating success in a current group is still greater than the zero mating success a male will experience when he spends any appreciable time alone before immigrating into another group (Alberts & Altmann, 1995). In addition, the early weeks or months in an unfamiliar group can also be a time of low mating success, both because a male may spend much time in the fights that establish mating access through high dominance status and because he will not yet obtain matings through 'friendships' (e.g., Smuts, 1985) or coalitions (e.g., Hall & DeVore, 1965; Noë & Sluijter, 1990; Noë; 1992) rather than dominance status. In other words, the opportunity costs are immediate and assured; the potential benefits are only potential and are future ones, i.e., discounted ones in economic terms.

   Immediate mortality costs are also entailed by time spent alone during dispersal, e.g., a twofold to tenfold difference among baboons in Amboseli (Alberts & Altmann, 1995a, Table 2), presumably primarily through predation. Therefore, when emigration entails some period of solitary living, as it commonly does, both mortality and mating opportunity costs of emigration will accrue.
2. Emigration potentially entails additional costs to both direct and indirect components of fitness derived from the change in social environment. Leaving a group will usually mean leaving relatives who may both benefit a male's survival and themselves gain survival benefits from his presence and behavior. Moreover, alliances or other strategies for interaction with non-kin will in general be more effective and more efficient in a group comprised of familiar individuals with whom an individual has established relationships than in one comprised of unfamiliar individuals. Therefore, even where densities of groups are high, levels of predation are low, and time spent living alone between groups is minimal – all situations that decrease the costs of emigration (Alberts & Altmann, 1995a) – emigration may still be costly. These emigration costs will often overshadow the potential advantage of dispersal that is gained by low male to female ratios in another group.

Can reasons be identified beyond these several costs of emigration that immigration does not more quickly and completely reduce the number of excess cycling females?

3. The third factor reducing dispersal results from the importance of individual differences, or heterogeneity among males, including the effects of male dominance rank on reproductive access. Because most groups have an excess of males most of the time, only if a male can join a group at a high dominance rank is he likely to have access to cycling females (see also Dunbar, Chapter 22). This is true even if the group into which he immigrates has fewer excess males than the group that he left. A second individual-specific effect is that for each male some groups with lower male excess or competition may be that male's group of birth, his natal group, in which some females are his mother or sisters, or a group in which some females are his mature daughters. Both situations will either entail the risk of inbreeding costs or reduce the apparent number of potential mates. Consequently, for many males, dispersal will not usually be worth its significant costs.

Ecological factors and some aspects of individual differences have recently been considered in a model of the processes underlying the distribution of the dispersing sex. A simple model of dispersal was proposed based on: (1) the survival differences between dispersing and staying (the mortality cost of dispersal); (2) the mating opportunity differences between dispersing and staying (the immediate reproductive costs of dispersal); and (3) the relationship of these to the time a dispersing male spends alone between groups (Alberts & Altmann, 1995a). Time spent alone was assumed to be inversely related to population density, which in turn reflects resource availability. In areas of high versus low predation, survival costs of dispersing will increase more steeply with time spent alone, that is, with decreasing density or food supply. Similarly, mating costs of dispersal will be minimal for males that spend little time alone but will be much more costly for successful males if time spent alone is great.

Complex and seemingly contradictory findings in the dispersal literature, e.g., regarding whether successful or unsuccessful males were more likely to disperse, emerge as simple predictions of this new model. Some parameters in the model, such as both survival values (those within groups and when solitary), will in general be the same during any given time period for all males of a single population, that is, primarily ecological in nature. Others, such as the current mating opportunities for different males in a group, will be at least partially a function of an individual male's age, tenure, or dominance status. Thus, the Alberts and Altmann model of dispersal is based not just on reproductive factors but also on ecological ones, and on individual differences among males as determinants of male dispersal, and, by implication, the ultimate distribution of males among groups in a population. That model was found to predict available reported differences between baboon populations/species. It remains to be tested more broadly among other populations or species, either of baboons or of other taxa, and the balance between intrapopulational and interpopulational variability in the relevant parameters will affect its value in exploring some apparent interpopulational differences. Finally, that model explicitly treated only one type of individual difference, dominance rank. It did not, for example, formally incorporate inbreeding constraints as they affect individual options and decision-making. Although treated primarily in isolation, the importance of inbreeding avoidance as a determinant of dispersal has long been recognized (e.g., Packer, 1979a; see also Alberts & Altmann, 1995a). As the authors note, the Alberts and Altmann model could readily be modified to incorporate inbreeding avoidance.

4. A fourth major factor that limits the match between the distribution of males and fertile females is that males are faced with the challenge of estimation and decision-making in the face of imperfect information. Even for baboons and other species in which the females have highly visible and informative sexual swellings, males outside a group have limited opportunity to assess the availability of cycling females in other groups; this will especially be the case for situations in which groups are at low densities and for situations in which predation is high. Males will have imperfect information, even regarding the simple number of females and males in other groups; information on the number of cycling females will be even less accessible. Researchers, too, rarely have simultaneous information on the distribution of males and cycling females in all potential dispersal groups of a population (nor is such information available from our research) to evaluate whether the males are, in fact, distributed about as well at any moment as predicted by an 'ideal free distribution' (Fretwell, 1972; Dunbar, Chapter 22). Moreover, the context of male decision-making is changing as the decision-making itself occurs. In particular, the number of cycling females in a group changes from month to month and is only modestly predicted by the number in

the previous month (see Fig. 20.5, bottom) or by any other simple variable. Females enter the pool of cycling females through maturation or, in most primates, through resumption of menstrual cycles at the end of post-partum amenorrhea. They leave it through pregnancy or death. The timing of each of these events has significant stochastic components. In addition, decisions by others are likely to compromise the effectiveness of any one individual's actions, and yet males probably make migration decisions independently. Both baboons and researchers may use an imperfect measure because that is the best information that is usually available. In order to incorporate the simultaneous consideration of opportunities in all groups within a population (see Dunbar, Chapter 22), the effects of simultaneous decision-making by many independent individuals, and the problem of decision-making in the face of incomplete information, analyses and models will in some cases be profitably borrowed from other fields, such as economics, and then integrated with those discussed above.

The final sections focus on two assumptions underlying most analyses of male and female grouping patterns.

Influences on female grouping patterns and the interest of females in male grouping patterns

In the traditional view, the distribution of females, like that of males, is also singly determined, but in the case of females the distribution is determined only by environmental factors, food and/or predation, not by reproductive access (but see Strier, 1994a; Hemelrijk & Luteijn, 1998). However, a number of actions by females not only affect their reproduction indirectly through effects on food availability but also directly affect their own reproduction and that of others. The most well-recognized action is that females of many mammalian species limit the number of mature or reproductive females through reproductive suppression; marmosets and tamarins are the main primate examples of such suppression. Clearly, the number of fertile females in a group will vary little or not at all in such species, and males would not be expected to favor groups with more females. Other behavior by females that can also affect reproduction by changing social environments has received less attention. Among primates, individual female dispersal and immigration occur particularly among some New World primates (surveyed and discussed by Strier, e.g., Chapter 7) but also among species that are primarily matrilocal (e.g., sifakas, *Propithecus verreauxi*: Kubzdela, 1997) in which reproductive constraints and opportunities are implicated for females. Even for species in which female dispersal is rare or non-existent, significant changes in female membership can occur. For cercopithecine females, including baboons, fissions or fusions of groups can be seen as the mechanism that is comparable to dispersal. Though not frequent, it occurs regularly and has profound effects on the reproductive environment of females. Further, it then secondarily affects the reproductive options available to males as well.

To what extent is it safe to assume that females will be indifferent to the number of males in their group? Females may garner higher levels of both paternal care and mate guarding in multi-male groups than in single-male groups. Interspecifically, paternal care is not simply predicted by paternity assurance across social/mating structures (Snowdon & Suomi, 1982), and males in single-male groups often forego mate-guarding (J. Altmann & S.C. Alberts, unpublished data). The potential benefits and costs of mate-guarding and paternal care have received little attention and are important to evaluate. Furthermore, females may prefer larger and thereby multi-male groups because such groups will provide them with the opportunity for mate choice, even more so if their cycles are synchronized. Alternatively, if variability in male quality is high, and if high quality is evaluated identically from every female's perspective (i.e., not a pairwise or otherwise female-specific phenomenon such as major histocompatibility complex haplotype) and is congruent with the outcome of male–male competition, then females might benefit from overdispersal rather than synchrony of cycles or benefit from single-male groups if such groups represent the product of greater male–male mate competition. In general, then, male grouping patterns will often have important reproductive consequences for females; it is unlikely that females have been passive in the face of those consequences (Eberhard, 1996).

Finally, group size also affects the familial and overall social environment in which females raise their young. Larger groups will increase the probability that offspring have peer playmates, have peers of both sexes, have more peers, and have more adults from whom care may be garnered in case of maternal incompetence, illness or even death. Although living in larger groups can also entail competitive costs, the range of likely potential compensatory advantages has not yet been evaluated. The role of females as interested and active determiners of their own reproductive destiny is just starting to be explored in a variety of domains (e.g., Smuts & Smuts, 1993; Gowaty, 1994). Its dynamic, ongoing role in affecting male grouping patterns is an intriguing and reasonable possibility.

Might multi-male groups be advantageous even to successful males?

Whereas single-male groups are ubiquitously assumed to be reproductively advantageous to males, this assumption is based on short-term perspectives and one that focuses solely on mating success. The assumption is likely to fail under a number of circumstances and particularly so when a male's full lifespan is the time unit of analysis and when offspring survival is included in reproductive success calculations rather than just offspring conception or even just mating success. Because male tenure is usually short in single-male primate populations and because tenure changes often entail the risk of infanticide, reproductive skew in single-male populations will not necessarily be greater than in multi-male populations. Moreover, because multi-male groups usually contain many more females than single-male groups, the most successful male in such groups has the potential to father more offspring than a male in a single-male group.

CONCLUSIONS

Environmental and reproductive factors are important to the grouping patterns of both males and females, historically and within extant populations. Intra-specific and intra-populational variability reveals the importance of a dynamic, multiple variable approach to identifying the processes at work. Distributions of males and females among groups are the result of action by many individuals for whom optimal positioning may change over time and space. The ideal group size and sex composition might be different for members of the two sexes and also among the various members of each sex-class as a function of age, dominance status and familial environment. Individual action will be constrained by the simultaneous actions of others and by the immediate options provided by current group sizes and compositions, requiring empirical and theoretical investigations at the level of populations as well as of groups and individuals. Little attention has focused on the question of why males of many species do not adopt the boom-or-bust lifestyle of forming all-male bands rather than living in bisexual groups that offer opportunities, however meager they may be, along a greater continuum. Males of the all-male-band species seem to be able to tolerate the presence of other males primarily in the absence of females who will be inaccessible to them as mates. The basis of this difference in tolerance and its species plasticity may hold important insights for understanding many species. Endocrinological and other physiological processes, perceptual and cognitive abilities, and other aspects of proximate mechanisms will ultimately facilitate or constrain individuals' abilities to capitalize on options that are potentially available to them, and, thereby, to determine the course of selection and the opportunity for evolutionary change.

ACKNOWLEDGMENTS

I am grateful to the Office of the President, Republic of Kenya, to the Kenya Wildlife Services, its Amboseli staff and Wardens, to the Institute of Primate Research, to R. Leakey, D. Western, J. Else, C.S. Bambra, and M. Isahakia, and to C. Mlay. Thanks go to S.L. Combes, D. Shimizu, and M. Nabong who provided assistance in data management, analysis, or manuscript preparation, and to the Amboseli fieldworkers who contributed to data collection, especially S. Alberts, P. Muruthi, R. Mututua, A. Samuels, S. Sayialel, and K. Warutere. Valuable comments on earlier drafts of the manuscript were provided by S. Alberts, S. Altmann, S.L. Combes, and two anonymous reviewers. For stimulating discussion of male grouping patterns and the opportunity to have that discussion, I am grateful to P. Kappeler, organizer, and the participants and sponsors of the conference on primate males. Major financial support for the long-term research has been provided at various times by NIMH15007, the Chicago Zoological Society, NSF IBN-9223335, IBN-9422013, and IBN-9729586.

21 • Why are male chimpanzees more gregarious than mothers? A scramble competition hypothesis

RICHARD W. WRANGHAM

INTRODUCTION

This chapter proposes an explanation for why male chimpanzees (*Pan troglodytes*) are more gregarious than females, and hence why male bonding among chimpanzees is more prevalent than female bonding. In all multi-male communities of chimpanzees, male bonding is evidenced by adult males associating and grooming together, as well as forming long-lasting coalitions used both within and between communities (e.g., Goodall, 1986). Female bonding, by contrast, is sporadic. It is barely present in East African sites (Gombe, Mahale, and Kibale), whereas coalitional or grooming bonds have been observed among females in captivity and, importantly, in two West African sites, Bossou and Taï (captivity: de Waal, 1978, 1982; Baker & Smuts, 1994; Bossou: Sugiyama, 1988; Sakura, 1994; Taï: Boesch, 1991, 1996a). Similarly, female bonobos (*Pan paniscus*) spend more time in parties of larger average size than female chimpanzees do, and have stronger bonds than female chimpanzees (Kano, 1992; White, 1992). Within-community bonds among chimpanzees are frequently, perhaps typically, formed among non-relatives or distant kin, as shown by the fact that male–male alliances are flexible (e.g., Nishida & Hosaka, 1996), as well as by direct genetic evidence (Goldberg & Wrangham, 1997). In the case of female–female bonds, the tendency for females to emigrate at adolescence means that bonds are normally among non-relatives both in chimpanzees and bonobos. Such studies make it clear that the more time that females spend together, the stronger are their bonds with each other. This is as expected from reciprocal altruism theory, which states that the frequency of reciprocal altruism among non-relatives (and, by implication, the strength of bonds) depends on the frequency of opportunities for interaction (Trivers, 1971).

Why, then, are female chimpanzees often less gregarious than males, spending more time alone, meeting fewer different individuals per day, and traveling in smaller parties (Goodall, 1986; see review below). Only a few other mammals share this pattern of relative male gregariousness, notably spider monkeys (*Ateles* spp.: Chapman, 1990a) and cheetahs (*Acinonyx jubatus*: Caro, 1994). (In this chapter 'relative male gregariousness' means 'relative to females of the same species'.) In none of these cases has a satisfactory explanation emerged for the sex difference, i.e., males being more, and females less, gregarious.

Cost–benefit explanations for male chimpanzees being more gregarious than females tend to imply that either the benefits or the costs of grouping differ between the sexes. Against this trend, it is possible that both benefits and costs differ importantly. However, sex differences in the cost of grouping have been the focus of attention, because hypotheses that invoke sex differences in the benefits have not yet explained species differences, and are therefore not compelling.

For example, there may be a sex difference in the benefit of grouping if males receive greater advantages than females from investing in bonds. The obvious male benefit is alliance support, which is important both in the defense of a territory and in competition for dominance rank (Nishida, 1979; Goodall, 1986; Wrangham, 1986). But why should this apply to chimpanzees and not to other xenophobic, fission–fusion species? One possibility is that there are ecological reasons why alliances are especially favored in males compared to females, such as high benefits to contest competition between groups (van Hooff & van Schaik, 1992). But no such reasons have been proposed. At present, no 'benefits' argument explains why chimpanzees have female alliances less than other fission–fusion species with intense inter-group contest competition, such as ruffed lemurs (*Varecia variegata*: Morland, 1991), or spotted hyenas (*Crocuta crocuta*: Frank, 1986; Smale *et al.*, 1997). Thus, there is not yet any empirical support for the hypothesis that with increasing party size, males gain benefits faster than females. This 'benefits hypothesis' is therefore not considered further here.

The 'cost hypothesis,' by contrast, attributes relative

male gregariousness to males having lower costs of grouping than females (and therefore being able to develop bonds more economically than females can: e.g., Wrangham, 1986; van Hooff & van Schaik, 1994). This hypothesis is prompted by a persistent correlation between food scarcity and reduction in size of chimpanzee parties (temporary subgroups). The food-scarcity/party-size correlation occurs not only with fruit abundance in the community range as a whole, but also with fruit availability at individual food patches (fruit trees, or fruit tree groves), and has been noted at all major study sites (Gombe, Tanzania: Goodall, 1986; Wrangham, 1986; Mahale, Tanzania: Nishida, 1974; Kanyawara, Uganda: Isabirye-Basuta, 1988; Wrangham *et al.*, 1991, 1992, 1996; Chapman *et al.*, 1995; Taï, Ivory Coast: Boesch, 1996a; Bossou, Guinea: Sakura, 1994). The correlation of food abundance and party size implies that as food becomes scarce, individuals benefit by dispersing away from each other. By comparison with species living in stable troops, therefore, chimpanzees appear to experience relatively intense feeding competition, so intense that it can lead to individuals traveling alone.

The reasons why competition should be particularly intense in chimpanzees have not been formally examined. However, comparative feeding data on troop-living monkeys and fission–fusion chimpanzees show that chimpanzees eat much more ripe fruit than monkeys do. This is true even when fruits are scarce, a time when monkeys are able to find a wide variety of alternative edible items (Wrangham *et al.*, 1998). The fact that chimpanzee foraging strategies are unusually focused on these high-quality food items, which occur in depletable patches, may thus contribute to chimpanzees being especially susceptible to competition.

But why should feeding competition have greater effects on female than on male chimpanzees? One suggestion is that in females, fitness may be more closely linked to feeding success than it is in males. Therefore, a given intensity of feeding competition has less impact on male than on female fitness (Wrangham, 1986). However, according to this proposal, greater male gregariousness should be found commonly in species subject to feeding competition, such as many fission–fusion species. In fact, however, fission–fusion species commonly have females who are at least as gregarious as males (e.g., primates: ruffed lemur: Morland, 1991; orangutans (*Pongo pygmaeus*): Rodman & Mitani, 1987, carnivores: spotted hyena: Frank, 1986; lion, *Panthera leo*: Schaller, 1972). This therefore suggests that there is something unusual about the form of competition faced by chimpanzees (apart from the possibility of bias due to the greater importance of food for female than for male reproductive success).

THE SCRAMBLE COMPETITION HYPOTHESIS

Feeding competition can take the form of either contest (interference) or scramble competition (van Hooff & van Schaik, 1992, 1994). Because male chimpanzees are socially dominant to females (Goodall, 1986; Takahata, 1990), it is easy to imagine that contest competition could cause females to experience high costs. Although chimpanzees rarely fight over plant foods, dominant individuals appear regularly to occupy the prime feeding sites (Goodall, 1986, p. 356), and among Gombe females there is evidence of substantial differences in reproductive rate, mediated by their dominance relationships (Pusey *et al.*, 1997). Even so, there is not yet any direct support for the importance of contest competition in forcing females to forage alone. Contest competition may prove important, but it is not considered further here.

I focus instead on scramble competition, the alternative source of sex differences in the costs of grouping. For frugivorous primates, it has recently been shown that between-species differences in the intensity of scramble competition explain much of the variance in group size (Janson & Goldsmith, 1995). This follows from the fact that for frugivores, their most important food items (ripe fruits) tend to be obtained from a series of discrete and finite food-patches. As a result, individuals traveling in relatively large groups must visit relatively more food patches (in order to satisfy their food requirements) and therefore incur travel costs (increased time and energy expenditure).

In comparisons between species, Janson & Goldsmith (1995) indexed the cost of grouping by the rate at which daily travel distance increases with group size (i.e., the 'relative ranging cost'). They found that relative ranging cost varies as a result of species differences in factors such as locomotor style or the distribution of food. Species with more intense scramble competition have higher relative ranging cost. They therefore live in smaller groups than species with low relative ranging cost, in order to prevent actual travel costs from increasing (see also Wrangham *et al.*, 1993).

In this chapter, similar logic is applied to sex differences in chimpanzee grouping. Hunt (1989) found that anestrous female chimpanzees (i.e., mainly mothers) travel more slowly than males. This means that when males and mothers travel the same distance, as they presumably do when in the same parties, mothers spend a longer time traveling than males. Time is an important resource for primates in general, and travel time may also predict total energy expenditure (Dunbar, 1988). I infer that mothers have a higher 'relative ranging cost'. By analogy with Janson and Goldsmith's (1995) argument, therefore, mothers can be expected to prefer smaller parties than males purely in order to avoid paying high travel costs (increased travel time compared to males). This is the scramble competition hypothesis for sex differences in gregariousness.

A strong form of the scramble competition hypothesis proposes that velocity differences between age–sex classes, through their effect on travel time, account for most of the variance in gregariousness; and, furthermore, that the specific level of gregariousness is such as to give all age–sex classes the same total travel time. To examine this idea, I first review variation in grouping among chimpanzee populations. Variation occurs in both the availability of same-sex associates for males and females in different communities, and in sex differences in gregariousness. I then attempt to test the hypothesis that a sex difference in velocity accounts for the sex difference in gregariousness.

METHODS

Fieldwork was carried out in Kibale National Park (Uganda) on the habituated chimpanzees of the Kanyawara community. For a description of the study site and methods, see Wrangham *et al.* (1996). Travel cost for parties of different size was assayed as distance covered per unit time. This was calculated by measuring travel distance (on a map) for each party, defined as a stable set of individuals. Thus, new parties were defined as occurring whenever there was a change in membership. Party size was summed as the total number of adult and adolescent individuals, excluding infants and dependent juveniles, recorded every 15 minutes. Data were tallied for all of the 2175 parties observed between April 1993 and March 1995. Fruit availability was assayed twice monthly from 1988 to 1996 by scoring the presence of ripe fruits on nine non-fig species ($n \geq 10$ individuals for each species). In a check for secular trends, I found that from 1988 to 1996, mean monthly party size tended to increase, probably influenced by the increasing habituation of peripheral females ($r^2 = 0.17$, $n = 101$ months, $p < 0.001$). The residual of party size on month was therefore used in analyses of the factors influencing party size. However, no values of statistical significance differed between analyses using the residual party size compared to actual party size. Ripe-fruit availability also showed a mild secular trend, perhaps because of increasing observer ability to detect fruit ripeness. Therefore, ripe-fruit availability was also measured as a residual on month, i.e., residual of the mean percentage of trees in fruit per month, for nine non-fig fruit-tree species.

RESULTS

Community size and the availability of partners

A chimpanzee community is a closed social group within which individuals have peaceful associations. In 11 different communities, representing 61 years of observation, the median community size was 41.2, containing 15.0 females (10.7 adults, 4.3 adolescents) and 9.3 males (6.3 adults, 3.0 adolescents). ('Adolescent females' means nulliparous females old enough to copulate; 'adolescent males' are as big as adolescent females, and still growing.) Thus, there are typically more females than males (mean ratio = 1.6 females per male), but both sexes have consistently high opportunities for social grouping and bonding.

Sexual dimorphism in grouping and bonding patterns: are females less social than males?

Female chimpanzees are consistently less gregarious than males, but there is variation among sites and among classes of female. The patterns are summarized in Table 21.1 and 21.2.

GOMBE

1. Grouping. The amount of time spent with at least one other adult was twice as high for adult males (73%) as for females without sexual swellings (35%). Such females also spent longer periods alone or only with dependent offspring, and shorter periods in parties, than adult males (Wrangham & Smuts, 1980). The most sociable females were nulliparous females who, after immigrating, spend much time with adult males even when they have no sexual swelling, whereas 'after giving birth to her first infant, the female typically becomes less gregarious' (Goodall, 1986). Only one female, Gigi, was as gregarious as adult males, and attended large gatherings as often as males. She differed from other females in having no

Table 21.1. *Adult and adolescent composition of chimpanzee communities*

			Males		Females			
Site	Community	Years	Adult	Adolescent	Adult	Adolescent	Total	F:M
Gombe	Kasekela 1	7	13.1	5.4	12.3	7.3	51.0	1.1
Budongo	Sonso	1	12.0	3.0	10.0	8.0	49.0	1.2
Mahale	M-group	9	9.6	9.3	35.8	10.9	95.0	2.5
Kibale	Kanyawara	9	9.0	4.0	12.8	4.3	41.2	1.3
Taï	Taï	2	7.0	?	27.0	?	76.0	3.9
Gombe	Kasekela 2	10	6.3	5.0	16.1	5.3	49.8	1.9
Mt Assirik	Mt Assirik	1	6.0	3.0	5.0	2.0	25.0	0.8
Gombe	Kahama	1	6.0	1.0	3.0	1.0	12.0	0.6
Kahuzi-Biega	C1	1	5.0	3.0	6.0	?	22.0	0.8
Mahale	K-group	18	3.8	0.8	10.7	2.6	25.6	2.9
Bossou	Bossou	3	2.0	0.5	6.7	1.3	19.3	3.4
Median			6.3	3.0	10.7	4.3	41.2	1.3

Note:

Data sources and years are: Gombe, Tanzania (Kasekela 1, before fission into Kasekela 2 and Kahama), 1965–71, Goodall (1986); Budongo, Uganda (Sonso), 1996, Reynolds (personal communication); Mahale, Tanzania (M-group), 1980–8, Nishida *et al.* (1990); Kibale, Uganda (Kanyawara), 1988–96, personal observation; Taï, Ivory Coast, 1987–9, Boesch (1996a); Gombe, Tanzania (Kasekela 2, after fission from Kahama community), 1972–83, Goodall, 1986; Mt Assirik, Senegal, 1979, Tutin *et al.* (1983); Gombe, Tanzania (Kahama), 1972, Wrangham (1975); Kahuzi-Biega, DR Congo (C1), 1991, Yamagiwa *et al.* (1996); Mahale, Tanzania (K-group), 1966–83, Nishida *et al.* (1985); Bossou, Guinea, 1977, 1980 and 1983, Sugiyama (1984).

offspring (Goodall, 1986). Two males became more solitary during their final one to two years before dying from old age, and unhealthy individuals were often solitary (Goodall, 1986).

2. Bonds. Whereas males form strong bonds, unrelated adult females do not, though they may form temporary associations (Goodall, 1986).

MAHALE

1. Grouping. Nishida (1968) reported that there was a clear sex difference in grouping patterns, with adult males of the K-group community being consistently more sociable than adult females. This was subsequently found in the M-group community also (e.g., Takahata, 1990). Although quantitative data are not available, among females without sexual swellings, mothers with dependent infants were stated to be less social than pregnant or cycling females (Takahata, 1990).
2. Bonding. According to Nishida (1990), females sometimes united in temporary coalitions against a new female immigrant, or against a young adult male. Immigrant females could also form rare, and not very effective, coalitions. This contrasted with the stronger social bonds formed among adult males, seen through proximity, grooming, and coalitions (Nishida, 1990; Nishida & Hosaka, 1996).

KIBALE

1. Grouping. At two sites within Kibale National Park, Kanyawara and Ngogo, a clear sex difference in grouping patterns has been found, with adult males being consistently more social than adult females (Wrangham *et al.*, 1992). Figure 21.1 shows how the sex difference is related to party size at Kanyawara. If the two sexes were equally gregarious, the sex ratio in large parties should reflect the sex ratio in the community. In fact, however, large parties commonly include all the community males, but have never been seen to include all the community females. Figure 21.1 illustrates this sex difference and also shows that adolescent (nulliparous) females were as gregarious as adult males. This suggests that the 'sex difference' in gregariousness is due not to gender but to parity.

Table 21.2. *Site differences in grouping and bonding*

	Gombe	Mahale	Kibale	Taï	Bossou
Grouping					
Females less social than males?	Y	Y	Y	Y	?
Mothers less social than other females?	Y	Y	Y	?	Y
Females mostly with males	N	N	N	Y	Y
Bonding					
Male bonding?	Y	Y	Y	Y	?
Female bonding?	N	N	N	Y	Y

Note:
See text for data.

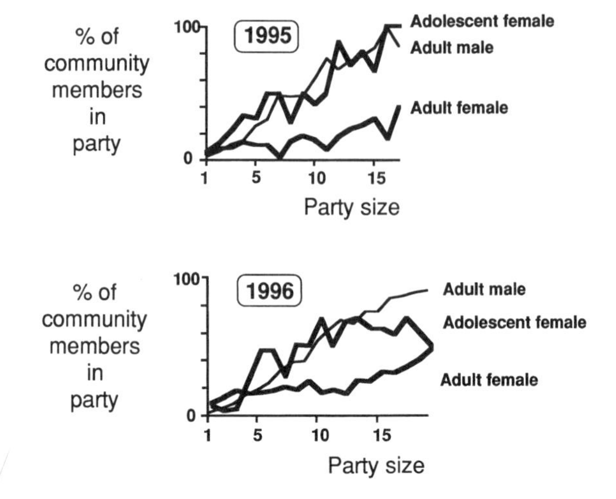

Fig. 21.1. Sex differences in gregariousness of Kanyawara chimpanzees. The proportion of community members in each of three age–sex classes is shown as a function of party size (adults and adolescents only). In the largest parties, a mean of over 90% of the community's adult males is present, whereas only about 40% of the community's adult females are present. This low degree of gregariousness by adult females contrasts to adolescent (nulliparous) females, who are as likely to be found in parties as males. Data source: all parties recorded in Kanyawara, initial composition only (i.e., each party scored only once per day, regardless of changes in membership). Sample sizes: 1995, $n = 313$ parties, 12 adult males, 1 adolescent female, 12 adult females; 1996, $n = 417$ parties, 13 adult males, 2 adolescent females, 14 adult females.

2. Bonding. In Kanyawara patterns of nearest neighbor association, association within parties, and grooming show matrix correlations indicating that males are more closely bonded than females (Wrangham *et al.*, 1992; Goldberg & Wrangham, 1997). Alliances tend to be found among the highest-ranking males, and males form coalitions in contests over rank and when guarding parous females within two to three days of detumescence (i.e., the peri-ovulatory period; R.W. Wrangham, unpublished data). Ngogo bonding patterns appear similar (D. Watts, personal communication). None of four male–male alliances in Kanyawara so far analyzed genetically has been between matrilineal relatives (Goldberg & Wrangham, 1997).

TAÏ

1. Grouping. In Taï, evidence indicating that females were less social than males comes from mixed parties, which were consistently male biased (0.54–1.34 males per female) in comparison to the community sex ratio (0.26 males per female) (Boesch, 1996a). However, there appears to be less of a sex difference in sociality than in Gombe, Mahale, and Kibale. For example, Boesch (1996a) concluded that 'females and males associate together for most of their time,' partly because females were rarely alone (4% of parties, compared to 6% for males).
2. Bonding. Boesch (1991) found that 17 out of 24 adult females in 1987 had 'at least one stable female associate with whom they shared food and formed coalitions to support each other in various contexts.' These friendships lasted for many years. Boesch (1991) stated that both male and female Taï chimpanzees form intrasexual bonds, and related this to the reduced sexual dimorphism in sociality compared to East African chimpanzees.

BOSSOU

1. Grouping. The Bossou community has had only one or two adult males, so sex differences in grouping are unknown. In Sakura's (1994) study, the adult male was one of the two most social individuals, compared to eight females, several of whom were relatively solitary. Mothers with nursing infants tended to spend more time alone than other females (Sakura, 1994).
2. Bonding. Male bonding and female bonding have both been recorded. Thus, based primarily on grooming frequency, Sugiyama (1988) found that female–female affiliative interactions occurred at higher than expected

rates, while male–male affiliation fluctuated but was less frequent than in Gombe, Mahale or Kibale.

The scramble competition hypothesis: velocity and sociality

FRUIT AVAILABILITY AND SEXUALLY ATTRACTIVE FEMALES BOTH INFLUENCE PARTY SIZE

Although there is consistent evidence that chimpanzee party size is correlated with fruit availability (see Introduction), party size is also related to other factors, such as danger (Sakura, 1994), hunting (Boesch & Boesch, 1989), and the number of females with maximal sexual swellings (Goodall, 1986, p. 158; see also Nishida, 1974; Sakura, 1994; Boesch, 1996a). If the number of sexually receptive females is positively correlated with fruit availability, the apparent correlation between party size and fruit availability could be spurious. However, Kanyawara data indicate that party size is influenced separately by both the number of maximally tumescent females and fruit availability.

First, party size (indexed as the residual of party size on month) was correlated both with the number of maximally tumescent females per month (e.g., parous females, $r^2=0.16$, $n=101$, $p<0.001$), and with fruit availability ($r^2=0.21$, $n=28$, $p<0.001$). Multiple regressions that used various different indices of these variables yielded a consistent result. Overall, party size was predicted best by a combination of both tumescent females and fruit availability ($r^2=0.28$, $n=92$, $p<0.0001$). Both independent variables were significantly correlated with the monthly residual of party size. Though the number of sexually receptive females was a greater influence (number of maximally tumescent females per month: partial $r=0.46$, $p<0.001$; fruit availability: partial $r=0.22$, $p<0.05$), fruit availability continued to predict party size even when female sexual availability was accounted for.

DAY-RANGE AND PARTY SIZE: DO CHIMPANZEES EXPERIENCE TRAVEL COSTS IN LARGER PARTIES?

Frugivorous primates that live in stable groups tend to travel further per day when in larger groups, apparently as a result of scramble competition (Janson & Goldsmith, 1995). Chimpanzees, in contrast, because they have a fission–fusion system, have the option of adjusting their party size so as to avoid this cost of scramble competition (increased travel distance) when they find themselves in a party that travels a long way. Nevertheless, they do not always do so. First, females (mostly mothers) and males in

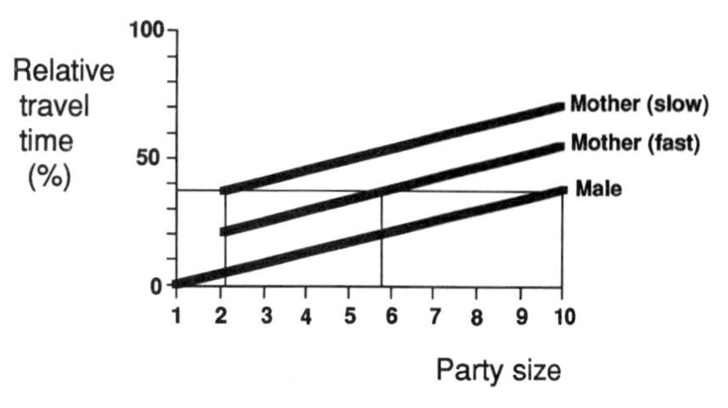

Fig. 21.2. Effect of sex difference in velocity on party size. The graph illustrates the scramble competition hypothesis. 'Relative travel time' is defined as the percentage increase in time spent traveling compared to a lone male, for individuals traveling in parties. Slopes show the rate of increase of travel distance for 2175 Kanyawara parties from 1993 to 1995 (see text). Mothers' travel times are higher than those of males when they are in parties, because mothers walk more slowly while covering the same distance as their companion males. Two elevations are shown for mothers, F min (fast) and F max (slow), respectively 16% and 32% higher than males, based on data in Table 21.3. showing that anestrous females are 16–32% slower than males. No points are shown for lone mothers, because they can choose to travel a shorter distance than males (and normally do so), and therefore are not constrained to have an increased travel time compared to males. On the assumption that mothers and males are constrained to the same maximal relative travel time, the sex difference in velocity causes mothers to optimize party size at a lower value than males. This is illustrated by males in a party of ten individuals, who have the same relative travel time as mothers in parties of 2.1 to 5.8 (depending on whether the mothers are slow or fast, respectively).

Gombe traveled significantly further per hour when in parties than when alone. The median increase was 15–28% (Wrangham & Smuts, 1980). Second, data from a two-year period at Kanyawara (1993–5) for 2175 different parties showed a small but non-significant increase in distance traveled per hour as parties grew from one to 15 ($r^2=0.17$, $n=15$, n.s.; see slope in Fig. 21.2). In combination with the evidence that party size is influenced by fruit availability, these data suggest that although chimpanzees generally respond to scramble competition by reducing party size, they nevertheless sometimes tolerate increased travel costs incurred by the presence of companions.

SEX DIFFERENCES IN SCRAMBLE COMPETITION: THE EFFECT OF VELOCITY

The costs of scramble competition include time, energy or risk, all of which increase with distance traveled. Time is

Table 21.3. *Velocities of lone or peripheral chimpanzees*

	Between fruit patch		Joining other chimpanzees	
	Velocity	*n*	Velocity	*n*
Adult male	3.2	128	3.1	106
Female (no sexual swelling)	2.7	133	2.3	65
Female (sexual swelling)	2.8	45	2.4	20

Note:
Velocities are in km/h, converted from feet/s. n = number of observations. Data are from eight males and ten females in Mahale Mountains National Park, Tanzania (Hunt, 1989, Table 5.23). All male–female differences were significantly different, assuming that individual differences were unimportant ($p < 0.001$).

easier to measure. If individuals vary in their mean velocity while traveling between fruit patches, lower velocities must cause a longer time spent traveling, and hence a higher cost of scramble competition. A critical question, therefore, is whether female chimpanzees travel more slowly than males.

The only data on chimpanzee velocities are those collected by Hunt (1989). Table 21.3 shows that lone males traveled some 16–32% faster than lone females, whether between fruit patches or parties of chimpanzees.

The specific effects on party size of this sex difference in velocity depend on how travel costs rise with party size, which itself depends on how willing individuals are to sustain extra travel as parties grow. We can estimate the rate of increase of travel distance with party size by using the actual slope of this relationship from the Kanyawara 1993–5 data. In Fig. 21.2, it is assumed that males and females experience an equal increase in travel distance as party size increases, that their difference in travel time is predicted solely by their difference in velocity, and that the travel cost (in time) that females will tolerate is the same as it is for males (see Table 21.3). According to this logic, Fig. 21.2 shows that for a male party size of ten, female party size is predicted to be between two and seven, depending on the estimate of sex difference in velocity. This corresponds well with actual differences in male and female party size (e.g., Wrangham & Smuts, 1980).

THE EFFECT OF MOTHERHOOD ON VELOCITY AND GROUPING

Hunt (1989) also studied how the presence of companions influenced a chimpanzee's velocity. He found that males traveled at the same mean velocity regardless of whether they were alone or in a party. Females without sexual swellings (i.e., mostly mothers with dependent infants) traveled slower than males regardless of the presence of companions. In practice, mothers with offspring tend to travel at the rear of a party, typically arriving at the next fruit tree several minutes after the males (personal observation). Therefore, at depletable resources, their slow travel means they lose feeding opportunities to their faster companions. On the other hand, females with sexual swellings (i.e., mostly females without dependent infants) traveled faster when in parties than when alone (3.1 km/h in parties, compared to 2.4 km/h alone, $p < 0.01$; Hunt, 1989). These data imply that females without infants were more able to modify their velocity, in relation to the presence of companions, than females with infants, and suggest that it is the presence of an infant that is particularly responsible for the slow travel of mothers.

The scramble competition hypothesis can therefore explain easily the strong tendency for mothers with infants to be less gregarious than other females (e.g., Fig. 21.1, Table 21.4). The demands of infant care presumably cause mothers to travel slowly, either because of delays caused by the infant traveling independently, or because infants are heavy (cf. Altmann & Samuels, 1992).

If low velocity rather than gender or dominance rank is responsible for asociality, it can be expected to cause unhealthy, wounded or senescent males to become asocial. In agreement with this prediction, Goodall (1986) notes several such cases. For example, two old adult males became as asocial as parous females during their last years of life.

Comparison with other species: why is relative male gregariousness so rare?

The scramble competition hypothesis for relative male gregariousness can be expressed in terms of three major assumptions. First, chimpanzees experience scramble competition so intense that individuals adjust their party size in order to keep travel distance low. Second, the time cost of

Table 21.4. *Travel time by sex*

Site	Females (%)	Males (%)	Source
Gombe	12–13	13.0	Wrangham & Smuts (1980)
Mahale	10.1	10.6	Hunt (1989)

Note:
Numbers show estimated percentage time traveling. Gombe data are based on 12 all-day observations for females, and 54 for males. Mahale data are from instantaneous scan samples (3683 female, 5282 male).

travel between fruit patches is higher for mothers than for other adults. Third, the total time that individuals can afford to spend traveling, given a particular set of ecological conditions, is the same for males and females. Can these three principles explain the distribution of relative male gregariousness in other species?

FISSION–FUSION SPECIES WITH ASOCIAL FEMALES

Spider monkeys are the only non-human primates other than chimpanzees in which females are clearly less gregarious than males (Symington, 1988a, 1988b; Chapman, 1990a). As with chimpanzees, they are ripe-fruit specialists that do little foraging on journeys between fruit trees, they have a fission–fusion society, and feeding competition has been invoked to explain both these features (Symington, 1988a, 1988b; Chapman, 1990a). Noting that females are often excluded from the best feeding sites, Symington (1988a) suggested that contest competition is more intense for females than for males. Chapman (1990a), on the other hand, implied that the level of competition is the same for the two sexes, but argued that females are more sensitive to its effects because their reproductive success is more closely tied to food acquisition than it is for males. These hypotheses have not yet been exhaustively tested.

In support of the scramble competition hypothesis, scramble competition has been documented among spider monkeys by a systematic increase of travel distance with party size (Symington, 1988a). Remarkably, indeed, Janson & Goldsmith (1995) found that the rate of increase of travel distance with party size in spider monkeys was one of the highest in their survey of primate species. While this supports the scramble competition hypothesis by showing that scramble competition was particularly intense, it is surprising because it means that, to a considerable degree, spider monkeys failed to lower party sizes in response to competition. Nevertheless, intense scramble competition is easily explicable because spider monkeys eat from depletable fruit patches, with little foraging between fruit patches (Chapman *et al.*, 1995).

The second expectation is that mothers are slower than males. Data on spider monkey velocities are not yet available to test this prediction but, like chimpanzees, females are known to have smaller core areas than males (Symington, 1988b). This raises the possibility that females also have shorter day-ranges than males.

The third expectation is not met: Symington (1988b) found that males spent more time traveling than females with infants (male 30%, mother 20%). This suggests that there are sex differences in their tolerance of total travel time, perhaps because the burden of carrying infants is high for spider monkeys, i.e., in this species, energy may be more limiting than time. Further data are therefore needed to understand this system, with its striking similarities to, but important differences from, chimpanzees.

Cheetahs are the only carnivore in which some adult males are more gregarious than females (Caro, 1994). Male coalitions travel together in small ranges where females are likely to occur. Many carnivores leave their young at dens, allowing mothers to travel without the burden of carrying their young or being delayed by them. Cheetah offspring, however, travel with their mothers until they are 18 months old. The travel velocities of cheetah are unknown, but if cheetahs follow the chimpanzee pattern, the scramble competition hypothesis predicts that larger parties travel further and/or have reduced food intake per capita, that mothers with young will be found to travel more slowly than adult males or females without young, and that the total travel time will be the same for mothers and gregarious males.

FISSION–FUSION SPECIES WITH SOCIAL FEMALES

Bonobos present a striking contrast to chimpanzees, because in spite of their close phylogenetic relationship and anatomical, locomotory and ecological similarity to chimpanzees, female bonobos are at least as gregarious as males, and female bonding occurs (White, 1988, 1992; Kano, 1992). Most attempts to solve this puzzle conclude that competition is relaxed in bonobos compared to chimpanzees. However, exactly how or why competition is relaxed among bonobos has been a matter of debate about the relative importance of the nature, density, and size of both preferred and fallback foods (see Wrangham *et al.*, 1996, for review).

The scramble competition hypothesis suggests a new dimension of solution. Based on inductive observations, Wrangham *et al.* (1996) proposed that the social ecology of chimpanzees and bonobos has diverged evolutionarily as a result of bonobos having greater access to a class of terrestrial herbs that allow 'feed-as-you-go foraging' ('H-THV,' or high-quality terrestrial herbaceous vegetation). In other words, chimpanzees traveling between fruit patches typically walk without feeding, whereas bonobos forage as they travel. Wrangham *et al.* (1996) suggested that this difference reduced the intensity of scramble competition for bonobos, but did not specify the mechanism.

In light of the hypothesis presented for chimpanzees, however, a clear prediction emerges (Fig. 21.3). If the scramble competition hypothesis applies, a key difference between bonobos and chimpanzees is that bonobos are adapted to a régime of feed-as-you-go foraging between fruit patches. This is responsible for reducing the intensity of scramble competition.

The logic is as follows. Bonobos forage as they travel because their environment provides abundant terrestrial herbs offering high-quality food; the result of their foraging as they travel is that the party travels at a lower velocity than non-foraging mothers would do on a journey between fruit patches; therefore, extra travel between fruit patches imposes the same time costs on females and males; therefore, there is no pressure on mothers to be non-gregarious. In addition, the fact that food is being obtained during travel between fruit patches lowers the cost of extra travel, so that parties are not subject to such an intense form of scramble competition as occurs in chimpanzees. This can therefore explain why bonobo parties are both more stable and sexually unbiased, compared to chimpanzees (Chapman *et al.*, 1994). A suite of behavioral changes follows easily from this divergence (Wrangham & Peterson, 1996). Bonobo data are therefore compatible with the scramble competition hypothesis, but further tests are needed.

FISSION–FUSION SPECIES WITH RELATIVELY SOCIAL FEMALES

Orangutans provide an instructive contrast to chimpanzees. Although these two apes are similar in being large-bodied frugivores subject to intense feeding competition, the slight sex difference in sociality in orangutans (females and small males are marginally more social than large males) is the reverse of the chimpanzee pattern (males are substantially more social than mothers) (Rodman & Mitani, 1987; Sugardjito *et al.*, 1987; van Schaik & van Hooff, 1996). To the extent that large males are arboreal, they are slower than females and small males, so that the contrast is easily explained by the scramble competition hypothesis.

STABLE-GROUP SPECIES WITH GREGARIOUS FEMALES BUT SCRAMBLE COMPETITION

Frugivorous monkeys living in stable groups experience scramble competition (Janson & Goldsmith, 1995). Since mothers in these species also carry their infants and thus face a travel burden not normally experienced by males, why are mothers as gregarious as males?

The scramble competition hypothesis predicts that, as in bonobos, the foraging style of frugivorous (and folivorous) monkeys reduces the mean velocity of group travel between fruit patches to a value at or below that at which mothers would travel if moving directly between fruit trees. Slow foraging between discrete food patches can occur for various reasons. Baboons (*Papio anubis*), for example, often travel slowly because they forage for grass as they walk, whereas arboreal cercopithecines search for insects while traveling. In some folivores, such as red colobus (*Colobus badius*), there is little feeding between major food patches, but the group moves slowly, perhaps because gut passage rate is slow, so that there is no need to reach the next food patch quickly.

These comparisons suggest that relative male gregariousness is rare because few species experience scramble competition so intense that the optimum realized velocity of mothers is below that of adult males. This complements the traditional hypotheses that in species with gregarious mothers, the benefits of grouping are higher than in chimpanzees and spider monkeys (whether in defense against infanticide, predation, or contest competition). Tests of the scramble competition hypothesis will be possible using data on scramble competition, velocities, and travel costs.

DISCUSSION

All analyses of chimpanzee social ecology agree that feeding competition plays an important role in regulating relative male gregariousness, but the specific mechanisms by which competition has its effects have remained uncertain. The scramble competition hypothesis suggests that a key feature is the lack of feed-as-you-go foraging between fruit trees. This places a premium on rapid travel between fruit patches. But, since mothers have a lower optimum velocity than males, they experience steeper increases in travel time as party size (and therefore travel distance) increases. Accordingly, optimum party size is lower for mothers than for males. The scramble competition hypothesis purports to explain not only sex differences, but also why it is specifically mothers and other slow individuals who tend to be less gregarious.

The scramble competition hypothesis also suggests that

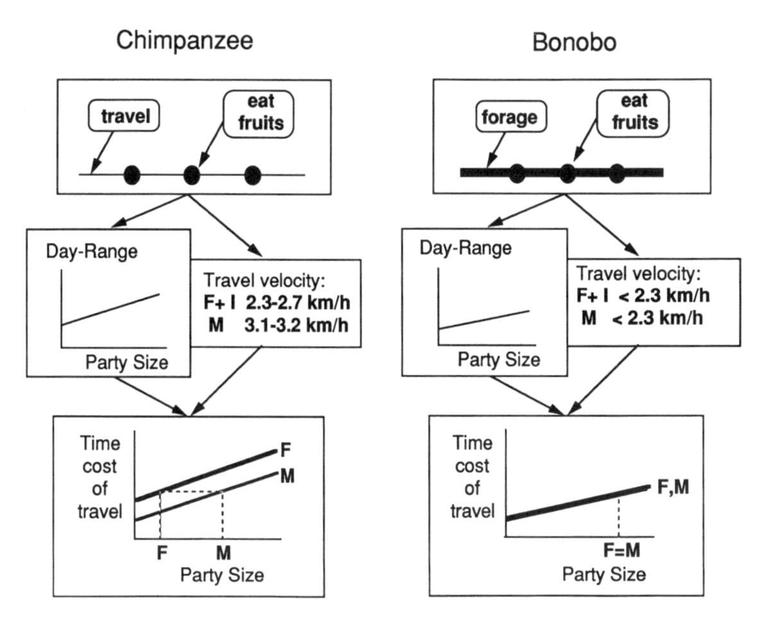

Fig. 21.3. Hypothesis for the ecological cause of differences in chimpanzee and bonobo grouping. Chimpanzees and bonobos both eat from widely separated fruit patches. In this model, chimpanzees traveling between fruit patches encounter no foods to slow them down. Scramble competition causes total travel distance (e.g., day-range) to rise with increasing party size, because parties often deplete fruit trees without satisfying all individuals. Velocities are lower for mothers (F + I = female plus infant) because of the time or energy burden of traveling with an infant. The party size at which travel time is held constant is therefore lower for mothers than for males, as in Fig. 20.2. Bonobos, by contrast, forage at small herb sites as they travel between fruit patches. This 'feed-as-you-go' pattern has two effects. First, by providing extra food, 'feeding-as-you-go' reduces the intensity of scramble competition, lowering both the elevation and slope of the party-size/day-range curve. Second, it means that travel velocity for the party is no higher than the optimal velocity of a female carrying an infant (here suggested to be the same for bonobos as it is for chimpanzees, i.e., 2.3–2.7 km/h). Accordingly, by this hypothesis, travel times for females and males rise equally slowly for both sexes, creating no pressure for a sex difference in gregariousness.

both males and females attempt to keep travel time between food patches below a threshold (the level of which needs to be explained). This postulated travel-time restriction may account for mothers having shorter day-ranges than males (Wrangham & Smuts, 1980; Goodall, 1986), a difference that has important social consequences in chimpanzees. Thus, longer day-ranges enable males to cover the community range more evenly than mothers, and to visit boundary areas more often (e.g., Chapman & Wrangham, 1993). The result is that males have lower costs of being involved in inter-community interactions (as long as they are in groups, i.e., safe!) and are more likely to encounter a variety of mothers, each restricted to a section of the community range. The sex differences in both day-range and gregariousness may therefore be understood as results of the importance of minimizing travel costs.

The scramble competition hypothesis also predicts that chimpanzee mothers have lower optimum velocities than males or females without offspring, and that optimum velocity is correlated with observed mean velocity. These are unsurprising predictions since infant carrying is known to be costly in other species (Altmann & Samuels, 1992). Tests are possible in the wild by observing the effects of different travel velocities on behavioral measures of fatigue, but would be easier in captivity using treadmills and physiological assays (cf. Taylor & Rowntree, 1973).

An alternative to the scramble competition hypothesis is that, in chimpanzees, the costs of motherhood are particularly high in comparison to other species. However, neonate mass and infant growth rate in chimpanzees are predicted well by maternal body mass (Ross & MacLarnon, 1995; Ross, 1998). Thus, there is no evidence in support of chimpanzees having an intrinsically high cost of motherhood. Another alternative is that the benefits of solitary life are particularly high for chimpanzee mothers. This is possible during the neonate's early weeks, due to the risk of infanticide from females or males (Goodall, 1986). However, even when infants are no longer vulnerable, mothers show the non-gregarious pattern, suggesting that this explanation has little to contribute.

Variation between populations may be explicable by the scramble competition hypothesis. As with bonobos, in chimpanzee sites where females are relatively gregarious (i.e., Taï and Bossou), the travel velocities of parties are expected to be no faster than those of mothers. A possible explanation is that foraging between fruit patches occurs more in these sites than in the East African sites (Gombe, Mahale, Kibale) where the sex differences are more pronounced. Ecological and behavioral data will allow tests of this hypothesis.

With respect to the evolution of male bonding, the implication of this analysis is that it has been permitted as a consequence of constraints on female bonding. Thus, if female bonding can evolve as a typical species feature, it favors female philopatry (or, as it should be called, philomatry: love of the motherland); selection for inbreeding avoidance in males then militates against male philopatry (Wrangham, 1980; van Hooff & van Schaik, 1994). In chimpanzees, however, the costs of motherhood and intense scramble competition constrain female gregariousness. Therefore, female bonding has not been favored sufficiently strongly for philomatry to evolve. Accordingly, males have been able to develop their cooperative potential by forming male bonds and lifelong relationships, leading to (male) philopatry.

In contrast to chimpanzees, spotted hyenas have fission–fusion systems with philomatry and female bonding (Frank, 1986; Smale *et al.*, 1997). The scramble competition hypothesis suggests that it is females who bond, rather than males, partly because in hyenas, the travel costs of motherhood are small (infants are denned and are never carried by the mother). This allows females to be as gregarious as males. Why intersexual conflict should be resolved in favor of female rather than male bonding remains an open question, but contest competition may be important (Frank, 1986).

Humans share with chimpanzees a number of relatively rare features, including fission–fusion social organization, more widespread and consequential male bonding than female bonding, and apparently longer day-ranges in males than in females (Rodseth *et al.*, 1991). Human foragers also seem likely to have regularly experienced intense scramble competition, to judge both from the high quality of the acceptable range of food items and evidence that during lean seasons, foraging units are small (e.g., Silberbauer, 1981). I have not found data on whether males are more gregarious than mothers, but it is clear that, as in other infant-carrying primates, the travel costs of motherhood are substantial (Blurton Jones *et al.*, 1989). The comparison suggests that intense scramble competition may have constrained the potential for female bonding in human foragers, as in chimpanzees. This hypothesis can be tested by generating models of pre-agricultural scramble competition, and by actualistic studies of human foraging practice.

This view suggests that in the hominid line, as in chimpanzees and spider monkeys, male bonding was able to emerge because female bonding was inhibited. Important benefits of male bonding are likely to have come from the same source in all three species, i.e., an arms race for male power (Alexander, 1987; Wrangham & Peterson, 1996). Accordingly, the scramble competition hypothesis suggests that human male bonding, male inter-group aggression, and patriarchy derive ultimately from our species being restricted to such high-quality foods that feed-as-you-go foraging was not favored.

ACKNOWLEDGMENTS

Thanks are due to Kevin Hunt and Mark Leighton for initial discussions, to Lisa Schweigler for data analysis, to Rachel Carmody and Andy Marshall for helpful comments, and to Jeanne Altmann, Peter Kappeler, and David Watts for their constructive reviews.

22 • Male mating strategies: a modeling approach

ROBIN I.M. DUNBAR

INTRODUCTION

It has often been suggested that the basic structure of primate societies is created by the females' decisions about how they should forage and group, with males simply mapping themselves onto the distribution patterns of the females (Goss-Custard *et al.*, 1972; Wrangham, 1980; Dunbar, 1988). In this chapter, I want to consider the decisions that the males make. I shall consider the male's problem as being essentially an optimal foraging problem in which he has to decide whether or not to join a specific female group (i.e., be social) once he has found one or to pursue a roving male strategy in which he searches continuously for receptive females with whom to mate, staying with the female(s) only long enough to ensure conception.

Emlen & Oring (1977) were perhaps the first to point out that male mating strategies could be explained in terms of males' abilities to monopolize females. They argued that males have a choice between group-defense polygamy and area-defense polygamy, with the male's ability to defend his patch (whether spatial or organismic) depending on the operational sex ratio (the number of reproductively active females per breeding male in the population). The operational sex ratio established the level of intrasexual competition within the population and this in turn was assumed to determine a male's ability to defend a mating resource patch. Which tactic males pursued would also depend on female foraging and ranging habits because these determined the size (or richness) of the resource patches.

This suggestion was developed into a formal model by Dunbar (1988), who argued that the substantive issue revolved around the likelihood of more than one female being in estrus (i.e., available for mating) on any given day. The conventional priority-of-access model (Altmann, 1962; Suarez & Ackerman, 1971) would imply that if only one female is in estrus, the dominant male will be able successfully to prevent other males from mating with her. Only when his attention is distracted by the presence of two or more females in estrus simultaneously will other males be able to gain access to receptive females.

Whereas the priority-of-access model considered this problem only with respect to multi-male groups, the same approach can be applied to the more general problem of whether males should be social or opt for a roving male strategy. It will only be worth a male's while joining a group of females if the number of females in estrus is more than the number of males already in the group who are more dominant than him. Thus, not only should the priority-of-access model predict the distribution of matings within a multi-male group, it should also predict whether or not males are social and, when males prefer the social option, how many males one should find in individual groups. Males should, in effect, distribute themselves around the habitat in direct proportion to the number of estrous females potentially available.

This chapter builds on a number of attempts to model the mating strategies of male mammals (Dunbar, 1984, 1988, 1995a, 1995b; Dunbar *et al.*, 1990; van Schaik & Dunbar, 1990; Srivastava & Dunbar, 1996). These models all revolve around a core of key issues, although each considers the problem in a specific context. The principal aim of this chapter is to try to pull together some of the main points that emerge from these models. In the first section, I present the basic model and its derivation. Since female dispersion is conceived as the principal driving force behind males' decisions, the second section explores the impact of female grouping patterns. Finally, I consider the effect that rivals have on the decisions that males make.

It is important to remember that the actual behavior of animals that we observe is the outcome of the animals' attempts to implement optimal strategies. As with all models, our concern here is with how animals should behave in the behavioral vacuum before anyone has actually implemented any decisions about what to do. All these models thus begin from first principles, and try to consider the bare bones of the situation shorn of all the complexities introduced by animals' strategic responses and counter-responses. Having established how the process should work in a vacuum, we can then ask whether this minimalist model is good enough to explain what we see. The interesting questions for future work are provided by those points at which the model does not explain all the variance in the data. These point us in the direction of hidden variables.

Note also that I shall not consider females' decisions at all in this context. This is not because I do not consider that females play a role in determining the mating system; rather, this is a heuristic device intended to allow us to examine each component of the system separately. For present purposes, I have considered females as essentially passive players in the mating system whose decisions are confined to deciding whom to associate with. Females' decisions about grouping patterns are presumed to drive the males' decisions, but not specifically to influence mating strategies. Once we have understood how the bare bones of the male component of the system works, we can then add in further layers of complication by introducing the females. Models that consider both sexes at the same time have been developed (e.g., for callitrichids: see Dunbar, 1995a, 1995b), but they are very complex.

A GENERAL MODEL OF MALE MATING STRATEGIES

The simplest situation from a male's point of view is to exclude all rivals from having access to the females with whom he wants to mate. All else being equal, the male would presumably prefer to be the only breeding male in the entire population, but pressure from rivals is unlikely to allow him to be able to get away with that, simply because, at some point, he will not be able to prevent other males gaining access to at least some of the females in the local population when he is defending other females elsewhere. The first issue we have to address here, then, is: how many females can a male monopolize?

The male's choice lies between 'defending' individual groups of females and 'defending' an area within which he has priority of access to all females present. (I use the term *defend* loosely here, to imply the ability to prevent rivals from mating with the females. Whether or not the male does that by forcibly excluding rivals is immaterial to the present argument. The issue here is whether the male can prevent other males from mating with the females under his purview.) In effect, the male has two choices: he can associate with one group of females all the time (the social strategy), or he can defend an area and search for individual females or groups of females (the roving male strategy).

The male's problem, then, is to decide whether he will achieve more conceptions by being social or by roving. A social male knows exactly how many conceptions he will achieve: as long as he is the only breeding male, he will sire his female's offspring every time she comes back into breeding condition. The more difficult problem is to determine how many sirings a roving male can expect to achieve during the equivalent time period. To determine this, we need to know two key components. The first is the size of the area that a male can defend (relative to the size of ranging area that a female needs to support herself and her offspring). The second is whether a roving male is likely to encounter the females within his territory often enough to be able to ensure fertilization.

We can estimate the first component by using the original Mitani & Rodman (1979) finding that territoriality in primates can be predicted from the relationship between home-range size and day journey length. They found that species (or populations) in which the typical day journey length was larger than the diameter of the home range were territorial. This relationship works because it reflects (albeit crudely) the animal's ability to shuttle backwards and forwards across its home range often enough to detect (and thus deter) intruders.

[It is worth pointing out (yet again) that the issue here is not how animals actually travel around their territories, but how the ratio of day journey length to territory size constrains the ability to monitor all parts of the boundary with sufficient frequency not to miss too many of the intruders. Neither the data themselves nor Mitani & Rodman (1979) give any indication of how often the animals actually do this (or over what length of time); they merely imply that, on average, it is equivalent to a degree of mobility that would allow the animals to cross their range at least once a day. A biologically more realistic analysis by Lowen & Dunbar (1995), which estimated the proportion of the boundary that could be searched each day by a randomly foraging group, demonstrated that, despite its simplicity, the Mitani–Rodman finding was a reasonable approximation. These points have been made repeatedly in the literature, but it seems they have not been well understood.]

We can use the Mitani–Rodman formula for defendability:

$$D = d/(4A/\pi)^{0.5} \quad [1]$$

(where D is the defendability index, d is the day journey length, and A is the area of the home range) to estimate the maximum area that a male could defend by setting $D=1$ (the minimum permissible value for successful territory defense) and inverting equation [1] to give:

$$A_{max} = 0.25\pi d^2$$

(Dunbar, 1988). This gives us the maximum possible area that a male could defend, given his existing day journey length (on the assumption that, since he can obviously manage to cover the observed distance while staying with a female group, then there is no reason to expect him not to be able to travel at least as far if he goes roving on his own). Note that this is the maximum area he could defend *given that he moves about his territory in a random way* and has no information about the arrival patterns of intruders. If intruders arrive in a non-random fashion (spatially or temporally) or there are natural boundaries that do not require defense (e.g., rivers or a forest edge), he may be able to concentrate his search effort accordingly and thus defend an even larger area (see, for example, Barrett, 1995). Although in principle possible, we should not be distracted by the specific tactics that individual populations have evolved to improve their performance over baseline: our concern here is to establish the baseline from which all species are working.

Once we know how big an area a male can defend, we need to know how many females he could expect to find within his territory. To determine this, we need to know how big an area a female needs to support herself and her dependent offspring from one year to the next. The easiest way of estimating this is to calculate the combined metabolic body weight of the group as a whole, and use this to estimate the area one female and her dependent offspring would need. (The metabolic body weight is the body weight raised to the power 0.75, following Kleiber's Law for basal metabolic weight: see Dunbar, 1988.) A simpler approximation may be obtained by determining the number of adult-equivalents in the group (counting all the dependent offspring of a single female as one adult-equivalent). Thus, in a pair-bonding species like gibbons, a female would require 67% of the original range area to support herself and her offspring. This value can then be used to determine the total number of females that would be found within the male's maximum defendable area (with due allowance made for the male's own feeding requirements).

Two other factors need to be taken into account. One is the fact that males who increase their area of control will inevitably do so only by driving other males out (or neutralizing in some other way their ability to gain access to receptive females). For a conventional territorial male (i.e., one that excludes all other males from his territory), this will increase the pressure from rivals trying to invade the territory. However, if all males opt for the roving strategy, sexual selection may lead to increased sexual dimorphism as males compete with each other for control over territory. There may well be a trade-off between a male's ability to exclude rivals and the costs of increased growth, but the evidence from most ungulate and primate species suggests that increased sexual dimorphism is an inevitable correlate of polygamy (Alexander *et al.*, 1979). Thus, the fact that some males become excluded as males start to defend larger territories is not a problem in this context, at least on an evolutionary time scale. For similar reasons, we are not here concerned with subordinate males' responses to this situation once it arises (i.e., the adoption of alternative or best-of-bad-job strategies: see Dunbar, 1982).

A more serious problem is that males who wander at random may miss some females during the fertile phase, and thus fail to fertilize them. In addition, they may encounter some females whom they have already fertilized on other occasions. Simply counting the number of females on a male's territory may overestimate the number that he can actually fertilize. We can correct for this by calculating the encounter rate with non-fertilized females (cf. van Schaik & Dunbar, 1990). This is given by:

$$E(f) = FgN_f\,(2Krd/A_{max}) \quad [2]$$

where $E(f)$ is the number of females that the male can expect to fertilize during the course of an average reproductive cycle (defined by the mean interbirth interval). This is made up of two components: the number of times that a roving male can expect to encounter groups of females, $N_f\,(2Krd/A)$, and the expected number of fertilizable females per group, Fg. The first component consists of two parts: the probability that a randomly searching male will encounter a female group during an average reproductive cycle calculated as the proportion of the total territory, A_{max} that he can search in a full reproductive cycle (given a daily search path of d km length and a search radius of r km either side of the search path and a reproductive cycle of length K days: $2Krd/A$), and the number of female groups in his territory (N_f). Similarly, the mean number of fertilizable females in a typical female group is the expectation of a Poisson process with parameter g (the probability that a female will be at risk of conception on any given day) and the mean number of females in a female group (F). I assume that a female is receptive (i.e., fertilizable and not merely in estrus, as conventionally defined) for about five days on each of three menstrual cycles during any given reproductive cycle (based on data for baboons: see Dunbar, 1988, p. 154), hence $g = 15/K$ (where K is the interbirth interval, in days).

The male's problem is that he does not know for sure when in the female's cycle ovulation will actually occur (even in those species that exhibit conspicuous sexual swellings). Although the lengths of both the menstrual cycle and the estrous phase vary between species of primates, the duration of the conception window (the period around ovulation when the female is at risk of being fertilized) almost certainly does not, because it is a function of the timing of ovulation and the viability of sperm and eggs and these are more or less constant across species. Since it is only the risk of conception that is at issue here, we can, at least as a first approximation, ignore those differences between species that do occur. I also assume that a male stays with a receptive female for some period of time after locating her in order to maximize the chances of fertilizing her.

Using the mean interbirth interval (or the length of the reproductive cycle) as our time base is convenient because it means that a social male who has exclusive mating access to the female(s) in his group will gain exactly one conception for each female in the group during this time period (irrespective of whether or not the females are reproductively synchronized). In other words, the expected payoff for social males is the same as the number of females in the group. Thus, the male's decision rule is:

Rove if $E(f) > F$, else be social. [3]

THE IMPACT OF FEMALE GROUP SIZE

The simplest situation to consider, and thus the obvious place to start, is that where females live alone (irrespective of whether or not they do so in overlapping ranges). The alternative options in this case reduce to a choice between monogamy and area-defense polygyny. These options are exemplified, respectively, by the monogamous primates (e.g., gibbons, callitrichids and small cebids) on the one hand and by the kinds of mating systems seen in orangutans (*Pongo pygmaeus*) and many nocturnal prosimians on the other.

Van Schaik & Dunbar (1990) used this model to decide whether male gibbons would, in principle, do better to stay with a single female (be monogamous) or pursue a roving male strategy. A similar analysis of callitrichids was carried out by Dunbar (1995a). In both cases, $F = 1$, the number of breeding females in the group. (In callitrichids, only one female normally breeds in the group even if there are more adult females present.) The observed day journey lengths suggested that, in most cases, the males of both taxa would be significantly better off pursuing a roving male strategy than by staying with individual females. That males in these species do not pursue a roving strategy implies that some other factor tips the balance in favor of monogamy. (Since the payoff differentials are far from marginal, it cannot be assumed that this lack of fit between observed and predicted behavior merely reflects errors of parameter estimation.) The magnitude of the payoff differential using a ten-day fertilization window is in the order of 4:1 in favor of roving, suggesting that the costs that males would incur by roving are worth more than three offspring per reproductive cycle. The equivalent using a five-day fertilization window is 2:1. In other words, some factor(s) that males cannot control when they leave a female imposes a mortality rate in excess of 50% on the offspring born to roving males. (Strictly speaking, this should be a mortality rate that is greater than twice that incurred by the offspring of monogamous males.) Van Schaik & Dunbar (1990) argued that this mortality factor was infanticide by roving males. In callitrichids, however, an additional contributory factor is likely to be the fact that females cannot cope with the energetic costs of rearing up to two sets of twins each year when they have to do this without the males' assistance.

When females live socially, the number of guaranteed fertilizations obtained by a social male will at some point exceed the number obtained by a roving male. The limiting case will occur when all the females in the local population (i.e., those living within the male's defendable range) live in the same group and the roving male loses fertilization opportunities because it takes him a long time to locate another group once he has left one. As before, however, a male should prefer to rove as long as inequality [3] holds (i.e., when he gets more fertilizations by roving than by staying with a group).

Chimpanzees provide a particularly convenient taxon on which to test this prediction because their fission–fusion social system provides a range of variation in female party sizes both across and within habitats. I will here consider only between-population-level data, but the same analysis

Table 22.1. *Ecological and demographic data for great ape populations*

Population	d^a (km)	A (km^2)	F	Mean female party size	$E(f)^b$	Males in female parties (%)	Sources
Chimpanzee populations							
Mahale (Tanzania)	(3.5)	15	10	2.9	3.5	10.9	Nishida (1968)
Kibale (Uganda)	(3.5)	14.9	14	<5	5.0	46	Ghiglieri (1984); Chapman & Wrangham (1993)
Tai (Ivory Coast)	(3.5)	27	9	c. 3–5	1.8	74.0	Boesch & Boesch (1989); Boesch (1996b)
Bossou (Guinea)	(2.0)	5.5	7	(3–5)	3.8	60.6	Sugiyama & Koman (1979)
Bonobo populations							
Wamba (Zaire)	2.0	58	c. 17	11.5	0.9	86	Kuroda (1979); Kano (1992)
Lomako (Zaire)	(2.0)	22	12	4.5	6.5	86.2	White (1989)
Gorilla populations							
Virunga (Rwanda)	0.4	8	2.0	2.0	0.1	100	Fossey & Harcourt (1977)
Lopé (Gabon)	1.1	21.7	2.5	2.5	0.2	100	Tutin (1996)
Orangutan populations							
Kutai (Borneo)	0.7	5	4	1	0.9	0	van Schaik & van Hooff (1996)
Tanjung Puting (Indonesia)	0.9	5?	4	1	1.1	0	van Schaik & van Hooff (1996)

Notes:

[a] Day journey length data are not available for values given in parentheses; in these cases, day journeys are assumed to be comparable to Gombe for common chimpanzees (except Bossou) and Wamba for bonobos. Bossou is assumed to have a smaller day journey length than other sites because of its small area.

[b] Detection distance of $r = 0.05$ km is assumed for all sites.

could, in principle, be applied to grouping patterns over time within a population. To broaden the appeal of the model, I also include data for the other two great apes (gorillas and orangutans).

Assuming that females are fertilizable for a period of five days in each of three menstrual cycles during any given reproductive cycle, and that the mean detection distance (i.e., the distance at which females are *on average* detected) is $d = 0.05$ km, a male's preference for staying with a female group will be proportional to the ratio of $E(f)/F$, the payoffs for roving and social males, respectively. The relevant data are given in Table 22.1 for all populations for which data are available. For these analyses, I make no assumptions about the size of areas that males can defend, but instead simply use the observed male home range size for the parameter A (equivalent to A_{max} in equation [2]). Figure 22.1 plots the observed proportion of adult males who were associated with female groups against the ratio of payoffs for all these populations. There is clearly an excellent correlation: males' preferences for traveling with female groups do seem to depend on the males' perceptions of whether or not they would do better by roving in search of other groups. The best-fit linear regression equation is:

$$\% \; males = 104.2 - 77.1 \; payoff$$
$$(r^2 = 0.78, F = 28.88, p < 0.001).$$

If the males were behaving optimally, we would expect the point of indifference in their behavior (i.e., the point at which they choose equally between the two strategies) to lie at the point where the payoffs are equal (i.e., where $E(f) = F$), as indicated by the cross-hairs in Fig. 22.1. In fact, the percentage of social males at $E(f)/F = 1$ is 27.0% rather than 50%. The main cause of the discrepancy is the two orangutan sites, where males should be somewhat more social than they actually are (the payoff ratios are ≈ 1). Since the data points as a whole are distributed in a very tight pattern, it seems unlikely that this deviation from optimality is due to an inappropriate analysis. The most likely explanations are:

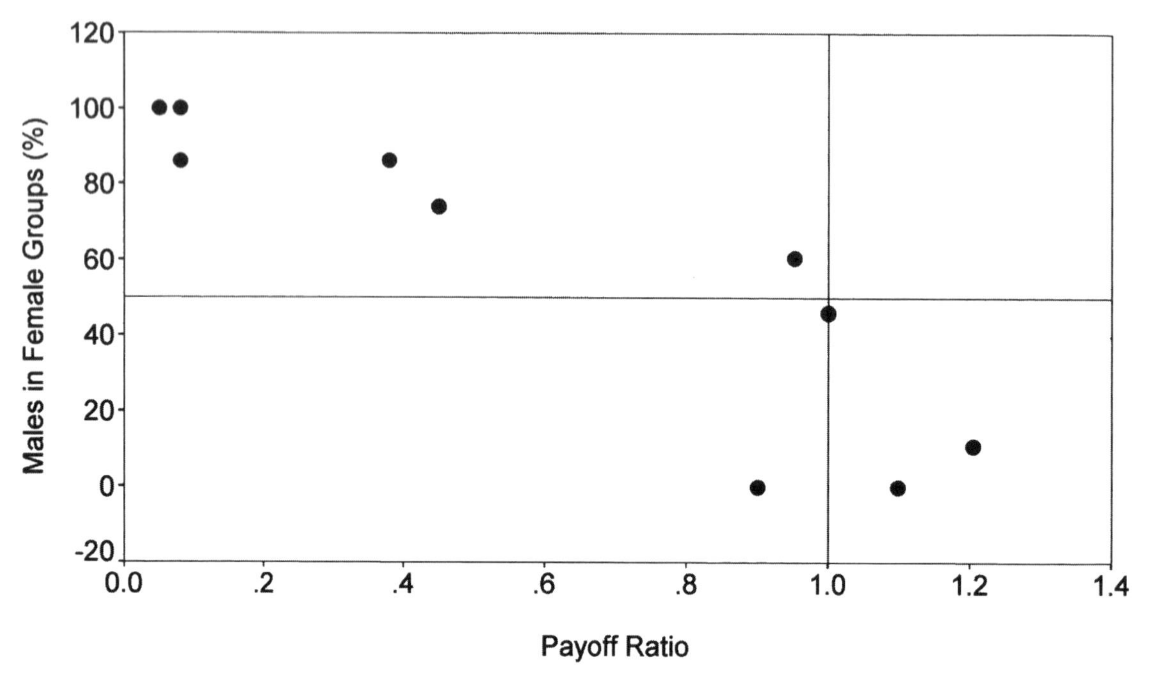

Fig. 22.1. The observed proportion of adult males who were associated with female groups plotted against the ratio of payoffs for social versus roving males for all populations for which there are relevant data. Source: Table 22.1.

(1) that the estimates for some parameters from the populations on the right-hand side of the graph are inaccurate (e.g., day journey lengths or association patterns at orangutan sites) and (2) that orangutan males are aberrant. If we omit the two orang sites, the fit is much improved:

$$\% \, males = 102.6 - 61.2 \, payoff$$
$$(r^2 = 0.88, F = 42.68, p < 0.001).$$

The equilibrium point at $E(f)/F = 1$ is 42.1% of males in female groups, a value that is not significantly different from 50% ($t_6 = 0.92, p > 0.50$).

Thus, the model seems to predict surprisingly well how males actually behave. This suggests that whatever females do in terms of their own mating strategies does not significantly alter the form of the problem that males seek to solve. An example of this may be the fact that female chimpanzees sometimes choose to join males (rather than it always being the other way around). This might seem to change radically the situation that the model seeks to describe. In reality, however, it has little impact on the model. In fact, the model is neutral as to how males find female groups: whether males attract females to them (as chimpanzee and orangutan males sometimes do with loud calls) or search actively for females (as prosimian males seem to do) is of no consequence for the model.

RIVALRY BETWEEN MALES

So far, the analysis has ignored the fact that males may compete with each other for access to females. If it pays males to be social (in a more or less permanent sense), then males should attempt to prevent other males joining the groups of females with whom they are associated. Following the Emlen–Oring hypothesis, a male's ability to keep rivals out will depend on how likely it is that more than one female is in estrus on any given day. As long as there is only one female in estrus, the male will be able to defend her and keep rivals out of the group. But if two females are in estrus, he may be unable to prevent rivals mating with one of them while he is preoccupied with the other.

Srivastava & Dunbar (1996) tested this prediction in Hanuman langurs (*Prebytis entellus*). They used a stepwise multiple regression, with the proportion of one-male groups in the population as the dependent variable, and a number of demographic, ecological, and climatic variables as independent variables. The best-fit equation was:

$$OMG = -15.3 + 10.3V + 0.3S$$

($n=21$, $r^2=0.71$), where OMG is the percentage of one-male groups, V is the number of dry months (average number of months with less than 50 mm of rainfall each year), and S is an index of breeding seasonality (calculated as the weighted number of months in the year when births occur: see Srivastava & Dunbar, 1996, for details). In this context, the number of dry months is a surrogate variable for the number of females in the group (the two variables are highly correlated, with the mean number of females per group increasing as the number of dry months increases: $r^2=0.575$, $F_{1,20}=9.88$, $p=0.005$).

We can test this hypothesis more directly by calculating the probability of two or more females cycling together on any given day for individual primate populations, and using this as an independent variable in a regression analysis to predict the number of males in groups. Mitani *et al.* (1996a) undertook such an analysis for a large sample of primates, but considered only the number of females in the group and the length of the breeding season as alternative independent variables in bivariate analyses. The substantive issue is not breeding seasonality *per se*, but the probability that two or more females will be cycling together (the probability of co-cycling females, $p(x \geq 2)$). We can calculate this from the binomial expansion:

$$p(x \geq 2) = \Sigma_{x=2} \binom{n}{x} p^x (1-p)^{n-x}$$

where p is the probability that an individual female will be at risk of conception on any given day in the year, and n is the number of females in the group. To estimate p, I once again assumed that females have a five–day conception window on each of three menstrual cycles during any given reproductive cycle (defined by the interbirth interval).

The demographic and reproductive data for the species sampled by Mitani *et al.* (1996a) are given in Table 22.2. Mitani *et al.* (1996a) included all species for which they were able to obtain demographic and birth seasonality data, irrespective of where these data came from. However, since both group composition and life history characteristics can vary significantly across a species' range, it is important that all data derive from the same population and time period. I have therefore corrected a number of values in that list. Mitani *et al.* (1996a), for example, give the mean number of females in some callitrichid species as two. Technically, this is correct: some species do have groups that contain several adult females. However, reproduction (and even puberty) is typically suppressed in all but one female in these species, so that, with only a few exceptions, the number of breeding females is actually one. Mitani *et al.* (1996a) also give male and female numbers for harems (reproductive units) in the case of hamadryas baboons (*Papio hamadryas*) and geladas (*Theropithecus gelada*). This is perhaps not unreasonable, but an alternative is to consider band size as the appropriate unit of analysis, since bands are equivalent to conventional groups for other primate species. I have preferred the latter option because it ensures consistency of social units across species (especially since the community rather than foraging party is used for chimpanzees and spider monkeys).

Because comparisons are being made between closely related species, I have carried out an independent contrasts analysis. I used the method of independent contrasts recommended by Harvey & Pagel (1991) (but without including branch lengths in the calculation) and Purvis's (1995) composite phylogeny for primates. All data were $\log_{10}$-transformed for statistical analysis; note that to avoid problems with log-transforms when $p=0$, I followed conventional practice in adding one to the value of all probabilities.

The results of a multiple regression with contrasts in the number of males as the dependent variable and contrasts in the number of females and in the probability of co-cycling females as the independent variables are shown in Table 22.3. The partial regression coefficient is significant only for the probability of co-cycling females. The constant is not significantly different from zero, as is required for an independent contrasts analysis. (Setting the regression through the origin reduces the significance levels slightly on both variables, but does not alter the outcome.)

Because the conception window could affect the results, I recalculated the probability of co-cycling females using a narrower conception window (three days in each menstrual cycle), but the results were identical. Once again, differences between species in the length of the menstrual cycle are less likely to be important than differences in the length of the conception window (which will vary much less between species).

We can check whether using bands rather than reproductive units for the gelada and hamadryas made any difference by using the results of this analysis to predict the numbers of males in the reproductive units of these two species. The best-fit bivariate regression equation for number of males versus probability of co-cycling females is:

Table 22.2. *Demographic and lifehistory data for species in the Mitani* et al. *(1996a) sample*

Species	Number of males	Number of females	IBI[a] (months)	Probability of co-cycling[b]
Lemur catta	4.5	4.0	12.0	0.00985
Callithrix jacchus	2.0	1.0 [c]	6.5	0.00000
Saguinus fuscicollis	1.9	1.0 [c]	10.2	0.00000
Saguinus mystax	2.0	1.0 [c]	11.5	0.00000
Saguinus oedipus	2.0	1.0 [c]	13.8	0.00000
Alouatta palliata	3.0	8.0	19.5	0.01661
Alouatta seniculus	1.5	2.5	16.6	0.00168
Ateles belzebuth	4.0	11.5	36.0	0.01097
Ateles paniscus	5.0	15.5	48.0	0.01110
Cebus apella	2.0	2.3	22.0	0.00072
Cebus capuchinus	5.5	4.0	19.0	0.00401
Cebus olivaceus	1.0	6.0	26.0	0.00527
Saimiri oerstedi	10.0	23.0	12.0[d]	0.24853
Saimiri sciureus	7.0	9.5	14.0	0.04306
Colobus guereza	1.7[e]	3.1[e]	21.4	0.03084
Procolobus badius	3.0	3.5	25.0	0.00114
Semnopithecus entellus	1.0	5.0	15.0	0.01039
Presbytis senex	1.0	6.0	23.5	0.00641
Cercocebus albigena	3.0	6.0	33.0	0.00331
Cercopithecus aethiops	3.0	4.3	16.0	0.00643
Cercopithecus ascanius	1.0	9.5	17.8	0.02768
Cercopithecus campbelli	1.0	4.0	12.0	0.00985
Cercopithecus mitis	1.0	18.0	21.5 [f]	0.06469
Cercopithecus neglectus	1.0	3.0	27.4	0.00099
Erythrocebus patas	1.0	12.5	14.6	0.06637
Macaca fascicularis	5.0 [g]	6.7 [g]	26.7 [g]	0.00627
Macaca mulatta	2.5	9.0	16.7	0.02805
Macaca radiata	7.0	9.0	12.0	0.05143
Macaca sinica	5.0	9.5	18.0	0.02711
Papio anubis	14.0	34.0	21.0	0.19378
Papio cynocephalus	8.0	13.0	21.0	0.03715
Papio ursinus	7.0	14.5	26.4	0.03000
Papio hamadryas [h]	15.7 (1)[j]	22.4 (1.8)[j]	24.0	0.07860
Theropithecus gelada [i]	34.0 (1.3)[j]	86.0 (4.1)[j]	24.0	0.53717
Gorilla gorilla	1.0	3.0	47.0	0.00034
Pan paniscus	8.0	8.0	60.0	0.00188
Pan troglodytes	10.0	35.0	72.0	0.02466

Notes:

[a] Interbirth interval

[b] Probability that two or more females will be in estrus on the same day.

[c] Mitani *et al.* (1996a) give 2.0 females for all callitrichid species, but since only one female is normally reproductively active in any one group, I use a value of one for the number of breeding females.

[d] Interbirth interval from S. Boinski (personal communication).

[e] Demographic data for Kibale, from Oates (1977).

[f] Interbirth interval from M. Cords (personal communication).

[g] Values from de Ruiter & van Hooff (1993).

[h] Demographic data for bands (from Sigg *et al.*, 1982).

[i] Demographic data for bands (from R. Dunbar, unpublished data for Sankaber population).

[j] Values in parentheses for reproductive units (following Mitani *et al.*, 1996a, but corrected from original sources). Note that the probability of two co-cycling females will be different in each case.

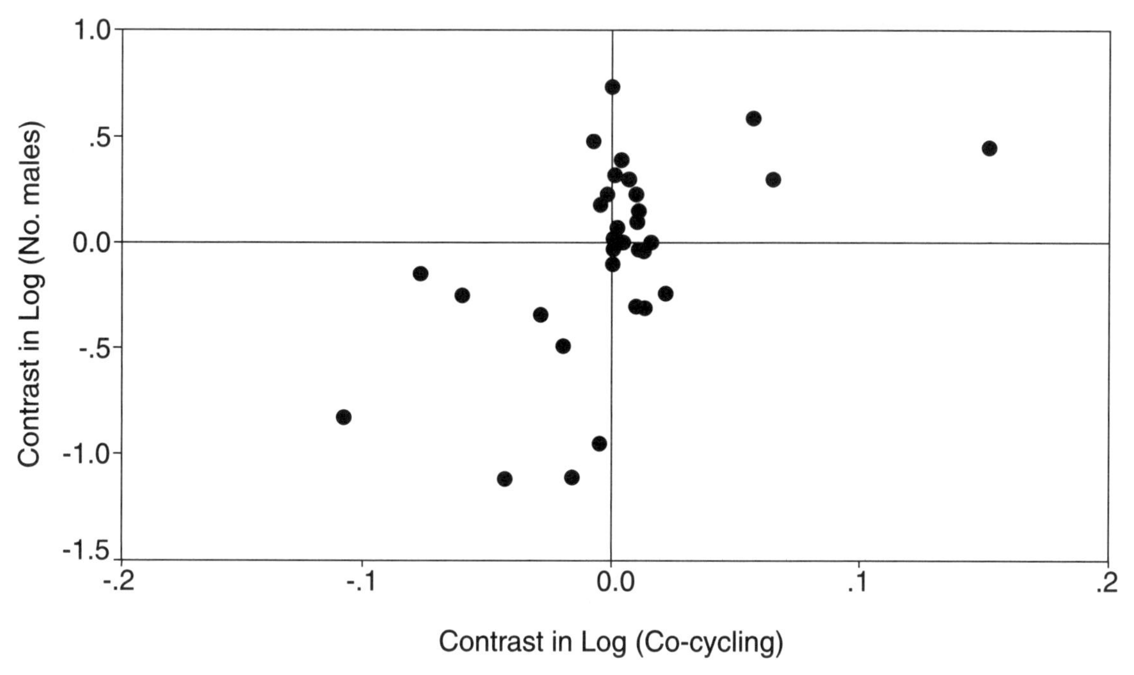

Fig. 22.2. Contrasts in the number of males in a group plotted against constrasts in the probability of co-cycling females (see text for details). Both variables were $\log_{10}$-transformed for analysis (with one being added to the probability of co-cycling females to avoid problems with logging values of $p=0$). Source: Table 22.2.

Table 22.3. *Multiple regression of factors influencing the number of males in primate groups: analysis based on independent contrasts of log_{10}-transformed variables*

Factor	Slope	t	P
Number of females	0.230	1.10	0.278
Probability of co-cycling	4.339	2.23	0.033
Intercept	−0.034	−0.53	0.603
ANOVA	$F_{2,33}=7.02, p=0.003, r^2=0.298$		

$$\text{Contrast } \log_{10}(\text{males}) = -0.052 + 5.61 \text{ contrast } \log_{10}(\text{co-cycling}) \qquad [4]$$

($r^2=0.27, p=0.001$). Substituting the appropriate probability of co-cycling females (based on the number of females per reproductive unit) into equation [4] yields predicted numbers of males of 0.90 for the gelada and 0.89 for the hamadryas: neither of these differs significantly from the observed values of 1.3 ($t=0.13, p>0.90$) and 1.0 ($t=-0.06, p>0.95$), respectively. Indeed, the observed values for reproductive units fit very neatly within the distribution of data points in Fig. 22.2. Thus, the probability of co-cycling females predicts the number of males in primate groups at both the population and the breeding unit scales. This in itself is very encouraging, since there is no intrinsic reason why there should be any difference between these two situations.

It is important to appreciate that it is not simply the number of females that dictates the number of males in primate groups, but rather the probability that there will be two or more females simultaneously in estrus on any given day. Because this is partly determined by the number of females in the group, an apparent correlation between the number of males and number of females is inevitable.

To find the switch point at which a male can no longer prevent a rival entering his group, we set $Y=\log_{10}(2)$ in equation [4] and solve for X. This yields a value for the probability of two or more females cycling together of 0.17 when the number of males is two. Thus, a male will be unable to prevent rivals joining his group when the probability of two or more females co-cycling is $p>0.17$. However, the male will only be able to guarantee keeping rivals out (i.e., the number of males in the group is $Y=1$) when the probability of co-cycling females is $p<0.039$. In any given species or population, this will be equivalent to a specific number of females, but in general there will be a trade-off between the length of the reproductive cycle and the number of females in the group: when the interbirth interval is long, males will be able to defend larger numbers of females.

CONCLUSIONS

I have shown that some simple principles allow us to explain a great deal of the observed variability in the number of males in primate groups. In effect, what we see seems to be simply the outcome of male mating strategies played out against a background of female grouping and life history patterns. It is in large part the variation in the latter (itself driven mainly by environmental factors and females' attempts to optimize their own reproductive opportunities) that is responsible for the diversity of grouping patterns observed in nature. Two points are worth emphasizing.

First, the same kinds of simple models can be used to explain male behavior at both the population (or species) and the local group level. What we see as species-typical patterns of behavior are, in fact, simply the outcome of individual males making their decisions about how to reproduce as effectively as they can. Specific constellations of parameter values lead to switch points at which systems can flip from one state (say monogamy) to another (roving male polygamy).

Second, despite the fact that they ignore what females seek to do in terms of maximizing their own mating opportunities, the models seem to be surprisingly successful at predicting what males do. This suggests that the well-recognized dichotomy in reproductive interests among mammals (with females being more concerned about rearing and males more concerned about mating) holds true, at least in the generality. This does not mean that females have no interest in whom they mate with or how often, but it does probably suggest that the search for good genes may be less important for female primates than it is in the case of, for example, peahens. Nonetheless, an important next step will be to develop these models to incorporate females' interests. Previous attempts to do this in the case of callitrichids (Dunbar, 1995a, 1995b) have again emphasized the importance of rearing conditions for females. But perhaps the important lesson in this case is the sheer complexity of the resulting analyses.

ACKNOWLEDGMENTS

I thank Charles Nunn, Richard Wrangham, and an anonymous reviewer for their helpful comments.

Part VI
Conclusions

23 • Understanding male primates

MICHAEL E. PEREIRA, TIMOTHY H. CLUTTON-BROCK & PETER M. KAPPELER

INTRODUCTION

This project was undertaken to clarify the causes and consequences of variation in the number of males residing in social groups of primates. The effort was inspired by a wealth of new data on social patterns within and among free-living species of primates, by new questions, especially about modes of male sexual coercion and of female mate choice, and by new methods of research, including unequivocal paternity determination and phylogenetic analysis of life history and behavioral traits. Comparison of results from recent fieldwork with diverse primates highlighted key factors underlying male and female behavioral decisions that conjointly determine male residence patterns and what sorts of things happen next. Collectively, contributions to the volume identify primary conceptual foci and methodological approaches for future research to illuminate more definitively how male strategies of reproduction and social relationship influence and are influenced by those of females, and how characteristic patterns of primate social grouping and reproductive success result from this interplay.

Our theoretical context is socioecology, a developing body of theory that understands social systems not as adaptations themselves but as by-products of natural and sexual selection operating among individual males and females to promote physical and behavioral traits that maximize reproductive success. Favored females best access resources that promote their own and offspring survival; favored males maximize the number of mates engaged over the course of their reproductive careers (Trivers, 1972; Emlen & Oring, 1977; Clutton-Brock, 1989). The socioecological perspective additionally emphasizes that sex-typical behavior, group size, and group composition are determined not only by resource distribution, predation risk and other ecological factors, but also by life history traits (e.g., Partridge & Endler, 1987; Dunbar, 1992; Geffen *et al.*, 1996; Kappeler, 1997a) and patterns of social interaction themselves. Especially important in this regard is to clarify the nature and mechanisms of conflicts of interest not only within but also between the sexes (Packer *et al.*, 1990; Smuts & Smuts, 1993; van Schaik, 1996; Davies, Chapter 2; Dunbar, Chapter 22).

PERMANENT MALE–FEMALE ASSOCIATION

Basic sexual selection theory predicts that male mammals will seek to maximize their number of mates either by roving among dispersed females (e.g., Dunbar, Chapter 22; Jarman, Chapter 3) or by monopolizing access to spatial clusters of females (e.g., Clutton-Brock *et al.*, 1982). Whereas most males desert females after mating to seek reproductive opportunities elsewhere (Clutton-Brock, 1989), male primates are typically found in permanent association with at least one adult female (Wrangham, 1987). All-male bands are common in few primates whereas many characteristically form multi-male–multi-female social groups (Smuts *et al.*, 1987). Continuous sexual access was initially suspected as the reason for this trait (Zuckerman, 1932), but primates' slow life histories and predictable periods of female fertility render this possibility unlikely (Harvey & Clutton-Brock, 1985; Hrdy & Whitten, 1987).

Just as interesting as toleration among males within groups including females is that female primates operate in ways that enable one or more males to associate with them. Males are often hungrier and more powerful than females and thus can increase feeding competition while presenting formidable added costs of social conflict. In some cases, males may reduce predation risk or improve access to resources contested between groups (van Schaik & van Hooff, 1983; Wrangham, 1987); but male–female association is observed across taxa that engage disparate predation risks and competitive regimes (Janson & Goldsmith, 1995).

Important perspective is provided by the recent extension of sexual selection theory to admit that, under appropriate conditions, males can be expected to attempt to coerce females to mate with them (Smuts & Smuts, 1993; Clutton-Brock & Parker, 1995b). Especially in slow-developing species like primates and where peak periods in male reproductive careers are short, infanticide is included among coercive male tactics (Hrdy, 1977; Watts, 1989; Pereira & Weiss 1991; Hrdy *et al.*, 1995; Steenbeek *et al.*, Chapter 12). In this context, male protection from male sexual coercion emerges as the factor with the broadest potential to explain male–female association, as risks of coercion should be widespread among primates and independent of basic ecology (Smuts & Smuts, 1993; Clutton-Brock & Parker, 1995b; cf. van Schaik & Kappeler, 1993; Brereton, 1995; van Schaik, 1996; Treves, 1998). Females that carry their infants, rather than 'park' them in the environment, should be especially vulnerable to infanticidal attacks, and this variation in maternal behavior predicts the incidence of permanent male–female association in primates (van Schaik & Kappeler, 1997) and also, to an extent, among non-primate mammals (van Schaik, Chapter 4; also Gubernick, 1994). Still unclear are the relative degrees to which male paternity benefits and female manipulations of male behavior (e.g., via synchronization or extension of estrus) have promoted the evolution of male–female association in primates (cf. Smuts, 1987; Dunbar, 1988; van Schaik *et al.*, 1999), and other benefits possibly obtained by males from permanent association with females remain to be investigated. Studies of non-primates with gregarious females will also be helpful in this regard (cf. Davies, Chapter 2; Jarman, Chapter 3).

MALE AND FEMALE PERSPECTIVES ON GROUP COMPOSITION

Across species, populations, and groups of primates, the number and proportion of co-resident adult males and females vary, with profound consequences for anatomy, reproduction, and social behavior (Clutton-Brock & Harvey, 1977a; Clutton-Brock *et al.*, 1977; Clutton-Brock, 1989; Hamilton & Bulger, 1992; Plavcan & van Schaik, 1997). From both male and female perspectives, a basic concern should be differences in conditions of existence between single-male and multi-male groups. Males should prefer to live in multi-male groups only when the number of mates per male is thereby increased somehow. Females, on the other hand, should prefer multi-male groups whenever these help them to access food or shelter or to avoid infanticide or other costs of aggression.

How do individual males in some groups or populations manage to monopolize access to several females (Clutton-Brock & Harvey, 1977a; Harvey & Harcourt, 1984)? Males' relations with one another are expected to be fundamentally competitive, commonly entailing agonistic conflict and social dominance relations. Males should generally try to exclude rivals from groups of females. However, females' spatial distributions or the temporal distributions of their estrous periods may intervene to preclude success for powerful males. The absolute number of spatially clumped females, in particular, appears to influence males' abilities to monopolize reproductive opportunities (Andelman, 1986; Dunbar, 1987, Chapter 22; Clutton-Brock, 1989; Altmann, 1990; Mitani *et al.*, 1996a). But, other factors are clearly involved, as many exceptions exist. Single marmoset and tamarin females often associate with several males (Heymann, Chapter 6), small groups of female lemurs are almost always accompanied by several males (Kappeler, Chapter 5), and some Hanuman langur males monopolize groups of more than 20 females (Sterck & van Hooff, Chapter 11).

Recent analyses of degrees of estrous overlap among female group mates revealed a positive association with number of males (Nunn, 1999). Seasonality of reproduction, however, is only a crude predictor of whether estrous periods are likely to overlap (Pereira, 1991; Dunbar, Chapter 22), and seasonal breeding may actually facilitate male monopolization of reproduction within groups because males are best able to sustain the necessary agonistic power over relatively short periods of time (e.g., Clutton-Brock *et al.*, 1982). Although only perfect estrus synchrony, not seasonality, can be assumed to distribute reproductive success widely among males (Pereira, 1998), it is notable that strong correlations between male rank and reproductive success in primates have rarely been demonstrated for seasonally breeding primates (Paul, 1997). Presumably, absolute size of female groups sets an upper limit on defendability by males, and degree of estrous synchrony modifies critical size. Female cohesiveness and environmental visibility are other factors that should influence males' abilities to monopolize access to fertile females (van Schaik & van Hooff, 1983).

In this context, it is important to emphasize that reproductive strategies and modes of competition among male primates remain poorly understood beyond simple patterns of dyadic dominance behavior. Do age-specific patterns of male mortality differ among single-male and multi-male species? What growth and maturational strategies are employed by males in different species? How common are solitary males and how do they promote their survival and reproductive success? Why are some young males philopatric while others disperse to join established groups or found new ones? Why are all-male bands so rare among primates and why do so few male primates collaborate to defend ranges to which females might be attracted? Despite recent insights into some of these questions in some species (cf. Henzi, 1988; Rowell, 1988b; Wickings & Dixson, 1992; Alberts & Altmann, 1995a, 1995b; Maggioncalda, 1995; Strier, 1996a; Altmann *et al.*, 1997; Borries, Chapter 13; Pope, Chapter 19; Watts, Chapter 15), detailed information on complete male life histories will be required from a broad range of taxa to achieve deep and unequivocal understanding.

Females are not just passive bystanders that end up in single-male or multi-male groups depending on male competitiveness and tolerance. They experience social and ecological costs and benefits from varying numbers of males and therefore can be expected to influence male number and behavior in a variety of ways (Eisenberg *et al.*, 1972). Multiple males should increase within-group feeding competition, but also might improve access to food contested between groups in certain cases (Crook & Gartlan, 1966; Wrangham, 1987). Little evidence for important effects of males on female resource acquisition has yet accumulated, however. Male aggression during intergroup conflict is variable across primates, not clearly linked to male number (Cheney, 1987), and generally aimed exclusively at other males (van Schaik *et al.*, 1992). Finally, the existence of so many multi-male taxa suggests that the energetic costs of this group structure are either unavoidable or not particularly strong for females (see also Wrangham, Chapter 21).

Multi-male group membership may also help to minimize predation risk, especially through dilution effects and added vigilance, and recent studies suggest that males in some species are better than females in detecting and repelling predators (van Schaik & van Noordwijk, 1989; van Schaik & Hörstermann, 1994; Koenig, 1998). *Per capita* payoffs are largest in small groups, however; so, unless males assume special positions (e.g., *Saimiri, Lemur catta*) or behavioral roles (e.g., above references), the importance of this effect of male number should diminish rapidly as groups exceed moderate sizes.

What social costs and benefits for females derive from male number? In most stable one-male groups, female mate choice should be constrained and a greater risk of infanticide should obtain (i.e., following takeovers; Watts, 1989; Sommer, 1994; cf. van Schaik, 1996), whereas the risk or intensity of other forms of sexual coercion may be reduced (Smuts & Smuts, 1993). That females in many multi-male groups evolved signals of reproductive state (e.g., sexual swellings) suggests that for them access to multiple mates confers important benefits (cf. van Schaik *et al.*, 1999). Females in some multi-male groups garner more male assistance in rearing young (Wright, 1990) and greater scope for friendships with males that provide support during conflicts (e.g., Smuts, 1985; Pereira & McGlynn, 1997; see also Barton, Chapter 9). Indeed, comparative data at several levels of taxonomic scale suggest that females in multi-male groups commonly compete over access to male services (Hemelrijk & Luteijn, 1998).

In sum, male and female primates inevitably experience conflicts of interest with members of their own and also the other sex regarding how many and which males should reside in their social group. What mechanisms do members of the two sexes have at their disposal to gain advantage in these evolutionary battles?

CONFLICTS OF INTEREST, REPRODUCTIVE SKEW, AND SOCIAL RELATIONSHIPS

Males can defend groups of females against invasion by rivals (Rowell, 1988a; Clutton-Brock, 1989) and can also aggressively discourage females from contacting other males (e.g., Stammbach, 1987). The latter tactic is most available to males that are larger than female counterparts, or endowed with more fearsome weaponry (e.g., Berenstain & Wade, 1983; Henzi *et al.*, 1998). Understanding of the relationships among mating system, dimorphism, and qualities of male–female relationship remains an area requiring more research (e.g., Barton, Chapter 9).

Depending on environmental and demographic conditions, however, males may benefit from multi-male group membership, via reduced risk of predation, cooperative defense of females, or both. Cooperative relations should be promoted by opportunities to collaborate with kin (Struhsaker & Pope, 1991; Goldberg & Wrangham, 1997; Struhsaker, Chapter 10); but, neither male philopatry nor kinship is necessary for male coalitions to form. Evidence is accumulating that related males commonly disperse together or rejoin in non-natal groups (e.g., Sussman, 1991; Pope, Chapter 19; Strier, Chapter 7) and unrelated males form important coalitions under certain circumstances (Pope, 1990; cf. Grinnell *et al.*, 1995; Davies, Chapter 2).

Given the prevalence of multi-male groups among primates, no single question should feature more prominently in future research than 'Who gets to reproduce?' More specifically, what is the nature of reproductive skew among males? In high-skew scenarios, one or just a few males reproduce; under low skew, reproduction is distributed more evenly among males. Models of evolutionarily stable skew assume that dominants control reproduction by subordinates (Vehrencamp, 1983a, 1983b; Keller & Reeve, 1994). Further, dominants are expected to modulate the reproduction yielded to subordinates as a function of how much they benefit from the subordinates' company, how much opportunity subordinates have to reproduce elsewhere, the probability that subordinates might overturn them in dominance, and the degree to which subordinates are genetically related. 'Staying' and 'peace' incentives (amounts of reproduction yielded) should decrease as relatedness increases because related subordinates benefit via kin selection for cooperating and should therefore require smaller inducements.

An alternative framework for understanding the distribution of male breeding success is provided by models that assume that dominant males cannot control reproduction in subordinates (e.g., Clutton-Brock, 1998; Reeve *et al.*, 1998). Under these conditions, related subordinates are as likely to be able to breed as unrelated ones, and there is no reason to suppose that the incidence of local breeding by subordinates should be inversely related to the chance that subordinates will disperse to breed elsewhere.

Primate societies offer one of the best opportunities for testing the assumptions and predictions of 'optimal skew' versus 'incomplete control' models. Does the presence of subordinate males enhance the breeding success of dominants or their tenure? Are subordinates that are allowed to share reproduction with the dominant male more likely to assist him or less likely to challenge him? Are breeding subordinates less likely to disperse to other groups? Do subordinates related to the dominant male achieve higher – or lower – reproductive success? There are many important questions that need answers.

Research on dominance relations among male primates offers many links to skew theory. Few subordinate male primates, for example, have appeared to their researchers to be at all powerless. In some cases, asymmetries in resource holding potential appear so slight, and weaponry so formidable, that males assiduously avoid escalated aggression, maintaining tensely tolerant relations (Preuschoft & Paul, Chapter 18). In others, subordinate males combine forces to overwhelm powerful residents (Borries, Chapter 13; Cords, Chapter 8; Pope, Chapter 19) or high-rankers (van Hooff, Chapter 16), and in some cases alliances between males shift specifically to overthrow or neutralize the power of despots (de Waal, 1982; Noë, 1992). Some elements of these social patterns may rely on cognitive capacities linked to primates' large forebrains and thus comprise mechanisms of multi-male group membership unique to large-brained species.

In some multi-male groups, reproductive skew among males appears to approach the mating structure of one-male groups. Age-graded hierarchies (Eisenberg *et al.*, 1972) explain part of this, where prime males are accompanied by younger followers, often sons, for whom costs of delayed breeding may be compensated by a safe haven for important social learning (e.g., Steenbeek *et al.*, Chapter 12; see also Janson & van Schaik, 1993) or later 'inheritance' of their groups (e.g., Watts, Chapter 15). Whether young males' reproductive activity is actually suppressed by dominants or, instead, simply reflects the youngsters' low-risk strategies remains to be determined for most species (Moore, 1984; Perret, 1992; Wickings & Dixson, 1992). Precise definition and determination of male adulthood are issues that loom large in this context (Pereira, 1993; Alberts & Altmann, 1995b).

Male reproductive skew seems less pronounced in many other multi-male primate scenarios (see, for example, Altmann *et al.*, 1997; de Ruiter & Geffen, 1998). Overviews of research on male dominance and mating success suggest only a weak general positive relationship (Bercovitch, 1991; Cowlishaw & Dunbar, 1991; de Ruiter & van Hooff, 1993). Further, many exceptions exist and studies finding no relationship are certainly under-reported. The interaction of dominance with other factors, such as seasonality and female choice, therefore requires close examination (cf. Smuts, 1987; Paul, 1997). Mechanisms that can skew male reproductive success or diminish skew should vary predictably, within and among taxa, in relation to adult sex ratios and other demographic and social features. These include mate guarding, male coalitions, elements of sperm competition, estrous synchrony or concealment, and other female adjustments (Berenstain & Wade, 1983; Birkhead & Møller, 1992; Alberts *et al.*, 1996; Eberhard, 1996; Harcourt, 1997; Watts, 1998a; Altmann, Chapter 20; Strier, Chapter 7).

Indeed, female strategies inevitably shape those of males, and not until we have amassed detailed data on female reproductive tactics for a range of primate taxa will we be in a favorable position to account for variability in social relations within or between the sexes. Females use a variety of behavioral and physiological mechanisms to choose their mates, thereby modulating actual and apparent reproductive skew among males. For primate females, this should have pivotal consequences for the number and identity of males that accompany them and for social relationships with and between those males (see, for example, van Schaik, Chapter 4).

Female primates have evolved a variety of precopulatory behavioral tactics that mitigate the potential for male harassment, including defensive coalitions with other females and protective males (Watts, 1989; Treves & Chapman, 1996; Palombit *et al.*, 1997; Sterck, 1997). Whereas nepotistic female-resident groups have been suggested to function generally as defensive coalitions against male harassment (Brereton, 1995), the behavior of individual females does not generally fit this interpretation well (Sterck *et al.*, 1997; Hemelrijk & Luteijn, 1998; Treves, 1998). Females can reduce the risk of infanticide, and generally increase their base of agonistic support, however, via multi-male group membership, and depending on food supply and environmental visibility, females may form groups of particular sizes in part to preclude their defense by just one or a few males (cf. van Schaik & van Hooff, 1983; Rowell, 1988a).

Synchronization and concealment of fertile periods can also reduce the potential for male monopolization (Andelman, 1986; Ims, 1990; Pereira & McGlynn, 1997), and sexual swellings, mating calls, and long follicular phases of estrous cycles are common among multi-male species (van Schaik *et al.*, 1999). Risk of infanticide may explain why females in some multi-male groups routinely engage in copulations with even extra-group males (e.g., Sauther, 1991; Richard, 1992). Experimental studies of female social relationships as a function of the number or tenure length of males would be of considerable value to investigate all these and related patterns.

Where females are free to migrate, they appear to time and target group transfers so as to minimize the infanticide risk (Sterck, 1997, 1998; Steenbeek *et al.*, Chapter 12), and they may regularly transfer not only into groups with the optimal number of males (e.g., Watts, Chapter 15), but also into those with males of particularly high quality (Sterck, 1997; Steenbeek *et al.*, Chapter 12; Struhsaker, Chapter 10). That this seems likely underscores that the development of useful measures of individual male quality remains an important challenge for future research on primates. Approaches to meeting this challenge would include innovative quantifications of male activity patterns, well-scheduled use of camera lenses that allow precise morphometric measurements to be made remotely, and direct measurements of body mass and lengths, skinfold thicknesses, hair quality, and other indices of condition at judicious intervals.

The most significant obstacle to learning more about male primates, however, may be post-copulatory or 'cryptic' mechanisms of female choice (Eberhard, 1996). Neither mounting with pelvic thrusting, nor ejaculation tells us that a copulating male is the sire of the infant produced at the appropriate later date, because even few ejaculations result in fertilization. Anatomical and physiological processes wholly under female control, along with several modes of competition among male gametes (Gomendio & Roldan, 1993), determine whether any given mate is likely to sire a female's next infant. Post-copulatory female control of paternity has received vastly insufficient attention generally in animal behavior (Birkhead & Møller, 1992; Eberhard, 1996), and virtually none in primatology, whereas, like male–male competition, female choice extends well past the moment of copulation. This is particularly important for primatologists because cryptic female choice is most likely to evolve where males can coerce female copulation (Packer & Pusey, 1983; Birkhead *et al.*, 1993; see also Borries, Chapter 13) and primates whose females typically mate with more than one male are those that show the greatest anatomical and behavioral evidence of post-copulatory choice (Dixson, 1987, 1991; Verrell, 1992; Eberhard, 1996).

Beyond adaptive modes of gregariousness, association with preferred males, movements that promote male conflict, and synchronization or extension of estrous cycles, then, females of a given species may selectively time copulations to occur at points of high or low fertility, adjust the timing of ovulation, extend or prematurely interrupt copulation, facilitate or hinder penile penetration, and facultatively transport, discharge or destroy a given male's sperm. By influencing their own level of stimulation, females modulate the likelihood of orgasm and, in turn, of sperm transport and chances of fertilization (Baker & Bellis, 1993). Ruling out stimulation effects, Troisi & Carosi (1998) observed orgasms in female Japanese macaques (*Macaca fuscata*) most often during matings between high-ranking males and low-ranking females and least often between low-ranking males and high-ranking females.

Mechanisms of female choice are numerous among animals, extending well beyond the behavioral and anatomical into the domains of chemical and immunological barriers to particular ejaculates (Birkhead & Møller, 1992; Roldan *et al.*, 1992; Eberhard, 1996). Collectively, the chapters in this volume distinctly highlight how little we know about female reproductive biology in primates. Investigators of guenons and langurs, for example, struggle to understand variation among populations and species, regarding proportions of groups that are multi-male and also proportions of group members that are male (e.g., Borries, Chapter 13; Cords, Chapter 8; Sterck & van Hooff, Chapter 11). Possible effects of seasonality are still obscured in part by lack of information about dispersion of estrus among females, and mechanisms underlying differential rates of conception per copulation among males remain mysterious for all primates (e.g., Borries, Chapter 13; Watts, Chapter 15).

Illumination of patterns of female choice among primates will pose the significant challenges of extensive hands-on investigation, often with naturalistic laboratory and enclosure social groups, and well-developed collaborations among biologists of different primary interests and training. The wedding of research in anatomy, physiology, and ecology offers immense scope for new research, however, and represents the essential path toward full illumination of primate evolution and development. Here again, we see that nothing will be more important to progress than coupling the long-term collection of behavioral data with unequivocal paternity determinations (cf. Davies, Chapter 2). This may be accomplished progressively more often via the new approach of recovering DNA from fecal samples.

LOOKING TOWARD THE FUTURE

Differences in basic biology and conditions of research have helped to generate contrasts in depth of understanding between relatively tractable research subjects like small birds (e.g., Davies, Chapter 2) and relatively intractable subjects like primates. These contrasts, however, also indicate what needs to be done to close this gap. The most significant sorts of advancement in our understanding of primate social behavior will be made only through the simultaneous study of larger numbers of groups, unambiguous quantification of fitness correlates (e.g., physical condition, rates of survivorship), variation in male quality, male care for immatures, and accumulation of long social and reproductive histories, including unequivocal paternity determinations. In addition, increased experimentation will be important, both under controlled laboratory conditions and also at field sites.

Armed with information about male reproductive skew, future work with multi-male primates, in particular, offers rich opportunities to illuminate strategies and behavioral mechanisms pertaining to male competition, tolerance, and co-operation (Noë & Sluijter, 1990; van Hooff & van Schaik, 1994; van Hooff, Chapter 16). Whereas dyadic agonistic competition has received appreciable attention (e.g., Plavcan & van Schaik, 1994; Leigh, 1995; Dixson, 1997; Harcourt, 1997), more complex questions about male social relations have been left almost entirely neglected. Tolerance and cooperative behavior have still primarily been studied in chimpanzees and other species showing male philopatry (de Waal, 1982; van Hooff & van Schaik, 1994), whereas other taxa offer great opportunities for research on reciprocity, mutualism, and negotiation among males (e.g., Noë & Hammerstein, 1994; Nunn, Chapter 17; Preuschoft & Paul, Chapter 18).

Increasing knowledge of female choice mechanisms, and how they are used to manipulate actual and apparent reproductive skew among males, will also expand our understanding of social relations, including those between males and females. In many species, members of the two sexes regularly affiliate, while sexual segregation is the rule in others throughout most of the year. In many taxa, particular males stand out as disproportionately integrated in the social lives of females. Sometimes they are current highest-rankers (Rowell & Dixson, 1975; Kappeler, 1993b; Boinski & Mitchell, 1994); in other cases they are recent sires (Smuts, 1985; van Noordwijk & van Schaik, 1988). Even among one-male groups of primates, female attention may be directed primarily toward the male (Kummer, 1968) or toward other females (Dunbar, 1983). Future research should link this variability to physiological, ecological, and social factors (cf. Barton, Chapter 9).

Studies of pair-living and non-gregarious primates should provide additional perspectives on basic aspects of male and female reproductive strategies. Other pair-living primates may turn out to be as flexible in their association and mating patterns as white-handed gibbons (Sommer & Reichard, Chapter 14), for example, so that some of the consequences arising from variation in the number of males should also be expected in these taxa. Species in which males and females do not permanently associate provide an important link to many other mammals, for which this is a typical form of social organization. The gathering of information on how males and females of these species manage their social relations, how males of some are able to prevent rivals from interacting with several dispersed females (Charles-Dominique, 1977; van Hooff, 1996), and why this is apparently not possible for males of others (Kappeler, 1997b) remains an important challenge for students of orangutans and nocturnal prosimians. Information on the spatial distribution of females and on the temporal distribution of their receptive periods may help to answer these fundamental questions (Emlen & Oring, 1977; Dunbar, Chapter 22).

In sum, the properties and dynamics of social systems are shaped not only by ecological factors. Life history traits, physiological mechanisms, and social patterns, especially those pertaining to sexual selection, can have both subtle and profound effects on group composition and the social behavior of individuals. More comparative information, on primates and non-primates, is needed to evaluate further the general importance of a variety of factors. Primate males play a more prominent role in shaping various aspects of primate social systems than is acknowledged by contemporary socioecological models. Clarification of the proximate and ultimate factors that determine association strategies, both with females and with other males, will strengthen socioecological theory, and further study of male social behavior will deepen our understanding of sexual conflict and behavioral strategies. More information on complete social and reproductive histories of males will be needed to achieve these goals.

References

Abegglen, J.J. 1984. *On Socialization in Hamadryas Baboons*. Lewisburg, PA: Bucknell University Press.

Abbott, D.H. 1984. Behavioral and physiological suppression of fertility in subordinate marmoset monkeys. *American Journal of Primatology*, **6**, 169–86.

Abbott, D.H., Barrett, J. & George, L.M. 1993. Comparative aspects of the social suppression of reproduction in female marmosets and tamarins. In: *Marmosets and Tamarins: Systematics, Behaviour, and Ecology*, ed. A.B. Rylands, pp. 152–63. Oxford: Oxford University Press.

Agoramoorthy, G. 1994a. Adult male replacement and social change in two troops of Hanuman langurs (*Presbytis entellus*) at Jodhpur, India. *International Journal of Primatology*, **15**, 225–38.

–1994b. An update on the long-term field research on red howler monkeys, *Alouatta seniculus*, at Hato Masaguaral, Venezuela. *Neotropical Primates*, **2**, 7–9.

Ahsan, F. 1995. Fighting between two females for a male in the Hoolock gibbon. *International Journal of Primatology*, **16**, 731–7.

Alatalo, R.V., Carlson, A., Lundberg, A. & Ulfstrand, S. 1981. The conflict between male polygamy and female monogamy: the case of the pied flycatcher. *American Naturalist*, **117**, 738–53.

Alberts, S.C. 1992. Maturation and dispersal in male baboons (*Papio cynocephalus*). PhD thesis, University of Chicago.

Alberts, S.C. & Altmann, J. 1995a. Balancing costs and opportunities: dispersal in male baboons. *American Naturalist*, **145**, 279–306.

–1995b. Preparation and activation: determinants of age at reproductive maturity in male baboons. *Behavioral Ecology and Sociobiology*, **36**, 397–406.

Alberts, S.C., Altmann, J. & Wilson, M. 1996. Mate guarding constrains foraging activity of male baboons. *Animal Behaviour* **51**, 1269–77.

Albignac, R., Fontenille, D., Maleyran, D. & Duvernoy, F. 1988. Evolution de l'organisation sociale et territoriale de *Propithecus verreauxi coquereli* pendant 6 ans, dans les forêts sèches du nord-ouest de Madagascar (Ankarafantsika). In: *L'Equilibre des Ecosystèmes Forestièrs à Madagascar: Actes d'un Séminaire International*, ed. L. Rakotovato, V. Barre & J. Sayer, IUCN, pp. 90–4. Cambridge: Page Bros Ltd.

Alcock, J., Jones, C.E. & Buchmann, S.L. 1977. Male mating strategies in the bee *Centris pallida* FOX (Anthophoridae: Hymenoptera). *American Naturalist*, **111**, 145–55.

Alexander, R.D. 1974. The evolution of social behavior. *Annual Review of Ecology and Systematics*, **5**, 325–83.

–1987. *The Biology of Moral Systems*. Hawthorne, NY: Aldine.

Alexander, R.D., Hoogland, J.L., Howard, R.D., Noonan, K.M. & Sherman, P.W. 1979. Sexual dimorphism and breeding systems in pinnipeds, ungulates, primates and humans. In: *Evolutionary Biology and Human Social Behaviour*, ed. N.A. Chagnon & W.A. Irons, pp. 402–35. North Scituate, MA: Duxbury.

Alfred, J.R.B. & Sati, J.P. 1991. On the first record of infanticide in the hoolock gibbon *Hylobates hoolock* in the wild. *Records of the Zoological Survey of India*, **89**, 319–21.

Altmann, J. 1990. Primate males go where the females are. *Animal Behaviour*, **39**, 193–5.

–1997. Mate choice and intrasexual reproductive competition: contributions to reproduction that go beyond acquiring more mates. In: *Feminism and Evolutionary Biology: Boundaries, Intersections, and Frontiers*, ed. P. Gowaty, pp. 320–33. New York: Chapman Hall.

Altmann, J. & Samuels, A. 1992. Costs of maternal care: infant-carrying in baboons. *Behavioral Ecology and Sociobiology*, **29**, 391–8.

Altmann, J., Alberts, S.C., Haines, S.A. *et al.* 1997. Behavior predicts genetic structure in a wild primate group. *Proceedings of the National Academy of Science of the United States of America*, **93**, 5797–801.

Altmann, J., Altmann, S.A., Hausfater, G. & McCuskey, S. 1977. Life history of yellow baboons: physical development, reproductive parameters, and infant mortality. *Primates*, **18**, 315–30.

Altmann, J., Hausfater, G. & Altmann, S.A. 1988. Determinants of reproductive success in savannah baboons (*Papio cynocephalus*). In: *Reproductive Success*, ed. T.H. Clutton-Brock, pp. 403–18. Chicago: University of Chicago Press.

Altmann, S.A. 1962. A field study of the sociobiology of the rhesus monkey, *Macaca mulatta*. *Annual Proceedings of the New York Academy of Sciences*, **102**, 338–435.

Altmann, S.A. & Altmann, J. 1970. *Baboon Ecology*. Chicago: University of Chicago Press.

Andelmann, S.J. 1986. Ecological and social determinants of cercopithecine mating patterns. In: *Ecological Aspects of Social Evolution: Birds and Mammals*, ed. D.I. Rubenstein & R.W. Wrangham, pp. 201–16. Princeton, NJ: Princeton University Press.

Anderson, C.M. 1981. Subtrooping in a chacma baboon (*Papio ursinus*) population. *Primates*, **22**, 445–58.

–1986. Predation and primate evolution. *Primates*, **27**, 15–39.

–1992. Male investment under changing conditions among chacma baboons at Suikerbosrand. *American Journal of Physical Anthropology*, **87**, 479–96.

Andrews, J.R. 1990. A preliminary survey of black lemurs, *Lemur macaco*, in North West Madagascar. Unpublished report.

Angst, W. & Thommen, D. 1977. New data and a discussion of infant killing in Old World monkeys and apes. *Folia Primatologica*, **27**, 198–229.

Anzenberger, G. 1992. Monogamous social systems and paternity in primates. In: *Paternity in Primates: Genetic Tests and Theories*, ed. R.D. Martin, A.F. Dixon & E.J. Wickings, pp. 203–24. Basel: S. Karger.

Arak, A. 1988. Callers and satellites in the natterjack toad: evolutionary stable decision rules. *Animal Behaviour*, **36**, 416–82.

Arbélot-Tracqui, V. 1983. Etude étho-écologique de deux primates prosimiens: *Lemur coronatus* et *Lemur fulvus sanfordi*. PhD thesis, University of Rennes.

Assink, P. & van Dijk, I. 1990. Social organization, ranging and density of *Presbytis thomasi* at Ketambe (Sumatra), and a comparison with other *Presbytis* species at several South-east Asian locations. MSc thesis, University of Utrecht.

Ayres, J.M. 1981. *Observacoes Sobre a Ecologia e o Comportamento dos Cuxius (*Chiropotes albinasus *e* Chiropotes satanus. *Cebidae. Primates)*. Instituto Nacional de Pesquisas da Amazonia: Manaus.

Bachmann, C. & Kummer, H. 1980. Male assessment of female choice in Hamadryas baboons. *Behavioral Ecology and Sociobiology*, **6**, 315–21.

Baker, A.J. 1991. Evolution of the social system of the golden lion tamarin. PhD thesis, University of Maryland.

Baker, A.J. & Dietz, J.M. 1996. Immigration in wild groups of golden lion tamarins (*Leontopithecus rosalia*). *American Journal of Primatology*, **38**, 47–56.

Baker, A.J., Dietz, J.M. & Kleiman, D.G. 1993. Behavioural evidence for monopolization of paternity in multi-male groups of golden lion tamarins. *Animal Behaviour*, **46**, 1091–103.

Baker, K.C. & Smuts, B.B. 1994. Social relationships of female chimpanzees: diversity between captive social groups. In: *Chimpanzee Cultures*, ed. R.W. Wrangham, W.C. McGrew, F.B.M. de Waal & P.G. Heltne, pp. 227–42. Cambridge, MA: Harvard University Press.

Baker, R.R. & Bellis, M.A. 1993. Human sperm competition: ejaculate manipulation by females and a function for the female orgasm. *Animal Behaviour*, **46**, 887–909.

Baldellou, M. & Henzi, S.P. 1992. Vigilance, predator detection and the presence of supernumerary males in vervet monkey troops. *Animal Behaviour*, **43**, 451–61.

Baldwin, J.D. & Baldwin, J.I. 1976. Vocalizations of howler monkeys (*Alouatta palliata*) in Southwestern Panama. *Folia Primatologica*, **26**, 81–108.

Barrett, L. 1995. Foraging strategies, ranging behaviour and territoriality among grey-cheeked mangabeys in Kibale forest, western Uganda. PhD thesis, University of London.

Barroso, C.M.L., Schneider, H., Schneider, M.P.C. *et al.* 1997. Update on the phylogenetic systematics of New World monkeys: further DNA evidence for placing the pygmy marmoset (*Cebuella*) within the genus *Callithrix*. *International Journal of Primatology*, **18**, 651–74.

Bartlett, T.Q. & Brockelman, W.Y. 1996. Gradual replacement of a male pairmate in white-handed gibbons (*Hylobates lar*) in Khao Yai National Park, Thailand. *American Journal of Physical Anthropology, Supplement*, **22**, 66.

Bartlett, T.Q., Sussman, R.W. & Cheverud, J.M. 1993. Infant killing in primates: a review of observed cases with specific reference to the sexual selection hypothesis. *American Anthropologist*, **95**, 958–90.

Barton, R.A. & Whiten, A. 1993. Feeding competition among female olive baboons, *Papio anubis*. *Animal Behaviour*, **46**, 777–89.

Barton, R.A., Whiten, A., Strum, S.C., Byrne, R.W. & Simpson, A.J. 1992. Habitat use and resource availability in baboons. *Animal Behaviour*, **43**, 831–44.

Barton, R.A., Byrne, R.W. & Whiten, A. 1996. Ecology, feeding competition and social structure in baboons. *Behavioral Ecology and Sociobiology*, **38**, 321–9.

Bateman, A.J. 1948. Intra-sexual selection in *Drosophila*. *Heredity*, **2**, 349–68.

Bates, R.H. 1988. Contra contractarianism: some reflections on the new institutionalism. *Politics and Society*, **16**, 387–401.

Becker, C.D. & Ginsberg, J.R. 1990. Mother–infant behaviour of wild Grevy's zebra: adaptations for survival in semi-desert East Africa. *Animal Behaviour*, **40**, 1111–18.

Beeson, M. 1985. The ecology and behaviour of blue monkeys (*Cercopithecus mitis nyasae*) on the Zomba Plateau, Malawi, in relation to bark-stripping of exotic softwoods. PhD thesis, University of Exeter.

Beeson, M., Tame, S., Keeming, E. & Lea, S.E.G. 1996. Food habits of guenons (*Cercopithecus* spp.) in Afro-montane forest. *African Journal of Ecology*, **34**, 202–10.

Bennett, E.L. 1983. The banded langur: ecology of a colobine in West Malaysian rain-forest. PhD thesis, Sidney Sussex College, Cambridge.

Berard, J.D., Nürnberg, P., Epplen, J.T. & Schmidtke, J. 1993. Male rank, reproductive behavior, and reproductive success in free-ranging rhesus macaques. *Primates*, **34**, 481–9.

Bercovitch, F.B. 1986. Male rank and reproductive activity in savanna baboons. *International Journal of Primatology*, **7**, 533–50.

–1987. Reproductive success in male savanna baboons. *Behavioral Ecology and Sociobiology*, **21**, 163–72.

–1988. Coalitions, cooperation and reproductive tactics among adult male baboons. *Animal Behaviour*, **36**, 1198–209.

–1991. Social stratification, social strategies, and reproductive success in primates. *Ethology and Sociobiology*, **12**, 315–33.

Berenstain, L. & Wade, T. 1983. Intrasexual selection and male mating strategies in baboons and macaques. *International Journal of Primatology*, **4**, 201–35.

Berger, J. 1986. *Wild Horses of the Great Basin: Social Competition and Population Size*. Chicago: Chicago University Press.

Bernstein, I.S. 1968. The lutong of Kuala Selangor. *Behaviour*, **32**, 1–16.

–1976. Dominance, aggression and reproduction in primate societies. *Journal of Theoretical Biology*, **60**, 459–72.

Biquand, S., Biquand-Guyot, V., Boug, A. & Gautier, J-P. 1992. Group composition in wild and commensal hamadryas baboons: a comparative study in Saudi Arabia. *International Journal of Primatology*, **13**, 533–43.

Birkhead, T.R. & Møller, A.P. (eds.) 1992. *Sperm Competition in Birds: Evolutionary Causes and Consequences*. London: Academic Press.

Birkhead, T.R., Møller, A.P. & Sutherland, W.J. 1993. Why do females make it so difficult for males to fertilize their eggs? *Journal of Theoretical Biology* **161**, 51–60.

Bishop, N.H. 1979. Himalayan langurs: temperate colobines. *Journal of Human Evolution*, **8**, 251–81.

Bishop, N., Hrdy, S.B., Teas, J. & Moore, J. 1981. Measures of human influence in habitats of South Asian monkeys. *International Journal of Primatology*, **2**, 153–67.

Blurton Jones, N.G., Hawkes, K. & O'Connell, J.F. 1989. Modelling and measuring costs of children in two foraging societies. In: *Comparative Socioecology: the Behavioural Ecology of Humans and Other Animals*, ed. V. Standen & R. Foley, pp. 367–90. Oxford: Blackwell Scientific Publications.

Boelkins, R.C. & Wilson, A.P. 1972. Intergroup social dynamics of the Cayo Santiago rhesus (*Macaca mulatta*) with special reference to changes in group membership by males. *Primates*, **13**, 125–39.

Böer, M. & Sommer, V. 1992. Evidence for sexually selected infanticide in captive *Cercopithecus mitis*, *Cercocebus torquatus*, and *Mandrillus leucophaeus*. *Primates*, **33**, 557–63.

Boesch, C. 1991. The effects of leopard predation on grouping patterns in forest chimpanzees. *Behaviour*, **117**, 220–42.

–1996a. Social grouping in Taï chimpanzees. In: *Great Ape Societies*, ed. W.C. McGrew, L.F. Marchant & T. Nishida, pp. 101–13. Cambridge: Cambridge University Press.

–1996b. The emergence of cultures among wild chimpanzees. In: *Evolution of Social Behaviour Patterns in Primates and Man*, ed. W.G. Runciman, J. Maynard Smith & R.I.M. Dunbar, pp. 251–68. Oxford: Oxford University Press.

Boesch, C. & Boesch, H. 1989. Hunting behavior of wild chimpanzees in the Tai National Park. *American Journal of Physical Anthropology*, **78**, 547–73.

Boese, G. 1975. Social behaviour and ecological considerations of West African baboons (*Papio papio*). In: *Socioecology and Psychology of Primates*, ed. R.H. Tuttle, pp. 205–30. Mouton: The Hague.

Boggess, J. 1980. Intermale relations and troop male membership changes in langurs (*Presbytis entellus*) in Nepal. *International Journal of Primatology*, **1**, 233–74.

Boinski, S. 1994. Affiliation patterns among male Costa Rican squirrel monkeys. *Behaviour*, **130**, 191–209.

Boinski, S. & Mitchell, C.L. 1994. Male residence and association patterns in Costa Rican squirrel monkeys (*Saimiri oerstedi*). *American Journal of Primatology*, **34**, 157–69.

Boone, J.L. 1992. Competition, conflict, and the development of social hierarchies. In: *Evolutionary Ecology and Human Behavior*, ed. E.A. Smith & B. Winterhalder, pp. 301–37. New York: Aldine.

Borries, C. 1996. Flexible langurs (*Presbytis entellus*): the influence of ecological factors on the social structure. *Primate Report*, **44**, 5.

–1997. Infanticide in seasonally breeding multimale groups of Hanuman langurs (*Presbytis entellus*) in Ramnagar (South Nepal). *Behavioral Ecology and Sociobiology*, **41**, 139–50.

Borries, C., Sommer, V. & Srivastava, A. 1991. Dominance, age, and reproductive success in free-ranging female Hanuman langurs (*Presbytis entellus*). *International Journal of Primatology*, **12**, 231–57.

Borries, C., Launhardt, K., Epplen, C. *et al.* 1999. DNA analyses support the hypothesis that infanticide is adaptive in larger monkeys. *Proceedings of the Royal Society, London, Series B*, **266**, 901–4.

Bourlière, F., Hunkeler, C. & Bertrand, M. 1970. Ecology and behavior of Lowe's guenon (*Cercopithecus campbelli lowei*) in the Ivory Coast. In: *Old World Monkeys*, ed. J.R. Napier & P.H. Napier, pp. 297–350. New York: Academic Press.

Boyd, R. & Silk, J.B. 1997. *How Humans Evolved.* New York, London: W.W. Norton.

Bradbury, J.W. & Vehrencamp, S.L. 1977. Social organization and foraging in Emballonurid Bats. III. Mating systems. *Behavioral Ecology and Sociobiology*, **2**, 1–17.

Brereton, A.R. 1995. Coercion-defence hypothesis: the evolution of primate sociality. *Folia Primatologica*, **64**, 207–14.

Briskie, J.V. 1992. Copulation patterns and sperm competition in the polygynandrous Smith's longspur. *Auk*, **109**, 563–75.

Brockelman, W.Y. 1984. Social behaviour of gibbons: introduction. In: *The Lesser Apes: Evolutionary and Behavioural Biology*, ed. H. Preuschoft, D.J. Chivers, W.Y. Brockelman & N. Creel, pp. 285–90. Edinburgh: Edinburgh University Press.

–1985. A gibbon pelt (*Hylobates lar entelloides*) from Khao Yai National Park, Saraburi Province, Thailand. *Natural History Bulletin of the Siam Society*, **33**, 55–7.

–1997. Extra males in gibbon groups. *Primate Report*, **48–2**, 14.

–1998. Study of tropical forest canopy height and cover using a point-intercept method. In: *Forest Biodiversity Research Monitoring and Modeling: Conceptual Background and Old World Case Studies*, ed. F. Dallmeier & J.A. Comiskey, pp. 521–31. Paris: UNESCO.

Brockelman, W.Y. & Gittins, S.P. 1984. Natural hybridization in the *Hylobates lar* species group: implications for speciations in gibbons. In: *The Lesser Apes: Evolutionary and Behavioural Biology*, ed. H. Preuschoft, D.J. Chivers, W.Y. Brockelman &

N. Creel, pp. 498–532. Edinburgh: Edinburgh University Press.

Brockelman, W.Y. & Srikosamatara, S. 1984. Maintenance and evolution of social structure in gibbons. In: *The Lesser Apes: Evolutionary and Behavioural Biology*, ed. H. Preuschoft, D.J. Chivers, W.Y. Brockelman & N. Creel, pp. 298–323. Edinburgh: Edinburgh University Press.

Brockelman, W.Y., Ross, B.A. & Pantuwatana, S. 1973. Social correlates of reproductive success in the gibbon colony on Ko Klet Kaeo, Thailand. *American Journal of Physical Anthropology*, **38**, 637–40.

Brockelman, W.Y., Reichard, U., Treesucon, U. & Raemaekers, J.J. 1998. Dispersal, pair formation and social structure in gibbons (*Hylobates lar*). *Behavioral Ecology and Sociobiology*, **42**, 329–39.

Brockman, D.K. 1994. Reproduction and mating system of Verreaux's sifaka, *Propithecus verreauxi*, at Beza Mahafaly, Madagascar. PhD thesis, Yale University.

Brockman, D.K., Whitten, P.L., Richard, A.F. & Schneider, A. 1998. Reproduction in free-ranging male *Propithecus verreauxi*: the hormonal correlates of mating and aggression. *American Journal of Physical Anthropology*, **105**, 137–51.

Bronikowski, A.M. & Altmann, J. 1996. Foraging in a variable environment: weather patterns and the behavioral ecology of baboons. *Behavioural Ecology and Sociobiology*, **39**, 11–25.

Bronikowski, A. & Webb, C. 1996. A critical examination of rainfall variability measures used in behavioral ecology studies. *Behavioral Ecology and Sociobiology*, **39**, 27–30.

Brotherton, P.N.M. & Manser, M.B. 1997. Female dispersion and the evolution of monogamy in the dik-dik. *Animal Behaviour*, **54**, 1413–24.

Brown, J.L. 1987. *Helping and Communal Breeding in Birds*. Princeton, NJ: Princeton University Press.

Buchanan-Smith, H.M. 1991. A field study on the red-bellied tamarin, *Saguinus l. labiatus*, in Bolivia. *International Journal of Primatology*, **12**, 259–76.

Buchanan-Smith, H.M. & Jordan, T.R. 1992. An experimental investigation of the pair bond in the Callithrichid monkey, *Saguinus labiatus*. *International Journal of Primatology*, **13**, 51–72.

Budnitz, N. & Dainis, K. 1975. *Lemur catta*: ecology and behavior. In: *Lemur Biology*, ed. I. Tattersall & R.W. Sussman, pp. 219–35. New York: Plenum Press.

Bulger, J.B. 1993. Dominance rank and access to estrous females in male savanna baboons. *Behaviour*, **127**, 67–103.

Burke, T., Davies, N.B., Bruford, M.W. & Hatchwell, B.J. 1989. Parental care and mating behaviour of polyandrous dunnocks *Prunella modularis* related to paternity by DNA fingerprinting. *Nature*, **338**, 249–51.

Busse, C.D. 1977. Chimpanzee predation as a possible factor in the evolution of red colobus monkey social organization. *Evolution*, **31**, 907–11.

–1984. Triadic interactions among male and infant chacma baboons. In: *Primate Paternalism*, ed. D.M. Taub, pp. 186–212. New York: van Nostrand Reinhold.

Butynski, T.M. 1988. Guenon birth seasons and correlates with rainfall and food. In: *A Primate Radiation: Evolutionary Biology of the African Guenons*, ed. A. Gautier-Hion, F. Bourlière, J.-P. Gautier & J. Kingdon, pp. 284–322. Cambridge: Cambridge University Press.

–1990. Comparative ecology of blue monkeys (*Cercopithecus mitis*) in high- and low-density subpopulations. *Ecological Monographs*, **60**, 1–26.

Byrne, R.W. 1981. Distance vocalizations of Guinea baboons (*Papio papio*) in Senegal: an analysis of function. *Behaviour*, **78**, 283–313.

Byrne, R.W., Whiten, A. & Henzi, S.P. 1987. One-male groups and intergroup interactions of mountain baboons. *International Journal of Primatology*, **8**, 615–33.

–1990. Social relationships of mountain baboons: leadership and affiliation in a non-female-bonded monkey. *American Journal of Primatology*, **20**, 313–29.

Caine, N.G. 1993. Flexibility and co-operation as unifying themes in *Saguinus* social organization and behaviour: the role of predation pressures. In: *Marmosets and Tamarins: Systematics, Behaviour, and Ecology*, ed. A.B. Rylands, pp. 200–19. Oxford: Oxford University Press.

Caldecott, J.O., Feistner, A.T. C. & Gadsby, E.L. 1996. A comparison of ecological strategies of pig-tailed macaques, mandrills and drills. In: *Evolution and Ecology of Macaque Societies*, ed. J.E. Fa & D.G. Lindburg, pp. 73–94. Cambridge: Cambridge University Press.

Cant, M. 1998. A model for the evolution of reproductive skew without reproductive suppression. *Animal Behaviour*, **55**, 163–9.

Caraco, T. 1979. Time budgeting and group size: a theory. *Ecology*, **60**, 611–17.

Caro, T.M. 1994. *Cheetahs of the Serengeti Plains: Group Living in an Asocial Species*. Chicago: University of Chicago Press.

Carpenter, C.R. 1940. A field study in Siam of the behavior and social relations of the gibbon (*Hylobates lar*). *Comparative Psychology Monographs*, **84**, 1–212.

Caryl, P.G. 1981. Escalated fighting and the war of nerves: game theory and animal combat. In: *Perspectives in Ethology*, Vol. 4, ed. P.G. Bateson & P.H. Klopfer, pp. 199–224. London: Plenum Press.

Castro Coronado, N.R. 1991. Behavioral ecology of two coexisting tamarin species (*Saguinus fuscicollis nigrifrons* and *Saguinus mystax mystax*, Callitrichidae, Primates) in Amazonian Peru. PhD thesis, Washington University, St Louis.

Chalmers, N.R. 1968. The visual and vocal communication of free-living mangabeys in Uganda. *Folia Primatologica*, **9**, 258–80.

Chapais, B. 1983. Reproductive activity in relation to male dominance and the likelihood of ovulation in male rhesus monkeys. *Behavioral Ecology and Sociobiology*, **12**, 215–28.

Chapman, C.A. 1990a. Association patterns of spider monkeys: the influence of ecology and sex on social organization. *Behavioral Ecology and Sociobiology*, **26**, 409–14.

–1990b. Ecological constraints on group size in three species of neotropical primates. *Folia Primatologica*, **55**, 1–9.

Chapman, C.A. & Chapman, L.J. 1990. Reproductive biology of captive and free-ranging spider monkeys. *Zoo Biology*, **9**, 1–9.

Chapman, C.A. & Wrangham, R.W. 1993. Range use of the forest chimpanzees of Kibale: implications for the understanding of chimpanzee social organization. *American Journal of Primatology*, **31**, 263–73.

Chapman, C.A., Fedigan, L.M., Fedigan, L. & Chapman, L.J. 1989. Post-weaning resource competition and sex ratio in spider monkeys. *Oikos*, **54**, 315–19.

Chapman, C.A., White, F.J. & Wrangham, R.W. 1994. Party size in chimpanzees and bonobos: a reevaluation of theory based on two similarly forested sites. In: *Chimpanzee Cultures*, ed. R.W. Wrangham, W.C. McGrew, F.B.M. de Waal & P.G. Heltne, pp. 41–57. Cambridge, MA: Harvard University Press.

Chapman, C.A., Wrangham, R.W. & Chapman, L.J. 1995. Ecological constraints on group size: an analysis of spider monkeys and chimpanzee subgroups. *Behavioral Ecology and Sociobiology*, **36**, 59–70.

Charles-Dominique, P. 1977. *Ecology and Behaviour of Nocturnal Primates*. New York: Columbia University Press.

Charnov, E.L. 1993. *Life History Invariants: Some Explorations of Symmetry in Evolutionary Ecology*. Oxford: Oxford University Press.

Cheney, D.L. 1981. Intergroup encounters among free-ranging vervet monkeys. *Folia Primatologica*, **35**, 124–46.

–1983. Proximate and ultimate factors related to the distribution of male migration. In: *Primate Social Relationships. An Integrated Approach*, ed. R.A. Hinde, pp. 241–9. Oxford: Blackwell.

–1987. Interactions and relationships between groups. In: *Primate Societies*, ed. B.B. Smuts, D.L. Cheney, R.M. Seyfarth, R.W. Wrangham & T.T. Struhsaker, pp. 267–81. Chicago: University of Chicago Press.

Cheney, D.L. & Seyfarth, R.M. 1977. Behaviour of adult and immature male baboons during inter-group encounters. *Nature*, **269**, 404–6.

–1981. Selective forces affecting the predator alarm calls of vervet monkeys. *Behaviour*, **76**, 25–61.

–1983. Non-random dispersal in free-ranging vervet monkeys: social and genetic consequences. *American Naturalist*, **122**, 392–412.

–1987. The influence of intergroup competition on the survival and reproduction of female vervet monkeys. *Behavioral Ecology and Sociobiology*, **21**, 375–86.

Cheney, D.L. & Wrangham, R.W. 1987. Predation. In: *Primate Societies*, ed. B.B. Smuts, D.L. Cheney, R.M. Seyfarth, R.W. Wrangham & T.T. Struhsaker, pp. 227–39. Chicago: University of Chicago Press.

Cheney, D.L., Seyfarth, R.M., Andelman, S.J. & Lee, P.C. 1988. Reproductive success in vervet monkeys. In: *Reproductive Success: Studies of Individual Variation in Contrasting Breeding Systems*, ed. T.H. Clutton-Brock, pp. 384–402. Chicago: University of Chicago Press.

Chesser, R.K. 1991a. Gene diversity and female philopatry. *Genetics*, **127**, 437–47.

–1991b. Influence of gene flow and breeding tactics on gene diversity within populations. *Genetics*, **129**, 573–83.

Cheverud, J.M., Dow, M.M. & Leutenegger, W. 1986. A phylogenetic autocorrelation analysis of sexual dimorphism in primates. *American Anthropologist*, **88**, 916–22.

Chism, J. & Rogers, W. 1997. Male competition, mating success, and female choice in a seasonally breeding primate (*Erythrocebus patas*). *Ethology*, **103**, 109–26.

Chism, J. & Rowell, T.E. 1986. Mating and residence patterns of male patas monkeys. *Ethology*, **72**, 31–9.

–1988. The natural history of patas monkeys. In: *A Primate Radiation: Evolutionary Biology of the African Guenons*, ed. A. Gautier-Hion, F. Bourlière, J.-P. Gautier, J. Kingdon, pp. 413–37. Cambridge: Cambridge University Press.

Chism, J., Rowell, T.E. & Olson, D. 1984. Life history patterns of female patas monkeys. In: *Female Primates: Studies by Women Primatologists*, ed. M.F. Small, pp. 175–90. New York: Alan R. Liss.

Chivers, D.J. 1971. The Malayan siamang. *Malayan Nature Journal*, **24**, 78–86.

–1977a. The feeding behaviour of siamang (*Hylobates syndactylus*). In: *Primate Ecology: Studies of Feeding and Ranging Behavior in Lemurs, Monkeys and Apes*, ed. T.H. Clutton-Brock, pp. 355–82. London: Academic Press.

–1977b. The lesser apes. In: *Primate Conservation*, ed. H.S.H. Prince Rainier III. of Monaco & G.H. Bourne, pp. 539–98. New York: Academic Press.

Chivers, D.J. & Raemaekers, J.J. 1980. Long-term changes in behaviour. In: *Malayan Forest Primates: Ten Years' Study in Tropical Rain Forest*, ed. D.J. Chivers, pp. 209–60. New York: Plenum Press.

–1986. Natural and synthetic diets of Malayan gibbons. In: *Primate Ecology and Conservation*, J.G. Else & P.C. Lee, pp. 39–56. Cambridge: Cambridge University Press.

Choudhury, A. 1996. A survey of Hoolock gibbon (*Hylobates hoolock*) in southern Assam, India. *Primate Report*, **44**, 77–85.

Clarke, J.L., Jones, M.E. & Jarman, P.J. 1989. A day in the life of a kangaroo: activities and movements of Eastern Grey kangaroos *Macropus giganteus* at Wallaby Creek. In: *Kangaroos, Wallabies and Rat-kangaroos*, ed. G. Grigg, P. Jarman & I. Hume, pp. 611–18. Sydney: Surrey Beatty & Sons.

–1995. Diurnal and nocturnal grouping and foraging behaviours of free-ranging Eastern Grey kangaroos. *Australian Journal of Zoology*, **43**, 519–29.

Clarke, M.R. & Zucker, E.L. 1994. Survey of the howling monkey population at La Pacifica: a seven-year follow-up. *International Journal of Primatology*, **15**, 61–73.

Clinton, W.L. & LeBoeuf, B.J. 1993. Sexual selection's effect on male life history and the pattern of male mortality. *Ecology*, **74**, 1884–92.

Clutton-Brock, T.H. 1972. Feeding and ranging behaviour of the red colobus monkey. PhD thesis, University of Cambridge.

–1974. Primate social organization and ecology. *Nature*, **250**, 539–42.

–1985. Size, sexual dimorphism, and polygyny in primates. In: *Size and Scaling in Primate Biology*, ed. W.L. Jungers, pp. 51–60. New York: Plenum Press.

–1989. Mammalian mating systems. *Proceedings of the Royal Society of London, Series B*, **236**, 339–72.

–1991. *The Evolution of Parental Care*. Princeton, NJ: Princeton University Press.

–1998. Reproductive skew, concessions and limited control. *Trends in Ecology and Evolution*, **13**, 288–92.

Clutton-Brock, T.H. & Albon, S.D. 1982. Parental investment in male and female offspring in mammals. In: *Current Problems in Sociobiology*, ed. King's College Sociobiology Group, pp. 223–48. Cambridge: Cambridge University Press.

Clutton-Brock, T.H. & Harvey, P.H. 1976. Evolutionary rules and primate societies. In: *Growing Points in Ethology*, ed. P.P.G. Bateson & R.A. Hinde. pp. 195–237. Cambridge: Cambridge University Press.

–1977a. Primate ecology and social organization. *Journal of Zoology, London*, **183**, 1–39.

–1977b. Species differences in feeding and ranging behaviour in primates. In: *Primate Ecology*, ed. T.H. Clutton-Brock, pp. 557–79. London: Academic Press.

Clutton-Brock, T.H. & Iason, G. 1986. Sex ratio variation in mammals. *Quarterly Review of Biology*, **61**, 339–74.

Clutton-Brock, T.H. & Parker, G.A. 1992. Potential reproductive rates and the operation of sexual selection. *Quarterly Review of Biology*, **67**, 437–56.

–1995a. Punishment in animal societies. *Nature*, **373**, 209–16.

–1995b. Sexual coercion in animal societies. *Animal Behaviour*, **49**, 1345–65.

Clutton-Brock, T.H. & Vincent, A.C. J. 1991. Sexual selection and the potential reproductive rates of males and females. *Nature*, **351**, 58–60.

Clutton-Brock, T.H., Harvey, P.H. & Rudder, B. 1977. Sexual dimorphism, socionomic sex ratio, and body weight in primates. *Nature*, **269**, 797–800.

Clutton-Brock, T., Guinness, F. & Albon, S. 1982. *Red Deer: Behavior and Ecology of Two Sexes*. Chicago: University of Chicago Press.

Colmenares, F. 1991. Greeting behaviour between male baboons: oestrous females, rivalry and negotiation. *Animal Behaviour*, **41**, 49–60.

–1992. Clans and harems in a colony of hamadryas and hybrid baboons: male kinship, familiarity and the formation of brother-teams. *Behaviour*, **121**, 61–94.

Colquhoun, I.C. 1993. The socioecology of *Eulemur macaco macaco*: a preliminary report. In: *Lemur Social Systems and Their Ecological Basis*, ed. P.M. Kappeler & J.U. Ganzhorn, pp. 11–23. New York: Plenum Press.

Colyn, M. 1991. L'importance zoogeographique du bassin du fleuve Zaïre pour la speciation: le cas des primates simiens. *Annales Sciences Zoologiques*, **264**, 1–250.

Cords, M. 1984. Mating patterns and social structure in redtail monkeys (*Cercopithecus ascanius*). *Zeitschrift für Tierpsychologie*, **64**, 313–29.

–1987a. Mixed-species association of *Cercopithecus* monkeys in the Kakamega Forest, Kenya. *University of California Publications in Zoology*, **117**, 1–109.

–1987b. Forest guenons and patas monkeys: male–male competition in one-male groups. In: *Primate Societies*, ed. B.B. Smuts, D.L. Cheney, R.M. Seyfarth, R.W. Wrangham & T.T. Struhsaker, pp. 98–111. Chicago: University of Chicago Press.

–1988. Mating systems of forest guenons: a preliminary review. In: *A Primate Radiation: Evolutionary Biology of the African Guenons*, ed. A. Gautier-Hion, F. Bourlière, J.-P. Gautier & J. Kingdon, pp. 323–39. Cambridge: Cambridge University Press.

Cords, M., Mitchell, B.J., Tsingalia, H.M. & Rowell, T.E. 1986. Promiscuous mating among blue monkeys in the Kakamega Forest, Kenya. *Ethology*, **72**, 214–26.

Coutinho, P.E.G. & Corrêa, H.K.M. 1995. Polygyny in a free-ranging group of buffy-tufted-ear marmosets, *Callithrix aurita*. *Folia Primatologica*, **65**, 25–9.

Cowlishaw, G. 1992. Song function in gibbons. *Behaviour*, **121**, 131–53.

–1994. Vulnerability to predation in baboon populations. *Behaviour*, **131**, 293–304.

–1995. Behavioural patterns in baboon group encounters: the role of resource competition and male reproductive strategies. *Behaviour*, **132**, 75–86.

–1997a. Alarm calling and implications for risk perception in a desert baboon population. *Ethology*, **103**, 384–94.

–1997b. Refuge use and predation risk in a desert baboon population. *Animal Behaviour*, **54**, 241–53.

Cowlishaw, G. & Dunbar, R.I.M. 1991. Dominance rank and mating success in male primates. *Animal Behaviour*, **41**, 1045–56.

Craig, J.L. 1984. Are communal pukeko caught in the prisoner's dilemma? *Behavioral Ecology and Sociobiology*, **14**, 147–50.

Craig, J.L. & Jamieson, I.G. 1990. Pukeko: different approaches and some different answers. In: *Cooperative Breeding in Birds*, ed. P.B. Stacey & W.D. Koenig, pp. 385–412. Cambridge: Cambridge University Press.

Crockett, C.M. 1984. Emigration by female red howler monkeys and the case for female competition. In: *Female Primates: Studies by Women Primatologists*, ed. M.F. Small, pp. 159–73. New York: Alan R. Liss.

–1985. Population studies of red howler monkeys (*Alouatta seniculus*). *National Geographic Research*, **1**(2), 264–73.

–1996. The relation between red howler monkey (*Alouatta

seniculus) troop size and population growth in two habitats. In: *Adaptive Radiations of Neotropical Primates*, ed. M.A. Norconk, A.L. Rosenberger & P.A. Garber, pp. 489–510. New York: Plenum Press.

Crockett, C.M. & Eisenberg, J.F. 1987. Howlers: variations in group size and demography. In: *Primate Societies*, ed. B.B. Smuts, D.L. Cheney, R.M. Seyfarth, R.W. Wrangham & T.T. Struhsaker, pp. 54–68. Chicago: University of Chicago Press.

Crockett, C.M. & Pope, T.R. 1988. Inferring patterns of aggression from red howler monkey injuries. *American Journal of Primatology*, **15**, 289–308.

–1993. Consequences of sex differences in dispersal for juvenile red howler monkeys. In: *Juvenile Primates: Life History, Development, and Behavior*, ed. M.E. Pereira & L.A. Fairbanks, pp. 104–18. New York: Oxford University Press.

Crockett, C.M. & Rudran, R. 1987a. Red howler monkey birth data I: seasonal variation. *American Journal of Primatology*, **13**, 347–68.

–1987b. Red howler monkey birth data II: interannual, habitat, and sex comparisons. *American Journal of Primatology*, **13**, 369–84.

Crockett, C.M. & Sekulic, R. 1984. Infanticide in red howler monkeys (*Alouatta seniculus*). In: *Infanticide: Comparative and Evolutionary Perspectives*, ed. G. Hausfater & S.B. Hrdy, pp. 173–91. New York: Aldine.

Croft, D.B. 1981. Behaviour of red kangaroos, *Macropus rufus* (Desmarest, 1822), in north-western New South Wales. *Australian Mammalogy*, **4**, 5–58.

–1989. Social organization of the Macropodoidea. In: *Kangaroos, Wallabies and Rat-kangaroos*, ed. G. Grigg, P. Jarman & I. Hume, pp. 505–25. Sydney: Surrey Beatty & Sons.

Crook, J.H. 1964. The evolution of social organization and visual communication in the weaver birds (Ploceinae). *Behaviour, Supplement*, **10**, 1–178.

–1968. The adaptive significance of avian social organizations. *Symposia of the Zoological Society of London*, **14**, 181–218.

–1970. The socio-ecology of primates. In: *Social Behavior in Birds and Mammals*, ed. J. Crook, pp. 103–66. London: Academic Press.

–1972. Sexual selection, dimorphism, and social organization in the primates. In: *Sexual Selection and the Descent of Man, 1871–1971*, ed. B.G. Campbell, pp. 231–81. Chicago: Aldine.

Crook, J.H. & Gartlan, J.C. 1966. Evolution of primate societies. *Nature*, **210**, 1200–3.

Curtin, S.H. 1980. Dusky and banded leaf monkeys. In: *Malayan Forest Primates: Ten Years' Study in Tropical Rain Forest*, ed. D.J. Chivers, pp. 107–45. New York: Plenum Press.

Curtin, S.H. & Chivers, D.J. 1978. Leaf-eating primates of peninsular Malaysia: the siamang and the dusky leaf monkey. In: *The Ecology of Arboreal Folivores*, ed. G.G. Montgomerey, pp. 441–64. Washington, DC: Smithsonian Institution Press.

Curtis, D. 1997. The mongoose lemur (*Eulemur mongoz*): a study in behavior and ecology. PhD thesis, Universität Zürich.

Dague, C. & Petter, J.-J. 1988. Observations sur le *Lemur rubriventer* dans son milieu naturel. In: *L'Equilibre des Ecosystèmes Forestièrs à Madagascar: Actes d'un Séminaire International*, ed. L. Rakotovao, V. Barre & J. Sayer, pp. 78–89. Cambridge: Page Bros Ltd.

Daniel, J.C. Jr 1970. Dormant embryos of mammals. *BioScience*, **20**, 411–15.

Darwin, C. 1871. *The Descent of Man and Selection in Relation to Sex*. London: John Murray.

Davies, A.G. 1984. *An Ecological Study of the Red Leaf Monkey* (Presbytis rubicunda) *in the Dipterocarp Forest of Northern Borneo*. Cambridge: Sidney Sussex College.

–1987. Adult male replacement and group formation in *Presbytis rubicunda*. *Folia Primatologica*, **49**, 111–14.

–1991. Seed-eating by red leaf monkeys (*Presbytis rubicunda*) in dipterocarp forest of Northern Borneo. *International Journal of Primatology*, **12**, 119–44.

Davies, A.G. & Oates, J.F. (eds.) 1994. *Colobine Monkeys: Their Ecology, Behaviour and Evolution*. Cambridge: Cambridge Unversity Press.

Davies, N.B. 1989. Sexual conflict and the polygamy threshold. *Animal Behaviour*, **38**, 226–34.

–1991. Mating systems. In: *Behavioural Ecology*, ed. J.R. Krebs & N.B. Davies, pp. 263–94. Oxford: Blackwell.

–1992. *Dunnock Behaviour and Social Evolution*. Oxford: Oxford University Press.

Davies, N.B. & Hatchwell, B.J. 1992. The value of male parental care and its influence on reproductive allocation by male and female dunnocks. *Journal of Animal Ecology*, **61**, 259–72.

Davies, N.B. & Lundberg, A. 1984. Food distribution and a variable mating system in the dunnock, *Prunella modularis*. *Journal of Animal Ecology*, **53**, 895–912.

Davies, N.B., Hatchwell, B.J., Robson, T. & Burke, T. 1992. Paternity and parental effort in dunnocks: how good are male chick-feeding rules? *Animal Behaviour*, **43**, 729–45.

Davies, N.B., Hartley, I.R., Hatchwell, B.J., Desrochers, A., Skeer, J. & Nebel, D. 1995. The polygynandrous mating system of the alpine accentor *Prunella collaris*. I. Ecological causes and reproductive conflicts. *Animal Behaviour*, **49**, 769–88.

Davies, N.B., Hartley, I.R., Hatchwell, B.J. & Langmore, N.E. 1996. Female control of copulations to maximize male help: a comparison of polygynandrous alpine accentors *Prunella collaris* and dunnocks *P. modularis*. *Animal Behaviour*, **51**, 27–47.

Dawson, G.A. 1977. Composition and stability of social groups of the tamarin, *Saguinus oedipus geoffroyi*, in Panama. Ecological and behavioral implications. In: *The Biology and Conservation of the Callitrichidae*, ed. D.G. Kleiman, pp. 23–37. Washington, DC: Smithsonian Institution Press.

–1979. The use of time and space by the Panamanian tamarin, *Saguinus oedipus*. *Folia Primatologica*, **31**, 253–84.

Dawson, G.A. & Dukelow, W.R. 1976. Reproductive characteristics of free-ranging Panamanian tamarins, *Saguinus oedipus geoffroyi*. *Journal of Medical Primatology*, **5**, 266–75.

Dawson, T.J. 1989. Diets of macropodoid marsupials: general patterns and environmental influences. In: *Kangaroos, Wallabies and Rat-kangaroos*, ed. G. Grigg, P. Jarman & I. Hume, pp. 129–42. Sydney: Surrey Beatty & Sons.

de la Torre, S., Campos, F. & de Vries, T. 1995. Home range and birth seasonality of *Saguinus nigricollis graellsi* in Ecuadorian Amazonia. *American Journal of Primatology*, **37**, 39–56.

de Ruiter, J.R. 1986. The influence of group size on predator scanning and foraging behaviour of wedge-capped capuchin monkeys (*Cebus olivaceus*). *Behaviour*, **98**, 240–58.

–1992. Capturing wild long-tailed macaques (*Macaca fascicularis*). *Folia Primatologica*, **59**, 89–104.

de Ruiter, J.R. & Geffen, E. 1998. Relatedness of matrilines, dispersing males and social groups in long-tailed macaques (*Macaca fascicularis*). *Proceedings of the Royal Society of London, Series B*, **265**, 79–87.

de Ruiter, J.R. & van Hooff, J.A.R.A.M. 1993. Male dominance rank and reproductive success in primate groups. *Primates*, **34**, 513–23.

de Ruiter, J.R., Scheffrahn, W., Trommelen, G.J.J.M., Uitterlinden, A.G., Martin, R.D. & van Hooff, J.A.R.A.M. 1992. Male social rank and reproductive success in wild long-tailed macaques: paternity exclusions by blood protein analysis and DNA fingerprinting. In: *Paternity in Primates: Genetic Tests and Theories*, ed. R.D. Martin, A.F. Dixson & E.J. Wickings, pp. 175–91. Basel: S. Karger.

de Ruiter, J.R., van Hooff, J.A.R.A.M. & Scheffrahn, W. 1994. Social and genetic aspects of paternity in wild long-tailed macaques (*Macaca fascicularis*). *Behaviour*, **129**, 203–24.

de Waal, F.B.M. 1977. The organization of agonistic relations within two captive groups of Java-monkeys (*Macaca fascicularis*). *Zeitschrift für Tierpsychologie*, **44**, 225–82.

–1978. Exploitative and familiarity-dependent support strategies in a colony of semi-free-living chimpanzees. *Behaviour*, **66**, 268–312.

–1982. *Chimpanzee Politics: Power and Sex Among the Apes*. New York: Harper and Row.

–1984. Sex differences in the formation of coalitions among chimpanzees. *Ethology and Sociobiology*, **5**, 239–55.

–1986. The integration of dominance and social bonding in primates. *Quarterly Review of Biology*, **61**, 459–79.

–1989a. Dominance 'style' and primate social organization. In: *Comparative Socioecology: the Behavioural Ecology of Humans and Other Mammals*, ed. V. Standen & R.A. Foley, pp. 243–63. Oxford: Blackwell Scientific Publications.

–1989b. Food sharing and reciprocal obligations among chimpanzees. *Journal of Human Evolution*, **18**, 433–59.

–1992. Coalitions as part of reciprocal relations in the Arnhem chimpanzee colony. In: *Coalitions and Alliances in Animals and Humans*, ed. A.H. Harcourt & F.B.M. de Waal, pp. 233–58. Oxford: Oxford University Press.

–1996a. Conflict as negotiation. In: *Great Ape Societies*, ed. W.C. McGrew, L.F. Marchant & T. Nishida, pp. 159–72. Cambridge: Cambridge University Press.

–1996b. *Good Natured: the Origins of Right and Wrong in Humans and Other Animals*. Cambridge, MA: Harvard University Press.

de Waal, F.B.M., van Hooff, J.A.R.A.M. & Netto, W.J. 1976. An ethological analysis of types of agonistic interaction in a captive group of Java-monkeys (*Macaca fascicularis*). *Primates*, **17**, 257–90.

Deag, J.M. & Crook, J.H. 1971. Social behaviour and 'agonistic buffering' in the wild Barbary macaque *Macaca sylvana* L. *Folia Primatologica*, **15**, 183–200.

Debyser, I.W.J. 1995. Prosimian juvenile mortality in zoos and primate centers. *International Journal of Primatology*, **16**, 889–907.

Decker, B.S. 1994a. Effects of habitat disturbance on the behavioral ecology and demographics of the Tana River red colobus (*Colobus badius rufomitratus*). *International Journal of Primatology*, **15**, 703–37.

–1994b. Endangered primates in the Selous Game Reserve and an imminent threat to their habitat. *Oryx*, **28**, 183–90.

Dennis, A.J. & Johnson, P.M. 1995. Musky rat-kangaroo *Hypsiprymnodon moschatus*. In: *The Mammals of Australia*, ed. R. Strahan, pp. 282–4. Sydney: Australian Museum & Reed Books.

Dietz, J.M. & Baker, A.J. 1993. Polygyny and female reproductive success in golden lion tamarins, *Leontopithecus rosalia*. *Animal Behaviour*, **46**, 1067–78.

Dietz, J.M., de Sousa, S.N.F. & da Silva, J.R.O. 1994. Population structure and territory size in golden-headed lion tamarins, *Leontopithecus chrysomelas*. *Neotropical Primates*, Supplement, **2**, 21–3.

DiFiore, A. & Rendall, D. 1994. Evolution of social organization: a reappraisal for primates by using phylogenetic methods. *Proceedings of the National Academy of Science of the United States of America*, **91**, 9941–5.

Digby, L.J. 1994. Social organization and reproductive strategies in a wild population of common marmosets (*Callithrix jacchus*). PhD thesis, University of California, Davis.

–1995. Infant care, infanticide, and female reproductive strategies in polygynous groups of common marmosets (*Callithrix jacchus*). *Behavioral Ecology and Sociobiology*, **37**, 51–61.

Digby, L.J. & Barreto, C.E. 1996. Activity and ranging patterns in common marmosets (*Callithrix jacchus*): implications for reproductive strategies. In: *Adaptive Radiations of Neotropical Primates*, ed. M.A. Norconk, A.L. Rosenberger & P.A. Garber, pp. 173–85. New York: Plenum Press.

Digby, L.J. & Ferrari, S.F. 1994. Multiple breeding females in free-ranging groups of *Callithrix jacchus*. *International Journal of Primatology*, **15**, 389–97.

Disotell, T.R. 1994. Generic level relationships of the papionini (Cercopithecoidea). *American Journal of Physical Anthropology*, **94**, 47–57.

–1996. The phylogeny of Old World monkeys. *Evolutionary Anthropology*, **5**, 18–24.

Dixson, A.F. 1983. Observations on the evolution and behavioural significance of 'sexual skin' in female primates. *Advances in the Study of Behavior*, **13**, 63–106.

–1987. Observations on the evolution of the genitalia and copulatory behaviour in male primates. *Journal of Zoology, London*, **213**, 423–43.

–1991. Sexual selection, natural selection and copulatory patterns in primates. *Folia Primatologica*, **57**, 96–101.

–1997. Evolutionary perspectives on primate mating systems and behavior. *Annals of the New York Academy of Sciences*, **807**, 42–61.

Dixson, A.F., Anzenberger, G., Monteiro da Cruz, M.A.O., Patel, I. & Jeffreys, A.J. 1992. DNA fingerprinting of free-ranging groups of common marmosets (*Callithrix jacchus jacchus*) in NE Brazil. In: *Paternity in Primates: Genetic Tests and Theories*, ed. R.D. Martin, A.F. Dixson & E.J. Wickings, pp. 192–202. Basel: S. Karger.

Donati, G., Lunardini, A. & Kappeler, P.M. 1999. Cathemeral activity of redfronted lemurs, *Eulemur fulvus rufus*, in Kirindy forest. In: *New Directions in Lemur Studies*, ed. J.U. Ganzhorn, S.M. Goodman & H. Rasamimanana, pp. 119–37. New York: Plenum Press. In press.

Drickamer, L.C. & Vessey, S.H. 1973. Group changing in free-ranging male rhesus monkeys. *Primates*, **14**, 359–68.

Dugatkin, L.A. 1997. *The Evolution of Cooperation*. New York: Oxford University Press.

Dunbar, R.I.M. 1974. Observations on the ecology and social organization of the green monkey, *Cercopithecus sabaeus*, in Senegal. *Primates*, **15**, 341–50.

–1978. Sexual behaviour and social relationships among gelada baboons. *Animal Behaviour*, **26**, 167–78.

–1982. Intraspecific variations in mating strategy. In: *Perspectives in Ethology*, Vol. 5, ed. P.H. Klopfer & P.P.G. Bateson, pp. 385–431. New York: Plenum Press.

–1983. Structure of gelada baboon reproductive units. II. The male's relationship with his females. *Animal Behaviour*, **31**, 565–75.

–1984. *Reproductive Decisions: an Economic Model of Gelada Baboon Social Strategies*. Princeton, NJ: Princeton University Press.

–1987. Habitat quality, population dynamics, and group composition in colobus monkeys (*Colobus guereza*). *International Journal of Primatology*, **8**, 299–329.

–1988. *Primate Social Systems*. Ithaca: Cornell University Press.

–1992. Time: a hidden constraint on the behavioural ecology of baboons. *Behavioral Ecology and Sociobiology*, **31**, 35–49.

–1995a. The mating system of callitrichid primates. I. Conditions for the coevolution of pair bonding and twinning. *Animal Behaviour*, **50**, 1057–70.

–1995b. The mating system of callitrichid primates. II. The impact of helpers. *Animal Behaviour*, **50**, 1071–89.

Dunbar, R.I.M. & Cowlishaw, G. 1992. Mating success in male primates: dominance rank, sperm competition and alternative strategies. *Animal Behaviour*, **44**, 1171–3.

Dunbar, R.I.M. & Dunbar, E.P. 1975. *Social Dynamics of Gelada Baboons*. Basel: S. Karger.

–1976. Contrasts in social structure among black and white colobus monkey groups. *Animal Behaviour*, **24**, 84–92.

Dunbar, R.I.M., Buckland, D. & Miller, D. 1990. Mating strategies of male feral goats: a problem in optimal foraging. *Animal Behaviour*, **40**, 653–67.

Eaton, R.L. 1978. Why some felids copulate so much: a model for the evolution of copulation frequency. *Carnivore*, **1**, 42–51.

Eberhard, W.G. 1996. *Female Control: Sexual Selection by Cryptic Female Choice*. Princeton: Princeton University Press.

Eisenberg, J.F. 1981. *The Mammalian Radiations: An Analysis of Trends in Evolution, Adaptation, and Behavior*. Chicago: University of Chicago Press.

Eisenberg, J.F., Muckenhirn, N.A. & Rudran, R. 1972. The relation between ecology and social structure in primates. *Science*, **176**, 863–74.

Elgar, M.A. 1989. Predator vigilance and group size in mammals and birds: a critical review of the empirical evidence. *Biological Reviews*, **64**, 13–33.

Ellefson, J.O. 1974. A natural history of white-handed gibbons in the Malayan Peninsular. In: *Gibbon and Siamang*, Vol. 3, *Natural History, Social Behavior, Reproduction, Vocalizations, Prehension*, ed. D.M. Rumbaugh, pp. 1–136. Basel: S. Karger.

Ely, J. & Kurland, J.A. 1989. Spatial autocorrelation, phylogenetic constraints, and the causes of sexual dimorphism in primates. *International Journal of Primatology*, **10**, 151–71.

Emerson, S.B. & Arnold, S.J. 1989. Intra- and interspecific relationships between morphology, performance, and fitness. In: *Complex Organismal Functions: Integration and Evolution in Vertebrates*, ed. D.B. Wake & G. Roth, pp. 295–314. New York: Wiley.

Emlen, S.T. 1982. The evolution of helping. II. The role of behavioral conflict. *American Naturalist*, **119**, 40–53.

–1997. Predicting family dynamics in social vertebrates. In: *Behavioural Ecology: An Evolutionary Approach*, 4th edn, ed. J.R. Krebs & N.B. Davies, pp. 228–53. Oxford: Blackwell Science.

Emlen, S.T. & Oring, L.W. 1977. Ecology, sexual selection and the evolution of mating systems. *Science*, **197**, 215–23.

Emlen, S.T. & Wrege, P.H. 1989. A test of alternative hypotheses for helping behaviour in white-fronted bee-eaters of Kenya. *Behavioral Ecology and Sociobiology*, **25**, 303–19.

Emmons, L.H. & Feer, F. 1990. *Neotropical Rainforest Mammals. A Field Guide*. Chicago: University of Chicago Press.

Encarnación, F. & Heymann, E.W. 1998. Body mass of wild *Callimico goeldii*. *Folia Primatologica*, **69**, 368–71.

Enomoto, T. 1974. The sexual behavior of Japanese macaques. *Journal of Human Evolution*, **3**, 351–72.

Epple, G. 1972. Social behavior of laboratory groups of *Saguinus fuscicollis*. In: *Saving the Lion Marmoset*, ed. D.D. Bridgewater, pp. 50–8. Wheeling, WV: The Wild Animal Propagation Trust.

–1975. The behavior of marmoset monkeys (Callitrichidae). In: *Primate Behavior*, Vol. 4, ed. L.A. Rosenblum, pp. 195–239. New York: Academic Press.

Fa, J.E. 1984. *The Barbary Macaque: a Case Study in Conservation*. New York: Plenum Press.

Faaborg, J. & Bednarz, J.C. 1990. Galapagos hawks and Harris hawks: divergent causes of sociality in two raptors. In: *Cooperative Breeding in Birds*, ed. P.B. Stacey & W.D. Koenig, pp. 357–83. Cambridge: Cambridge University Press.

Faaborg, J., Parker, P.G., DeLay, L. *et al.* 1995. Confirmation of cooperative polyandry in the Galapagos hawk *Buteo galapagoensis*. *Behavioral Ecology and Sociobiology*, **36**, 83–90.

Fairbanks, L.A. 1993. Juvenile vervet monkeys: establishing relationships and practicing skills for the future. In: *Juvenile Primates: Life History, Development, and Behavior*, ed. M.E. Pereira & L.A. Fairbanks, pp. 211–27. New York: Oxford University Press.

Fairgrieve, C. 1993. The comparative ecology of blue monkeys (*Cercopithecus mitis stuhlmanni*) in logged and un-logged forest, Budongo Forest Reserve, Uganda. Unpublished report.

–1995. Infanticide and infant eating in the blue monkey (*Cercopithecus mitis stuhlmanni*) in the Budongo Forest Reserve, Uganda. *Folia Primatologica*, **64**, 69–72.

Fedigan, L. & Fedigan, L.M. 1988. *Cercopithecus aethiops*: a review of field studies. In: *A Primate Radiation: Evolutionary Biology of the African Guenons*, ed. A. Gautier-Hion, F. Bourlière, J.-P. Gautier & J. Kingdon, pp. 389–411. Cambridge: Cambridge University Press.

Fedigan, L.M. 1983. Dominance and reproductive success. *Yearbook of Physical Anthropology*, **26**, 91–129.

–1993. Sex differences and intersexual relations in adult white-faced capuchins (*Cebus capucinus*). *International Journal of Primatology*, **14**, 853–77.

Fedigan, L.M. & Baxter, M.J. 1984. Sex differences and social organization in free-ranging spider monkeys (*Ateles geoffroyi*). *Primates*, **25**, 279–94.

Fedigan, L.M. & Rose, L.M. 1995. Interbirth interval variation in three sympatric species of neotropical monkey. *American Journal of Primatology*, **37**, 9–24.

Fedigan, L.M. Rose, L.M. & Avila, R.M. 1996. See how they grow: tracking capuchin monkey (*Cebus capucinus*) populations in a regenerating Costa Rican dry forest. In: *Adaptive Radiations of Neotropical Primates*, ed. M.A. Norconk, A.L. Rosenberger & P.A. Garber, pp. 289–307. New York: Plenum Press.

Feh, C. 1990. Long-term paternity data in relation to different aspects of rank for Camargue stallions, *Equus caballus*. *Animal Behaviour*, **40**, 995–6.

Felsenstein, J. 1985. Phylogenies and the comparative method. *American Naturalist*, **125**, 1–15.

Ferrari, S.F. 1992. The care of infants in a wild marmoset (*Callithrix flaviceps*) group. *American Journal of Primatology*, **26**, 109–18.

Ferrari, S.F. & Lopes Ferrari, M.A. 1989. A re-evaluation of the social organization of the Callitrichidae, with reference to the ecological differences between genera. *Folia Primatologica*, **52**, 132–47.

Ferris, C.F. & Potegal, M. 1988. Vasopressin receptor blockade in the anterior hypothalamus suppresses aggression in hamsters. *Physiology and Behavior*, **44**, 235–9.

Flannery, T.F., Martin, R. & Szalay, A. 1996. *Tree Kangaroos: a Curious Natural History*. Sydney: Reed Books.

Fontaine, R. 1981. The uakaris, genus *Cacajao*. In: *Ecology and Behavior of Neotropical Primates*, ed. A. Coimbra-Filho & R.A. Mittermeier, pp. 443–93. Rio de Janeiro: Academia Brasiliera de Ciencias.

Ford, S.M. 1994. Evolution of sexual dimorphism in body weight in platyrrhines. *American Journal of Primatology*, **34**, 221–44.

Fossey, D. 1984. Infanticide in mountain gorillas (*Gorilla gorilla beringei*) with comparative notes on chimpanzees. In: *Infanticide: Comparative and Evolutionary Aspects*, ed. G. Hausfater & S.B. Hrdy, pp. 217–35. New York: Aldine.

Fossey, D. & Harcourt, A.H. 1977. Feeding ecology of free-ranging mountain gorillas (*Gorilla gorilla beringei*). In: *Primate Ecology*, ed. T.H. Clutton-Brock, pp. 415–47. London: Academic Press.

Frank, L.G. 1986. Social organization of the spotted hyaena (*Crocuta crocuta*). I. Demography. *Animal Behaviour*, **35**, 1500–9.

Freed, B. 1996. Co-occurrence among crowned lemurs (*Lemur coronatus*) and Sanford's lemurs (*Lemur fulvus sanfordi*) of Madagascar. PhD thesis, Washington University St Louis, MO.

French, J.A. & Snowdon, C.T. 1981. Sexual dimorphism in responses to unfamiliar intruders in the tamarin, *Saguinus oedipus*. *Animal Behaviour*, **29**, 822–9.

Fretwell, S.D. 1972. *Populations in a Seasonal Environment*. Monographs in Population Biology (5). Princeton, NJ: Princeton University Press.

Fukuda, F. 1982. Relationships between age and troop shifting in male Japanese monkeys. *Japanese Journal of Ecology*, **32**, 491–8.

Galat, G. & Galat-Luong, A. 1978. *Les Effectifs des Bandes et les Strategies d'Occupation de l'Espace chez le Singe Vert* (Cercopithecus aethiops sabaeus), *au Senegal: Methodes d'Étude et Resultats Preliminaires*. Abidjan: ORSTOM, Centre d'Adiopodoumé.

–1985. La communauté des primates diurnes de la forêt de Tai, Cote d'Ivoire. *Review of Ecology* (*La Terre et la Vie*), **40**, 3–32.

Galat-Luong, A. 1975. Notes préliminaires sur l'écologie de *Cercopithecus ascanius schmidti* dans les environs de Bangui (R.C.A.). *Review of Ecology* (*La Terre et la Vie*), **122**, 288–97.

Galdikas, B.M.F. 1979. Orangutan adaptation at Tanjung Puting Reserve: mating and ecology. In: *The Great Apes*, ed. D.A. Hamburg & E.R. McCown, pp. 195–233. New York: Academic Press

–1985a. Adult male sociality and reproductive tactics among orangutans at Tanjung Puting. *Folia Primatologica*, **45**, 9–24.

–1985b. Subadult male orangutan sociality and reproductive behavior at Tanjung Puting. *American Journal of Primatology*, **8**, 87–99.

–1988. Orangutan diet, range, and activity at Tanjung Puting, Central Borneo. *International Journal of Primatology*, **9**, 1–35.

Garber, P.A. 1993. Feeding ecology and behaviour of the genus

Saguinus. In: *Marmosets and Tamarins. Systematics, Behaviour, and Ecology*, ed. A.B. Rylands, pp. 273–95. Oxford: Oxford University Press.

–1994. Phylogenetic approach to the study of tamarin and marmoset social systems. *American Journal of Primatology*, **34**, 199–219.

–1997. One for all and breeding for one: cooperation and competition as a tamarin reproductive strategy. *Evolutionary Anthropology*, **5**, 187–99.

Garber, P.A. & Leigh, S.R. 1997. Ontogenetic variation in small-bodied New World primates: implications for patterns of reproduction and infant care. *Folia Primatologica*, **68**, 1–22.

Garber, P.A., Moya, L. & Malaga, C. 1984. A preliminary field study of the moustached tamarin monkey (*Saguinus mystax*) in northeastern Peru: questions concerned with the evolution of a communal breeding system. *Folia Primatologica*, **42**, 17–32.

Garber, P.A., Encarnacion, F., Moya, L. & Pruetz, J.D. 1993a. Demographic and reproductive patterns in moustached tamarin monkeys (*Saguinus mystax*): implications for reconstructing platyrrhine mating systems. *American Journal of Primatology*, **29**, 235–54.

Garber, P.A., Pruetz, J.D. & Isaacson, J. 1993b. Patterns of range use, range defense, and intergroup spacing in moustached tamarin monkeys (*Saguinus mystax*). *Primates*, **34**, 11–25.

Garber, P.A., Moya, L., Pruetz, J.D. & Ique, C. 1996. Social and seasonal influences on reproductive biology in male moustached tamarins (*Saguinus mystax*). *American Journal of Primatology*, **38**, 29–46.

Gartlan, J.S. 1968. Structure and function in primate society. *Folia Primatologica*, **8**, 89–120.

Gartlan, J.S. & Gartlan, S.C. 1973. Quelques observations sur les groupes exclusivement males chez *Erythrocebus patas*. *Annales de la Faculté des Sciences de Cameroon*, **12**, 121–44.

Gatinot, B.L. 1975. Ecologie d'un colobe bai (*Colobus badius temmincki*, Kuhn 1820) dans un milieu marginal au Senegal. PhD thesis, Université de Paris VI.

Gautier, J.-P. 1994. Quelques caracteristiques écologiques du singe des marais: *Allenopithecus nigroviridis* Lang 1923. *Review of Ecology (La Terre et la Vie)*, **40**, 331–42.

Gautier-Hion, A. 1970. L'organisation sociale d'une bande de talapoins (*Miopithecus talapoin*) dans le nord-est du Gabon. *Folia Primatologica*, **12**, 116–41.

–1971. L'écologie du talapoin du Gabon. *Review of Ecology* (*La Terre et la Vie*), **25**, 427–90.

Gautier-Hion, A. & Gautier, J.-P. 1974. Les associations polyspécifiques des Cercopitheques du plateau de M'passa, Gabon. *Folia Primatologica*, **22**, 134–77.

–1978. Le singe de Brazza: une stratégie originale. *Zeitschrift für Tierpsychologie*, **46**, 84–104.

Gautier-Hion, A., Bourlière, F., Gautier, J.-P. & Kingdon, J. (eds.) 1988. *A Primate Radiation. Evolutionary Biology of the African Guenons*. Cambridge: Cambridge University Press.

Geffen, E., Gompper, M.E., Gittleman, J.L., Luh, H.K., Macdonald, D.W. & Wayne, R.K. 1996. Size, life-history traits, and social organization in the Canidae: a reevaluation. *American Naturalist* **147**, 140–60.

Ghiglieri, M.P. 1984. Feeding ecology and sociality of chimpanzees in Kibale Forest, Uganda. In: *Adaptations for Foraging in Nonhuman Primates*, ed. P.S. Rodman & J. Cant, pp. 161–94. New York: Columbia University Press.

Gittins, S.P. 1980. Territorial behavior in the agile gibbon. *International Journal of Primatology*, **1**, 381–99.

Gittleman, J.L. 1986. Carnivore life history patterns: allometric, phylogenetic, and ecological associations. *American Naturalist*, **127**, 744–71.

Gittleman, J., Anderson, C., Kot, M. & Luh, H.-K. 1996. Phylogenetic lability and rates of evolution: a comparison of behavioral, morphological and life history traits. In: *Phylogenies and the Comparative Method in Animal Behavior*, ed. E. Martins, pp. 166–205. Oxford: Oxford University Press.

Glander, K.E. 1980. Reproduction and population growth in free-ranging mantled howling monkeys. *American Journal of Physical Anthropology*, **53**, 25–36.

–1992. Dispersal patterns in Costa Rican mantled howling monkeys. *International Journal of Primatology*, **13**, 415–36.

Glenn, M.E. 1997. Group size and group composition of the mona monkey (*Cercopithecus mona*) on the island of Grenada, West Indies. *American Journal of Primatology*, **43**, 167–73.

Goldberg, T.A. & Wrangham, R.W. 1997. Genetic correlates of social behaviour in wild chimpanzees: evidence from mitochondrial DNA. *Animal Behaviour*, **54**, 559–70.

Goldizen, A.W. 1987a. Facultative polyandry and the role of infant-carrying in wild saddle-back tamarins (*Saguinus fuscicollis*). *Behavioral Ecology and Sociobiology*, **20**, 99–109.

–1987b. Tamarins and marmosets: communal care of offspring. In: *Primate Societies*, ed. B.B. Smuts, D.L. Cheney, R.M. Seyfarth, R.W. Wrangham & T.T. Struhsaker, pp. 34–43. Chicago: University of Chicago Press.

–1988. Tamarin and marmoset mating systems: unusual flexibility. *Trends in Ecology and Evolution*, **3**, 36–40.

–1989. Social relationships in a cooperatively polyandrous group of tamarins (*Saguinus fuscicollis*). *Behavioral Ecology and Sociobiology*, **24**, 79–89.

–1990. A comparative perspective on the evolution of tamarin and marmoset social systems. *International Journal of Primatology*, **11**, 63–83.

Goldizen, A.W. & Terborgh, J. 1989. Demography and dispersal patterns of a tamarin population: possible causes of delayed breeding. *American Naturalist*, **134**, 208–24.

Goldizen, A.W., Terborgh, J., Cornejo, F., Porras, D.T. & Evans, R. 1988. Seasonal food shortage, weight loss, and the timing of births in saddle-back tamarins (*Saguinus fuscicollis*). *Journal of Animal Ecology*, **57**, 893–901.

Goldizen, A.W., Mendelson, J., van Vlaardingen, M. & Terborgh, J. 1996. Saddle-back tamarin (*Saguinus fuscicollis*) reproductive strategies: evidence from a thirteen-year study of a marked population. *American Journal of Primatology*, **38**, 57–83.

Goldizen, A.W., Putland, D.A. & Goldizen, A.R. 1998. Variable mating patterns in Tasmanian native hens *Gallinula mortierii*: correlates of reproductive success. *Journal of Animal Ecology*, **67**, 307–17.

Gomendino, M. & Roldan, E.R.S. 1993. Mechanisms of sperm competition: linking physiology and behavioural ecology. *Trends in Ecology and Evolution*, **8**, 95–100.

Goodall, J. 1965. Chimpanzees of the Gombe Stream Reserve. In: *Primate Behaviour*, ed. I. de Vore, pp. 425–73. New York: Holt, Rinehart and Winston.

–1983. Population dynamics during a 15-year period in one community of free-living chimpanzees in the Gombe National Park. *Zeitschrift für Tierpsychologie*, **61**, 1–60.

–1986. *The Chimpanzees of Gombe: Patterns of Behavior.* Cambridge, MA: Harvard University Press.

Goodman, S.M., O'Connor, S. & Langrand, O. 1993. A review of predation on lemurs: implications for the evolution of social behavior in small, nocturnal primates. In: *Lemur Social Systems and Their Ecological Basis*, ed. P.M. Kappeler & J.U. Ganzhorn, pp. 51–66. New York: Plenum Press.

Goss-Custard, J., Dunbar, R.I.M. & Aldrich-Blake, F.P.G. 1972. Survival, mating and rearing strategies in the evolution of primate social structure. *Folia Primatologica*, **17**, 1–19.

Gould, L. 1996a. Male–female affiliative relationships in naturally occurring ringtailed lemurs (*Lemur catta*) at the Beza-Mahafaly Reserve, Madagascar. *American Journal of Primatology*, **39**, 63–78.

–1996b. Vigilance behavior during the birth and lactation season in naturally occurring ring-tailed lemurs (*Lemur catta*) at the Beza-Mahafaly Reserve, Madagascar. *International Journal of Primatology*, **17**, 331–47.

–1997. Intermale affiliative behavior in ringtailed lemurs (*Lemur catta*) at the Beza-Mahafaly Reserve, Madagascar. *Primates*, **38**, 15–30.

Gould, L., Fedigan, L.M. & Rose, L.M. 1997. Why be vigilant? The case of the alpha animal. *International Journal of Primatology*, **18**, 401–14.

Gouzoules, S. & Gouzoules, H. 1987. Kinship. In: *Primate Societies*, ed. B.B. Smuts, D.L. Cheney, R.M. Seyfarth, R.W. Wrangham & T.T. Struhsaker, pp. 299–305. Chicago: University of Chicago Press.

Gowaty, P.A. 1981. An extension of the Orians-Verner-Willson model to account for mating systems besides polygyny. *American Naturalist*, **118**, 851–9.

–1994. Architects of sperm competition. *Trends in Ecology and Evolution*, **9**, 160–2.

–1996. Battles of the sexes and origins of monogamy. In: *Partnerships in Birds. The Study of Monogamy*, ed. J.M. Black, pp. 21–52. Oxford: Oxford University Press.

Graur, D. 1993. Molecular phylogeny and the higher classification of eutherian mammals. *Trends in Ecology and Evolution*, **8**, 141–7.

Green, K.M. 1981. Preliminary observations on the ecology and behavior of the capped langur, *Presbytis pileatus*, in the Madhupur Forest of Bangladesh. *International Journal of Primatology*, **2**, 131–51.

Green, S.M. 1981. Sex differences and age gradations in vocalizations of Japanese and lion-tailed monkeys (*Macaca fuscata* and *Macaca silenus*). *American Zoologist*, **21**, 165–83.

Greenwood, P.J. 1980. Mating systems, philopatry, and dispersal in birds and mammals. *Animal Behaviour*, **28**, 1140–62.

Grether, G.F., Palombit, R.A. & Rodman, P.S. 1992. Gibbon foraging decisions and the marginal value model. *International Journal of Primatology*, **13**, 1–17.

Grinnell, J., Packer, C. & Pusey, A. 1995. Cooperation in male lions: kinship, reciprocity or mutualism? *Animal Behaviour* **49**, 95–105.

Gross, M.R. 1985. Disruptive selection for alternative life histories in salmon. *Nature*, **313**, 47–8.

–1996. Alternative reproductive strategies and tactics: diversity within sexes. *Trends in Ecology and Evolution*, **11**, 92–8.

Gubernick, D. 1994. Biparental care and male–female relations in mammals. In: *Infanticide and Parental Care*, ed. S. Parmigiani & F. vom Saal, pp. 427–63. Chur, Switzerland: Harwood.

Gupta, A.K. 1996. A note on a review of the taxonomic status of Phayre's langur (*Trachypithecus phayrei*) in Tripura, North-East India. *Folia Primatologica*, **69**, 22–7.

Gurmaya, K.J. 1986. Ecology and behavior of *Presbytis thomasi* in Northern Sumatra. *Primates*, **27**, 151–72.

Haimoff, E.H. 1984. Acoustic and organizational features of gibbon songs. In: *The Lesser Apes: Evolutionary and Behavioural Biology*, ed. H. Preuschoft, D.J. Chivers, W.Y. Brockelman & N. Creel, pp. 333–53. Edinburgh: Edinburgh University Press.

Haimoff, E.H., Yang, X-J., He, S.-J. & Chen, N. 1986. Census and survey of black-crested gibbons (*Hylobates concolor concolor*) in Yunnan Province, People's Republic of China. *Folia Primatologica*, **46**, 205–14.

Hall, K.R.L. 1965. Behaviour and ecology of the wild patas monkeys, *Erythrocebus patas*, in Uganda. *Journal of Zoology, London*, **148**, 15–87.

Hall, K.R.L. & DeVore, I. 1965. Baboon social behavior. In: *Primate Behavior: Field Studies of Monkeys and Apes*, ed. I. DeVore, pp. 53–110. New York: Holt, Rinehart and Winston.

Hall, K.R.L. & Gartlan, J.S. 1965. Ecology and behaviour of the vervet monkey, *Cercopithecus aethiops*, Lolui Island, Lake Victoria. *Proceedings of the Royal Society of London, Series B*, **145**, 37–57.

Hamilton, W.D. 1964. The genetical evolution of social behaviour, I, II. *Journal of Theoretical Biology*, **7**, 1–52.

–1971. Geometry for the selfish herd. *Journal of Theoretical Biology*, **31**, 295–311.

Hamilton, W.J. III. & Bulger, J. 1992. Facultative expression of behavioral differences between one-male and multimale savanna baboon groups. *American Journal of Primatology*, **28**, 61–71.

Hampton, J.K. Jr, Hampton, S.H. & Landwehr, B.T. 1966.

Observations on a successful breeding colony of the marmoset, *Oedipomidas oedipus*. *Folia Primatologica*, **4**, 265–87.

Hanby, J.P. 1974. Male–male mounting in Japanese macaques (*Macaca fuscata*). *Behaviour*, **49**, 152–96.

Hand, J.L. 1986. Resolution of social conflicts: dominance, egalitarianism, spheres of dominance, and game theory. *Quarterly Review of Biology*, **61**, 201–20.

Hannon, S.J., Mumme, R.L., Koenig, W.D. & Pitelka, F.A. 1985. Replacement of breeders and within-group conflict in the cooperatively breeding acorn woodpecker. *Behavioral Ecology and Sociobiology*, **17**, 303–12.

Harada, Y. & Iwasa, Y. 1996. Female mate preference to maximize paternal care: a two-step game. *American Naturalist*, **147**, 996–1027.

Harcourt, A.H. 1978. Strategies of emigration and transfer by primates, with particular reference to gorillas. *Zeitschrift für Tierpsychologie*, **48**, 401–20.

–1979a. Social relationships between adult male and female mountain gorillas in the wild. *Animal Behaviour*, **27**, 325–42.

–1979b. Contrasts between male relationships in wild gorilla groups. *Behavioral Ecology and Sociobiology*, **5**, 39–49.

–1981. Intermale competition and the reproductive behavior of the great apes. In: *Reproductive Biology of the Great Apes*, ed. C.A. Graham, pp. 301–18. New York: Academic Press.

–1995. Sexual selection and sperm competition in primates: what are male genitalia good for? *Evolutionary Anthropology*, **4**, 121–9.

–1997. Sperm competition in primates. *American Naturalist*, **149**, 189–94.

Harcourt, A.H. & de Waal, F.B.M. (eds.) 1992. *Coalitions and Alliances in Humans and other Animals*. New York: Oxford University Press.

Harcourt, A.H. & Gardiner, J. 1994. Sexual selection and genital anatomy of male primates. *Proceedings of the Royal Society of London, Series B*, **255**, 47–53.

Harcourt, A.H. & Stewart, K.J. 1981. Gorilla male relationships: can differences during immaturity lead to contrasting reproductive tactics in adulthood? *Animal Behaviour*, **29**, 206–10.

Harcourt, A.H., Fossey, D. & Sabter-Pi, J. 1981a. Demography of *Gorilla gorilla*. *Journal of Zoology, London*, **195**, 215–53.

Harcourt, A.H., Harvey, P.H., Larson, S.G. & Short, R.V. 1981b. Testis weight, body weight, and breeding system in primates. *Nature*, **293**, 55–7.

Harcourt, A.H., Purvis, A. & Liles, L. 1995. Sperm competition: mating system, not breeding season, affects testes size of primates. *Functional Ecology*, **9**, 468–76.

Hardie, S.M. & Buchanan-Smith, H.M. 1997. Vigilance in single- and mixed-species groups of tamarins (*Saguinus labiatus* and *Saguinus fuscicollis*). *International Journal of Primatology*, **18**, 217–34.

Hardin, R. 1982. *Collective Action*. Baltimore: Johns Hopkins University Press.

Harding, R.S.O. & Olson, D.K. 1986. Patterns of mating among male patas monkeys (*Erythrocebus patas*) in Kenya. *American Journal of Primatology*, **11**, 343–58.

Harrington, J.E. 1975. Field observations of social behavior of *Lemur fulvus fulvus* E. Geoffroy 1812. In: *Lemur Biology*, ed. I. Tattersall & R.W. Sussman, pp. 259–79. New York: Plenum Press.

–1978. Diurnal behavior of *Lemur mongoz* at Ampijoroa. *Folia Primatologica*, **29**, 291–302.

Harrison, M.J.S. 1983. Territorial behavior in the green monkey, *Cercopithecus sabaeus*: seasonal defense of local food supplies. *Behavioral Ecology and Sociobiology*, **12**, 85–94.

Hartley, I.R. & Davies, N.B. 1994. Limits to cooperative polyandry in birds. *Proceedings of the Royal Society of London, Series B*, **257**, 67–73.

Hartley, I.R., Davies, N.B., Hatchwell, B.J., Desrochers, A., Nebel, D. & Burke, T. 1995. The polygynandrous mating system of the alpine accentor *Prunella collaris*. II. Multiple paternity and parental effort. *Animal Behaviour*, **49**, 789–803.

Harvey, P.H. & Clutton-Brock, T.H. 1985. Life history variation in primates. *Evolution* **39**, 559–81.

Harvey, P.H. & Harcourt, A.H. 1984. Sperm competition, testes size, and breeding system in primates. In: *Sperm Competition and the Evolution of Animal Mating* Systems, ed. R.L. Smith, pp. 589–600. New York: Academic Press.

Harvey, P.H. & Pagel, M.D. 1991. *The Comparative Method in Evolutionary Biology*. Oxford: Oxford University Press.

Harvey, P.H. & Zammuto, R.M. 1985. Patterns of mortality and age at first reproduction in natural populations of mammals. *Nature*, **315**, 319–20.

Harvey, P.H., Kavanaugh, M. & Clutton-Brock, T.H. 1978. Sexual dimorphism in primate teeth. *Journal of Zoology, London*, **186**, 475–85.

Harvey, P.H., Martin, R.D. & Clutton-Brock, T.H. 1987. Life histories in comparative perspective. In: *Primate Societies*, ed. B.B. Smuts, D.L. Cheney, R.M. Seyfarth, R.W. Wrangham & T.T. Struhsaker, pp. 181–96. Chicago: University of Chicago Press.

Harvey, P.H., Promislow, D.E.L. & Read, A.F. 1989. Causes and correlates of life history differences among mammals. In: *Comparative Socioecology*, ed. V. Standen & R.A. Foley, pp. 305–18. Oxford: Blackwell Science Publications.

Hashimoto, C., Furuichi, T. & Takenaka, O. 1996. Matrilineal kin relationship and social behavior of wild bonobos (*Pan paniscus*): sequencing the D-loop region of mitochondrial DNA. *Primates*, **37**, 305–18.

Hasselquist, D., Bensch, S. & von Schantz, T. 1996. Correlation between male song repertoire, extra-pair paternity and offspring survival in the great reed warbler. *Nature*, **381**, 229–32.

Hatchwell, B.J. & Davies, N.B. 1990. Provisioning of nestlings of dunnocks *Prunella modularis* in pairs and trios: compensation reactions by males and females. *Behavioral Ecology and Sociobiology*, **27**, 199–209.

Hauser, M.D. 1986. Male responsiveness to infant distress calls in free-ranging vervet monkeys. *Behavioral Ecology and Sociobiology*, **19**, 65–71.

Hauser, M.D. & Marler, P. 1993. Food-associated calls in rhesus macaques (*Macaca mulatta*): II. Costs and benefits of call production and suppression. *Behavioral Ecology*, **4**, 206–12.

Hausfater, G. & Hrdy, S.B. 1984. *Infanticide. Comparative and Evolutionary Perspectives*. New York: Aldine.

Hausfater, G., Altmann, J. & Altmann, S.A. 1982. Long-term consistency of dominance relations among female baboons (*Papio cynocephalus*). *Science*, **217**, 752–5.

Hawkes, K. 1992. Sharing and collective action. In: *Evolutionary Ecology and Human Behavior*, ed. E.A. Smith & B. Winterhalder, pp. 269–300. New York: Aldine.

Hawkes, K., Rogers, A.R. & Charnov, E.L. 1995. The male's dilemma: increased offspring production is more paternity to steal. *Evolutionary Ecology*, **9**, 662–77.

Hayssen, V., van Tienhoven, A. & van Tienhoven, A. 1993. *Asdell's Patterns of Mammalian Reproduction: a Compendium of Species-Specific Data*. Ithaca, NY: Comstock Publishing Associates.

Heathcote, C.F. 1987. Grouping of eastern grey kangaroos in open habitat. *Australian Wildlife Research*, **14**, 343–8.

Heer, L. 1996. Cooperative breeding by alpine accentors *Prunella collaris*: polygynandry, territoriality and multiple paternity. *Journal für Ornithologie*, **137**, 35–51.

Heinsohn, R. & Packer, C. 1995. Complex cooperative strategies in group-territorial lions. *Science*, **269**, 1260–2.

Hemelrijk, C.K. & Ek, A. 1991. Reciprocity and interchange of grooming and 'support' in captive chimpanzees. *Animal Behaviour*, **41**, 923–35.

Hemelrijk, C. & Luteijn, M. 1998. Philopatry, male presence and grooming reciprocation among female primates: a comparative perspective. *Behavioral Ecology and Sociobiology* **42**, 207–16.

Henzi, S.P. 1985. Genital signalling and the coexistence of male vervet monkeys (*Cercopithecus aethiops pygerythrus*). *Folia Primatologica*, **45**, 129–47.

–1988. Many males do not a multimale troop make. *Folia Primatologica*, **51**, 165–8.

Henzi, S.P. & Lawes, M. 1987. Breeding season influxes and the behaviour of adult male Samango monkeys (*Cercopithecus mitis albogularis*). *Folia Primatologica*, **48**, 125–36.

–1988. Strategic responses of male Samango monkeys (*Cercopithecus mitis*) to a decline in the number of receptive females. *International Journal of Primatology*, **9**, 479–95.

Henzi, S.P. & Lucas, J.W. 1980. Observations on the inter-troop movement of adult vervet monkeys (*Cercopithecus aethiops*). *Folia Primatologica*, **33**, 220–35.

Henzi, S.P. & Lycett, J.E. 1993. Male induced fission as the determinant of mountain baboon population structure. *Primate Eye*, **48**, 7–8.

–1995. Population-structure, demography, and dynamics of mountain baboons: an interim report. *American Journal of Primatology*, **35**, 155–63.

Henzi, S.P., Lycett, J.E. & Piper, S.E. 1997a. Fission and troop size in a mountain baboon population. *Animal Behaviour*, **53**, 525–35.

Henzi, S.P., Lycett, J.E. & Weingrill, T. 1997b. Cohort size and the allocation of social effort by female mountain baboons. *Animal Behaviour*, **54**, 1235–43.

–1998. Mate guarding and risk assessment by male mountain baboons during inter-troop encounters. *Animal Behaviour*, **55**, 1421–8.

Hershkovitz, P. 1977. *Living New World Monkeys (Platyrrhini)*, Vol. 1. Chicago: University of Chicago Press.

Heymann, E.W. 1996. Social behavior of wild moustached tamarins, *Saguinus mystax*, at the Estación Biológica Quebrada Blanco, Peruvian Amazonia. *American Journal of Primatology*, **38**, 101–13.

–1997. Aspekte der Verhaltensbiologie von Schnurrbarttamarinen, *Saguinus mystax* (Callitrichinae; Cebidae; Primates), in ökologischer, evolutionärer und phylogenetischer Perspektive. Habilitation thesis, University of Giessen.

Heymann, E.W. & Soini, P. in press. Offspring number in pygmy marmosets, *Cebuella pygmaeus*, in relation to group size and the number of adult males. *Behavioral Ecology and Sociobiology* **46**.

Hill, D.A. 1987. Social relationships between adult male and female rhesus macaques: 1. Sexual consortships. *Primates*, **28**, 439–56.

–1994. Affiliative behaviour between adult males of the genus *Macaca*. *Behaviour*, **130**, 293–308.

Hill, D.A. & van Hooff, J.A.R. A.M. 1994. Affiliative relationships between males in groups of nonhuman primates: a summary. *Behaviour*, **130**, 143–9.

Hilty, S.L. & Brown, W.L. 1986. *A Guide to the Birds of Colombia*. Princeton, NJ: Princeton University Press.

Hinde, R.A. 1983. A conceptual framework. In: *Primate Social Relationships: an Integrated Approach*, ed. R.A. Hinde, pp. 1–7. Oxford: Blackwell.

Hochberg, Y. 1988. A sharper Bonferroni procedure for multiple tests of significance. *Biometrika*, **75**, 800–2.

Hoffmann, H.U. 1993. *Eine Monographie zu den Weißhandgibbons* Hylobates lar *mit ergänzenden Angaben zu den verwandten Arten*. Zoologische Dokumentationsstelle Saarbrücken: Neuweiler.

Hogg, J.T. & Forbes, S.H. 1997. Mating in bighorn sheep: frequent male reproduction via a high-risk 'unconventional' tactic. *Behavioral Ecology and Sociobiology*, **41**, 33–48.

Hohmann, G. 1989. Group fission in Nilgiri langurs (*Presbytis johnii*). *International Journal of Primatology*, **10**, 441–54.

Honer, O.P., Leumann, L. & Noë, R. 1997. Dyadic associations of red colobus and Diana monkey groups in the Taï National Park, Ivory Coast. *Primates*, **38**, 281–91.

Hood, L.C. 1994. Infanticide among ringtailed lemurs (*Lemur catta*) at Berenty Reserve, Madagascar. *American Journal of Primatology*, **33**, 65–9.

Hood, L.C. & Jolly, A. 1995. Troop fission in female *Lemur catta* at Berenty Reserve, Madagascar. *International Journal of Primatology*, **16**, 997–1015.

Hoogland, J.L. 1998. Why do female Gunnison's prairie dogs copulate with more than one male? *Animal Behaviour*, **55**, 351–9.

Horrocks, J. & Hunte, W. 1983. Maternal rank and offspring rank in vervet monkeys: an appraisal of the mechanisms of rank acquisition. *Animal Behaviour*, **31**, 772–82.

Hörstermann, M. 1995. Geschlechtsspezifische und individuelle Verhaltensunterschiede von Rotstirnmakis (*Eulemur fulvus rufus*) in Interaktionen zwischen Gruppen (in Gefangenschaft). MSc thesis, University of Würzburg.

Horsup, A.B. 1996. The behavioural ecology of the allied rock-wallaby, *Petrogale assimilis*. PhD thesis, James Cook University, Townsville.

Houston, A.I., Gasson, C.E. & McNamara, J.M. 1997. Female choice of matings to maximize parental care. *Proceedings of the Royal Society of London, Series B*, **264**, 173–9.

Hrdlicka, A. 1925. Weight of the brain and of the internal organs in American monkeys. *American Journal of Physical Anthropology*, **8**, 201–11.

Hrdy, S.B. 1977. *The Langurs of Abu. Female and Male Strategies of Reproduction*. Cambridge, MA: Harvard University Press.

–1979. Infanticide among animals: a review, classification, and examination of the implications for the reproductive strategies of females. *Ethology and Sociobiology*, **1**, 13–40.

–1981. *The Woman that Never Evolved*. Cambridge: Harvard University Press.

Hrdy, S.B. & Whitten, P.L. 1987. Patterning of sexual activity. In: *Primate Societies*, ed. B.B. Smuts, D.L. Cheney, R.M. Seyfarth, R.W. Wrangham & T.T. Struhsaker, pp. 370–84. Chicago: University of Chicago Press.

Hrdy, S.B., Janson, C.H. & van Schaik, C.P. 1995. Infanticide: let's not throw out the baby with the bath water. *Evolutionary Anthropology*, **3**, 151–4.

Hume, I.D., Jarman, P.J., Renfree, M.B. & Temple-Smith, P.D. 1989. Macropodidae. In: *Fauna of Australia*, ed. D.W. Walton & B.J. Richardson, pp. 679–715. Canberra: Australian Government Publishing Service.

Hunkeler, C., Bourlière, F. & Bertrand, M. 1972. Le comportement de la Mone de Lowe (*Cercopithecus campbelli lowei*). *Folia Primatologica*, **17**, 218–36.

Hunt, K.D. 1989. Positional behavior in *Pan troglodytes* at the Mahale mountains and the Gombe Stream National Parks, Tanzania. PhD thesis, University of Michigan.

Hussmann, S. 1996. Wachsamkeitsverhalten bei freilebenden Larvensifakas (*Propithecus v. verreauxi* Grandidier, 1867): Sind die Männchen wachsamer als die Weibchen und kann dies als „Dienstleistung" gegenüber den Weibchen gedeutet werden? MSc thesis, University of Würzburg.

Ims, R.A. 1988. Spatial clumping of sexually receptive females induces space sharing among male voles. *Nature*, **335**, 541–3.

–1990. The ecology and evolution of reproductive synchrony. *Trends in Ecology and Evolution* **5**, 135–40.

Inoue, M., Mitsunaga, F., Nozaki, M. *et al.* 1993. Male dominance rank and reproductive success in an enclosed group of Japanese macaques: with special reference to post-conception mating. *Primates*, **34**, 503–11.

Irion, G. 1978. Soil fertility in the African rain forest. *Naturwissenschaften*, **65**, 515–19.

Isaac, R.M., Walker, J.M. & Thomas, S.H. 1984. Divergent evidence on free-riding: an experimental examination of possible explanations. *Public Choice*, **43**, 113–49.

Isabirye-Basuta, G. 1988. Food competition among individuals in a free-ranging chimpanzee community in Kibale Forest, Uganda. *Behaviour*, **105**, 135–47.

Isbell, L.A. 1994. Predation on primates: ecological patterns and evolutionary consequences. *Evolutionary Anthropology*, **3**, 61–71.

Isbell, L.A., Cheney, D.L. & Seyfarth, R.M. 1991. Group fusions and minimum group sizes in vervet monkeys (*Cercopithecus aethiops*). *American Journal of Primatology*, **25**, 57–65.

Itani, I. 1972. A preliminary essay on the relationship between social organization and incest avoidance in nonhuman primates. In: *Primate Socialization*, ed. F.E. Poirier, pp. 165–71. New York: Random House.

Itani, J. 1977. Evolution of primate social structure. *Journal of Human Evolution*, **6**, 235–43.

–1980. Social structure of African great apes. *Journal of Reproductive Fertility*, Supplement, **28**, 33–41.

Iwasa, Y. & Harada, Y. (1998). Female mate preference to maximize paternal care. II. Female competition leads to monogamy. *American Naturalist*, **151**, 367–82.

Izawa, K. 1978. A field study of the ecology and behavior of the black-mantled tamarin (*Saguinus nigricollis*). *Primates*, **19**, 241–74.

–1994a. Group division of wild black-capped capuchins. *Field Studies of New World Monkeys, La Macarena, Colombia*, **9**, 5–14.

–1994b. Social changes within a group of wild black-capped capuchins, IV. *Field Studies of New World Monkeys, La Macarena, Colombia*, **9**, 15–21.

Izawa, K. & Lozano, M.H. 1991. Social changes within a group of red howler monkeys (*Alouatta seniculus*) III. *Field Studies of New World Monkeys, La Macarena, Colombia*, **5**, 1–16.

–1994. Social changes within a group of red howler monkeys (*Alouatta seniculus*) V. *Field Studies of New World Monkeys, La Macarena, Colombia*, **9**, 33–9.

Jacobs, S.C., Larson, A. & Cheverud, J.M. 1995. Phylogenetic relationships and orthogenetic evolution of coat color among tamarins (genus *Saguinus*). *Systematic Biology*, **44**, 515–32.

Jamieson, I.G. 1997. Testing reproductive skew models in a communally breeding bird, the pukeko *Porphyrio porphyrio*. *Proceedings of the Royal Society of London, Series B*, **264**, 335–40.

Jamieson, I.G., Quinn, J.S., Rose, P.A. & White, B.N. 1994. Shared paternity among non-relatives is a result of an egalitarian mating system in a communally breeding bird, the pukeko. *Proceedings of the Royal Society of London, Series B*, **257**, 271–7.

Janson, C.H. 1984. Female choice and mating system of the brown capuchin monkey *Cebus apella* (Primates: Cebidae). *Zeitschrift für Tierpsychologie*, **65**, 177–200.

–1985. Aggressive competition and individual food consumption in wild brown capuchin monkeys (*Cebus apella*). *Behavioural Ecology and Sociobiology*, **18**, 125–38.

–1986. The mating system as a determinant of social evoluiton in capuchin monkeys (*Cebus*). In: *Primate Ecology and Conservation*, Vol. 2, ed. J.G. Else & P.C. Lee, pp. 169–79. Cambridge: Cambridge University Press.

–1988. Intraspecific food competition and primate social structure: a synthesis. *Behaviour*, **105**, 1–17.

–1992. Evolutionary ecology of primate social structure. In: *Evolutionary Ecology and Human Behavior*, ed. E.A. Smith & B. Winterhalder, pp. 95–130. New York: Aldine.

Janson, C.H. & Goldsmith, M.L. 1995. Predicting group size in primates: foraging costs and predation risks. *Behavioral Ecology*, **6**, 326–36.

Janson, C.H. & van Schaik, C.P. 1988. Recognising the many faces of primate food competition: methods. *Behaviour*, **105**, 165–86.

–1993. Ecological risk aversion in juvenile primates: slow and steady wins the race. In *Juvenile Primates: Life History, Development and Behavior*, ed. M. Pereira & L. Fairbanks, pp. 57–74. New York: Oxford University Press.

Jaremovic, R. 1984. Space and time related behaviour in eastern grey kangaroos (*Macropus giganteus* Shaw). PhD thesis, University of New South Wales, Australia.

Jarman, P.J. 1974. The social organization of antelope in relation to their ecology. *Behaviour*, **48**, 215–56.

–1983. Mating system and sexual dimorphism in large, terrestrial, mammalian herbivores. *Biological Reviews*, **58**, 485–520.

–1987. Group size and activity in eastern grey kangaroos. *Animal Behaviour*, **35**, 1044–50.

–1989a. Sexual dimorphism in Macropodoidea. In: *Kangaroos, Wallabies and Rat-kangaroos*, ed. G. Grigg, P. Jarman & I. Hume, pp. 433–47. Sydney: Surrey Beatty & Sons.

–1989b. On being thick-skinned: dermal shields in large, mammalian herbivores. *Biological Journal of the Linnean Society*, **36**, 169–91.

–1991. Social behavior and organization in the Macropodoidea. *Advances in the Study of Behavior*, **20**, 1–50.

–1994. Individual behaviour and social organization of kangaroos. In: *Animal Societies: Individuals, Interactions and Organization*, ed. P.J. Jarman & A. Rossiter, pp. 70–85. Kyoto, Japan: Kyoto University Press.

Jarman, P.J. & Coulson, G.M. 1989. Dynamics and adaptiveness of grouping in macropods. In: *Kangaroos, Wallabies and Rat-kangaroos*, ed. G. Grigg, P. Jarman & I. Hume, pp. 527–47. Sydney: Surrey Beatty & Sons.

Jarman, P.J. & Kruuk, H. 1996. Phylogeny and spatial organization in mammals. In: *Comparison of Marsupial and Placental Behaviour*, ed. D.B. Croft & U. Ganslosser, pp. 80–101. Fürth: Filander Verlag.

Jarman, P.J. & Southwell, C.N. 1986. Grouping, associations and reproductive strategies in Eastern grey kangaroos. In: *Ecological Aspects of Social Evolution*, ed. D.I. Rubenstein & R.W. Wrangham, pp. 399–428. Princeton, NJ: Princeton University Press.

Jarman, P.J. & Wright, S.M. 1993. Macropod studies at Wallaby Creek. IX. Exposure and responses of Eastern grey kangaroos to dingoes. *Wildlife Research*, **20**, 833–43.

Jarman, P.J., Johnson, C.N., Southwell, C.J. & Stuart-Dick, R.I. 1987. Macropod studies at Wallaby Creek. I. The area and animals. *Australian Wildlife Research*, **14**, 1–14.

Johns, A.D. 1983. *Ecological Effects of Selective Logging in a West Malaysian Rain Forest*. Cambridge: Cambridge University Press.

–1985. Behavioral responses of two Malaysian primates (*Hylobates lar* and *Presbytis melalophos*) to selective logging: vocal behavior, territoriality, and nonemigration. *International Journal of Primatology*, **6**, 423–33.

Johnson, C.N. 1987. Relationships between mother and infant red-necked wallabies (*Macropus rufogriseus banksianus*). *Ethology*, **74**, 1–20.

–1989. Social interactions and reproductive tactics in red-necked wallabies (*Macropus rufogriseus banksianus*). *Journal of Zoology, London*, **217**, 267–80.

Jolly, A. 1966. *Lemur Behavior*. Chicago: University of Chicago Press.

–1967. Breeding synchrony in wild *Lemur catta*. In: *Social Communication among Primates*, ed. S.A. Altman, pp. 3–14. Chicago: University of Chicago Press.

–1972. Troop continuity and troop spacing in *Propithecus verreauxi* and *Lemur catta* at Berenty (Madagascar). *Folia Primatologica*, **17**, 335–62.

–1984. The puzzle of female feeding priority. In: *Female Primates: Studies by Women Primatologists*, ed. M.F. Small, pp. 197–215. New York: Alan R. Liss.

–1985. *The Evolution of Primate Behavior*, 2nd edn. New York: MacMillan Publishing.

–1998. Pair-bonding, female aggression, and the evolution of lemur societies. *Folia Primatologica*, **69**, 1–13.

Jolly, A., Gustafson, H., Oliver, W.L.R. & O'Connor, S.M. 1982. *Propithecus verreauxi* population and ranging at Berenty, Madagascar, 1975 and 1980. *Folia Primatologica*, **39**, 124–44.

Jones, C.B. 1980. The functions of status in the mantled howler monkey, *Alouatta palliata*: intraspecific competition for group membership in a folivorous Neotropical primate. *Primates*, **21**, 389–405.

Jones, K.C. 1983. Inter-troop transfer of *Lemur catta* males at Berenty, Madagascar. *Folia Primatologica*, **40**, 145–60.

Jones, W.T. & Bush, B.B. 1988. Movement and reproductive behavior of solitary male redtail guenons (*Cercopithecus ascanius*). *American Journal of Primatology*, **14**, 203–22.

Jorde, L.B. & Spuhler, J.N. 1974. A statistical analysis of selected aspects of primate demography, ecology, and social behavior. *Journal of Anthropological Research*, **30**, 199–223.

Jurke, M.H. & Pryce, C.R. 1994. Parental and infant behaviour during early periods of infant care in Goeldi's monkey, *Callimico goeldii*. *Animal Behaviour*, **48**, 1095–112.

Kano, T. 1992. *The Last Ape: Pygmy Chimpanzee Behaviour and Ecology*. Stanford, CA: Stanford University Press.

Kappeler, P.M. 1990. Social status and scent marking behaviour in *Lemur catta*. *Animal Behaviour*, **40**, 774–6.

–1993a. Sexual selection and lemur social systems. In: *Lemur Social Systems and Their Ecological Basis*, ed. P.M. Kappeler & J.U. Ganzhorn, pp. 223–40. New York: Plenum Press.

–1993b. Variation in social structure: the effects of sex and kinship on social interactions in three lemur species. *Ethology*, **93**, 125–45.

–1996. Causes and consequences of life-history variation among strepsirhine primates. *American Naturalist*, **148**, 868–91.

–1997a. Determinants of primate social organization: comparative evidence and new insights from Malagasy lemurs. *Biological Reviews*, **72**, 111–51.

–1997b. Intrasexual selection and testis size in strepsirhine primates. *Behavioral Ecology*, **8**, 10–19.

–1998. To whom it may concern: the transmission and function of chemical signals in *Lemur catta*. *Behavioral Ecology and Sociobiology*, **42**, 411–21.

–1999a. Convergence and nonconvergence in primate social systems. In: *Primate Communities*, ed. J. Fleagle, C. Janson & K. Reid, pp. 158–70. Cambridge: Cambridge University Press.

–1999b. Lemur social structure and convergence in primate socioecology. In: *Comparative Primate Socioecology*, ed. P.C. Lee, pp. 273–99. Cambridge: Cambridge University Press.

–1999c. Primate socioecology: new insights from males. *Naturwissenschaften*, **85**, 18–29.

Kappeler, P.M. & Ganzhorn, J.U. (eds.) 1993. *Lemur Social Systems and Their Ecological Basis*. New York: Plenum Press.

Kappeler, P.M. & Heymann, E.W. 1996. Nonconvergence in the evolution of primate life history and socio-ecology. *Biological Journal of the Linnean Society*, **59**, 297–326.

Kauffman, S.A. 1993. *The Origins of Order*. New York: Oxford University Press.

Kaufmann, J. 1965. A three-year study of mating behavior in a free-ranging band of rhesus monkeys. *Ecology*, **46**, 500–12.

Kavanagh, M. 1981. Variable territoriality among tantalus monkeys in Cameroon. *Folia Primatologica*, **36**, 76–98.

–1983. Birth seasonality in *Cercopithecus aethiops*: a social advantage from synchrony? In: *Perspectives in Primate Biology*, ed. P.K. Seth, pp. 89–98. New Delhi: Today and Tomorrow.

Kawai, M., Dunbar, R.I.M., Ohsawa, H. & Mori, U. 1983. Social organization of gelada baboons: social units and definitions. *Primates*, **24**, 13–24.

Keane, B., Dittus, W.P.J. & Melnick, D.J. 1997. Paternity assessment in wild groups of toque macaques, *Macaca sinica*, at Polonnaruwa, Sri Lanka, using molecular markers. *Molecular Ecology*, **6**, 267–82.

Keller, L. & Reeve, H.K. 1994. Partitioning of reproduction in animal societies. *Trends in Ecology and Evolution*, **9**, 98–102.

Kempenaers, B., Verheyen, G.R. & Dhondt, A.A. 1997. Extrapair paternity in the blue tit *Parus caeruleus*: female choice, male characteristics and offspring quality. *Behavioral Ecology*, **8**, 481–92.

Keuroghlian, A. 1990. Observations on the behavioral ecology of the black lion tamarin (*Leontopithecus chrysopygus*) at Caetetus Reserve, Sao Paulo, Brazil. MSc thesis, West Virginia University, Morgantown.

Kingdon, J. 1971. *East African Mammals*, Vol. 1. London, New York: Academic Press.

Kingsley, S. 1982. Causes of non-breeding and the development of secondary sexual characteristics in the male orang-utan: a hormonal study. In: *The Orang-utan, its Biology and Conservation*, ed. L.E.M. de Boer, pp. 215–29. The Hague: Junk.

Kinzey, W.G. 1987. A primate model for human mating systems. In: *The Evolution of Human Behavior: Primate Models*, ed. W.G. Kinzey, pp. 105–14. New York: State University of New York Press.

Kirkpatrick, R.C. 1997. Ecology and behavior in snub-nosed and douc langurs. In: *The Natural History of the Snub-Nosed and Douc Langurs*, ed. N. Jablonski. Singapore: World Scientific Press.

Kirkpatrick, R.C., Long, Y.T., Zhong, T. & Xiao, L. 1998. Social organization and range use in the Yunnan snub-nosed monkey *Rhinopithecus bieti*. *International Journal of Primatology*, **19**, 13–51.

Kirsch, J.A.W., Lapointe, F.-J. & Springer, M.S. 1997. DNA-hybridisation studies of marsupials and their implications for metatherian classification. *Australian Journal of Zoology*, **45**, 211–80.

Kleiman, D.G. 1977. Monogamy in mammals. *Quarterly Review of Biology*, **52**, 39–69.

Koenig, A. 1995. Group size, composition, and reproductive success in wild common marmosets (*Callithrix jacchus*). *American Journal of Primatology*, **35**, 311–17.

–1998. Visual scanning by common marmosets (*Callithrix jacchus*): functional aspects and the special role of adult males. *Primates*, **39**, 85–90.

Koenig, A., Borries, C., Chalise, M.K. & Winkler, P. 1997. Ecology, nutrition, and timing of reproductive events in an Asian primate, the Hanuman langur (*Presbytis entellus*). *Journal of Zoology, London*, **243**, 215–35.

Koenig, A., Beise, J., Chalise, M.K. & Ganzhorn, J.U. 1998. When females should contest for food – testing hypotheses about resource density, distribution, size, and quality with Hanuman langurs (*Presbytis entellus*). *Behavioral Ecology and Sociobiology*, **42**, 225–37.

Koenig, W.D. & Mumme, R.L. 1987. *Population Ecology of the Cooperatively Breeding Acorn Woodpecker*. Princeton, NJ: Princeton University Press.

Koenig, W.D. & Pitelka, F.A. 1979. Relatedness and inbreeding avoidance: counterploys in the communally nesting acorn woodpecker. *Science*, **206**, 1103–5.

–1981. Ecological factors and kin selection in the evolution of cooperative breeding in birds. In: *Natural Selection and Social Behavior: Recent Research and New Theory*, ed. R.D. Alexander & D. Tinkle, pp. 261–80. New York: Chiron Press.

Koenig, W.D., Pitelka, F.A., Carmen, W.J., Mumme, R.L. & Stanback, M.T. 1992. The evolution of delayed dispersal in cooperative breeders. *Quarterly Review of Biology*, **67**, 111–50.

Komdeur, J. 1996. Influence of helping and breeding experience on reproductive performance in the Seychelles warbler: a translocation experiment. *Behavioral Ecology*, **7**, 326–33.

Komers, P.E. & Brotherton, P.N.M. 1997. Female space use is the best predictor of monogamy in mammals. *Proceedings of the Royal Society of London, Series B*, **264**, 1261–70.

Kool, K.M. 1989. Behavioural ecology of the silver leaf monkey (*Trachypithecus auratus sondaicus*) in the Pangandaran Nature Reserve, West Java, Indonesia. PhD thesis, University of New South Wales, Australia.

–1993. The diet and feeding behavior of the silver leaf monkey (*Trachypithecus auratus sondaicus*) in Indonesia. *International Journal of Primatology*, **14**, 667–700.

Koyama, N. 1988. Mating behavior of ring-tailed lemurs (*Lemur catta*) at Berenty, Madagascar. *Primates*, **29**, 163–75.

–1991. Troop division and inter-troop relationships of ring-tailed lemurs (*Lemur catta*) at Berenty, Madagascar. In: *Proceedings of the XIII Congress of the International Primatological Society*, ed. A. Ehara, T. Kimura, O. Takenaka & M. Iwamoto, pp. 173–6. Amsterdam: Elsevier.

Kraus, C. 1997. Fortpflanzungstaktiken männlicher Larvensifakas (*Propithecus verreauxi verreauxi* Grandidier 1867). MSc thesis, University of Göttingen.

Krebs, J.R. & Davies, N.B. 1984. *Behavioural Ecology: an Integrated Approach*. Oxford: Blackwell Scientific Publications.

Kubzdela, K. 1997. Sociodemography in diurnal primates: the effects of group size and female dominance rank on intra-group spatial distribution, feeding competition, female reproductive success, and female dispersal patterns in white sifaka, *Propithecus verreauxi verreauxi*. PhD thesis, University of Chicago.

Kuester, J. & Paul, A. 1986. Male–infant relationships in semi-free-ranging Barbary macaques (*Macaca sylvanus*) of Affenberg Salem/FRG: testing the 'male care' hypothesis. *American Journal of Primatology*, **10**, 315–27.

–1989. Reproductive strategies of subadult Barbary macaque males at Affenberg Salem. In: *Sociobiology of Reproductive Strategies in Animals*, ed. A.E.C. Rasa, C. Vogel & E. Voland, pp. 93–109. London: Chapman & Hall.

–1992. Influence of male competition and female mate choice on male mating success in Barbary macaques (*Macaca sylvanus*). *Behaviour*, **120**, 192–217.

–1997. Group fission in Barbary macaques (*Macaca sylvanus*) at Affenberg Salem. *International Journal of Primatology*, **18**, 941–66.

–1999. Male migration in Barbary macaques (*Macaca sylvanus*) at Affenberg Salem. *International Journal of Primatology*, **20**, 85–106.

Kuester, J., Paul, A. & Arnemann, J. 1995. Age-related and individual differences of reproductive success in male and female Barbary macaques. *Primates*, **36**, 461–76.

Kummer, H. 1968. *Social Organization of Hamadryas Baboons: a Field Study*. Chicago: University of Chicago Press.

–1980. From laboratory to desert and back: a social system of hamadryas baboons. *Animal Behaviour*, **32**, 965–71.

Kummer, H. & Abegglen, J.J. 1978. Gesellschaftsordnung bei Mantelpavianen. In: *Die Psychologie des 20. Jahrhunderts*, Vol. 6, *Lorenz und die Folgen*, ed. R.A. Stamm & H. Zeier, pp. 163–76. Zürich: Kindler.

Kummer, H., Götz, W. & Angst, W. 1974. Triadic differentiation: an inhibitory process protecting pair bonds in baboons. *Behaviour*, **49**, 62–87.

Kummer, H., Banaja, A.A., Abo-Khatwa, A.N. & Gandour, A.M. 1981. A survey of hamadryas baboons in Saudi Arabia. *Fauna of Saudi Arabia*, **3**, 441–71.

–1985. Differences in social behaviour between Ethiopian and Arabian hamadryas baboons. *Folia Primatologica*, **45**, 1–8.

Kuroda, S. 1979. Grouping of the pygmy chimpanzees. *Primates*, **20**, 161–83.

Kvarnemo, C. & Ahnesjö, I. 1996. The dynamics of operational sex ratios and competition for mates. *Trends in Ecology and Evolution*, **11**, 404–8.

Lack, D. 1968. *Ecological Adaptations for Breeding in Birds*. London: Methuen.

Lamprey, V. 1998. A review of the mating system of captive cotton-top tamarins. *International Zoo News*, **46**, 16–19.

Langmore, N.E., Davies, N.B., Hatchwell, B.J. & Hartley, I.R. 1996. Female song attracts males in the alpine accentor *Prunella collaris*. *Proceedings of the Royal Society of London, Series B*, **263**, 141–6.

Launhardt, K. 1998. Paarungserfolg und Reproduktionserfolg männlicher Hanuman Languren (*Presbytis entellus*) in Ramnagar, Südnepal. PhD thesis, University of Göttingen.

Launhardt, K. & Borries, C. 1997. Male reproductive success in seasonally breeding Hanuman langurs (*Presbytis entellus*). *Primate Report*, **48**, 25–6.

Lawes, M.J. 1990. The socioecology and conservation of the samango monkey (*Cercopithecus mitis erythrarchus*) in Natal. PhD thesis, University of Natal.

Lawes, M.J. & Henzi, S.P. 1995. Inter-group encounters in blue monkeys: how territorial must a territorial species be? *Animal Behaviour*, **49**, 239–42.

Lawes, M.J. & Piper, S.E. 1992. Activity patterns in free-ranging samango monkeys (*Cercopithecus mitis erythrarchus* Peters, 1852) at the southern range limit. *Folia Primatologica*, **59**, 186–202.

Lawes, M.J., Henzi, S.P. & Perrin, M.R. 1990. Diet and feeding behaviour of samango monkeys (*Cercopithecus mitis labiatus*) in Ngoye Forest, South Africa. *Folia Primatologica*, **54**, 57–69.

Laws, J.W. & Vonder Haar Laws, J. 1984. Social interactions among adult male langurs (*Presbytis entellus*) at Rajaji Wildlife Sanctuary. *International Journal of Primatology*, **5**, 31–50.

Lee, P.C. 1994. Social structure and evolution. In: *Behaviour and Evolution*, ed. P.J.B. Slater & T.R. Halliday, pp. 266–337. Cambridge: Cambridge University Press.

–1996. The meaning of weaning: growth, lactation, and life history. *Evolutionary Anthropology*, **5**, 87–96.

Lee, P.C., Majluf, P. *et al.* 1991. Growth, weaning and maternal investment from a comparative perspective. *Journal of Zoology, London*, **225**, 99–114.

Leigh, S. 1995. Socioecology and the ontogeny of sexual size dimorphism in anthropoid primates. *American Journal of Physical Anthropology*, **97**, 339–56.

Leighton, D.R. 1987. Gibbons: territoriality and monogamy. In: *Primate Societies*, ed. B.B. Smuts, D.L. Cheney, R.M. Seyfarth, R.W. Wrangham & T.T. Struhsaker, pp. 135–45. Chicago: University of Chicago Press.

Leutenegger, W. 1978. Scaling of sexual dimorphism in body size and breeding system in primates. *Nature*, **272**, 610–11.

Lindburg, D.G. 1969. Rhesus monkeys: mating season mobility of adult males. *Science*, **166**, 1176–8.

Liu, Z., Zhang, Y., Jiang, H. & Southwick, C. 1989. Population structure of *Hylobates concolor* in Bawanglin Nature Reserve, Hainan, China. *American Journal of Primatology*, **19**, 247–54.

Lloyd, L.E., McDonald, B.E. & Crampton, E.W. 1978. *Fundamentals of Nutrition*. San Francisco: W.H. Freeman and Company.

Lloyd, P.H. & Rasa. O.A.E. 1989. Status, reproductive success, and fitness in Cape Mountain zebra. *Behavioral Ecology and Sociobiology*, **25**, 411–20.

Lorenz, K. 1963. *Das sogenannte Böse. Zur Naturgeschichte der Aggression*. Wien: Borotha-Schoeler.

Lowen, C. & Dunbar, R.I.M. 1995. Territory size and defendability in primates. *Behavioral Ecology and Sociobiology*, **35**, 347–54.

Loy, J. 1971. Estrous behaviour in free-ranging rhesus monkeys (*Macaca mulatta*). *Primates*, **12**, 1–31.

Macdonald, D.W. (ed.) 1984. *The Encyclopaedia of Mammals*. London: George Allen & Unwin Hyman.

Macedonia, J.M. 1993. Adaptation and phylogenetic constraints in the antipredator behavior of ringtailed and ruffed lemurs. In: *Lemur Social Systems and Their Ecological Basis*, ed. P.M. Kappeler & J.U. Ganzhorn, pp. 67–84. New York: Plenum Press.

MacKinnon, J. 1974. The behaviour and ecology of wild orang utans (*Pongo pygmaeus ABELII*). *Animal Behaviour*, **22**, 3–74.

MacKinnon, J.R. & MacKinnon, K.S. 1984. Territoriality, monogamy and song in gibbons and tarsiers. In: *The Lesser Apes: Evolutionary and Behavioural Biology*, ed. H. Preuschoft, D.J. Chivers, W.Y. Brockelman & N. Creel, pp. 291–7. Edinburgh: Edinburgh University Press.

Maddison, W.P. 1990. A method for testing the correlated evolution of two binary characters: are gains or losses concentrated on certain branches of a phylogenetic tree? *Evolution*, **44**, 539–57.

Maddison, W.P. & Maddison, D.R. 1992. *MacClade*. Sunderland, MA: Sinauer Associates.

Maggioncalda, A.N. 1995. The socioendocrinology of orangutan growth, development, and reproduction: an analysis of endocrine profiles of juvenile, developing adolescent, developmentally arrested adolescent, adult, and aged captive male orangutans. PhD thesis, Duke University, Durham, NC.

Magrath, R.D. & Whittingham, L.A. 1997. Subordinate males are more likely to help if unrelated to the breeding female in cooperatively breeding white-browed scrubwrens. *Behavioral Ecology and Sociobiology*, **41**, 185–92.

Manley, G.H. 1986. Through the territorial barrier: harem accretion in *Presbytis senex*. In: *Primate Ontogeny, Cognition and Social Behaviour*, Vol. III, ed. J.G. Else & P.C. Lee, pp. 363–70. Cambridge: Cambridge University Press.

Manson, J.H. 1992. Measuring female mate choice in Cayo Santiago rhesus macaques. *Animal Behaviour*, **44**, 405–16.

–1996. Male dominance and mount series duration in Cayo Santiago rhesus macaques. *Animal Behaviour*, **51**, 1219–31.

Marler, P. 1968. Aggregation and dispersal: two functions in primate communication. In: *Primates: Studies in Adaptation and Variability*, ed. P.C. Jay, pp. 420–38. New York: Holt, Rinehart and Winston.

–1972. Vocalizations of East African monkeys II. Black and white colobus. *Behaviour*, **42**, 175–97.

Marsh, C.W. 1979a. Comparative aspects of social organization in the Tana River red colobus, *Colobus badius rufomitratus*. *Zeitschrift für Tierpsychologie*, **51**, 337–63.

–1979b. Female transfer and mate choice among Tana River red colobus. *Nature*, **281**, 568–9.

Martin, R.D. 1990. *Primate Origins and Evolution. A Phylogenetic Reconstruction*. London: Chapman and Hall.

–1992. Female cycles in relation to paternity in primate societies. In: *Paternity in Primates: Genetic Tests and Theories*, ed. R.D. Martin, A.F. Dixon & E.J. Wickings, pp. 238–74. Basel: S. Karger.

Martin, R.D. & MacLarnon, A.M. 1985. Gestation period, neonatal size, and maternal investment in placental mammals. *Nature*, **313**, 220–3.

Martin, R.D., Dixson, A.F. & Wickings, E.J. (eds.) 1992. *Paternity in Primates: Genetic Tests and Theories: Implications of Human DNA Fingerprinting*. Basel: S. Karger.

Martin, R.D., Willner, L.A. & Dettling, A. 1994. The evolution of sexual size dimorphism in primates. In: *The Differences Between the Sexes*, ed. R.V. Short & E. Balaban, pp. 159–200. New York: Cambridge University Press.

Marwell, G. & Ames, R.E. 1981. Economists free ride, does anyone else? Experiments on the provision of public goods, IV. *Journal of Public Economics*, **15**, 295–310.

Masataka, N. 1981. A field study of the social behavior of Goeldi's monkeys (*Callimico goeldii*) in north Bolivia. I. Group composition, breeding cycle, and infant development. In: *Kyoto University Overseas Research Reports of New World Monkeys*, pp. 23–32.

Maynard Smith, J. 1974. The theory of games and the evolution of animal conflicts. *Journal of Theoretical Biology*, **47**, 209–21.

–1976. Evolution and the theory of games. *American Scientist*, **64**, 41–5.

–1977. Parental investment – a prospective analysis. *Animal Behaviour*, **25**, 1–9.

McCann, C. 1933. Notes on the colouration and habits of the white-browed gibbon or Hoolock (*Hylobates hoolock* Harlan). *Journal of the Bombay Natural History Society*, **36**, 395–405.

McFarland, M.J. 1986. Ecological determinants of fission-fusion sociality in *Ateles* and *Pan*. In: *Primate Ecology and Conservation*, ed. J.G. Else & P.C. Lee, pp. 181–90. Cambridge: Cambridge University Press.

McGraw, S. 1994. Census, habitat preference, and polyspecific associations of six monkeys in the Lomako Forest, Zaire. *American Journal of Primatology*, **34**, 295–307.

McGrew, W.C., Marchant, L.F. & Nishida, T. (eds.) 1996. *Great Ape Societies*. Cambridge: Cambridge University Press.

McLean, I.G., Lundie-Jenkins, G., Jarman, P.J. & Kean, L.E. 1993. Copulation and associated behaviour in the rufous hare-wallaby, *Lagorchestes hirsutus*. *Australian Mammalogy*, **16**, 77–9.

McMillan, C.A. 1989. Male age, dominance, and mating success among rhesus macaques. *American Journal of Physical Anthropology*, **80**, 83–9.

McNeilage, A.D. 1995. Mountain gorillas in the Virunga volcanoes: ecology and carrying capacity. PhD thesis, University of Bristol.

Megantara, E.N. 1989. Ecology, behavior and sociality of *Presbytis femoralis* in Eastcentral Sumatra. In: *Comparative Primatology Monographs*, Vol. 2, ed. A. Ehara & S. Kawamura, pp. 171–301. Padjadjaran: University of Padjadjaran, India.

Mehlman, P.T. 1986. Male intergroup mobility in a wild population of the Barbary macaque (*Macaca sylvanus*), Ghomaran Rif Mountains, Morocco. *American Journal of Primatology*, **10**, 67–81.

–1989. Comparative density, demography, and ranging behavior of Barbary macaques (*Macaca sylvanus*) in marginal and prime conifer habitats. *International Journal of Primatology*, **10**, 269–92.

Meikle, D.B. & Vessey, S.H. 1981. Nepotism among rhesus monkey brothers. *Nature*, **294**, 160–1.

Melnick, D.J. & Pearl, M.C. 1987. Cercopithecines in multimale groups: genetic diversity and population structure. In: *Primate Societies*, ed. B.B. Smuts, D.L. Cheney, R.M. Seyfarth, R.W. Wrangham & T.T. Struhsaker, pp. 121–34. Chicago: University of Chicago Press.

Melnick, D.J., Pearl, M.C. & Richard, A.F. 1984. Male migration and inbreeding avoidance in wild rhesus monkeys. *American Journal of Primatology*, **7**, 229–43.

Ménard, N. & Vallet, D. 1993. Population dynamics of *Macaca sylvanus* in Algeria: an 8-year study. *American Journal of Primatology*, **30**, 101–18.

–1996. Demography and ecology of Barbary macaques (*Macaca sylvanus*) in two different habitats. In: *Evolution and Ecology of Macaque Societies*, ed. J.E. Fa & D.G. Lindburg, pp. 106–31. Cambridge: Cambridge University Press.

Ménard, N., Scheffrahn, W., Vallet, D., Zidane, C. & Reber, C. 1992. Application of blood protein electrophoresis and DNA fingerprinting to the analysis of paternity and social characteristics of wild Barbary macaques. In: *Paternity in Primates: Genetic Tests and Theories*, ed. R.D. Martin, A.F. Dixson & E.J. Wickings, pp. 153–74. Basel: S. Karger.

Merenlender, A.M. 1993. The effects of sociality on the demography and genetic structure of *Lemur fulvus rufus* (polygamous) and *Lemur rubriventer* (monogamous) and the conservation implications. PhD thesis, University of Rochester, NY.

Mesnick, S.L. 1997. Sexual alliances: evidence and evolutionary implications. In: *Feminism and Evolutionary Biology: Boundaries, Intersections, and Frontiers*, ed. P.A. Gowaty, pp. 207–60. New York: Chapman and Hall.

Meyers, D. 1993. The effects of resource seasonality on behavior and reproduction in the golden-crowned sifaka (*Propithecus tattersalli*, Simons, 1988) in three Malagasy forests. PhD thesis, Duke University, Durham, NC.

Millar, C.D., Anthony, I., Lambert, D.M. *et al.* 1994. Patterns of reproductive success determined by DNA fingerprinting in a communally breeding ocean bird. *Biological Journal of the Linnean Society*, **52**, 31–48.

Milton, K. 1985. Mating patterns of woolly spider monkeys. *Brachyteles arachnoides:* implications for female choice. *Behavioral Ecology and Sociobiology*, **17**, 53–9.

Mitani, J.C. 1985a. Sexual selection and adult male orangutan long calls. *Animal Behaviour*, **33**, 272–83.

–1985b. Mating behaviour of male orangutans in the Kutai Game Reserve, Indonesia. *Animal Behaviour*, **33**, 392–402.

–1990. Demography of agile gibbons (*Hylobates agilis*). *International Journal of Primatology*, **11**, 411–24.

Mitani, J.C. & Rodman, P.S. 1979. Territoriality: the relation of ranging pattern and home range size to defendability, with an analysis of territoriality among primate species. *Behavioral Ecology and Sociobiology*, **5**, 241–51.

Mitani, J.C., Gros-Louis, J. & Manson, J.H. 1996a. Number of males in primate groups: comparative tests of competing hypotheses. *American Journal of Primatology*, **38**, 315–32.

Mitani, J.C., Gros-Louis, J. & Richards, A.F. 1996b. Sexual dimorphism, the operational sex ratio, and the intensity of male competition in polygynous primates. *American Naturalist*, **147**, 966–80.

Mitani, M. 1991. Niche overlap and polyspecific associations among sympatric cercopithecids in the Campo Animal Reserve, southwestern Cameroon. *Primates*, **32**, 137–51.

Mitchell, C.L. 1994. Migration alliances and coalitions among adult male South American squirrel monkeys (*Saimiri sciureus*). *Behaviour*, **130**, 169–90.

Mitchell, C.L., Boinski, S. & van Schaik, C.P. 1991. Competitive regimes and female bonding in two species of squirrel monkeys (*Saimiri oerstedi* and *S. sciureus*). *Behavioral Ecology and Sociobiology*, **28**, 55–60.

Mittermeier, R.A., Rylands, A.B., Coimbra-Filho, A.F. & da Fonseca, G.A.B. (eds.) 1988. *Ecology and Behavior of Neotropical Primates*, Vol. 2. Washington, DC: World Wildlife Fund.

Moehlman, P.D. & Hofer, H. 1996. Cooperative breeding, reproductive suppression, and body mass in canids. In: *Cooperative Breeding in Mammals*, ed. N.G. Solomon & J.A. French, pp. 76–128. Cambridge: Cambridge University Press.

Møller, A.P. 1988. Ejaculate quality, testes size and sperm competition in primates. *Journal of Human Evolution*, **17**, 479–88.

Monard, A. & Duncan, P. 1996. Consequences of natal dispersal in female horses. *Animal Behaviour*, **52**, 565–79.

Moore, J. 1984. Female transfer in primates. *International Journal of Primatology*, **5**, 537–89.

–1985. Demography and sociality in primates. PhD thesis, Harvard University, Cambridge, MA.

–1992. Dispersal, nepotism, and primate social behavior. *International Journal of Primatology*, **13**, 361–78.

–1993. Inbreeding and outbreeding in primates: what's wrong with 'the dispersing sex'? In: *The Natural History of Inbreeding and Outbreeding. Theoretical and Empirical Perspectives*, ed. N.W. Thornhill, pp. 392–426. Chicago: University of Chicago Press.

Moore, J. & Ali, R. 1984. Are dispersal and inbreeding avoidance related? *Animal Behaviour*, **32**, 94–112.

Moraes, P.L.R., Carvalho, O. Jr & Strier, K.B. 1998. Population variation in patch and party size in muriquis (*Brachyteles arachnoides*). *International Journal of Primatology*, **19**, 325–37.

Morin, T.A., Moore, J.J., Chakraborty, R., Jin, L., Goodall, J. & Woodruff, D.S. 1994. Kin selection, social structure, gene flow, and the evolution of chimpanzees. *Science*, **265**, 1193–201.

Morland, H.S. 1991. Preliminary report on the social organization of ruffed lemurs (*Varecia variegata variegata*) in a Northeast Madagascar rain forest. *Folia Primatologica*, **56**, 157–61.

–1993. Reproductive activity of ruffed lemurs (*Varecia variegata variegata*) in a Madagascar rain forest. *American Journal of Physical Anthropology*, **91**, 71–82.

Mturi, F.A. 1991. The feeding ecology and behaviour of the red colobus monkey (*Colobus badius kirkii*). PhD thesis, University of Dar es Salaam, Tanzania.

Muckenhirn, N.A. 1972. *Leaf Eaters and Their Predators in Ceylon: Ecological Roles of Grey Langurs* Presbytis entellus *and Leopards*. College Park, MD: University of Maryland.

Mukherjee, R.P. & Saha, S.S. 1974. The golden langurs (*Presbytis geei* Khajuria, 1956) of Assam. *Primates*, **15**, 327–40.

Mulder, R.A. & Langmore N.E. 1993. Dominant males punish short-term defection by helpers in superb fairy wrens. *Animal Behaviour*, **45**, 830–3.

Mulder, R.A., Dunn, P.O., Cockburn, A., Lazenby-Cohen, K.A. & Howell, M.J. 1994. Helpers liberate female fairy wrens from constraints on extra-pair mate choice. *Proceedings of the Royal Society of London, Series B*, **255**, 223–9.

Mumme, R.L., Koenig, W.D. & Pitelka, F.A. 1983. Reproductive competition in the communal acorn woodpecker: sisters destroy each other's eggs. *Nature*, **306**, 583–4.

–1988. Costs and benefits of joint nesting in the acorn woodpecker. *American Naturalist*, **131**, 654–77.

Muroyama, Y. 1994. Exchange of grooming for allomothering in female patas monkeys. *Behaviour*, **128**, 103–19.

Mutschler, T., Feistner, A. & Nievergelt, C. 1998. Preliminary field data on group size, diet and activity in the Alaotran genlte lemur, *Hapalemur griseus alaotrensis*. *Folia Primatologica*, **69**, 325–30.

Nakagawa, N. 1989. Activity budget and diet of patas monkeys in the Kala Maloue National Park, Cameroon: a preliminary report. *Primates*, **30**, 27–34.

–1995. A case of infant kidnapping and allomothering by members of a neighboring group in patas monkeys. *Folia Primatologica*, **64**, 62–8.

Nakamura, M. 1998a. Multiple mating and cooperative breeding in polygynandrous alpine accentors. I. Competition among females for males. *Animal Behaviour*, **55**, 259–76.

–1998b. Multiple mating and cooperative breeding in polygynandrous alpine accentors. II. Dominance and male mating tactics. *Animal Behaviour*, **55**, 277–90.

–1990. Cloacal protuberance and copulatory behaviour of the alpine accentor *Prunella collaris*. *Auk*, **107**, 284–95.

National Research Council, USA. 1981. *Techniques for the Study of Primate Population Ecology*. Washington, DC: National Academy Press.

Neudenberger, J. 1993. Monogamie als Paarungssystem: eine Fallstudie am Weißhandgibbon (*Hylobates lar*) im Khao Yai Nationalpark, Thailand. MSc thesis, University of Göttingen.

Neville, M.K. 1968. Male leadership change in a free-ranging troop of Indian rhesus monkeys (*Macaca mulatta*). *Primates*, **9**, 13–27.

Newton, P.N. 1986. Infanticide in an undisturbed forest population of Hanuman langurs, *Presbytis entellus*. *Animal Behaviour*, **34**, 785–9.

–1987. The social organization of forest Hanuman langurs (*Presbytis entellus*). *International Journal of Primatology*, **8**, 199–232.

–1988. The variable social organization of Hanuman langurs (*Presbytis entellus*), infanticide, and the monopolization of females. *International Journal of Primatology*, **9**, 59–77.

Newton, P.N. & Dunbar, R.I.M. 1994. Colobine monkey society. In: *Colobine Monkeys: Their Ecology, Behaviour and Evolution*, ed. A.G. Davies & J.F. Oates, pp. 311–46. Cambridge: Cambridge University Press.

Neyman, P.F. 1977. Aspects of the ecology and social organization of free-ranging cotton-top tamarins (*Saguinus oedipus*) and the conservation status of the species. In: *The Biology and Conservation of the Callitrichidae*, ed. D.G. Kleiman, pp. 39–71. Washington, DC: Smithsonian Institution Press.

Nikolei, J. & Borries, C. 1997. Sex differential behavior of immature Hanuman langurs (*Presbytis entellus*) in Ramnagar, South Nepal. *International Journal of Primatology*, **18**, 415–37.

Nishida, T. 1968. The social group of wild chimpanzees in the Mahale mountains. *Primates*, **9**, 167–224.

–1974. Ecology of wild chimpanzees. In: *Human Ecology*, ed. R.

Ohtsuka, J. Tanaka & T. Nishida, pp. 15–60. Tokyo: Kyoritsu-suppan.
–1979. The social structure of the chimpanzees of the Mahale mountains. In: *The Great Apes*, ed. D.A. Hamburg & E.R. McCown, pp. 73–121. Menlo Park, CA: Benjamin/Cummings.
–1990. A quarter century of research in the Mahale mountains: an overview. In: *The Chimpanzees of the Mahale Mountains. Sexual and Life History Strategies*, ed. T. Nishida, pp. 3–35. Tokyo: University of Tokyo Press.
Nishida, T. & Hiraiwa-Hasegawa, M. 1987. Chimpanzees and bonobos: cooperative relationships among males. In: *Primate Societies*, ed. B.B. Smuts, D.L. Cheney, R.M. Seyfarth, R.W. Wrangham & T.T. Struhsaker, pp. 165–77. Chicago: University of Chicago Press.
Nishida, T. & Hosaka, K. 1996. Coalition strategies among adult male chimpanzees of the Mahale mountains, Tanzania. In: *Great Ape Societies*, ed. W.C. McGrew, L.F. Marchant & T. Nishida, pp. 114–34. Cambridge: Cambridge University Press.
Nishida, T., Hiraiwa-Hasegawa, M., Hasegawa T. & Takahata, Y. 1985. Group extinction and female transfer in wild chimpanzees in the Mahale National Park, Tanzania. *Zeitschrift für Tierpsychologie*, **67**, 284–301.
Nishida, T., Takasaki, H. & Takahata, Y. 1990. Demography and reproductive profiles. In: *The Chimpanzees of the Mahale Mountains. Sexual and Life History Strategies*, ed. T. Nishida, pp. 63–97. Tokyo: University of Tokyo Press.
Nishimura, A. 1990. Mating behavior of woolly monkeys, *Lagothrix lagotricha*, at La Macarena, Colombia (II): mating relationships. *Field Studies of New World Monkeys, La Macarena, Colombia*, **3**, 7–12.
–1992. Mating behaviors of woolly monkeys, *Lagothrix lagoricha*, at La Macarena, Colombia (III): reproductive parameters viewed from a longterm study. *Field Studies of New World Monkeys, La Macarena, Colombia*, **7**, 1–7.
–1994. Social interaction patterns of woolly monkeys (*Lagothrix lagotricha*): a comparison among the atelins. *The Science and Engineering Review of Doshisha University*, **35**, 235–54.
Noë, R. 1986. Lasting alliances among adult male savannah baboons. In: *Primate Ontogeny, Cognition and Social Behaviour*, ed. J.G. Else & P.C. Lee, pp. 381–92. Cambridge: Cambridge University Press.
–1990. A veto game played by baboons: a challenge to the use of the prisoner's dilemma as a paradigm for reciprocity and cooperation. *Animal Behaviour*, **39**, 78–90.
–1992. Alliance formation among male baboons: shopping for profitable partners. In: *Coalitions and Alliances in Animals and Humans*, ed. A.H. Harcourt & F.B.M. de Waal, pp. 285–321. Oxford: Oxford University Press.
Noë, R. & Hammerstein, P. 1994. Biological markets: supply and demand determine the effect of partner choice in cooperation, mutualism and mating. *Behavioral Ecology and Sociobiology*, **35**, 1–11.
–1995. Biological markets. *Trends in Ecology and Evolution*, **10**, 336–9.
Noë, R. & Sluijter, A.A. 1990. Reproductive tactics of male savanna baboons. *Behaviour*, **113**, 117–70.
–1995. Which adult male savanna baboons form coalitions? *International Journal of Primatology*, **16**, 77–105.
Noë, R., van Schaik, C.P. & van Hooff, J.A.R. A.M. 1991. The market effect: an explanation for pay-off asymmetries among collaborating animals. *Ethology*, **87**, 97–118.
Norconk, M., Rosenberger, A. & Garber, P. (eds.) 1996. *Adaptive Radiations of Neotropical Primates*. New York: Plenum Press.
Norikoshi, K. & Koyama, N. 1974. Group shifting and social organization among Japanese macaques. In: *Proceedings from the Symposia of the Fifth Congress of the International Primatological Society*, ed. S. Kondo, M. Kawai, A. Ehara & S. Kawamura, pp. 43–61. Tokyo: Japan Science Press.
Nowak, R.M. 1991. *Walker's Mammals of the World*, Vol. 2, 5th edn. Baltimore, MD: Johns Hopkins University Press.
Nunn, C.L. 1996. Testing the function of between-group conflict in ringtailed lemurs (*Lemur catta*) using playbacks of extra-group vocalizations. *American Journal of Physical Anthropology* Supplement, **22**, 179.
–1999. The number of males in primate social groups: a comparative test of the socioecological model. *Behavioral Ecology and Sociobiology*, **46**, 1–13.
Nunn, C.L. & Lewis, R.J. 1999. Cooperation and collective action in animal behaviour. In: *Economics in Nature*, ed. R. Noë, J.A.R.A.M. van Hooff & P. Hammerstein, pp. 00–00. Cambridge: Cambridge University Press.
O'Brien, T.G. 1991. Female–male social interactions in wedge-capped capuchin monkeys: benefits and costs of group living. *Animal Behaviour*, **41**, 555–67.
O'Brien, T.G. & Robinson, J.G. 1993. Stability of social relationships in female wedge-capped capuchin monkeys. In: *Juvenile Primates: Life History, Development, and Behavior*, ed. M.E. Pereira & L.A. Fairbanks, pp. 197–210. New York: Oxford University Press.
O'Connell, S.M. & Cowlishaw, G. 1994. Infanticide avoidance, sperm competition and mate choice: the function of copulation calls in female baboons. *Animal Behaviour*, **48**, 687–94.
O'Connor, S.M. 1987. The effect of human impact on the vegetation and the consequences to primates in two riverine forests, Southern Madagascar. PhD thesis, University of Cambridge.
Oates, J.F. 1977. The guereza and its food. In: *Primate Ecology*, ed. T.H. Clutton-Brock, pp. 276–323. London: Academic Press.
–1996. *African Primates. Status Survey and Conservation Action Plan*. Gland, Switzerland: IUCN.
Ogawa, H. 1995. Bridging behavior and other affiliative interactions among male Tibetan macaques (*Macaca thibetana*). *International Journal of Primatology*, **16**, 707–29.
Ohsawa, H., Inoue, M. & Takenaka, O. 1993. Mating strategy and reproductive success of male patas monkeys (*Erythrocebus patas*). *Primates*, **34**, 533–44.
Oi, T. 1990. Population organization of wild pig-tailed macaques (*Macaca nemestrina*) in West Sumatra. *Primates*, **31**, 15–31.

–1996. Sexual behaviour and mating system of the wild pig-tailed macaque in West Sumatra. In: *Evolution and Ecology of Macaque Societies*, ed. J.E. Fa & D.G. Lindburg, pp. 342–68. Cambridge: Cambridge University Press.

Olson, M. 1965. *The Logic of Collective Action*. Cambridge, MA: Harvard University Press.

Olson, M. & Zeckhauser, R. 1966. An economic theory of alliances. *Review of Economics and Statistics*, **48**, 266–79.

Oring, L.W. 1982. Avian mating systems. In: *Avian Biology*, Vol. 6, ed. D.S. Farner, J.R. King & K.C. Parkes, pp. 1–92. New York: Academic Press.

Ostner, J. 1998. Soziale Konsequenzen der Kathemeralität bei Rotstirnmakis (*Eulemur fulvus rufus*). MSc thesis, University of Göttingen.

Ostrom, E. 1990. *Governing the Commons: the Evolution of Institutions for Collective Action*. Cambridge: Cambridge University Press.

Overdorff, D.J. 1988. Preliminary report on the activity cycle and diet of the red-bellied lemur (*Lemur rubriventer*) in Madagascar. *American Journal of Primatology*, **16**, 143–53.

–1996. Ecological correlates to social structure in two lemur species in Madagascar. *American Journal of Physical Anthropology*, **100**, 487–506.

–1998. Are *Eulemur* species pair-bonded? Social organization and mating strategies in *Eulemur fulvus rufus* from 1988–1995 in southeast Madagascar. *American Journal of Physical Anthropology*, **105**, 153–66.

Overdorff, D.J. & Rasmussen, M.A. 1995. Determinants of night-time activity in 'diurnal' lemurid primates. In: *Creatures of the Dark*, ed. L. Alterman, G.A. Doyle & M.K. Izard, pp. 61–74. New York: Plenum Press.

Overdorff, D.J., Strait, S.G. & Telo, A. 1997. Seasonal variation in activity and diet in a small-bodied folivorous primate, *Hapalemur griseus*, in southeastern Madagascar. *American Journal of Primatology*, **43**, 211–23.

Owens, I. & Thompson, D. 1994. Sex differences, sex ratios and sex roles. *Proceedings of the Royal Society of London, Series B*, **258**, 93–9.

Packer, C. 1975. Male transfer in olive baboons. *Nature*, **255**, 219–20.

–1977. Reciprocal altruism in olive baboons. *Nature*, **265**, 441–3.

–1979a. Inter-troop transfer and inbreeding avoidance in *Papio anubis*. *Animal Behaviour*, **27**, 1–36.

–1979b. Male dominance and reproductive activity in *Papio anubis*. *Animal Behaviour*, **27**, 37–45.

–1985. Dispersal and inbreeding avoidance. *Animal Behaviour*, **33**, 676–8.

Packer, C. & Pusey, A.E. 1983. Adaptations of female lions to infanticide by incoming males. *American Naturalist*, **121**, 716–28.

–1985. Asymmetric contests in social mammals: respect, manipulation and age-specific aspects. In: *Evolution. Essays in Honour of John Maynard Smith*, ed. P.J. Greenwood, P.H. Harvey & M. Slatkin, pp. 173–86. Cambridge: Cambridge University Press.

Packer, C., Scheel, D. & Pusey, A.E. 1990. Why lions form groups: food is not enough. *American Naturalist* **136**, 1–19.

Pagel, M. 1994a. Detecting correlated evolution on phylogenies: a general method for the comparative analysis of discrete characters. *Proceedings of the Royal Society of London, Series B*, **255**, 37–45.

–1994b. The evolution of conspicuous oestrus advertisement in Old World monkeys. *Animal Behaviour*, **47**, 1333–41.

Palombit, R.A. 1993. Lethal territorial aggression in a white-handed gibbon. *American Journal of Primatology*, **31**, 311–18.

–1994a. Extra-pair copulations in a monogamous ape. *Animal Behaviour*, **47**, 721–3.

–1994b. Dynamic pair bonds in hylobatids: implications regarding monogamous social systems. *Behaviour*, **128**, 65–101.

–1997. Inter- and intraspecific variation in the diets of sympatric Siamang (*Hylobates syndactylus*) and lar gibbons (*Hylobates lar*). *Folia Primatologica*, **68**, 321–37.

Palombit, R.A., Seyfarth, R.M. & Cheney, D.L. 1997. The adaptive value of 'friendships' to female baboons: experimental and observational evidence. *Animal Behaviour*, **54**, 599–614.

Parker, G.A. 1970. Sperm competition and its evolutionary consequences in the insects. *Biological Reviews*, **45**, 525–67.

–1979. Sexual selection and sexual conflict. In: *Sexual Selection and Reproductive Competition in Insects*, ed. M.S. Blum & N.A. Blum, pp. 123–66. New York: Academic Press.

–1984. Sperm competition and the evolution of animal mating strategies. In: *Sperm Competition and the Evolution of Animal Mating Systems*, ed. R.L. Smith, pp. 1–60. New York: Academic Press.

Parmigiani, S. & vom Saal, F.S. 1994. *Infanticide and Parental Care*. London: Harwood Academic Publishers.

Parsons, P.E. & Taylor, C.R. 1977. Energetics of brachiation versus walking: a comparison of a suspended and inverted pendulum mechanism. *Physiological Zoology*, **50**, 182–8.

Partridge, L. & Endler, J.A. 1987. Life history constraints on sexual selection. In *Sexual Selection: Testing the Alternatives*, ed. J.W. Bradbury & M.B. Andersson, pp. 265–77. New York: J. Wiley.

Paul, A. 1989. Determinants of male mating success in a large group of Barbary macaques (*Macaca sylvanus*) at Affenberg Salem. *Primates*, **30**, 461–76.

–1997. Breeding seasonality affects the association between dominance and reproductive success in non-human male primates. *Folia Primatologica*, **68**, 344–9.

Paul, A. & Kuester, J. 1985. Intergroup transfer and incest avoidance in semifree-ranging Barbary macaques (*Macaca sylvanus*) at Salem (FRG). *American Journal of Primatology*, **8**, 317–22.

–1988. Life history patterns of Barbary macaques (*Macaca sylvanus*) at Affenberg Salem. In: *Ecology and Behavior in Food-Enhanced Primate Groups*, ed. J.E. Fa & C.H. Southwick, pp. 199–228. New York: Alan Liss.

Paul, A., Kuester, J. & Arnemann, J. 1992. DNA fingerprinting reveals that infant care by male Barbary macaques is not parental investment. *Folia Primatologica*, **58**, 93–8.

Paul, A., Kuester, J., Timme, J. & Arnemann, J. 1993. The association between rank, mating effort, and reproductive success in male Barbary macaques (*Macaca sylvanus*). *Primates*, **34**, 491–502.

Paul, A., Kuester, J. & Arnemann, J. 1996. The sociobiology of male–infant interactions in Barbary macaques, *Macaca sylvanus*. *Animal Behaviour*, **51**, 155–70.

Penzhorn, B.L. 1984. A long-term study of social organization and behaviour of cape mountain zebras, *Equus zebra zebra*. *Zeitschrift für Tierpsychologie*, **64**, 97–146.

Pereira, M.E. 1991. Asynchrony within estrous synchrony among ringtailed lemurs (Primates: Lemuridae). *Physiology and Behavior*, **49**, 47–52.

–1993. Agonistic interaction, dominance relation, and ontogenetic trajectories in ringtailed lemurs. In: *Juvenile Primates: Life History, Development, and Behavior*, ed. M.E. Pereira & L.A. Fairbanks, pp. 285–305. New York: Oxford University Press.

–1995. Development and social dominance among group-living primates. *American Journal of Primatology*, **37**, 143–75.

–1998. One male, two males, three males, more. *Evolutionary Anthropology*, **7**, 39–45.

Pereira, M.E. & Kappeler, P.M. 1997. Divergent systems of agonistic relationship in lemurid primates. *Behaviour*, **134**, 225–74.

Pereira, M.E. & McGlynn, C.A. 1997. Special relationships instead of female dominance for redfronted lemurs, *Eulemur fulvus rufus*. *American Journal of Primatology*, **43**, 239–58.

Pereira, M.E. & Weiss M.L. 1991. Female mate choice, male migration, and the threat of infanticide in ringtailed lemurs. *Behavioral Ecology and Sociobiology*, **28**, 141–52.

Pereira, M.E., Strohecker, R., Cavigelli, S., Hughes, C. & Pearson, D. 1999. Metabolic tactics in Lemuridae and implications for social behavior. In: *New Directions in Lemur Studies*, ed. J.U. Ganzhorn, S.M. Goodman & H. Rasamimanana, pp. 93–118. New York: Plenum Press.

Peres, C.A. 1993. Anti-predation benefits in a mixed-species group of Amazonian tamarins. *Folia Primatologica*, **61**, 61–76.

–1994. Diet and feeding ecology of gray woolly monkeys (*Lagothrix lagotricha cana*) in Central Amazonia: comparisons with other atelines. *International Journal of Primatology*, **15**, 333–72.

–1996. Use of space, spatial group structure, and foraging group size of gray woolly monkeys (*Lagothrix lagotricha cana*) at Urucu, Brazil: a review of the Atelinae. In: *Adaptive Radiations of Neotropical Primates*, ed. M.A. Norconk, A.L. Rosenberger & P.A. Garber, pp. 467–88. New York: Plenum Press.

Perret, M. 1992. Environmental and social determinants of sexual function in the male lesser mouse lemur (*Microcebus murinus*). *Folia Primatologica*, **59**, 1–25.

Perry, S. 1998. Male–male social relationships in wild white-faced capuchins, *Cebus capucinus*. *Behaviour*, **135**, 139–72.

Phillips-Conroy, J.E., Jolly, C.J., Nystrom, P. & Hemmalin, H.A. 1992. Migration of male hamadryas baboons into anubis groups in the Awash National Park, Ethiopia. *International Journal of Primatology*, **13**, 455–76.

Piper, W.H. & Slater, G. 1993. Polyandry and incest avoidance in the cooperative stripe-backed wren of Venezuela. *Behaviour*, **124**, 227–47.

Platt, J.R. 1964. Strong inference. *Science*, **146**, 347–52.

Plavcan, J.M. & van Schaik, C.P. 1992. Intrasexual competition and canine dimorphism in anthropoid primates. *American Journal of Physical Anthropology*, **87**, 461–77.

–1994. Canine dimorphism. *Evolutionary Anthropology*, **2**, 208–14.

–1997. Intrasexual competition and body weight dimorphism in anthropoid primates. *American Journal of Physical Anthropology*, **103**, 37–68.

Podzuweit, D. 1994. Sozio-Ökologie weiblicher Hanuman Languren (*Presbytis entellus*) in Ramnagar, Südnepal. PhD thesis, University of Göttingen.

Podzuweit, D., Winkler, P. & Borries, C. 1992. Mating success and infant defense: keys to male migration patterns in troops of langurs (*Presbytis entellus*) in Nepal? *XIVth Congress of the International Primatological Society, Strasbourg, IPS*.

Poirier, F.E. 1969. The Nilgiri langur (*Presbytis johnii*) troop: its composition, structure, function and change. *Folia Primatologica*, **10**, 20–47.

–1970. The Nilgiri langur (*Presbytis johnii*) of South India. In: *Primate Behavior. Developments in Field and Laboratory Research*, ed. L.A. Rosenblum, pp. 251–383. New York: Academic Press.

–1972. The St. Kitts green monkey (*Cercopithecus aethiops sabaeus*): ecology, population dynamics, and selected behavioral traits. *Folia Primatologica*, **17**, 20–55.

Pollock, J.I. 1975. Field observations on *Indri indri*: a preliminary report. In: *Lemur Biology*, ed. I. Tattersall & R.W. Sussman, pp. 287–311. New York: Plenum Press.

Pook, A.G. & Pook, G. 1981. A field study of the socio-ecology of the Goeldi's monkey (*Callimico goeldii*) in northern Bolivia. *Folia Primatologica*, **35**, 288–312.

Pope, T.R. 1989. The influence of mating system and dispersal patterns on the genetic structure of red howler monkey populations. PhD thesis, University of Florida.

–1990. The reproductive consequences of male cooperation in the red howler monkey: paternity exclusion in multi-male and single-male troops using genetic markers. *Behavioral Ecology and Sociobiology*, **27**, 439–46.

–1992. The influence of dispersal patterns and mating system on genetic differentiation within and between populations of the red howler monkey (*Alouatta seniculus*). *Evolution*, **46**, 1112–28.

–1996a. Influence of social dynamics on mtDNA diversity in red howler monkey populations. *American Journal of Physical Anthropology*, Supplement, **22**, 188.

–1996b. Socioecology, population fragmentation, and patterns of genetic loss in endangered primates. In: *Conservation Genetics: Case Histories from Nature*, ed. J.C. Avise & J.L. Hamrick, pp. 119–59. New York: Chapman and Hall.

–1998. Effects of demographic change on group kin structure and gene dynamics of red howling monkey populations. *Journal of Mammalogy*, **79**, 692–712.

Porter, C.A., Page, S.L., Czelusniak, J. *et al.* 1997. Phylogeny and evolution of selected primates as determined by sequences of the epsilon-globin locus and 5′ flanking regions. *International Journal of Primatology*, **18**, 261–95.

Powell, G.V.N. 1974. Experimental analysis of the social value of flocking by starlings (*Sturnus vulgaris*) in relation to predation and foraging. *Animal Behaviour*, **22**, 501–5.

Preuschoft, H., Chivers, D.J., Brockelman, W.Y. & Creel, N. (eds.) 1984. *The Lesser Apes: Evolutionary and Behavioural Biology*. Edinburgh: Edinburgh University Press.

Preuschoft, S. 1999. Are primates behaviorists? Formal dominance, cognition, and free floating rationales. *Journal of Comparative Psychology*, **113**, 91–5.

Preuschoft, S., Paul, A. & Kuester, J. 1998. Dominance styles of female and male Barbary macaques (*Macaca sylvanus*). *Behaviour*, **135**, 731–55.

Price, E.C. 1990. Infant carrying as a courtship strategy of breeding male cotton-top tamarins. *Animal Behaviour*, **40**, 784–6.

Price, E.C. & McGrew, W.C. 1991. Departures from monogamy in colonies of captive cotton-top tamarins. *Folia Primatologica*, **57**, 16–27.

Prince, R.I.T. 1995. Banded hare-wallaby *Lagostrophus fasciatus*. In: *The Mammals of Australia*, ed. R. Strahan, pp. 406–8. Sydney: Australian Museum & Reed Books.

Puertas, P., Encarnación, F., Aquino, R. & García, J.E. 1995. Análisis poblacional del pichico pecho anaranjado *Saguinus labiatus*, en el sur oriente peruano. *Neotropical Primates*, **3**, 4–7.

Purvis, A. 1995. A composite estimate of primate phylogeny. *Philosophical Transactions of the Royal Society of London, Series B*, **348**, 405–21.

Purvis, A. & Rambaut, A. 1994. *Comparative Analysis by Independent Contrasts (CAIC), Version 2*. Oxford: Oxford University.

Pusey, A.E. 1979. Intercommunity transfer of chimpanzees in Gombe National Park. In: *The Great Apes*, ed. D.A. Hamburg & E.R. McCown, pp. 465–79. Menlo Park, CA: Benjamin/Cummings.

–1992. The primate perspective on dispersal. In: *Animal Dispersal: Small Mammals as a Model*, ed. N.C. Stenseth & W.Z. Lidicker Jr, pp. 243–59. London: Chapman & Hall.

Pusey, A.E. & Packer, C. 1987. Dispersal and philopatry. In: *Primate Societies*, ed. B.B. Smuts, D.L. Cheney, R.M. Seyfarth, R.W. Wrangham & T.T. Struhsaker, pp. 250–66. Chicago: University of Chicago Press.

–1994. Infanticide in lions: consequences and counterstrategies. In: *Infanticide and Parental Care*, ed. S. Parmigiani & F.S. vom Saal, pp. 277–99. London: Harwood Academic Publishers.

Pusey, A.E., Williams, J. & Goodall, J. 1997. The influence of dominance rank on the reproductive success of female chimpanzees. *Science*, **277**, 828–31.

Quinn, J.F. & Dunham, A.E. 1983. On hypothesis testing in ecology and evolution. *American Naturalist*, **122**, 602–17.

Quris, R. 1976. Données comparatives sur la socioécologie de huit espèces de Cercopithecidae vivant dans une meme zone de forêt primitive périodiquement inondée (N-E Gabon). *Review of Ecology* (*La Terre et la Vie*), **30**, 193–209.

Quris, R., Gautier, J.-P., Gautier, J.Y. & Gautier-Hion, A. 1981. Organisation spatio-temporelle des activités individuelles et sociales dans une troupe de *Cercopithecus cephus*. *Review of Ecology* (*La Terre et la Vie*), **35**, 37–53.

Rabenold, P.P., Rabenold, K.N., Piper, W.H., Haydock, J. & Zack, S.W. 1990. Shared paternity revealed by genetic analysis in cooperatively breeding tropical wrens. *Nature*, **348**, 538–40.

Rabinowitz, A.T. 1989. The density and behavior of large cats in a dry tropical forest mosaic in Huai Kha Khaeng Wildlife Sanctuary, Thailand. *The Natural History Bulletin of the Siam Society*, **37**, 235–51.

Raemaekers, J.J. & Chivers, D.J. 1980. Socio-ecology of Malayan forest primates. In: *Malayan Forest Primates: Ten Years' Study in Tropical Rain Forest*, ed. D.J. Chivers, pp. 279–315. New York: Plenum Press.

Raemaekers, J.J. & Raemaekers, P.M. 1984a. The Ooaa duet of the gibbon (*Hylobates lar*): a group call which triggers other groups to respond in kind. *Folia Primatologica*, **42**, 209–15.

Raemaekers, J.J. & Raemaekers, P.M. 1984b. Vocal interaction between two male gibbons, *Hylobates lar*. *The Natural History Bulletin of the Siam Society*, **32**, 95–106.

Raemaekers, J.J., Aldrich-Blake, F.P.G. & Payne, J.B. 1980. The forest. In: *Malayan Forest Primates. Ten Years' Study in Tropical Rain Forest*, ed. D.J. Chivers, pp. 29–61. New York: Plenum Press.

Raemaekers, J.J., Raemaekers, P.M. & Haimoff, E.H. 1984. Loud calls of the gibbon (*Hylobates lar*): repertoir, organization and context. *Behaviour*, **91**, 146–89.

Rajanathan, R. & Bennett, E.L. 1990. Notes on the social behaviour of wild proboscis monkeys (*Nasalis larvatus*). *Malayan Nature Journal*, **44**, 35–44.

Rajpurohit, L.S. 1995. Temporary splitting or subgrouping in male bands of Hanuman langurs, *Presbytis entellus*, around Jodhpur, western India. *Mammalia*, **59**, 3–8.

Rajpurohit, L.S. & Sommer, V. 1991. Sex differences in mortality among langurs (*Presbytis entellus*) of Jodhpur, Rajastan. *Folia Primatologica*, **56**, 17–27.

–1993. Juvenile male emigration from natal one-male troops in Hanuman langurs. In: *Juvenile Primates: Life History, Development, and Behavior*, ed. M.E. Pereira & L.A. Fairbanks, pp. 86–103. New York: Oxford University Press.

Rajpurohit, L.S., Sommer, V. & Mohnot, S.M. 1995. Wanderers between harems and bachelor bands: male Hanuman langurs (*Presbytis entellus*) at Jodhpur in Rajasthan. *Behaviour*, **132**, 255–99.

Rakotoarisoa, S.V. 1994. Etude des influences des facteurs externes sur la structure de la population de *Propithecus verreauxi verreauxi* dans la réserve privée de Berenty et ses interets pédagogiques et éducationnels. Memoire de CAPEN, Université d'Antananarivo.

Read, A.F. & Nee, S. 1995. Inference from binary comparative data. *Journal of Theoretical Biology*, **173**, 99–108.

Reed, C., O'Brien, T.G. & Kinnaird, M.F. 1997. Male social behavior and dominance hierarchy in the Sulawesi crested black macaque (*Macaca nigra*). *International Journal of Primatology*, **18**, 247–60.

Reeve, H.K., & Ratnieks, F.L.W. 1993. Queen–queen conflict in polygynous societies: mutual tolerance and reproductive skew. In: *Queen Number and Sociality in Insects*, ed. L. Keller, pp. 45–85. Oxford: Oxford University Press.

Reeve, H.K., Emlen, S.T. & Keller, L. 1998. Reproductive sharing in animal societies: reproductive incentives or incomplete control by dominant breeders? *Behavioral Ecology*, **9**, 267–78.

Reichard, U. 1995. Extra-pair copulations in a monogamous gibbon (*Hylobates lar*). *Ethology*, **100**, 99–112.

–1996. Sozial- und Fortpflanzungsverhalten von Weisshandgibbons (*Hylobates lar*): Eine Freilandstudie im thailändischen Khao Yai Regenwald. PhD thesis, University of Göttingen.

–1998. Sleeping sites, sleeping places, and pre-sleep behavior of gibbons (*Hylobates lar*). *American Journal of Primatology*, **46**, 35–62.

Reichard, U. & Sommer, V. 1997. Group encounters in wild gibbons (*Hylobates lar*): agonism, affiliation, and the concept of infanticide. *Behaviour*, **134**, 1135–74.

Reinhardt, V., Reinhardt, A., Bercovitch, F.B. & Goy, R.W. 1986. Does intermale mounting function as a dominance demonstration in rhesus monkeys? *Folia Primatologica*, **47**, 55–60.

Relethford, J.H. 1996. *The Human Species: an Introduction to Biological Anthropology*, 3rd edn. Mountain View, CA: Mayfield Publications.

Reyer, H.-U. 1990. Pied kingfishers: ecological causes and reproductive consequences of cooperative breeding. In: *Cooperative Breeding in Birds*, ed. P.B. Stacey & W.D. Koenig, pp. 529–57. Cambridge: Cambridge University Press.

Richard, A.F. 1974a. Intra-specific variation in the social organization and ecology of *Propithecus verreauxi*. *Folia Primatologica*, **22**, 178–207.

–1974b. Patterns of mating in *Propithecus verreauxi verreauxi*. In: *Prosimian Biology*, ed. R.D. Martin, G.A. Doyle & A.C. Walker, pp. 49–74. London: Duckworth.

–1985. Social boundaries in a Malagasy prosimian, the sifaka (*Propithecus verreauxi*). *International Journal of Primatology*, **6**, 553–68.

–1987. Malagasy prosimians: female dominance. In: *Primate Societies*, ed. B.B. Smuts, D.L. Cheney, R.M. Seyfarth, R.W. Wrangham & T.T. Struhsaker, pp. 25–33. Chicago: University of Chicago Press.

–1992. Aggressive competition between males, female-controlled polygyny and sexual monomorphism in a Malagasy primate, *Propithecus verreauxi*. *Journal of Human Evolution*, **22**, 395–406.

Richard, A.F. & Dewar, R.E. 1991. Lemur ecology. *Annual Review of Ecology and Systematics*, **22**, 145–75.

Richard, A.F., Rakotomanga, P. & Schwartz, M. 1991. Demography of *Propithecus verreauxi* at Beza Mahafaly, Madagascar: sex ratio, survival, and fertility. *American Journal of Physical Anthropology*, **84**, 307–22.

–1993. Dispersal by *Propithecus verreauxi* at Beza Mahafaly, Madagascar: 1984–1991. *American Journal of Primatology*, **30**, 1–20.

Ridley, M. 1986. The number of males in a primate troop. *Animal Behaviour*, **34**, 1848–58.

Rigamonti, M.M. 1993. Home range and diet in red ruffed lemurs (*Varecia variegata rubra*) on the Masoala peninsula, Madagascar. In: *Lemur Social Systems and Their Ecological Basis*, ed. P.M. Kappeler & J.U. Ganzhorn, pp. 25–39. New York: Plenum Press.

Rijksen, H.D. 1978. *A Field Study on Sumatran Orang Utans (Pongo pygmaeus abelii*, Lesson 1827)*: Ecology, Behaviour and Conservation*. Wageningen: Veenman & Zonen.

Robbins, D., Chapman, C.A. & Wrangham, R.W. 1991. Group size and stability: why do gibbons and spider monkeys differ? *Primates*, **32**, 301–5.

Robbins, M.M. 1995. A demographic analysis of male life history and social structure of mountain gorillas. *Behaviour*, **132**, 21–47.

–1996. Male–male interactions in heterosexual and all-male wild mountain gorilla groups. *Ethology*, **102**, 942–65.

Roberts, G. 1996. Why individual vigilance declines as group size increases. *Animal Behaviour*, **51**, 1077–86.

Robertshaw, J.D. & Harden, R.H. 1985. The ecology of the dingo in north-eastern New South Wales. III. Analysis of bone fragments in dingo scats. *Australian Wildlife Research*, **12**, 163–71.

–1986. The ecology of the dingo in north-eastern New South Wales. IV. Prey selection by dingoes and its effect on the major prey species, the swamp wallaby, *Wallabia bicolor* (Desmarest). *Australian Wildlife Research*, **13**, 141–63.

–1989. Predation on Macropodoidea: a review. In: *Kangaroos, Wallabies and Rat-kangaroos*, ed. G. Grigg, P. Jarman & I. Hume, pp. 735–53. Sydney: Surrey Beatty & Sons.

Robinson, J.G. & Janson, C.H. 1987. Capuchins, squirrel monkeys, and atelines: socioecological convergence with Old World primates. In: *Primate Societies*, ed. B.B. Smuts, D.L. Cheney, R.M. Seyfarth, R.W. Wrangham & T.T. Struhsaker, pp. 69–82. Chicago: University of Chicago Press.

Robinson, J.G. & O'Brien, T.G. 1991. Adjustment in birth sex ratio in wedge-capped capuchin monkeys. *American Naturalist*, **138**, 1173–86.

Robinson, J.G., Wright, P.C. & Kinzey, W.G. 1987. Monogamous cebids and their relatives: intergroup calls and spacing. In: *Primate Societies*, ed. B.B. Smuts, D.L. Cheney, R.M. Seyfarth, R.W. Wrangham & T.T. Struhsaker, pp. 44–53. Chicago: University of Chicago Press.

Rodman, P.S. & Mitani, J.C. 1987. Orangutans: sexual dimorphism in a solitary species. In: *Primate Societies*, ed. B.B. Smuts, D.L.

Cheney, R.M. Seyfarth, R.W. Wrangham & T.T. Struhsaker, pp. 146–54. Chicago: University of Chicago Press.

Rodseth, L., Wrangham, R.W., Harrigan, A.M. & Smuts, B.B. 1991. The human community as a primate society. *Current Anthropology*, **32**, 221–54.

Roldan, E.R.S., Gomendio, M. & Vitullo, A.D. 1992. The evolution of eutherian spermatozoa and underlying selective forces: female selection and sperm competition. *Biological Reviews*, **67**, 551–93.

Rose, L.M. 1994. Benefits and costs of resident males to females in white-faced capuchins, *Cebus capucinus*. *American Journal of Primatology*, **32**, 235–48.

Rose, L.M. & Fedigan, L.M. 1995. Vigilance in white-faced capuchins, *Cebus capucinus*, in Costa Rica. *Animal Behaviour*, **49**, 63–70.

Rosenberger, A.L. & Strier, K.B. 1989. Adaptive radiation of the ateline primates. *Journal of Human Evolution*, **18**, 717–50.

Ross, C. 1991. Life history patterns of New World monkeys. *International Journal of Primatology*, **12**, 481–502.

–1993. Predator mobbing by an all-male band of Hanuman langurs (*Presbytis entellus*). *Primates*, **34**, 105–7.

–1998. Primate life histories. *Evolutionary Anthropology*, **6**, 54–63.

Ross, C. & MacLarnon, A. 1995. Ecological and social correlates of maternal expenditure on infant growth in haplorhine primates. In: *Motherhood in Human and Nonhuman Primates*, ed. C.R. Pryce, R.D. Martin & D. Skuse, pp. 37–46. Basel: S. Karger.

Rothe, H. 1975. Some aspects of sexuality and reproduction in groups of captive common marmosets (*Callithrix jacchus*). *Zeitschrift für Tierpsychologie*, **37**, 255–73.

Rowe, N. 1996. *The Pictorial Guide to the Living Primates*. East Hampton, NY: Pogonias Press.

Rowell, T.E. 1972. Toward a natural history of the talapoin monkeys in Cameroon. *Annales de la Faculté des Sciences de Cameroon*, **10**, 121–34.

–1973. Social organization of wild talapoin monkeys. *American Journal of Physical Anthropology*, **38**, 593–8.

–1982. The breeding season of talapoin monkeys. *National Geographic Society Research Reports*, **14**, 577–83.

–1988a. Beyond the one-male group. *Behaviour*, **104**, 189–201.

–1988b. The social system of guenons, compared with baboons, macaques and mangabeys. In: *A Primate Radiation: Evolutionary Biology of the African Guenons*, ed. A. Gautier-Hion, F. Bourlière, J.-P. Gautier & J. Kingdon, pp. 439–51. Cambridge: Cambridge University Press.

–1994. Choosy or promiscuous – it depends on the time scale. In: *Current Primatology: Selected Proceedings of the XIVth Congress of the International Primatological Society*, Vol. 2, *Social Development, Learning and Behaviour*, ed. J.J. Roder, B. Thierry, J.R. Anderson & N. Herrenschmidt, pp. 11–18. Strasbourg: Université Louis Pasteur.

–1996. Book review: perspectives in ethology. *International Journal of Primatology*, **17**, 297–8.

Rowell, T.E. & Chism, J. 1986. The ontogeny of sex differences in the behavior of patas monkeys. *International Journal of Primatology*, **7**, 83–107.

Rowell, T. E. & Dixson, A.F. 1975. Changes in social organization during the breeding season of wild talapoin monkeys. *Journal of Reproduction and Fertility*, **43**, 419–34.

Rubenstein, D.I. 1986. Ecology and sociality in horses and zebras. In: *Ecological Aspects of Social Evolution: Birds and Mammals*, ed. D.I. Rubenstein & R.W. Wrangham, pp. 282–302. Princeton, NJ: Princeton University Press.

Rubenstein, D.I. & Wrangham, R.W. (eds.) 1986. *Ecological Aspects of Social Evolution: Birds and Mammals*. Princeton, NJ: Princeton University Press.

Rudran, R. 1973a. Adult male replacement in one-male troops of purple-faced langurs (*Presbytis senex senex*) and its effect on population structure. *Folia Primatologica*, **19**, 166–92.

–1973b. The reproductive cycles of two subspecies of purple-faced langurs (*Presbytis senex*) with relation to environmental factors. *Folia Primatologica*, **19**, 41–60.

–1979. The demography and social mobility of a red howler (*Alouatta seniculus*) population in Venezuela. In: *Vertebrate Ecology in the Northern Neotropics*, ed. J.F. Eisenberg, pp. 107–26. Washington, DC: Smithsonian Institution Press.

Ruhiyat, Y. 1983. Socio-ecological study of *Presbytis aygula* in West Java. *Primates*, **24**, 344–59.

Rutberg, A.T. 1990. Inter-group transfer in Assateague pony mares. *Animal Behaviour*, **40**, 945–52.

Ruxton, G.D. & van der Meer, J. 1997. Policing: it pays the strong to protect the weak. *Trends in Ecology and Evolution*, **12**, 250–1.

Ryder, O.A. & Massena, A.R. 1988. A case of male infanticide in *Equus przewalskii*. *Applied Animal Behavior Science*, **21**, 187–90.

Rylands, A.B. 1982. The behaviour and ecology of three species of marmosets and tamarins (Callitrichidae, Primates) in Brazil. PhD thesis, University of Cambridge.

–1993. The ecology of the lion tamarins, *Leontopithecus*: some intrageneric differences and comparisons with other callitrichids. In: *Marmosets and Tamarins. Systematics, Behaviour, and Ecology*, ed. A.B. Rylands, pp. 296–313. Oxford: Oxford University Press.

–1996. Habitat and the evolution of social and reproductive behavior in Callitrichidae. *American Journal of Primatology*, **38**, 5–18.

Rylands, A.B., Coimbra-Filho, A.F. & Mittermeier, R.A. 1993. Systematics, geographic distribution, and some notes on the conservation status of the Callitrichidae. In: *Marmosets and Tamarins. Systematics, Behaviour, and Ecology*, ed. A.B. Rylands, pp. 262–72. Oxford: Oxford University Press.

Sakura, O. 1994. Factors affecting party size and composition of chimpanzees (*Pan troglodytes verus*) at Bossou, Guinea. *International Journal of Primatology*, **15**, 167–83.

Salafsky, N.N. 1988. The foraging patterns and socioecology of the Kelasi (*Presbytis rubicunda*). MSc thesis, Harvard College, Cambridge, MA.

Samuelson, P.A. 1954. The pure theory of public expenditure. *Review of Economics and Statistics*, **36**, 387–9.

Sánchez, S., Peláez, F., Gil-Bürmann, C. & Kaumanns, W. 1999. Costs of infant carrying in the cotton-top tamarin (*Saguinus oedipus*). *American Journal of Primatology*, **48**, 99–111.

Sandler, T. 1992. *Collective Action*. Ann Arbor: University of Michigan Press.

Sauther, M.L. 1991. Reproductive behavior of free-ranging *Lemur catta* at Beza Mahafaly Special Reserve, Madagascar. *American Journal of Physical Anthropology*, **84**, 463–77.

Savage, A., Giraldo, L.H., Soto, L.H. & Snowdon, C.T. 1996a. Demography, group composition, and dispersal in wild cotton-top tamarin (*Saguinus oedipus*) groups. *American Journal of Primatology*, **38**, 85–100.

Savage, A., Snowdon, C.T., Giraldo, L.H. & Soto, L.H. 1996b. Parental care patterns and vigilance in wild cotton-top tamarins (*Saguinus oedipus*). In: *Adaptive Radiations of Neotropical Primates*, ed. M.A. Norconk, A.L. Rosenberger & P.A. Garber, pp. 187–99. New York: Plenum Press.

Schaller, G.B. 1963. *The Mountain Gorilla*. Chicago: University of Chicago Press.

–1972. *The Serengeti Lion*. Chicago: University of Chicago Press.

Schilling, A. 1979. Olfactory communication in prosimians. In: *The Study of Prosimian Behavior*, ed. G.A. Doyle & R.D. Martin, pp. 461–542. New York: Academic Press.

Schlichte, H.J. 1978. The ecology of two groups of blue monkeys, *Cercopithecus mitis stuhlmanni*, in an isolated habitat of poor vegetation. In: *The Ecology of Arboreal Folivores*, ed. G.G. Montgomery, pp. 507–15. Washington, DC: Smithsonian Institution Press.

Schneider, G. 1906. Ergebnisse zoologischer Forschungsreisen in Sumatra. *Zoologische Jahrbücher. Abteilung Systematik, Geographie und Biologie der Tiere*, **23**, 1–172.

Schneider, H. & Rosenberger, A.L. 1996. Molecules, morphology, and platyrrhine systematics. In: *Adaptive Radiations of Neotropical Primates*, ed. M.A. Norconk, A.L. Rosenberger & P.A. Garber, pp. 3–19. New York: Plenum Press.

Schürmann, C.L. 1982. Mating behaviour of wild orangutans. In: *The Orangutan: Its Biology and Conservation*, ed. L. de Boer, pp. 271–86. The Hague: W. Junk.

Schürmann, C.L. & van Hooff, J.A.R. A.M. 1986. Reproductive strategies of the orang-utan: new data and a reconsideration of existing sociosexual models. *International Journal of Primatology*, **7**, 265–87.

Seebeck, J.H. & Rose, R.W. 1989. Potoroidae. In: *Fauna of Australia*, Vol. 1B, *Mammalia*, ed. D.W. Walton & B.J. Richardson, pp. 716–39. Canberra: Australian Government Publishing Service.

Sekulic, R. 1982a. Behavior and ranging patterns of a solitary female red howler (*Alouatta seniculus*). *Folia Primatologica*, **38**, 217–32.

–1982b. The function of howling in red howler monkeys (*Alouatta seniculus*). *Behaviour*, **81**, 38–54.

–1983a. Male relationships and infant deaths in red howler monkeys (*Alouatta seniculus*). *Zeitschrift für Tierpsychologie*, **61**, 185–202.

–1983b. The effect of female call on male howling in red howler monkeys (*Alouatta seniculus*). *International Journal of Primatology*, **4**, 291–305.

Seyfarth, R.M. 1976. Social relationships among adult female baboons. *Animal Behaviour*, **24**, 917–38.

Seyfarth, R.M. & Cheney, D.L. 1984. Grooming, alliances and reciprocal altruism in vervet monkeys. *Nature*, **308**, 541–3.

Shields, W.M. 1987. Dispersal and mating systems: investigating their causal connections. In: *Mammalian Dispersal Patterns: the Effects of Social Structure on Population Genetics*, ed. B.D. Chepko-Sade & Z.T. Halpin, pp. 3–24. Chicago: University of Chicago Press.

Short, R.V. 1979. Sexual selection and its component parts, somatic and genital selection as illustrated by man and the great apes. *Advances in the Study of Behaviour*, **9**, 131–58.

Sicchar, L.A. & Heymann, E.W. 1992. Preliminary observations on external signs of estrus in *Saguinus mystax* (Callitrichidae). *Laboratory Primate Newsletter*, **31**, 4–6.

Sicotte, P. 1993. Inter-group encounters and female transfer in mountain gorillas: influence of group composition on male behavior. *American Journal of Primatology*, **30**, 21–36.

–1994. Effect of male competition on male–female relationships in bi-male groups of mountain gorillas. *Ethology*, **97**, 47–64.

Siegel, S. & Castellan, N.J. 1988. *Nonparametric Statistics for the Behavioral Sciences*. New York: McGraw-Hill.

Siex, K.S. & Struhsaker, T.T. 1999. Ecology of the Zanzibar red colobus monkey: demographic variability and habitat stability. *International Journal of Primatology*, **20**, 163–92.

Sigg, H. & Falett, J. 1985. Experiments of possession and property in hamadryas baboons (*Papio hamadryas*). *Animal Behaviour*, **33**, 978–84.

Sigg, H., Stolba, A., Abegglen, J.-J. & Dasser, V. 1982. Life history of hamadryas baboons: physical development, infant mortality, reproductive parameters and family relationships. *Primates*, **23**, 473–87.

Silberbauer, G.B. 1981. *Hunter and Habitat in the Central Kalahari Desert*. Cambridge: Cambridge University Press.

Silk, J.B. 1992. Patterns of intervention in agonistic contests among male bonnet macaques. In: *Coalitions and Alliances in Animals and Humans*, ed. A.H. Harcourt & F.B.M. de Waal, pp. 215–32. Oxford: Oxford University Press.

–1994. Social relationships of male bonnet macaques: male bonding in a matrilineal society. *Behaviour*, **130**, 271–91.

Sillén-Tullberg, B. & Møller, A.P. 1993. The relationship between concealed ovulation and mating systems in anthropoid primates: a phylogenetic analysis. *American Naturalist*, **141**, 1–25.

Skorupa, J.P. 1988. The effects of selective timber harvesting on rainforest primates in Kibale Forest, Uganda. PhD thesis, University of California, Davis.

Smale, L., Nunes, S. & Holekamp, K.E. 1997. Sexually dimorphic dispersal in mammals: patterns, causes and consequences. *Advances in the Study of Behaviour*, **26**, 181–250.

Small, M. 1990. Promiscuity in Barbary macaques (*Macaca sylvanus*). *American Journal of Primatology*, **20**, 267–82.

Smith, A.C. 1997. Comparative ecology of saddleback (*Saguinus fuscicollis*) and moustached (*Saguinus mystax*) tamarins. PhD thesis, University of Reading.

Smith, A.P. & Ganzhorn, J.U. 1996. Convergence in community structure and dietary adaptations in Australian possums and gliders and Malagasy lemurs. *Australian Journal of Ecology*, **21**, 31–46.

Smith, D.G. 1981. The association between rank and reproductive success of male rhesus monkeys. *American Journal of Primatology*, **1**, 83–90.

–1993. A 15-year study of the association between dominance rank and reproductive success of male rhesus macaques. *Primates*, **34**, 471–80

Smith, E.O. 1992. Dispersal in sub-saharan baboons. *Folia Primatologica*, **59**, 177–85.

Smith, R.J. & Jungers, W.L. 1997. Body mass in comparative primatology. *Journal of Human Evolution*, **32**, 523–59.

Smuts, B.B 1985. *Sex and Friendship in Baboons*. Chicago: Aldine.

–1987. Sexual competition and mate choice. In: *Primate Societies*, ed. B.B. Smuts, D.L. Cheney, R.M. Seyfarth, R.W. Wrangham & T.T. Struhsaker, pp. 385–99. Chicago: University of Chicago Press.

Smuts, B.B. & Gubernick, D.J. 1992. Male–infant relationships in nonhuman primates: paternal investment or mating effort? In: *Father–Child Relations: Cultural and Biosocial Contexts*, ed. B.S. Hewlett, pp. 1–30. New York: Aldine.

Smuts, B.B. & Smuts, R.W. 1993. Male aggression and sexual coercion of females in nonhuman primates and other mammals: evidence and theoretical implications. *Advances in the Study of Behavior*, **22**, 1–63.

Smuts, B.B., Cheney, D.L., Seyfarth, R.M., Wrangham R.W. & Struhsaker T.T. (eds.) 1987. *Primate Societies*. Chicago: University of Chicago Press.

Snowdon, C.T. & Suomi, S.J. 1982. Paternal behavior in primates. In: *Child Nurturance*, Vol.3*: Studies of Development in Nonhuman Primates*, ed. H. Fitzgerald, J. Mullins & P. Gage, pp. 63–108. New York: Plenum Press.

Soini, P. 1987a. Sociosexual behavior of a free-ranging *Cebuella pygmaea* (Callitrichidae, Platyrrhini) troop during postpartum estrus of its reproductive female. *American Journal of Primatology*, **13**, 223–30.

–1987b. Ecology of the saddle-back tamarin *Saguinus fuscicollis illigeri* on the Rio Pacaya, northeastern Peru. *Folia Primatologica*, **49**, 11–32.

–1988. The pygmy marmoset, *Cebuella*. In: *Ecology and Behavior of Neotropical Primates*, Vol. 2, ed. R.A. Mittermeier, A.B. Rylands, A.F. Coimbra-Filho & G.A.B. Fonseca, pp. 79–129. Washington, DC: World Wildlife Fund.

–1990. Ecología y dinámica poblacional de pichico común *Saguinus fuscicollis* (Callitrichidae, Primates). In: *La Primatología en el Perú. Investigaciones Primatológicas (1973–1985)*, ed. Dirección General Forestal y de Fauna, Instituto Veterinario de Investigaciones Tropicales y de Altura & Organización Panamericana de Salud, pp. 202–53. Lima: Imprenta Propaceb.

–1995a. Ecologia del coto mono (*Alouatta seniculus*). In: *Reporte Pacaya-Samiria: Investigaciones en la Estacion Biologica Cahuana*, ed. P. Soini, A. Tovar & U. Valdez, pp. 373–84. Iquitos: Fundacion Peruana par la Conservacion de la Naturaleza.

–1995b. El huapo (*Pithecia monachus*): Dinamica poblacional y organizacion social. In: *Reporte Pacaya-Samiria: Investigaciones en la Estacion Biologica Cahuana*, ed. P. Soini, A. Tovar & U. Valdez, pp. 289–302. Iquitos: La Fundacion Peruana para la Conservacion de la Naturaleza.

Soini, P. & Cóppula, M. 1981. Ecología y dinámica poblacional de pichico *Saguinus fuscicollis* (Primates, Callitrichidae). *Informe de Pacaya*, **4**, 1–43.

Soini, P. & de Soini, M. 1990a. Un estudio y saca experimental de *Cebuella pygmaea*. In: *La Primatología en el Perú. Investigaciones Primatológicas (1973–1985)*, ed. Dirección General Forestal y de Fauna, Instituto Veterinario de Investigaciones Tropicales y de Altura & Organización Panamericana de Salud, pp. 104–21. Lima: Imprenta Propaceb.

–1990b. Distribución geográfica y ecología poblacional de *Saguinus mystax* (Primates, Callitrichidae). In: *La Primatología en el Perú. Investigaciones Primatológicas (1973–1985)*, ed. Dirección General Forestal y de Fauna, Instituto Veterinario de Investigaciones Tropicales y de Altura & Organización Panamericana de Salud, pp. 272–313. Lima: Imprenta Propaceb.

Solomon, N.G. & French, J.A. (eds.) 1996. *Cooperative Breeding in Mammals*. Cambridge: Cambridge University Press.

Sommer, V. 1988. Male competition and coalitions in langurs (*Presbytis entellus*) at Jodhpur, Rajastan, India. *Human Evolution*, **3**, 261–78.

–1989. *Die Affen. Unsere wilde Verwandtschaft*. Hamburg: GEO.

–1994. Infanticide among the langurs of Jodhpur: testing the sexual selection hypothesis with a long-term record. In: *Infanticide and Parental Care*, ed. S. Parmigiani & F.S. vom Saal, pp. 155–98. London: Harwood Academic Publishers.

Sommer, V. & Rajpurohit, L.S. 1989. Male reproductive success in harem troops of Hanuman langurs (*Presbytis entellus*). *International Journal of Primatology*, **10**, 293–317.

Southwell, C.J. 1984. Variability on grouping in the eastern grey kangaroo, *Macropus giganteus*. II. Dynamics of group formation. *Australian Wildlife Research*, **11**, 437–49.

Sozou, P.D. & Houston, A.I. 1994. Parental effort in a mating system involving two males and two females. *Journal of Theoretical Biology*, **171**, 251–66.

Sprague, D.S. 1992. Life history and male intertroop mobility among Japanese macaques (*Macaca fuscata*). *International Journal of Primatology*, **13**, 437–54.

Srikosamatara, S. 1984. Ecology of pileated gibbons in south-east Thailand. In: *The Lesser Apes: Evolutionary and Behavioural Biology*, ed. H. Preuschoft, D.J. Chivers, W.Y. Brockelman & N. Creel, pp. 498–532. Edinburgh: Edinburgh University Press.

Srikosamatara, S. & Brockelman, W.Y. 1987. Polygyny in a group of pileated gibbons via a familial route. *International Journal of Primatology*, **8**, 389–93.

Srivastava, A. & Dunbar, R.I.M. 1996. The mating system of Hanuman langurs: a problem in optimal foraging. *Behavioral Ecology and Sociobiology*, **39**, 219–26.

Stacey, P.B. 1979. Habitat saturation and communal breeding in the acorn woodpecker. *Animal Behaviour*, **27**, 1153–66.

Stacey, P.B. & Ligon, J.D. 1987. Territory quality and dispersal options in the acorn woodpecker, and a challenge to the habitat-saturation model of cooperative breeding. *American Naturalist*, **130**, 654–76.

Stammbach, E. 1987. Desert, forest and montane baboons: multi-level-societies. In: *Primate Societies*, ed. B.B. Smuts, D.L. Cheney, R.M. Seyfarth, R.W. Wrangham & T.T. Struhsaker, pp. 112–20. Chicago: University of Chicago Press.

Stanford, C.B. 1991a. Social dynamics of intergroup encounters in the capped langur (*Presbytis pileata*). *American Journal of Primatology*, **25**, 35–47.

–1991b. The capped langur in Bangladesh: behavioral ecology and reproductive tactics. *Contributions to Primatology*, **26**, 1–179.

–1995. The influence of chimpanzee predation on group size and anti-predator behaviour in red colobus monkeys. *Animal Behaviour*, **49**, 577–87.

–1999. *Chimpanzee and Red Colobus: the Ecology of Predator and Prey*. Cambridge, MA: Harvard University Press.

Starin, E.D. 1994. Philopatry and affiliation among red colobus. *Behaviour*, **130**, 253–70.

Steenbeek, R. 1996. What a maleless group can tell us about the constraints on female transfer in Thomas's langurs (*Presbytis thomasi*). *Folia Primatologica*, **67**, 169–81.

–1999. Female choice and male coercion in wild Thomas's langurs (*Presbytis thomasi*). PhD thesis, University of Utrecht.

Steenbeek, R., Piek, R.C., van Buul, M. & van Hooff, J.A.R.A.M. 1999. Vigilance in wild Thomas's langurs (*Presbytis thomasi*): the importance of infanticide risk. *Behavioral Ecology and Sociobiology*, **45**, 137–50.

Sterck, E.H.M. 1995. Females, foods and fights: a socioecological comparison of the sympatric Thomas langur and long-tailed macaque. PhD thesis, University of Utrecht.

–1997. Determinants of female dispersal in Thomas langurs. *American Journal of Primatology*, **42**, 179–98.

–1998. Female dispersal, social organization, and infanticide in langurs: are they linked to human disturbance? *American Journal of Primatology*, **44**, 235–54.

Sterck, E.H.M. & Steenbeeck, R. 1997. Female dominance relationships and food competition in the sympatric Thomas langur and long-tailed macaque. *Behaviour*, **134**, 749–74.

Sterck, E.H.M., Watts, D.P. & van Schaik, C.P. 1997. The evolution of social relationships in nonhuman primates. *Behavioral Ecology and Sociobiology*, **41**, 291–309.

Sterling, E.J. 1994. Evidence for nonseasonal reproduction in wild aye-ayes (*Daubentonia madagascariensis*). *Folia Primatologica*, **62**, 46–53.

Stevens, E.F. 1990. Instability of harems of feral horses in relation to season and presence of subordinate stallions. *Behaviour*, **112**, 149–61.

Stevenson, P.R., Quiñoes, M.J. & Ahumada, J.A. 1994. Ecological strategies of woolly monkeys (*Lagothrix lagotricha*) at Tinigua National Park, Colombia. *American Journal of Primatology*, **32**, 123–40.

Stewart, K.J. & Harcourt, A.H. 1987. Gorillas: variation in female social relationships. In: *Primate Societies*, ed. B.B. Smuts, D.L. Cheney, R.M. Seyfarth, R.W. Wrangham & T.T. Struhsaker, pp. 155–64. Chicago: University of Chicago Press.

Stolba, A. 1979. Entscheidungsfindung in Verbänden von *Papio hamadryas*. PhD thesis, University of Zürich.

Strahan, R. (ed.) 1995. *The Mammals of Australia*. Sydney: Australian Museum & Reed Books.

Strier, K.B. 1990. New World primates, new frontiers: insights from the woolly spider monkey, or muriqui (*Brachyteles arachnoides*). *International Journal of Primatology*, **11**, 7–19.

–1992a. Atelinae adaptations: behavioral strategies and ecological constraints. *American Journal of Physical Anthropology*, **88**, 515–24.

–1992b. *Faces in the Forest: the Endangered Muriqui Monkeys of Brazil*. New York: Oxford University Press.

–1994a. Myth of the typical primate. *Yearbook of Physical Anthropology*, **37**, 233–71.

–1994b. Brotherhoods among atelins: kinship, affiliation, and competition. *Behaviour*, **130**, 151–67.

–1996a. Male reproductive strategies in New World primates. *Human Nature*, **7**, 105–23.

–1996b. Reproductive ecology of female muriquis. In: *Adaptive Radiations of Neotropical Primates*, ed. M.A. Norconk, A.L. Rosenberger & P.A. Garber, pp. 511–32. New York: Plenum Press.

–1996c. Viability analyses of an isolated population of muriqui monkeys (*Brachyteles arachnoides*): implications for primate conservation and demography. *Primate Conservation*, **14/15**, 43–52.

–1997a. Behavioral ecology and conservation biology of primates and other animals. *Advances in the Study of Behavior*, **26**, 101–58.

–1997b. Mate preferences of wild muriqui monkeys (*Brachyteles arachnoides*): reproductive and social correlates. *Folia Primatologica*, **68**, 120–33.

–1999a. Why is female kin bonding so rare: comparative sociality of New World primates. In: *Primate Socioecology*, ed. P.C. Lee, pp. 300–19. Cambridge: Cambridge University Press.

–1999b. Predicting primate responses to 'stochastic' demographic events. *Primates*.

–1999c. The atelines. In: *Comparative Primate Behavior*, ed. A. Fuentes & P. Dolhinow. New York: McGraw Hill.

Strier, K.B., Mendes, F.D.C., Rímoli, J. & Rímoli, A.O. 1993. Demography and social structure in one group of muriquis (*Brachyteles arachnoides*). *International Journal of Primatology*, **14**, 513–26.

Struhsaker, T.T. 1967. Social structure among vervet monkeys (*Cercopithecus aethiops*). *Behaviour*, **29**, 83–121.

–1969. Correlates of ecology and social organization among African cercopithecines. *Folia Primatologica*, **11**, 80–118.

–1975. *The Red Colubus Monkey*. Chicago: University of Chicago Press.

–1977. Infanticide and social organization in the redtail monkey (*Cercopithecus ascanius schmidti*) in the Kibale Forest, Uganda. *Zeitschrift für Tierpsychologie*, **45**, 75–84.

–1988. Male tenure, multi-male influxes, and reproductive success in redtail monkeys (*Cercopithecus ascanius*). In: *A Primate Radiation: Evolutionary Biology of the African Guenons*, ed. A. Gautier-Hion, F. Bourlière, J.-P. Gautier & J. Kingdon, pp. 340–63. Cambridge: Cambridge University Press.

–1997. *Ecology of an African Rain Forest: Logging in Kibale and the Conflict between Conservation and Exploitation*. Gainesville: University of Florida Press.

–1999. The effects of predation and habitat quality on the socioecology of African monkeys: lessons from the islands of Bioko and Zanzibar. In: *Old World Monkeys*, ed. P. Whitehead & C. Jolly. New York: Cambridge University Press.

Struhsaker, T.T. & Gartlan, J.S. 1970. Observations on the behaviour and ecology of the patas monkey (*Erythrocebus patas*) in the Waza Reserve, Cameroon. *Journal of Zoology, London*, **161**, 49–63.

Struhsaker, T.T. & Leakey, M. 1990. Prey selectivity by crowned hawk-eagles on monkeys in the Kibale Forest, Uganda. *Behavioral Ecology and Sociobiology*, **26**, 435–43.

Struhsaker, T.T. & Leland, L. 1979. Socioecology of five sympatric monkey species in the Kibale forest, Uganda. *Advances in the Study of Behavior*, **9**, 159–228.

–1985. Infanticide in a patrilineal society of red colobus monkeys. *Zeitschrift für Tierpsychologie*, **69**, 89–132.

–1987. Colobines: infanticide by adult males. In: *Primate Societies*, ed. B.B. Smuts, D.L. Cheney, R.M. Seyfarth, R.W. Wrangham & T.T. Struhsaker, pp. 83–97. Chicago: University of Chicago Press.

Struhsaker, T.T. & Pope, T.R. 1991. Mating system and reproductive success: a comparison of two African forest monkeys (*Colobus badius* and *Cercopithecus ascanius*). *Behaviour*, **117**, 182–205.

Strum, S.C. 1987. *Almost Human. A Journey into the World of Baboons*. New York: Random House.

Stuart-Dick, R.I. 1987. Parental investment and rearing schedules in the eastern grey kangaroo. PhD thesis, University of New England, Armidale, Australia.

Stuart-Dick, R.I. & Higginbottom, K.B. 1989. Strategies of parental investment in macropodoids. In: *Kangaroos, Wallabies and Rat-kangaroos*, ed. G. Grigg, P. Jarman & I. Hume, pp. 571–92. Sydney: Surrey Beatty & Sons.

Suarez, B. & Ackerman, D.R. 1971. Social dominance and reproductive behaviour in male rhesus monkeys. *American Journal of Physical Anthropology*, **35**, 219–22.

Sugardjito, J., te Boekhorst, I.J.A. & van Hooff, J.A.R. A.M. 1987. Ecological constraints on the grouping of wild orang-utans (*Pongo pygmaeus*) in the Gunung Leuser National Park, Sumatra, Indonesia. *International Journal of Primatology*, **8**, 17–41.

Sugardjito, J., van Schaik, C.P., van Noordwijk, M.A. & Mitrasetia, T. 1989. Population status of the Simeulue monkey (*Macaca fascicularis fusca*). *American Journal of Primatology*, **17**, 197–207.

Sugg, D.W., Chesser, R.K., Dobson, F.S. & Hoogland, J.L. 1996. Population genetics meets behavioral ecology. *Trends in Ecology and Evolution*, **11**, 338–42.

Sugiyama, Y. 1966. An artificial social change in a Hanuman langur troop (*Presbytis entellus*). *Primates*, **7**, 41–72.

–1976. Life history of male Japanese monkeys. *Advances in the Study of Behavior*, **7**, 255–84.

–1984. Population dynamics of wild chimpanzees at Bossou, Guinea, between 1976 and 1983. *Primates*, **25**, 391–400.

–1988. Grooming interactions among adult chimpanzees at Bossou, Guinea, with special reference to social structure. *International Journal of Primatology*, **9**, 393–407.

Sugiyama, Y. & Koman, J. 1979. Social structure and dynamics of wild chimpanzees at Bossou, Guinea. *Primates*, **20**, 323–39.

Sugiyama, Y. & Ohsawa, H. 1975. Life history of male Japanese macaques at Ryozenyama. In: *Contemporary Primatology*, ed. S. Kondo, M. Kawai & A. Ehara, pp. 407–10. Basel: S. Karger.

Sugiyama, Y., Yoshiba, K. & Pathasarathy, M.D. 1965. Home range, mating season, male group and inter-troop relations in Hanuman langurs (*Presbytis entellus*). *Primates*, **6**, 73–106.

Supriatna, J., Manullang, B.O. & Soekara, E. 1986. Group composition, home range, and diet of the maroon leaf monkey (*Presbytis rubicunda*) at Tanjung Puting reserve, Central Kalimantan, Indonesia. *Primates*, **27**, 185–90.

Sussman, R.W. 1974. Ecological distinctions of sympatric species of *Lemur*. In: *Prosimian Biology*, ed. R.D. Martin, G.A. Doyle & A.C. Walker, pp. 75–108. London: Duckworth.

–1991. Demography and social organization of free-ranging *Lemur catta* in the Beza Mahafaly Reserve, Madagascar. *American Journal of Physical Anthropology*, **84**, 43–58.

–1992. Male life history and intergroup mobility among ringtailed lemurs (*Lemur catta*). *International Journal of Primatology*, **13**, 395–413.

Sussman, R.W. & Garber, P.A. 1987. A new interpretation of the social organization and mating system of the Callitrichidae. *International Journal of Primatology*, **8**, 73–92.

Sussman, R.W. & Richard, A.F. 1974. The role of aggression among diurnal prosimians. In: *Primate Aggression, Territoriality and Xenophobia*, ed. R.L. Holloway, pp. 49–76. New York: Academic Press.

Sussman, R.W., Cheverud, J.M. & Bartlett, T.Q. 1995. Infant killing as an evolutionary strategy: reality or myth? *Evolutionary Anthropology*, **3**, 149–51.

Symington, M.M. 1988a. Food competition and foraging party size in the black spider monkey (*Ateles paniscus chamek*). *Behaviour*, **105**, 117–34.

–1988b. Demography, ranging patterns, and activity budgets of black spider monkeys (*Ateles paniscus chamek*) in the Manu National Park, Peru. *American Journal of Primatology*, **15**, 45–67.

–1990. Fission-fusion social organization in *Ateles* and *Pan*. *International Journal of Primatology*, **11**, 47–61.

Szekely, T., Webb, J.N., Houston, A.I. & McNamara, J.M. 1996. An evolutionary approach to offspring desertion in birds. *Current Ornithology*, **13**, 271–330.

Takahata, Y. 1990. Adult males' social relations with adult females. In: *The Chimpanzees of the Mahale Mountains. Sexual and Life History Strategies*, ed. T. Nishida, pp. 133–48. Tokyo: University of Tokyo Press.

Tanaka, J. 1965. Social structure of nilgiri langurs. *Primates*, **6**, 107–22.

Tattersall, I. 1977. Ecology and behavior of *Lemur fulvus mayottensis* (Primates, Lemuriformes). *Anthropological Papers of the American Museum of Natural History*, **54**, 425–82.

–1978. Behavioral variation in *Lemur mongoz*. In: *Recent Advances in Primatology*, Vol. III, ed. D.A. Chivers & K.A. Joysey, pp. 127–32. London: Academic Press.

Taub, D.M. 1980. Female choice and mating strategies among wild Barbary macaques (*Macaca sylvanus* L.). In: *The Macaques: Studies in Ecology, Behavior and Evolution*, ed. D.G. Lindberg, pp. 287–344. New York: van Nostrand Reinhold.

–1982. Sexual behavior of wild Barbary macaque males (*Macaca sylvanus*). *American Journal of Primatology*, **2**, 109–13.

–1984. Male caretaking behavior among wild Barbary macaques (*Macaca sylvanus*). In: *Primate Paternalism*, ed. D.M. Taub, pp. 20–55. New York: van Nostrand Reinhold.

Taylor, C.R. & Rowntree, V.J. 1973. Running on two or on four legs: which consumes more energy? *Science*, **179**, 186–7.

te Boekhorst, I.J.A. 1991. Social structure of three great ape species: an approach based on field data and individual oriented models. PhD thesis, University of Utrecht.

te Boekhorst, I.J.A. & Hogeweg, P. 1994. Self-structuring in artificial 'CHIMPS' offers new hypotheses for male grouping in chimpanzees. *Behaviour*, **130**, 229–52.

Templeton, A.R. 1980. The theory of speciation via the founder principle. *Genetics*, **94**, 1011–38.

Tenaza, R.R. & Hamilton, W.J. III 1971. Preliminary observations of the Mentawai islands gibbon, *Hylobates klossii*. *Folia Primatologica*, **15**, 201–11.

Terborgh, J. 1983. *Five New World Primates. A Study in Comparative Ecology*. Princeton, NJ: Princeton University Press.

Terborgh, J. & Janson, C.H. 1986. The socioecology of primate groups. *Annual Review of Ecology and Systematics*, **17**, 111–36.

Thorington, R.W. Jr & Groves, C.P. 1970. List of genera of *Presbytis*. In: *Old World Monkeys*, ed. J.R. Napier & P.H. Napier, pp. 640–3. London: Academic Press.

Tilson, R.L. 1979. On the behavior of Hoolock gibbons (*Hyobates hoolock*) during different seasons in Assam, India. *Journal of the Bombay Natural History Society*, **76**, 1–16.

–1981. Family formation strategies of Kloss's gibbons. *Folia Primatologica*, **35**, 259–87.

Treesucon, U. 1984. Social development of young gibbons (*Hylobates lar*) in Khao Yai National Park, Thailand. MSc thesis, Mahidol University, Bangkok.

Treves, A. 1998. Primate social systems: conspecific threat and coercion-defense hypotheses. *Folia Primatologica*, **69**, 81–8.

Treves, A. & Chapman, C.A. 1996. Conspecific threat, predation avoidance, and resource defense: implications for grouping in langurs. *Behavioral Ecology and Sociobiology*, **39**, 43–53.

Trivers, R.L. 1971. The evolution of reciprocal altruism. *Quarterly Review of Biology*, **46**, 35–57.

–1972. Parental investment and sexual selection. In: *Sexual Selection and the Descent of Man*, ed. B. Campbell, pp. 136–79. Chicago: Aldine.

Troisi, A. & Carosi, M. 1998. Female orgasm rate increases with male dominance in Japanese macaques. *Animal Behaviour* **56**, 1261–6.

Tsingalia, H.M. & Rowell, T.E. 1984. The behaviour of adult male blue monkeys. *Zeitschrift für Tierpsychologie*, **64**, 253–68.

Tutin, C.E.G. 1979. Mating patterns and reproductive strategies in a community of wild chimpanzees (*Pan troglodytes schweinfurthii*). *Behavioral Ecology and Sociobiology*, **6**, 29–38.

–1996. Ranging and social structure of lowland gorillas in the Lope Reserve, Gabon. In: *Great Ape Societies*, ed. W.C. McGrew, L.F. Marchant & T. Nishida, pp. 58–70. Cambridge: Cambridge University Press.

Tutin, C.E.G., McGrew, W.C. & Baldwin, P.J. 1983. Social organization of savanna-dwelling chimpanzees, *Pan troglodytes verus*, at Mt. Assirik, Senegal. *Primates*, **24**, 154–73.

Tuttle, R.H. 1986. *Apes of the World: Their Social Behavior, Communication, Mentality, and Ecology*. Park Ridge, NJ: Noyes Publications.

Uhde, N.L. 1997. Das Raubfeindrisiko bei Gibbons (*Hylobates lar*). Eine sozioökologische Studie im Regenwald des Khao Yai National Park. MSc thesis, University of Göttingen.

Uhde, N.L. & Sommer, V. 1998. The importance of predation risk for gibbon behaviour and evolution. *Folia Primatologica*, **69**, 224.

Valladares-Padua, C.B. 1993. The ecology, behavior and conservation of the black lion tamarin (*Leontopithecus chrysopygus* Mikan, 1823). PhD thesis, University of Florida, Gainesville.

van Hooff, J.A.R.A.M. 1996. The orangutan: a social outsider. In: *The Neglected Ape*, ed. R.D. Nadler, B.F.M. Galdikas, L.K. Sheeran & N. Rosen, pp. 153–62. New York: Plenum Press.

van Hooff, J.A.R.A.M. & van Schaik, C.P. 1992. Cooperation in competition: the ecology of primate bonds. In: *Coalitions and Alliances in Humans and Other Animals*, ed. A.H. Harcourt & F.B.M. de Waal, pp. 357–89. Oxford: Oxford University Press.

–1994. Male bonds: affiliative relationships among nonhuman primate males. *Behaviour*, **130**, 309–37.

van Noordwijk, M.A. 1985. Sexual behaviour of Sumatran long-tailed macaques (*Macaca fascicularis*). *Zeitschrift für Tierpsychologie*, **70**, 277–96.

van Noordwijk, M.A. & van Schaik, C.P. 1985. Male migration and rank acquisition in wild long-tailed macaques (*Macaca fascicularis*). *Animal Behaviour*, **33**, 849–61.

–1987. Competition among adult female long-tailed macaques, *Macaca fascicularis*. *Animal Behavior*, **36**, 577–89.

–1988. Male careers in Sumatran long-tailed macaques (*Macaca fascicularis*). *Behaviour*, **107**, 24–43.

–1999. The effects of dominance rank and group size on female lifetime reproductive success in wild long-tailed macaques. *Macaca fascicularis*. *Primates*, **40**, 109–34.

van Rhijn, J.G. 1985. A scenario for the evolution of social organization in ruffs *Philomachus pugnax* and other Charadriiform species. *Ardea*, **73**, 25–37.

van Rhijn, J.G. & Vodegel, R. 1980. Being honest about one's intentions: an evolutionary stable strategy for animal conflicts. *Journal of Theoretical Biology*, **85**, 623–41.

van Roosmalen, M.G.M. 1980. *Habitat Preferences, Diet, Feeding Strategy, and Social Organization of the Black Spider Monkey* (*Ateles p. paniscus* Linnaeus 1758) *in Surinam*. Arnhem: Rijksinstituut voor Natuurbeheer.

van Schaik, C.P. 1983. Why are diurnal primates living in groups? *Behaviour*, **87**, 120–44.

–1986. Phenological changes in a Sumatran rain forest. *Journal of Tropical Ecology*, **2**, 327–47.

–1989. The ecology of social relationships amongst female primates. In: *Comparative Socioecology. The Behavioural Ecology of Humans and Other Mammals*, ed. V. Standen & R.A. Foley, pp. 195–218. Oxford: Blackwell Scientific Publications.

–1992. Sex-biased juvenile mortality in primates: a reply to Hauser and Harcourt. *Folia Primatologica*, **58**, 53–5.

–1996. Social evolution in primates: the role of ecological factors and male behaviour. *Proceedings of the British Academy*, **88**, 9–31.

van Schaik, C.P. & de Visser, J.A.G. M. 1990. Fragile sons or harassed daughters? Sex differences in mortality among juvenile primates. *Folia Primatologica*, **55**, 10–23.

van Schaik, C.P. & Dunbar, R.I.M. 1990. The evolution of monogamy in large primates: a new hypothesis and some crucial tests. *Behaviour*, **115**, 30–62.

van Schaik, C.P. & Hörstermann, M. 1994. Predation risk and the number of adult males in a primate group: a comparative test. *Behavioral Ecology and Sociobiology*, **35**, 261–72.

van Schaik, C.P. & Kappeler, P.M. 1993. Life history, activity period and lemur social systems. In: *Lemur Social Systems and Their Ecological Basis*, ed. P.M. Kappeler & J.U. Ganzhorn, pp. 241–60. New York: Plenum Press.

–1996. The social systems of gregarious lemurs: lack of convergence with anthropoids due to evolutionary disequilibrium? *Ethology*, **102**, 915–41.

–1997. Infanticide risk and the evolution of male–female association in primates. *Proceedings of the Royal Society of London, Series B*, **264**, 1687–94.

van Schaik, C.P. & Mirmanto, E. 1985. Spatial variation in the structure and litterfall of a Sumatran rain forest. *Biotropica*, **17**, 196–205.

van Schaik, C.P. & Paul, A. 1997. Male care in primates: does it ever reflect paternity? *Evolutionary Anthropology*, **5**, 152–6.

van Schaik, C.P. & van Hooff, J.A.R. A.M. 1983. On the ultimate causes of primate social systems. *Behaviour*, **85**, 91–117.

–1996. Toward an understanding of the orangutan's social system. In: *Great Ape Societies*, ed. W.C. McGrew, L.F. Marchant & T. Nishida, pp. 3–15. Cambridge: Cambridge University Press.

van Schaik, C.P. & van Noordwijk, M.A. 1985. Evolutionary effect of the absence of felids on the social organization of the macaques on the island of Simeulue (*Macaca fascicularis fusca*, Miller 1903). *Folia Primatologica*, **44**, 138–47.

–1986. The hidden costs of sociality: intra-group variation in feeding strategies in Sumatran long-tailed macaques (*Macaca fascicularis*). *Behaviour*, **99**, 296–315.

–1988. Scramble and contest in feeding competition among female long-tailed macaques (*Macaca fascicularis*). *Behaviour*, **105**, 77–98.

–1989. The special role of male *Cebus* monkeys in predation avoidance and its effect on group composition. *Behavioral Ecology and Sociobiology*, **24**, 265–76.

van Schaik, C.P., van Noordwijk, M.A., Wasone, M.A. & Sitriono, E. 1983. Party size and early detection of predators in Sumatran rain forest primates. *Primates*, **24**, 211–21.

van Schaik, C.P., Assink, P.R. & Salafsky, N. 1992. Territorial behavior in Southeast Asian langurs: resource defense or mate defense? *American Journal of Primatology*, **26**, 233–42.

van Schaik, C.P., van Noordwijk, M.A. & Nunn, C.L. 1999. Sex and social evolution in primates. In: *Comparative Primate Socioecology*, ed. P.C. Lee, pp. 204–31. Cambridge: Cambridge University Press.

Vandenbergh, J.G. & Vessey, S.H. 1968. Seasonal breeding of free-ranging rhesus monkeys and related ecological factors. *Journal of Reproduction and Fertility*, **15**, 71–9.

Vedder, A.L. 1984. Movement patterns of a group of free-ranging mountain gorillas (*Gorilla gorilla beringei*) and their relationship to food availability. *American Journal of Primatology*, **7**, 73–88.

Vehrencamp, S.L. 1983a. A model for the evolution of despotic versus egalitarian societies. *Animal Behaviour*, **31**, 667–82.

–1983b. Optimal degree of skew in cooperative societies. *American Zoologist*, **23**, 327–35.

Vehrencamp, S.L. & Bradbury, J.W. 1984. Mating systems and ecology. In: *Behavioural Ecology: an Evolutionary Approach*, ed. J.R. Krebs & N.B. Davies, pp. 251–78. Sunderland, MA: Sinauer.

Vellayan, S. 1981. The nutritive value of *Ficus* in the diet of lar gibbon (*Hylobates lar*). *Malaysian Applied Biology*, **10**, 177–81.

Verrell, P.A. 1992. Primate penile morphologies and social systems: further evidence for an association. *Folia Primatologica*, **59**, 114–20.

Vick, L.G. & Pereira, M.E. 1989. Episodic targeting aggression and the histories of *Lemur* social groups. *Behavioral Ecology and Sociobiology*, **25**, 3–12.

Vitousek, P.M. & Sanford, J.R.L. 1986. Nutrient cycling in moist tropical forest. *Annual Review of Ecology and Systematics*, **17**, 137–67.

Vogel, C. & Loch, H. 1984. Reproductive parameters, adult-male replacements, and infanticide among free-ranging langurs (*Presbytis entellus*) at Jodhpur (Rajasthan), India. In: *Infanticide. Comparative and Evolutionary Perspectives*, ed. G. Hausfater & S.B. Hrdy, pp. 237–55. New York: Aldine.

Wahome, J.M., Rowell, T.E. & Tsingalia, H.M. 1993. The natural history of de Brazza's monkey in Kenya. *International Journal of Primatology*, **14**, 445–66.

Walker, L.V. 1995. Mate choice in female eastern grey kangaroos *Macropus giganteus*. PhD thesis, University of New England, Armidale, Australia.

–1996. Female mate-choice. In: *Comparison of Marsupial and Placental Behaviour*, ed. D.B. Croft & U. Ganslosser, pp. 208–25. Fürth: Filander Verlag.

Walker, T.W. & Syers, J.K. 1976. The fate of phosphorus during pedogenesis. *Geoderma*, **15**, 1–19.

Walters, J.R. 1987. Transition to adulthood. In: *Primate Societies*, ed. B.B. Smuts, D.L. Cheney, R.M. Seyfarth, R.W. Wrangham & T.T. Struhsaker, pp. 358–69. Chicago: University of Chicago Press.

Walters, J.R. & Seyfarth, R.M. 1987. Conflict and cooperation. In: *Primate Societies*, ed. B.B. Smuts, D.L. Cheney, R.M. Seyfarth, R.W. Wrangham & T.T. Struhsaker, pp. 306–17. Chicago: University of Chicago Press.

Wang, Z., Ferris, C.F. & de Vries, G.J. 1994. Role of arginine vasopressin intervention in paternal behavior in prairie voles (*Microtus ochrogaster*). *Proceedings of the National Academy of Sciences of the United States of America*, **91**, 400–4.

Waser, P.M. 1976. *Cercocebus albigena*: site attachment, avoidance, and intergroup spacing. *American Naturalist*, **110**, 911–35.

–1988. Resources, philopatry, and social interactions among mammals. In: *The Ecology of Social Behavior*, ed. C.N. Slobodchikoff, pp. 109–30. San Diego: Academic Press.

Waser, P.M. & Wiley, R.H. 1980. Mechanisms and evolution of spacing in animals. In: *Handbook of Behavioral Neurobiology*, ed. P. Marler & J.G. Vandenbergh, pp. 159–223. New York: Plenum Press.

Watson, S., Ward, J., Izard, K. & Stafford, D. 1996. An analysis of birth sex ratio bias in captive prosimian species. *American Journal of Primatology*, **38**, 303–14.

Watts, D.P. 1984. Composition and variability of mountain gorillas' diets in the central Virungas. *American Journal of Primatology*, **7**, 323–56.

–1985. Relations between group size and composition and feeding competition in mountain gorilla groups. *Animal Behaviour*, **33**, 72–85.

–1989. Infanticide in mountain gorillas: new cases and a reconsideration of the evidence. *Ethology*, **81**, 1–18.

–1990a. Ecology of gorillas and its relationship to female transfer in mountain gorillas. *International Journal of Primatology*, **11**, 21–45.

–1990b. Mountain gorilla life histories, reproductive competition, and sociosexual behavior and some implications for captive husbandry. *Zoo Biology*, **9**, 185–200.

–1991. Strategies of habitat use by mountain gorillas. *Folia Primatologica*, **56**, 1–16.

–1992. Social relationships of immigrant and resident female mountain gorillas. I. Male–female relationships. *American Journal of Primatology*, **28**, 159–81.

–1994a. Social relationships of immigrant and resident female mountain gorillas, II. Relatedness, residence, and relationships between females. *American Journal of Primatology*, **32**, 13–30.

–1994b. Agonistic relationships between female mountain gorillas (*Gorilla gorilla beringei*). *Behavioral Ecology and Sociobiology*, **34**, 347–58.

–1995. Post-conflict in wild mountain gorillas (Mammalia, Hominoidea). II. Redirection, side direction, and consolation. *Ethology*, **100**, 158–74.

–1996. Comparative socio-ecology of gorillas. In: *Great Ape Societies*, ed. W.C. McGrew, L.F. Marchant & T. Nishida, pp. 16–28. Cambridge: Cambridge University Press.

–1997. Agonistic interventions in wild mountain gorilla groups. *Behaviour*, **134**, 23–57.

–1998a. Coalitionary mate guarding by male chimpanzees at Ngogo, Kibale National Park, Uganda. *Behavioral Ecology and Sociobiology*, **44**, 43–56.

–1998b. Long-term habitat use by mountain gorillas (*Gorilla gorilla beringei*), 1. Consistency, variation, and home range size and stability. *International Journal of Primatology*, **19**, 651–80.

Watts, D.P. & Pusey, A.E. 1993. Behavior of juvenile and adolescent great apes. In: *Juvenile Primates: Life History, Development, and Behavior*, ed. M.E. Pereira & L.A. Fairbanks, pp. 148–67. New York: Oxford University Press.

Weber, A.W. & Vedder, A.L. 1983. Population dynamics of the Virunga gorillas: 1959–78. *Biological Conservation*, **26**, 341–66.

Weitzel, V. & Groves, C.P. 1985. The nomenclature and taxonomy of the colobine monkeys of Java. *International Journal of Primatology*, **6**, 399–409.

Wesche, K. 1997. A classification of a tropical *Shorea robusta* forest stand in southern Nepal. *Phytocoenologia*, **27**, 103–18.

Westneat, D.F., Sherman, P.W. & Morton, M.L. 1990. The ecology and evolution of extra-pair copulations in birds. *Current Ornithology*, **7**, 331–69.

White, F.J. 1988. Party composition and dynamics in *Pan paniscus*. *International Journal of Primatology*, **9**, 179–93.

–1989. Ecological correlates of pygmy chimpanzee social structure. In: *Comparative Socioecology. The Behavioural Ecology of Humans and Other Mammals*, ed. V. Standen & R. Foley, pp. 151–64. Oxford: Blackwell.

–1991. Social organization, feeding ecology, and reproductive strategy of ruffed lemurs, *Varecia variegata*. In: *Proceedings of the XIII Congress of the International Primatological Society*, ed. A. Ehara, T. Kimura, O. Takenaka & M. Iwamoto, pp. 81–4. Amsterdam: Elsevier.

–1992. Pygmy chimpanzee social organization: variation with party size and between study sites. *American Journal of Primatology*, **26**, 203–14.

Whiten, A., Byrne, R.W., Henzi, S.P. 1987. The behavioral ecology of mountain baboons. *International Journal of Primatology*, **8**, 367–88.

Whitesides, G.H. 1981. Community and population ecology of non-human primates in the Douala-Edea forest reserve. MSc thesis, Johns Hopkins University, Baltimore.

–1989. Interspecific associations of Diana monkeys, *Cercopithecus diana*, in Sierra Leone, West Africa: biological significance or chance? *Animal Behaviour*, **37**, 760–76.

Whitington, C. 1990. Seed dispersal by white-handed gibbons (*Hylobates lar*) in Khao Yai National Park, Thailand. MSc thesis, Mahidol University, Bangkok.

Whitten, P.L. 1982. Female reproductive strategies among vervet monkeys. PhD thesis, Harvard University.

–1984. Competition among female vervet monkeys. In: *Female Primates: Studies by Women Primatologists*, ed. M.F. Small, pp. 127–40. New York: Alan R. Liss.

–1987. Infants and adult males. In: *Primate Societies*, ed. B.B. Smuts, D.L. Cheney, R.M. Seyfarth, R.W. Wrangham & T.T. Struhsaker, pp. 343–57. Chicago: University of Chicago Press.

Whittingham, L.A., Dunn, P.O. & Magrath, R.D. 1997. Relatedness, polyandry and extra-group paternity in the cooperatively-breeding white-browed scrubwren *Sericornis frontalis*. *Behavioral Ecology and Sociobiology*, **40**, 261–70.

Wickings, E. & Dixson, A. 1992. Testicular function, secondary sexual development, and social status in male mandrills (*Mandrillus sphinx*). *Physiology and Behaviour*, **52**, 909–16.

Wickler, W. & Seibt, U. 1983. Monogamy: an ambiguous concept. In: *Mate Choice*, ed. P.P.G. Bateson, pp. 33–50. Cambridge: Cambridge University Press.

Wielgus, R.B. & Bunnell, F.L. 1995. Tests of hypotheses for sexual segregation in grizzly bears. *Journal of Wildlife Management*, **59**, 552–60.

Wiley, R.H. & Rabenold, K.R. 1984. The evolution of cooperative breeding by delayed reciprocity and queueing for favorable social positions. *Evolution*, **38**, 609–21.

Wilson, A.P. & Boelkins, R.C. 1970. Evidence for seasonal variation in aggressive behaviour in *Macaca mulatta*. *Animal Behaviour*, **18**, 719–24.

Wilson, E.O. 1975. *Sociobiology*. Cambridge, MA: Belknap Press.

Wilson, J.M., Stewart, P.D., Ramangason, G.S., Denning, A.M. & Hutchings, M.S. 1989. Ecology and conservation of the crowned lemur, *Lemur coronatus*, at Ankarana, N. Madagascar. *Folia Primatologica*, **52**, 1–26.

Winkler, P., Podzuweit, D. & Borries, C. 1993. Zur Toleranz gezwungen – oder warum Hanuman Languren nicht nur in Harems leben. In: *Evolution und Anpassung. Warum die Vergangenheit die Gegenwart erklärt*, ed. E. Voland, pp. 94–103. Stuttgart: Hirzel.

Winslow, J., Hastings, N., Carter, C.S., Harbaugh, C. & Insel, T.R. 1993. A role for central vasopressin in pair bonding in monogamous prairie voles. *Nature*, **365**, 545–8.

Winter, J.W. 1996. Australasian possums and Madagascan lemurs: behavioural comparison of ecological equivalents. In: *Comparison of Marsupial and Placental Behaviour*, ed. D.B. Croft & U. Ganslosser, pp. 263–92. Fürth: Filander Verlag.

Wittenberger, J.F. 1980. Group size and polygamy in social mammals. *American Naturalist*, **115**, 197–222.

Wolf, K.E. 1984. Reproductive competition among co-resident male silvered leaf-monkeys (*Presbytis cristata*). PhD thesis, Yale University.

Wolf, K.E. & Fleagle, J.G. 1977. Adult male replacement in a group of silvered leaf-monkeys (*Presbytis cristata*) at Kuala Selangor, Malaysia. *Primates*, **18**, 949–55.

Wrangham, R.W. 1975. Behavioral ecology of chimpanzees in Gombe National Park, Tanzania. PhD thesis, University of Cambridge.

–1977. Aspects of feeding and social behaviour in gelada baboons. Unpublished report to the Science Research Council, UK.

–1979. On the evolution of ape social systems. *Social Science Information*, **18**, 334–68.

–1980. An ecological model of female-bonded primate groups. *Behaviour*, **75**, 262–300.

–1986. Ecology and social relationships in two species of chimpanzee. In: *Ecological Aspects of Social Evolution: Birds and Mammals*, ed. D.I. Rubenstein & R.W. Wrangham, pp. 352–78. Princeton, NJ: Princeton University Press.

–1987. Evolution of social structure. In: *Primate Societies*, ed. B.B. Smuts, D.L. Cheney, R.M. Seyfarth, R.W. Wrangham & T.T. Struhsaker, pp. 282–96. Chicago: University of Chicago Press.

Wrangham, R.W. & Peterson, D. 1996. *Demonic Males: Apes and the Origins of Human Violence*. New York: Houghton Mifflin.

Wrangham, R.W. & Rubenstein, D.I. 1986. Social evolution in birds and mammals. In: *Ecological Aspects of Social Evolution. Birds and Mammals*, ed. D.I. Rubenstein & R.W. Wrangham, pp. 452–70. Princeton, NJ: Princeton University Press.

Wrangham, R.W. & Smuts, B.B. 1980. Sex differences in the behavioural ecology of chimpanzees in the Gombe National Park, Tanzania. *Journal of Reproduction and Fertility*, Supplement, **28**, 13–31.

Wrangham, R.W., Conklin, N.L., Chapman, C.A. & Hunt, K.D. 1991. The significance of fibrous foods for Kibale Forest chimpanzees. *Philosophical Transactions of the Royal Society of London, Series B*, **334**, 171–8.

Wrangham, R.W., Clark, A.P. & Isabirye-Basuta, G. 1992. Female social relationships and social organization of the Kibale Forest chimpanzees. In: *Topics in Primatology.* Vol. 1*: Human Origins*, ed. T. Nishida, W.C. McGrew, P. Marler, M. Pickford & F.B.M. de Waal, pp. 81–98. Tokyo: University of Tokyo Press.

Wrangham, R.W., Gittleman, J.L. & Chapman, C.A. 1993. Constraints on group size in primates and carnivores: population density and day-range as assays of exploitation competition. *Behavioral Ecology and Sociobiology*, **32**, 199–209.

Wrangham, R.W., Chapman, C.A., Clark-Arcadi, A.P. & Isabirye-Basuta, G. 1996. Social ecology of Kanyawara chimpanzees: implications for understanding the costs of great ape groups. In: *Great Ape Societies*, ed. W.C. McGrew, L.F. Marchant & T. Nishida, pp. 45–57. Cambridge: Cambridge University Press.

Wrangham, R.W., Conklin-Brittain, N.L. & Hunt, K.D. 1998. Dietary response of chimpanzees and cercopithecines to seasonal variation in fruit abundance: I. Antifeedants. *International Journal of Primatology*, **19**, 949–70.

Wright, P.C. 1984. Biparental care in *Aotus trivirigatus* and *Callicebus moloch*. In: *Female Primates: Studies by Women Primatologists*, ed. M.F. Small, pp. 59–75. New York: Alan R. Liss.

–1989. Comparative ecology of three sympatric bamboo lemurs in Madagascar. *American Journal of Physical Anthropology*, **78**, 327.

–1990. Patterns of paternal care in primates. *International Journal of Primatology*, **11**, 89–102.

–1995. Demography and life history of free-ranging *Propithecus diadema edwardsi* in Ranomafana National Park, Madagascar. *International Journal of Primatology*, **16**, 835–54.

–1998. Impact of predation risk on the behaviour of *Propithecus diadema edwardsi* in the rain forest of Madagascar. *Behaviour*, **135**, 483–512.

Wright, P.C., Heckscher, S.K. & Dunham, A.E. 1997. Predation on Milne-Edward's sifaka (*Propithecus diadema edwardsi*) by the fossa (*Cryptoprocta ferox*) in the rain forest of Southeastern Madagascar. *Folia Primatologica*, **68**, 34–43.

Wright, S. 1980. Genic and organismic selection. *Evolution*, **34**, 825–42.

Wright, S.M. 1993. Observations on the behaviour of male eastern grey kangaroos when attacked by dingoes. *Wildlife Research*, **20**, 845–9.

Yamagiwa, J. 1983. Diachronic changes in two eastern lowland gorilla groups (*Gorilla gorilla graueri*) in the Mt. Kahuzi region, Zaïre. *Primates*, **24**, 174–83.

–1985. Socio-sexual factors of troop fission in wild Japanese monkeys (*Macaca fuscata yakui*) on Yakushima Island, Japan. *Primates*, **26**, 105–20.

–1987a. Intra- and inter-group interactions of an all-male group of Virunga mountain gorillas (*Gorilla gorilla beringei*). *Primates*, **28**, 1–30.

–1987b. Male life history and the social structure of wild mountain gorillas (*Gorilla gorilla beringei*). In: *Evolution and Coadaptation in Biotic Communities*, ed. S. Kawano, J.H. Connell & T. Hidaka, pp. 31–51. Tokyo: University of Tokyo Press.

Yamagiwa, J., Maruhashi, T., Yumoto, T. & Mwanza, N. 1996. Dietary and ranging overlap in sympatric gorillas and chimpanzees in Kahuzi-Biega National Park, Zaïre. In: *Great Ape Societies*, ed. W.C. McGrew, L.F. Marchant & T. Nishida, pp. 82–98. Cambridge: Cambridge University Press.

Yeager, C.P., Kirkpatrick, R.C. & Craig, R. 1998. Asian colobine social structures: ecological and evolutionary constraints. *Primates*, **39**, 147–55.

Yoder, A.D., Cartmill, M., Ruvolo, M., Smith, K. & Vilgalys, R. 1996. Ancient single origin for Malagasy primates. *Proceedings of the National Academy of Sciences of the United States of America*, **93**, 5122–6.

Young, L.J. 1997. Species differences in V_1a receptor gene expression in monogamous and nonmonogamous voles: behavioral consequences. *Behavioral Neuroscience*, **111**, 599–605.

Zuberbühler, K., Noë, R. & Seyfarth, R.M. 1997. Diana monkey long-distance calls: messages for conspecifics and predators. *Animal Behaviour*, **53**, 589–604.

Zuckerman, S. 1932. *The Social Life of Monkeys and Apes*. London: Routledge & Kegan Paul.

Index

access to females 23, 31, 55, 60, 69, 72, 120, 130, 146, 154, 173, 183, 222, 264
adaptation 4, 22, 41, 47, 55, 189, 203, 219, 271
adult sex ratio 5, 27, 55, 58, 108, 111, 116, 146, 179, 206, 208, 238, 274
affiliation 4, 85, 102, 106, 210, 212, 234, 253
age-graded group 4, 120, 126, 130, 136, 142, 169, 178, 187, 222, 274
aggression 31, 35, 58, 95, 103, 110, 126, 131, 138, 144, 166, 175, 183, 195, 207, 215, 220, 226, 234, 272
all-male band 6, 108, 120, 131, 169, 238, 247, 271
Allenopithecus 84, 91, 96
alliance 7, 19, 57, 78, 97, 102, 107, 132, 142, 178, 189, 206, 219, 230, 234, 244, 248, 252, 274
Alouatta 5, 19, 41, 56, 72, 80, 120, 131, 142, 188, 197, 219, 226, 233
alpha male 12, 19, 27, 29, 72, 78, 153, 156, 195
alternative strategies 42, 169, 172
altruism 11, 189, 248
Aotus 64, 232,
arboreal 4, 22, 42, 94, 106, 155, 160, 256
asynchrony 15, 187
Ateles 73, 82, 188, 248, 255, 265

baboon, *see Papio*
bachelor 170
behavioral ecology 106, 119, 159
between-group conflict 59, 62, 73, 120, 129, 131, 137, 143, 170, 198
bimaturism 190
birth seasonality 93, 120, 127, 187, 265
body size 47, 56, 64, 72, 82, 171, 174, 214, 232
bonding 6, 14, 32, 42, 71, 102, 165, 188, 206, 219, 248, 258
Brachyteles 72, 77, 188
breeding season 5, 37, 84, 88, 184, 198, 228, 265
breeding seasonality 37, 93, 200, 265
breeding system 11, 19, 159, 178, 226

Cacajao 233
Callimico 64, 71
Callithrix 66, 69, 166, 227
canine size 55
Carnivora 37, 47, 100, 249, 255
Cebuella 65, 70
Cebus 72, 78, 233
Cercopithecus 34, 43, 84, 91, 100, 146, 155
chimpanzee, *see Pan*
Chiropotes 233
coalition 5, 11, 17, 28, 31, 33, 97, 103, 108, 110, 117, 120, 129, 143, 161, 175, 183, 189, 206, 216, 220, 223, 227, 244, 251, 273
collective action 130, 192, 201, 205
Colobus 5, 42, 108, 131, 143, 188, 256
comparative methods 192, 198
consortship 26, 28, 34
contest competition 46, 61, 102, 140, 184, 195, 199, 206, 215, 248, 255
cooperation 61, 72, 131, 158, 188, 190, 195, 219, 221, 276
cooperative breeding 11, 178, 226, 232
copulation 7, 13, 18, 26, 28, 32, 45, 69, 77, 85, 108, 118, 136, 155, 160, 164, 173, 178, 183, 202, 214, 216, 222, 227, 231, 275
cycling females 149, 156, 241, 265

daily path length 70, 232
delayed implantation 37, 46
demography 55, 61, 72, 78, 86, 99, 113, 120, 164, 170, 220, 226, 236, 265, 273
dimorphism 55, 64, 100, 102, 104, 106, 132, 185, 190, 206, 215, 230, 250, 261, 273
dispersal 11, 17, 29, 41, 43, 45, 63, 72, 78, 80, 108, 111, 132, 146, 148, 152, 154, 159, 165, 169, 174, 179, 185, 189, 199, 219, 222, 226, 229, 236, 244
display 15, 94, 108, 208
DNA fingerprinting 7, 13, 35, 156, 164, 221, 276
dominance 3, 6, 12, 26, 28, 30, 40, 55, 61, 69, 85, 97, 102, 106, 132, 146, 148, 152, 156, 169, 175, 178, 183, 194, 197, 205, 207, 214, 220, 227, 234, 244, 248, 254, 272, 274
duet 159, 163, 169

ecology 4, 21, 46, 63, 100, 102, 121, 159, 170, 185, 205, 256, 272, 276
economics 193, 246
egalitarian relationship 77, 82, 178, 184, 186, 206, 214
emigration 112, 118, 146, 149, 155, 174, 220, 226, 244
Erythrocebus 34, 43, 84, 91, 93, 95, 100, 155, 199
escalation 206, 215
estrous synchrony 60, 272, 274
estrus 23, 25, 29, 32, 36, 42, 44, 68, 87, 93, 186, 190, 205, 208, 222, 259, 262, 267, 276
Eulemur 57, 59, 60
eutherian mammals 21, 39
evolution 4, 11, 16, 32, 35, 41, 48, 55, 71, 97, 99, 106, 119, 167, 191, 219, 231, 258, 272
extra-troop males 146, 148, 155, 157

female alliances 97, 104, 248
female choice 6, 16, 58, 60, 78, 169, 187, 190, 275
female dispersal 41, 108, 132, 154, 169, 220, 224, 233, 246
female interests 59, 60
female strategies 71, 77, 97, 107, 187, 202, 274
female transfer 58, 97, 103, 170, 176, 178
female–female competition 33, 62, 103, 106, 161, 164
fertility 15, 86, 167, 170, 187, 271, 275
fission–fusion groups 62, 98, 100, 103, 109, 117, 137, 206, 233, 246, 248, 255, 262
fitness 11, 16, 19, 120, 128, 171, 176, 183, 187, 196, 205, 223, 244, 249, 276
follower male 43, 100, 106, 143, 170, 175, 274
food competition 131, 134, 144, 161, 166
food distribution 169, 240

game theory 14, 16, 192, 203
gelada, *see Theropithecus*
gibbon, *see Hylobates*
Gorilla 41, 56, 106, 120, 131, 143, 169, 187, 197, 263
gregariousness 170, 248, 250, 254, 275
grooming 23, 26, 57, 62, 97, 102, 106, 134, 139, 163, 177, 195, 210, 248
group cohesion 3, 60, 99, 103
group composition 4, 7, 27, 56, 58, 63, 84, 120, 148, 226, 233, 239, 243, 265, 271, 277
group formation 221, 229, 233
group membership 59, 130, 137, 155, 169, 185, 214, 222, 224, 229, 273
group size 5, 22, 27, 45, 56, 60, 66, 73, 78, 91, 100, 101, 103, 110, 116, 124, 134, 144, 171, 186, 197, 220, 228, 232, 236, 246, 262
group transfer 6, 58, 108, 275
growth 22, 24, 47, 67, 257, 261, 272

habitat quality 104, 111, 117, 124
Hapalemur 59
harassment 16, 29, 34, 45, 57, 60, 102, 123, 132, 141, 144, 178, 202, 275

harem 3, 23, 30, 44, 56, 98, 143, 183, 206, 265
helper 11, 16, 48, 65, 70, 178, 223, 226, 230
hierarchy 26, 28, 30, 69, 85, 103, 108, 234
home range 23, 27, 95, 109, 111, 120, 123, 132, 137, 147, 165, 170, 198, 223, 228, 233, 260, 263
howler monkey, *see Alouatta*
human 3, 108, 110, 116, 123, 165, 192, 193, 195, 258
Hylobates 64, 159, 161, 233, 261, 276

immigration 40, 59, 71, 74, 77, 118, 120, 143, 149, 151, 155, 170, 195, 220, 223, 227, 233, 245
inbreeding 19, 77, 146, 155, 165, 184, 245, 258
inclusive fitness 6, 131, 143, 183, 187, 189, 196, 219, 223, 231
independent contrasts 37, 64, 198, 265
individual difference 118, 245
infant care 4, 38, 42, 45, 65, 69, 230, 254
infant carrying 16, 39, 64, 67
infant sex ratio 73, 77, 78, 80, 82
infanticide 5, 6, 19, 32, 34, 59, 60, 102, 107, 128, 131, 143, 162, 166, 172, 176, 187, 190, 193, 202, 205, 222, 231, 247, 256, 262, 272, 275
intergroup encounters 41, 95, 109, 196
intrasexual competition 55, 205, 219, 220, 226, 229, 233
intraspecific variability 185

juvenile 21, 24, 35, 57, 91, 108, 110, 112, 131, 134, 147, 164, 170, 175, 208, 223, 233, 250

kinship 19, 106, 131, 188, 195, 198

lactation 36, 41, 47, 87, 205
Lagothrix 72, 74, 77, 82
langur, *see Presbytis*
Lemur 39, 57, 59, 146, 155, 273
Leontopithecus 69, 219, 226
life history 4, 21, 35, 37, 40, 42, 72, 81, 130, 170, 178, 219, 234, 271
litter mass 65, 67
long-term study 6, 40, 63, 95, 117, 160, 220, 232
longevity 187
loud calls 91, 127, 137, 143, 196, 200, 264

Macaca 34, 41, 43, 99, 146, 155, 169, 183, 197, 200, 205, 215, 275
Macropus 22, 25
male alliances/coalitions 5, 17, 33, 108, 110, 117, 132, 142, 178, 189, 223, 227, 234, 248, 252, 255, 273
male influx 43, 85, 146, 151, 156, 242
male philopatry 6, 72, 73, 82, 169, 185, 188, 219, 233, 258, 273, 276
male quality 118, 246, 276
male service 60, 273
male tenure 61, 82, 135, 142, 148, 151, 155, 178, 247
male transfer 58, 136
male–female association 34, 40, 41, 43, 47, 160, 170, 271
male–female relationship 45, 60, 106, 177, 188, 273
male–male competition 6, 28, 55, 58, 61, 74, 106, 118, 154, 157, 167, 191, 205, 246, 275
male–male relationship 144, 188
Mandrillus 43, 99, 106
marsupial 21
mate defense 198
mate guarding 13, 15, 64, 69, 169, 227, 274
mating share 12, 19
mating strategy 23, 27, 71, 120, 189, 259, 264
mating success 29, 31, 61, 78, 108, 146, 156, 179, 216, 223, 243, 247
mating system 6, 11, 14, 19, 22, 27, 30, 33, 64, 69, 77, 100, 106, 169, 205, 219, 226, 229, 231, 234, 260, 273
Miopithecus 84, 90, 96, 106
model 14, 16, 55, 71, 83, 97, 100, 104, 163, 171, 186, 192, 200, 202, 234, 236, 240, 245, 259, 263, 274
monogamy 12, 16, 70, 159, 219, 232, 262, 267
monopolization 5, 12, 60, 64, 70, 121, 125, 156, 183, 189, 197, 206, 227, 272, 275
mortality 6, 29, 41, 55, 57, 63, 74, 78, 81, 110, 113, 115, 117, 136, 167, 170, 176, 179, 221, 224, 229, 240, 244, 262, 272
mounting 26, 28, 210, 275
multi-male group 4, 11, 19, 34, 40, 44, 58, 72, 85, 90, 97, 99, 106, 120, 127, 130, 175, 183, 227, 237, 246, 247, 273
multiple mating 11, 16
mutualism 195, 198, 276

natal dispersal 42, 148, 155, 157, 159, 165, 228
neighboring group 4, 17, 35, 62, 77, 84, 95, 132, 143, 148, 156, 161, 164, 178, 223, 233
non-resident males 84, 87, 89, 156
number of females 5, 23, 56, 60, 65, 77, 86, 90, 93, 97, 101, 104, 115, 118, 123, 127, 146, 151, 157, 170, 183, 186, 228, 240, 245, 253, 259, 261, 265
number of males 3, 6, 16, 34, 45, 55, 59, 65, 70, 72, 78, 83, 88, 97, 101, 112, 120, 124, 148, 155, 169, 176, 186, 192, 197, 200, 205, 228, 236, 240, 265, 271

one-male group 3, 84, 99, 103, 106, 120, 125, 127, 130, 142, 147, 156, 172, 176, 183, 186, 196, 264, 273, 276
operational sex ratio 55, 62, 174, 185, 259
opportunity costs 16, 244
orang-utan, *see Pongo*

pair-bond 14, 30, 46, 60, 165, 166
Pan 33, 97, 108, 110, 114, 117, 169, 188, 248, 255
Papio 43, 85, 96, 97, 107, 120, 131, 143, 146, 157, 169, 174, 178, 185, 188, 196, 205, 206, 238, 256, 265
parental care 15, 164
party size 116, 248, 250, 253
patas monkey, *see Erythrocebus*
paternal care 16, 32, 56, 59, 61, 64, 69, 177, 231, 246
paternity 11, 13, 14, 16, 35, 40, 45, 60, 62, 68, 71, 100, 164, 173, 176, 190, 222, 227, 231, 246, 271, 275
patrilineal society 74, 108
payoff 16, 169, 171, 194, 262
philopatry 6, 72, 73, 82, 97, 165, 169, 185, 188, 219, 233, 273, 276
phylogenetic analyses 7, 99
phylogenetic inertia 4, 38, 219, 234
phylogeny 37, 96, 110, 118, 198, 200, 265
Pithecia 232, 233
platyrrhine 38, 42, 72, 78, 219, 226, 231
polyandry 11, 18, 64, 70, 160, 163, 168, 190
polygamy 160, 164, 219, 232, 259, 261, 267
polygynandry 11, 12, 14, 15, 64, 160, 169
polygyny 12, 15, 19, 33, 64, 71, 160, 164, 169, 206, 220, 228, 262
Pongo 31, 33, 190, 191, 249, 256, 262, 276
population density 81, 103, 113, 120, 127, 178, 220, 224, 233, 245
population growth 179, 219, 222, 224, 230
population sex ratio 5, 58
predation 3, 5, 11, 19, 22, 24, 46, 59, 67, 93, 100, 101, 103, 110, 113, 115, 127, 136, 141, 144, 161, 169, 187, 203, 229, 236, 244, 256, 271, 273
predator 4, 5, 22, 24, 31, 33, 40, 43, 47, 59, 62, 68, 91, 94, 100, 104, 110, 120, 123, 127, 130, 141, 144, 161, 186, 193, 196, 231, 273
Presbytis 3, 35, 49, 106, 120, 124, 130, 143, 146, 178, 185, 187, 196, 264, 272, 275
priority of access 55, 61, 85, 108, 164, 183, 208, 214, 260
Procolobus 108, 113
promiscuity 34, 45, 189, 190
Propithecus 59
Prunella 11, 15
punishment 195

rainfall 70, 99, 103, 123, 265
range defense 94, 95
rank 25, 27, 29, 34, 40, 102, 128, 138, 146, 148, 152, 156, 165, 171, 177, 183, 194, 206, 214, 245, 254, 275
receptivity 28, 45, 200
reciprocity 189, 195, 276
reproductive quiescence 36, 37, 44
reproductive skew 6, 19, 69, 184, 194, 197, 247, 273, 274
reproductive success 6, 12, 14, 17, 28, 31, 55, 60, 63, 95, 117, 131, 143, 157, 159, 166, 174, 183, 191, 205, 216, 229, 247, 255, 271
resource distribution 219, 271
ritualized submission 208, 212
roving male 23, 30, 62, 128, 166, 259, 260, 271

Saguinus 16, 64, 166, 219, 226, 233
Saimiri 73, 78, 106, 188, 233, 273
scramble competition 131, 142, 144, 184, 187, 195, 198, 248, 253, 256
seasonal breeding 30, 56, 60, 95, 146, 155, 200, 205, 272
secondary dispersal 80, 132, 146, 149, 155

sex difference 55, 57, 190, 248, 253
sexual coercion 6, 7, 32, 102, 104, 107, 169, 178, 188, 195, 202, 205, 271, 273
sexual conflict 11, 14, 16, 35, 202, 258, 277
sexual dimorphism 6, 22, 55, 100, 102, 104, 106, 132, 185, 190, 206, 215, 230, 250, 261
sexual selection 3, 44, 55, 62, 64, 187, 261, 271, 277
sexual swelling 7, 15, 34, 45, 100, 245, 250, 254, 262, 273, 275
social evolution 4, 55, 63, 64, 71, 202
social organization 3, 7, 21, 25, 42, 45, 55, 64, 71, 97, 99, 102, 106, 108, 119, 146, 154, 159, 162, 258, 276
social structure 4, 84, 91, 93, 98, 99, 102, 104, 107, 130, 159, 161, 166, 185, 232
socioecology 3, 4, 27, 33, 55, 61, 67, 97, 99, 107, 170, 192, 202, 271
solitary individuals 5, 6, 22, 23, 26, 31, 39, 42, 46, 62, 108, 120, 130, 142, 169, 178, 220, 224, 238, 251, 272
sperm competition 7, 11, 26, 61, 69, 100, 185, 198, 200, 206, 216, 274
spider monkey, *see Ateles*
status 18, 26, 27, 28, 43, 61, 63, 84, 163, 166, 170, 174, 207, 222, 231, 244, 247
submission 208, 210, 212
subordination 29, 207, 208
supernumerary males 60, 62, 86, 89, 95, 186
suppression 62, 227, 233, 246
swamp monkey, *see Allenopithecus*
synchrony 5, 13, 15, 60, 61, 93, 187, 203, 246, 272, 274

take-over 17, 18, 41, 81, 121, 126, 128, 131, 135, 142, 163, 186, 224
talapoin monkey, *see Miopithecus*
tenure length 61, 148, 156, 171, 178, 275
terrestriality 22, 31, 94, 97, 100, 256
territoriality 22, 30, 95, 108, 159, 165, 192, 196, 203, 220, 232, 260
testes 55, 61, 69, 100, 132, 147, 198, 199, 227
Theropithecus 43, 99, 100, 143, 265
tolerance 110, 118, 127, 130, 139, 144, 176, 184, 186, 208, 215, 228, 233, 247, 255, 273, 276
Trachypithecus 121, 123, 124, 127
trade-off 16, 164, 167, 186, 187, 190, 207
transfer 6, 7, 58, 62, 97, 99, 103, 108, 156, 169, 176, 178, 220, 223, 230, 275
travel costs 249

vaginal plug 26
Varecia 42, 57, 59, 60, 248
vigilance 59, 100, 130, 136, 141, 161, 186, 197, 203, 231, 273
violence 206
visibility 3, 7, 147, 155, 272, 275

war of attrition 207, 216
weaning 25, 28, 35, 37, 40, 57, 132, 136, 147, 226
wounds 110, 134, 141, 206, 215